Other Titles in This Series

45 **George M. Bergman and Adam O. Hausknecht,** Cogroups and co-rings in categories of associative rings, 1996

44 **J. Amorós, M. Burger, K. Corlette, D. Kotschick, and D. Toledo,** Fundamental groups of compact Kähler manifolds, 1996

43 **James E. Humphreys,** Conjugacy classes in semisimple algebraic groups, 1995

42 **Ralph Freese, Jaroslav Ježek, and J. B. Nation,** Free lattices, 1995

41 **Hal L. Smith,** Monotone dynamical systems: an introduction to the theory of competitive and cooperative systems, 1995

40.2 **Daniel Gorenstein, Richard Lyons, and Ronald Solomon,** The classification of the finite simple groups, number 2, 1995

40.1 **Daniel Gorenstein, Richard Lyons, and Ronald Solomon,** The classification of the finite simple groups, number 1, 1994

39 **Sigurdur Helgason,** Geometric analysis on symmetric spaces, 1993

38 **Guy David and Stephen Semmes,** Analysis of and on uniformly rectifiable sets, 1993

37 **Leonard Lewin, Editor,** Structural properties of polylogarithms, 1991

36 **John B. Conway,** The theory of subnormal operators, 1991

35 **Shreeram S. Abhyankar,** Algebraic geometry for scientists and engineers, 1990

34 **Victor Isakov,** Inverse source problems, 1990

33 **Vladimir G. Berkovich,** Spectral theory and analytic geometry over non-Archimedean fields, 1990

32 **Howard Jacobowitz,** An introduction to CR structures, 1990

31 **Paul J. Sally, Jr. and David A. Vogan, Jr., Editors,** Representation theory and harmonic analysis on semisimple Lie groups, 1989

30 **Thomas W. Cusick and Mary E. Flahive,** The Markoff and Lagrange spectra, 1989

29 **Alan L. T. Paterson,** Amenability, 1988

28 **Richard Beals, Percy Deift, and Carlos Tomei,** Direct and inverse scattering on the line, 1988

27 **Nathan J. Fine,** Basic hypergeometric series and applications, 1988

26 **Hari Bercovici,** Operator theory and arithmetic in H^∞, 1988

25 **Jack K. Hale,** Asymptotic behavior of dissipative systems, 1988

24 **Lance W. Small, Editor,** Noetherian rings and their applications, 1987

23 **E. H. Rothe,** Introduction to various aspects of degree theory in Banach spaces, 1986

22 **Michael E. Taylor,** Noncommutative harmonic analysis, 1986

21 **Albert Baernstein, David Drasin, Peter Duren, and Albert Marden, Editors,** The Bieberbach conjecture: Proceedings of the symposium on the occasion of the proof, 1986

20 **Kenneth R. Goodearl,** Partially ordered abelian groups with interpolation, 1986

19 **Gregory V. Chudnovsky,** Contributions to the theory of transcendental numbers, 1984

18 **Frank B. Knight,** Essentials of Brownian motion and diffusion, 1981

17 **Le Baron O. Ferguson,** Approximation by polynomials with integral coefficients, 1980

16 **O. Timothy O'Meara,** Symplectic groups, 1978

15 **J. Diestel and J. J. Uhl, Jr.,** Vector measures, 1977

14 **V. Guillemin and S. Sternberg,** Geometric asymptotics, 1977

13 **C. Pearcy, Editor,** Topics in operator theory, 1974

(See the AMS catalog for earlier titles)

MATHEMATICAL
Surveys and Monographs

Volume 45

Cogroups and Co-rings in Categories of Associative Rings

George M. Bergman
Adam O. Hausknecht

American Mathematical Society
Providence, Rhode Island

The first author was supported by National Science Foundation contracts MCS 77-03719, MCS 82-02632, DMS 85-02330, DMS 90-01234, DMS 92-41325, and DMS 93-03379.

1991 *Mathematics Subject Classification*. Primary 16B50, 17B99, 18A40, 18D35; Secondary 08B99, 13K05, 14L17, 16S10, 16N40, 16W10, 16W30, 17A99, 20J15, 20M50.

ABSTRACT. This book studies representable functors among well-known varieties of algebras. All such functors from associative rings over a fixed ring R to each of the categories of abelian groups, associative rings, Lie rings, and to several others are determined. Results are also obtained on representable functors on varieties of groups, semigroups, commutative rings, and Lie algebras.

Library of Congress Cataloging-in-Publication Data

Bergman, George M., 1943–
Cogroups and co-rings in categories of associative rings / George M. Bergman, Adam O. Hausknecht.
p. cm. — (Mathematical surveys and monographs, ISSN 0076-5376; v. 45)
Includes bibliographical references (p. –) and index.
ISBN 0-8218-0495-2 (alk. paper)
1. Associative rings. 2. Categories (Mathematics) 3. Functor theory. I. Hausknecht, Adam O. II. Title. III. Series: Mathematical surveys and monographs; no. 45.
QA251.5.B465 1996
512′.55–dc20

96-147
CIP

∞ The paper used in this book is acid-free and falls within the guidelines
established to ensure permanence and durability.
♻ Printed on recycled paper.

10 9 8 7 6 5 4 3 2 1 01 00 99 98 97 96

To

Sylvia and Lester Bergman

and to

Rita, Morrissa, and Elizabeth Hausknecht

CONTENTS

Chapter I. Introduction

Chapter II. Review of coalgebras and representable functors

Chapter III. Representable functors from rings to abelian groups

Chapter IV. Digressions on semigroups, etc.

Chapter V. Representable functors from algebras over a field to rings

Chapter VI. Representable functors from k-rings to rings

Chapter VII. Representable functors from rings to general groups and semigroups

Chapter VIII. Representable functors on categories of commutative associative algebras

Chapter IX. Representable functors on categories of Lie algebras

Chapter X. Multilinear algebra of representable functors on k-**Ring**[1]

Chapter XI. Directions for further investigation

Dependence of sections

All parts of these notes assume the material of Chapter II; or more precisely, the first four sections thereof. The statement of the main result of Chapter III, Theorem 13.15, is also assumed in all subsequent Chapters except VIII. (Partial exceptions: the last three sections of Chapter IV do not require this result, though they assume familiarity with the notational approach of §12; nor, for the most part, do §39, §§59-63 or §65.)

The Chapters after III are mutually independent, except that Chapter V is assumed in Chapters VI and IX, and in §57 and §64; and its first section, §23, in Chapter X as well.

Within each Chapter, later sections generally assume the earlier ones. Notable exceptions are Chapter IV, where the last three sections are independent of the first five, Chapter VII, where the last section is self-contained, and the final chapter, which breaks into largely self-contained sequences {57, 58}, {59, 65}, {60, 61, 62} and {63, 64}.

There are minor dependencies not noted above (for instance, some of the definitions, though not the results, of §§28-29 are called on at various places). Thus, the reader following a shortcut through the text based on the above notes will occasionally have to backtrack, using an internal cross-reference or the index to locate the material required.

CHAPTER I.

Introduction.

§0. General prerequisites.

In this work we shall assume familiarity with the elementary concepts of noncommutative ring theory, including that of a tensor product of bimodules, and with the basic terminology of category theory, including ''natural transformations'' (morphisms of functors). General references for ring theory are [**115**], [**48**], [**49**], for category theory, [**122**].

More specialized results and concepts which are important in this work, in particular, those of algebra and coalgebra objects in a category, and of representable algebra-valued functors, are developed in Chapter II. Below, in §1, we give a brief motivated sketch of some of this background material, followed in §2 by a summary of the main results of Chapters III, V, and VI, and some indication of the subject matter of the remaining chapters.

§1. Introductory sketch – what are coalgebras, and why?

Clearly, the following is a very basic type of construction in algebra: Given two categories of algebraic objects **C** and **D** (e.g., commutative rings, and groups), we construct from each object S of **C** an object $V(S)$ of **D** by taking for the elements of $V(S)$ all X-tuples of elements of S (X a fixed set) that satisfy a fixed system Y of equations in their coordinates, and defining the operations of $V(S)$ by formulas (in precise language, ''terms'') in these coordinates under the operations of S. Examples include ''forgetful functors'', where $X = 1$, $Y = \varnothing$ and the operations of $V(S)$ are a subset (possibly empty) of the operations of S, as well as less trivial constructions, such as the formal-power-series-ring functor from rings to rings, and the functor SL_n from commutative rings to groups. Note that to establish that one has such a construction, one must verify that the operations so defined will always take a family of X-tuples satisfying the equations Y to an X-tuple again satisfying these equations, and that the resulting objects $V(S)$ will satisfy the identities or other conditions defining membership in **D**.

A key tool in studying such functors is the following observation. Let categories **C** and **D**, and a construction V based, as above, on families X and Y be given. Then if **C** is a reasonable category of algebraic objects (e.g., a *variety of algebras,* definition recalled in §6 below), we can form an object of **C** presented by X and Y as *generators and relations*:

(1.1) $$R = <X \mid Y>.$$

The universal property of R says that for every X-tuple v of elements of an object S of **C** satisfying the relations Y, there exists a unique homomorphism $R \to S$ carrying the distinguished generating X-tuple of elements of R to v. In other words, the set of elements of the object $V(S)$ can be identified with the hom-set $\mathbf{C}(R, S)$. So at the set-level, the construction V is "encoded" in the object R of **C**.

Consider for example the functor GL_n taking every associative unital ring S to the group of all invertible $n \times n$ matrices over S. So described, this construction does not fit our formulation, since the property of invertibility is not a set of equations in the entries of a matrix. But because of the *uniqueness of inverses*, to give an invertible $n \times n$ matrix P over S is equivalent to giving two $n \times n$ matrices P and Q satisfying $PQ = QP = I$, and these matrix relations comprise a system of equations in the entries of P and Q. The group operations on invertible matrices are given by ring-theoretic expressions in the coordinates, so up to natural isomorphism, GL_n is indeed a construction of the sort just discussed. The object encoding this construction is the ring presented by $2n^2$ generators p_{ij}, q_{ij} $(i, j = 1, \ldots, n)$ and the $2n^2$ relations constituting the matrix equations $PQ = I$, $QP = I$:

(1.2) $$R = <p_{ij}, q_{ij} \mid PQ = QP = I>.$$

In general, category-theorists call a covariant functor V from any category **C** to **Set** *representable* if there exists an object R of **C** (a *representing object*) such that V is isomorphic to $\mathbf{C}(R, -)$ as functors $\mathbf{C} \to \mathbf{Set}$. Dually, a *contravariant functor* from a category **C** to **Set** is called representable if it is isomorphic to a functor of the form $\mathbf{C}(-, R)$ for some object R of **C**.

Now the representing object (1.1) nicely encodes the description of the underlying *sets* of the algebras we are constructing; but we are interested in *operations* as well. Is there a way to extend the above approach so as to also encode a description of these?

There is!

Consider again the example of GL_n. The representing object (1.2) for this construction may be described as a ring R with a *universal* invertible $n \times n$ matrix P over it. Since group multiplication is a binary operation, let us also form the ring with a universal *pair* of invertible $n \times n$ matrices, P^λ and P^ρ (the superscripts marking the intended *left* and *right* factors in the product we intend to form). This ring will be a *coproduct* (§5 below) of two copies of R, and we will write it $R^\lambda \amalg R^\rho$. Matrix multiplication in $\mathrm{GL}_n(R^\lambda \amalg R^\rho)$, applied to the universal pair of invertible matrices P^λ, P^ρ over this ring, gives a new invertible matrix, $P^\lambda P^\rho$, and by the universal property of R, that matrix in turn corresponds to a morphism $\mathbf{m}$: $R \to R^\lambda \amalg R^\rho$.

This morphism $\mathbf{m}$ can be shown to "encode" the operation of multiplying arbitrary invertible matrices over arbitrary rings; one likewise gets ring homomorphisms $\mathbf{i}$ and $\mathbf{e}$ encoding the inverse-matrix and identity-matrix operations. (Namely, $\mathbf{i}$: $R \to R$ is the ring homomorphism interchanging the entries of P and the corresponding entries of Q, and $\mathbf{e}: R \to \mathbf{Z}$ is the homomorphism sending the entries of both P and Q to the corresponding entries of the identity matrix.) The resulting system $(R, \mathbf{m}, \mathbf{i}, \mathbf{e})$ is what is called a *cogroup* in the category of rings (with $\mathbf{m}$, $\mathbf{i}$, and $\mathbf{e}$ its *co-operations*), and this cogroup is said to *represent* the functor GL_n: **Ring** $\to$ **Group**, just as R is said to represent the set-valued construction $S \mapsto |\mathrm{GL}_n(S)|$.

In fact, we shall see in §8 below that if V is a representable functor from a reasonable category of algebras $\mathbf{C}$ to **Set**, with representing object R, then the following are equivalent:

(a) a way of giving the sets $V(S)$ operations, by fixed formulas in the "coordinates", so as to satisfy some set of identities Σ,

(b) any *functorial* way of giving these sets operations that satisfy the identities of Σ,

(c) a system of *co-operations* on the object R which satisfy the *coidentities* corresponding to the members of Σ (concepts to be made precise in §8).

Let us next recall the category-theoretic concept of *adjoint functor* (details in §5 below).

The standard introductory example is the construction associating to every set X the *free group* $F(X)$ on X, i.e., the group with a universal X-tuple of elements. If we write U: **Group** $\to$ **Set** for the underlying set functor, and $U(G)^X$ for the set of X-tuples of elements of $U(G)$, then the universal property of $F(X)$ is equivalent to saying that it represents the set-valued functor $G \mapsto U(G)^X$. Rewriting $U(-)^X$ as $\mathbf{Set}(X, U(-))$, this condition becomes an isomorphism of functors **Group** $\to$**Set**,

$$\mathbf{Group}(F(X), -) \cong \mathbf{Set}(X, U(-)). \tag{1.3}$$

The bijection (1.3) is functorial in both variables (the one in **Set**, which we have so far regarded as constant and written X, and the one in **Group**, written "$-$" above). Generally, given categories $\mathbf{C}$, $\mathbf{D}$ and functors $\mathbf{C} \underset{F}{\overset{V}{\rightleftarrows}} \mathbf{D}$, one calls an isomorphism

$$\mathbf{C}(F(-), -) \cong \mathbf{D}(-, V(-)) \tag{1.4}$$

as functors $\mathbf{D}^{\mathrm{op}} \times \mathbf{C} \to \mathbf{Set}$ an *adjunction* between V and F; F is called the *left adjoint* of V, and V the *right adjoint* of F. Further familiar examples are the abelianization functor **Group** $\to$ **Ab**, which is the left adjoint of the inclusion functor **Ab** $\to$ **Group**, and the construction of the universal enveloping algebra of a Lie algebra, which is the left adjoint to the "commutator-brackets" construction taking associative algebras over a field k to Lie algebras over k.

In fact, after representable functors, the next commonest sort of constructions in algebra are probably functors arising as left adjoints.

But left adjoints to what?

Freyd ([**69**], summarized in §8 below) shows that a functor between varieties of algebras, $V: \mathbf{C} \to \mathbf{D}$, has a left adjoint if and only if V is *representable* in the sense we have discussed, i.e., by a co-$\mathbf{D}$-object of $\mathbf{C}$. For example, we saw above that the functor GL_n is representable. Thus it has a left adjoint; this is the functor taking every group G to the ring R_G with a universal $n \times n$ matrix representation of G (straightforward to construct by generators and relations).

Thus, the representable functors are doubly important, as a class of fundamental constructions, and also as the context in which another fundamental type of construction becomes possible. Hence for two given varieties of algebras $\mathbf{C}$ and $\mathbf{D}$, it is of interest to study, and where possible, describe completely the class of all representable functors from $\mathbf{C}$ to $\mathbf{D}$; or what is equivalent, and often more convenient, the class of all co-$\mathbf{D}$ objects of $\mathbf{C}$.

In algebraic geometry, cogroups in the category of commutative k-algebras are studied under the name *affine group schemes*. (We shall discuss these in Chapter VIII; they include the "classical groups".) In algebraic topology, cogroups in the category of topological spaces, with homotopy-classes of maps for morphisms are considered [**178**, §1.6]. For example, the fact that the n-sphere S^n has such a structure for $n>0$ is the "reason" the sets $\pi_n(X)$, for X a topological space, may be made into groups in a functorial way.

The main new results of the present work, sketched in the next section, concern coalgebras which represent functors from categories of associative rings to categories whose objects have abelian group structures – abelian groups themselves, rings, etc.. Incidentally, the concept called a coalgebra here should not be confused with the concept given that name by Hopf algebraists; the two are related, but neither is a case of the other. (We shall discuss the latter objects briefly in the last paragraph of §9.1 below, and more extensively in §43.)

§2. Overview of results.

Let k be an associative ring with 1, and let k-$\mathbf{Ring}^1$ denote the category whose objects are associative rings S with 1 given with unital homomorphisms $k \to S$, and whose morphisms are the obvious commuting triangles. (Thus, when k is commutative, these include the *k-algebras*.) For any *k-bimodule* M, let $k\langle M\rangle$ denote the tensor k-ring

$$k\langle M\rangle \ = \ k \oplus M \oplus (M \otimes_k M) \oplus \dots .$$

The construction of $k\langle M\rangle$ is the left adjoint of the forgetful functor k-$\mathbf{Ring}^1 \to k$-$\mathbf{Bimod}$. That is, for every k-ring S, if we loosely denote the underlying k-bimodule of S by the same symbol S (writing now as ring-theorists rather than category-theorists), we have

(2.1) $$k\text{-}\mathbf{Ring}^1(k\langle M\rangle, S) \ \cong \ k\text{-}\mathbf{Bimod}(M, S).$$

Fixing M, and regarding this as a *functor in* S, we see from the right-hand-side of (2.1) that the values of this functor have natural structures of *abelian group* (since bimodule homomorphisms can be added and subtracted), and from the left-hand side that this functor is *representable*. The fundamental result of this work, proved in §§10-13, is that *every representable functor from* k-$\mathbf{Ring}^1$ *to abelian groups has the form* (2.1) *for some* M. This yields a contravariant equivalence between the category of such representable functors and the category k-$\mathbf{Bimod}$. (The reader might stop to verify at this point that the "underlying additive group" functor from k-rings to abelian groups is of this form, and think about how to describe the functor corresponding to a bimodule M presented by an arbitrary system of generators and relations.)

To study *ring-valued* functors on k-$\mathbf{Ring}^1$, we need to understand *bilinear maps* among these abelian-group-valued functors. Suppose F, G, H are three representable functors from k-$\mathbf{Ring}^1$ to abelian groups, with representing objects $k{<}L{>}$, $k{<}M{>}$, $k{<}N{>}$. A morphism

$$m\colon\ F \times G \to H$$

as set-valued functors corresponds to a k-ring homomorphism

$$\mathbf{m}\colon\ k{<}N{>} \to k{<}L{>} \amalg k{<}M{>} \cong k{<}L \oplus M{>}.$$

It is natural to call m *bilinear* if for each k-ring S, the map $m(S)$: $F(S) \times G(S) \to H(S)$ is a bilinear map of abelian groups. We shall see that this is the case if and only if the representing ring-homomorphism $\mathbf{m}$ carries the generating subbimodule $N \subseteq k{<}N{>}$ into the "bilinear component" $L \otimes M \oplus M \otimes L \subseteq k{<}L \oplus M{>}$.

It follows, on taking $L = M = N$, that representable functors

$$F\colon\ k\text{-}\mathbf{Ring}^1 \to \text{not necessarily associative rings} \tag{2.2}$$

correspond to k-bimodules M given with homomorphisms $\mathbf{m}\colon M \to M \otimes M \oplus M \otimes M$. (In context, the latter bimodule will be written $M^\lambda \otimes M^\rho \oplus M^\rho \otimes M^\lambda$.) Restrictions on the right hand side of (2.2), such as associativity, can now be translated into conditions on this bimodule map $\mathbf{m}$.

The structure theory that develops from this point on is easiest to sketch if we assume k a field, let our k-rings be k-algebras, and restrict attention to functors represented by *finitely generated* k-algebras. The bimodule M corresponding to such a functor (2.1) is then a finite-dimensional k-vector-space. If we write

$$A = M^*$$

for its finite-dimensional dual space, then, first of all, the functor from k-algebras S to abelian groups described by (2.1) takes the form

$$S \mapsto \mathbf{Mod}_k(M, S) \cong S \otimes_k A,$$

and secondly, a linear map $\mathbf{m}\colon M \to M \otimes M \oplus M \otimes M$ corresponds by duality to a linear map $\beta\colon (A \otimes A) \oplus (A \otimes A) \to A$, that is, to a pair of bilinear "multiplications" on A. By studying this situation under various hypotheses, we shall obtain characterizations including the following:

(i) Any representable functor F with finitely generated representing object from associative unital k-algebras to *associative unital rings* has the form $S \mapsto (S \otimes_k A') \times (S \otimes_k A'')^{\mathrm{op}}$, where A' and A'' are finite-dimensional associative unital k-algebras, and $(\ )^{\mathrm{op}}$ denotes the opposite-ring construction.

(ii) Any representable functor F with finitely generated representing object from associative unital k-algebras to *Lie rings* has the form $S \mapsto (S \otimes_k A)_{\mathrm{Lie}}$, where A is a finite-dimensional associative(!) nonunital k-algebra, and the subscript means that after making $S \otimes_k A$ an associative ring in the standard way, we then make this a Lie ring via commutator brackets.

(iii) If $\operatorname{char} k \neq 2$, then any representable functor F with finitely generated representing object from associative unital k-algebras to unital *Jordan rings* similarly has the form $S \mapsto (S \otimes_k A)_{\text{Jordan}}$, where A is as in (ii), except that it is now assumed unital, and the associative unital ring $S \otimes_k A$ is made a Jordan ring using the operation $(a, b) = ab+ba$. (So in particular, a representable functor from associative unital k-algebras to unital Jordan rings can only yield *special* Jordan rings.)

If we remove the assumption that our representing algebra be finitely generated, but keep k a field and continue to work with k-algebras, we get the same three characterizations, but with the finite-dimensional algebras A replaced by *linearly compact* algebras, and tensor products by *completed* tensor products (§§24-25). The corresponding results in the still more general context of an arbitrary ring k and representable functors on the category of all k-rings are formulated and proved in §28.

The material sketched above constitutes about one third of this work. The remainder includes results, often not so complete, on functors from associative rings to other categories such as abelian semigroups, nonabelian groups, and general semigroups (§§16-18 and §§33-39), a chapter on representable functors on varieties of Lie algebras (Chapter IX) the main results of which are deduced from our results on associative algebras, some results on representable functors on varieties of semigroups, including a characterization of all representable endofunctors of the variety of semigroups with neutral element (§§20-21), and a survey of some features of the much-studied subject of representable functors on varieties of commutative associative algebras (Chapter VIII).

A minor theme that comes up several times, though it was not part of the original plan of this work, is the question of when semigroup-valued representable functors must have idempotent constants. (Every finite nonempty semigroup has an idempotent element, so the question is when representable functors behave like finite objects in this respect.) The answer seems to depend in a delicate way on the domain category.

Open questions are noted throughout the presentation of the above topics; still others are discussed in the final chapter.

§3. Results in the literature.

We know of only a few works in the literature where the category of *all* representable functors between given varieties of algebras is determined. The earliest is Kan [**106**], where simple characterizations are obtained for both comonoid and monoid objects in the category of groups. We shall recover Kan's structure theorem for comonoids as a consequence (Corollary 21.1(i)) of our results on representable functors from semigroups to semigroups.

Berstein [**31**] notes that Kan's result is relevant to the study of the fundamental groups of group and cogroup objects in the category whose objects are topological spaces and whose morphisms are homotopy classes of maps, since a group or cogroup structure on such a space X induces a group or cogroup structure on the fundamental group of X. He goes on to determine, for the purpose of similarly studying cohomology rings of loop spaces, all cogroups in the category of *connected graded* associative algebras. This result will be developed in §34. A result of a similar nature in Hopf algebra theory is the Milnor-Moore theorem, discussed briefly at the end of §43. E. Katz [**109**] shows that Kan's characterization of comonoids also holds in the category of *topological* groups (which falls outside the scope of this work).

In [**28**] (and under more general hypotheses on k, [**81**] or [**80**, Chapter 2]), all sufficiently "good" co-k-algebra structures on the free associative k-algebra on one generator were determined, and used to obtain a description of the automorphism class group of the category of associative k-algebras. In [**82**] a description of all representable functors from associative k-algebras to not necessarily abelian groups will be obtained; we preview this description in §33 below. Some specialized results obtained in [**23**] are also briefly sketched at the end of §21 below.

The concept of a cogroup in the category of *operator algebras* has been introduced as a tool in the study of probability distributions [**193**], [**194**].

§4. Notes on this book; acknowledgements.

We use throughout this work a common numbering system for proclamations (theorems, corollaries, etc.), definitions, numbered displays, and occasionally other items. The nth among the items of all of these sorts in section m is numbered $m.n$. The reason is that it is easier to quickly locate, say, Theorem 13.15 when it lies between display (13.14) and Definition 13.16 than if it were Theorem 13.1 sitting between display (13.8) and Definition 13.3. A consequence, however, is that the statement of Theorem $m.n$ often contains display $(m.n{+}1)$.

Results labeled as exercises (mostly occurring in sections where background material is developed or sketched) represent digressions from the needs of the immediate context, which are nonetheless instructive or of independent interest. Occasionally, an exercise may be referred back to for an alternate proof of a result, or an alternate reference for a standard fact.

The new results presented in this work are due variously to the two authors; specifically, the key result of Chapter III is based on Chapter 1 of Hausknecht's doctoral thesis [**80**], and §§17-18 and 35 are collaborative, while most of the other new results are Bergman's.

The organization and writing of this work was done by Bergman. While working on this book, he was also working on [**24**], and there has been a certain amount of borrowing of text back and forth between the material reviewed in Chapter II below and parts of the later chapters of [**24**]. §§9.5-9.6 of [**24**] are also based on §20 and some of §21 of this work.

For notes on dependence among Chapters and Sections of this work, see p.iii.

The authors are indebted to Keith Kearnes, Ka-Hin Leung, Thomas Lopes and Mike May, who gave Chapters I-IV a "trial run" in a Topics in algebra course at Berkeley in Fall 1986 and pointed out many errata and unclear points; to P.M. Cohn, Mike May, and Perry McDonnell for additional corrections, and to I. Berstein, P.M. Cohn, K. Goodearl, M. Hazewinkel, F.W. Lawvere, J.-L. Loday,

S. Montgomery, M. Takeuchi, and R. Wisbauer for helpful references to the literature. Bergman's Summer support under NSF contracts MCS 77-03719, MCS 82-02632, DMS 85-02330, DMS 90-01234, DMS 92-41325 and DMS 93-03379 helped make it possible to complete this project. We are also indebted to Edward Moy, then of U. C. Berkeley's Academic Computing Services, for his enhancements and bug-fixing of the typesetting program `troff`, with which this book was prepared, and to Sarah Donnelly of the AMS for her ready help and advice when we were reformatting the text for the Surveys and Monographs series.

The authors can be reached by e-mail as

gbergman@math.berkeley.edu   ahausknecht@UMassD.edu

Readers who note any errata to this volume are encouraged to send this information to Bergman, either by regular mail or by e-mail. We would also welcome information on existing or future results closely related to the material covered here, including the open questions we mention. We plan to make a list of errata and related information available on-line through Bergman's home-page on the WWW server http://math.berkeley.edu/ .

CHAPTER II.

Review of coalgebras and representable functors.

§5. Category-theoretic formulations of universal properties, and some other matters.

This section reviews some standard concepts of category theory. The reader familiar with these can skim to the next section. References both for the background assumed and the concepts presented here are [**122**] and [**24**, Chapters 6-7].

If X and Y are objects of a category $\mathbf{C}$, we shall write $\mathbf{C}(X, Y)$ for the set of morphisms of $\mathbf{C}$ from X to Y. A composite of morphisms $X \xrightarrow{f} Y \xrightarrow{g} Z$ will be written gf (not fg); i.e., we use the order of composition compatible with writing maps on the left of their arguments. (Note that in this situation, we may refer to gf as obtained from f by composing with g on the *left*, but of the diagram $X \xrightarrow{f} Y \xrightarrow{g} Z$ as obtained from $X \xrightarrow{f} Y$ by adding the arrow g on the *right*.) The identity morphism of an object X of a category will be written 1_X; the identity *functor* of a category $\mathbf{C}$ will be written $\mathrm{Id}_{\mathbf{C}}$. **Set** will denote the category of sets. (Starting in §25, we will write 1_R for the unit element of a ring R; we will point out this transition at the time.)

We shall write both covariant and contravariant functors in "covariant form"; that is, a *contravariant* functor from $\mathbf{A}$ to $\mathbf{B}$ will be written $F\colon \mathbf{A}^{\mathrm{op}} \to \mathbf{B}$, with F covariant. The category of all covariant functors from $\mathbf{A}$ to $\mathbf{B}$ will often be written $\mathbf{B}^{\mathbf{A}}$. We shall use the term "morphism of functors" where many authors say "natural transformation". We shall use $\cong$ to denote isomorphism (of objects, functors, etc.), $\simeq$ for equivalence of categories.

Suppose F is any covariant functor from a category $\mathbf{C}$ to **Set**, R an object of $\mathbf{C}$, and x a member of the set $F(R)$. Then for each morphism of $\mathbf{C}$ with domain R, $u \in \mathbf{C}(R, S)$, we get an element $F(u)(x) \in F(S)$. Thus, x determines for each object S a set-map $\mathbf{C}(R, S) \to F(S)$, and this assignment is easily seen to be a morphism of functors $Y_x\colon \mathbf{C}(R, -) \to F(-)$ in $\mathbf{Set}^{\mathbf{C}}$. Note that the element $x \in F(R)$ can be recovered from Y_x, as $Y_x(1_R)$. Using this method of going from morphisms back to elements, it is straightforward to verify the first assertion of the following statement; the second is true by symmetry:

5.1 Yoneda's Lemma. *Let* $\mathbf{C}$ *be a category, and* $R \in \mathrm{Ob}(\mathbf{C})$.

(i) *Given* F *in* $\mathbf{Set}^{\mathbf{C}}$, *every morphism* $\mathbf{C}(R, -) \to F$ *is of the form* Y_x *for a unique* $x \in F(R)$.

(ii) *Given* G *in* $\mathbf{Set}^{(\mathbf{C}^{\mathrm{op}})}$, *every morphism* $\mathbf{C}(-, R) \to G$ *is induced in a like manner by a unique* $x \in G(R)$. □

If we think of a covariant functor from $\mathbf{C}$ to $\mathbf{Set}$ as a system of sets indexed by the objects of $\mathbf{C}$, given with a system of maps among these sets indexed by the morphisms of $\mathbf{C}$, then (i) above can be thought of as saying that in the class $\mathbf{Set}^{\mathbf{C}}$ of such systems, the system $\mathbf{C}(R, -)$ is "free, on one generator in the component indexed by R", that free generator being $1_R \in \mathbf{C}(R, R)$; and (ii) says the same for $\mathbf{C}(-, R)$ in $\mathbf{Set}^{(\mathbf{C}^{\mathrm{op}})}$.

If for F in (i) above we take another covariant hom-functor, $\mathbf{C}(S, -)$, we get the result that $\mathbf{C}(S, R)$ is in natural bijective correspondence with $\mathbf{Set}^{\mathbf{C}}(\mathbf{C}(R, -), \mathbf{C}(S, -))$. It is easy to check that this correspondence associates to an element $f \in \mathbf{C}(S, R)$ the operation of composing with f on the right; so Yoneda's Lemma tells us that each morphism of functors $\mathbf{C}(R, -) \to \mathbf{C}(S, -)$ is given by composition on the right with a unique such f. This yields a contravariant isomorphism of $\mathbf{C}$ with a full subcategory of $\mathbf{Set}^{\mathbf{C}}$, the "Yoneda embedding"

$$\mathbf{C}^{\mathrm{op}} \to \mathbf{Set}^{\mathbf{C}},$$

taking each object R to the functor $\mathbf{C}(R, -)$. By the dual argument, (ii) yields a *covariant* full embedding

$$\mathbf{C} \to \mathbf{Set}^{(\mathbf{C}^{\mathrm{op}})},$$

which takes R to $\mathbf{C}(-, R)$, and takes a morphism $f\colon R \to S$ to the operation of composing with f on the *left*. We now give names to the classes of functors which, up to isomorphism, form the images of these two embeddings:

Definition 5.2. *Let* $\mathbf{C}$ *be a category.*

(i) *A covariant functor* $F\colon \mathbf{C} \to \mathbf{Set}$ *is called* representable *if there exists an isomorphism* $i\colon \mathbf{C}(R, -) \cong F$ *for some* $R \in \mathrm{Ob}(\mathbf{C})$. *The object* R *is called a* representing object *for* F.

(ii) *A contravariant functor* $G\colon \mathbf{C}^{\mathrm{op}} \to \mathbf{Set}$ *is likewise called representable if there exists an isomorphism* $i\colon \mathbf{C}(-, S) \cong G$ *for some* representing object $S \in \mathrm{Ob}(\mathbf{C})$.

Note that to give a representation i of F as in (i), it suffices to specify the representing object R and the element $x = i(1_R) \in F(R)$, since by Yoneda's Lemma this determines the isomorphism i. Putting it another way, to represent F is equivalent to finding an object R with a *universal* element $x \in F(R)$. The reader new to these concepts should write down and prove a lemma establishing the equivalence of the representability of F as defined above, and the existence of an object and element with a "universal property" of the sort familiar from descriptions of concepts such as that of a free group.

We also see from the Yoneda embedding that if a functor is representable, any two representing objects are canonically isomorphic; i.e., up to canonical isomorphism, representing objects are unique. For brevity, we shall not explicitly state similar uniqueness conditions in subsequent results in these introductory sections.

The next concept, that of *adjoint functors*, introduced by D. M. Kan in [**107**], offers an embarrassment of important equivalent characterizations, of which we give a few in the next Proposition. We shall mainly use (i) and (ii). We include (ii*) for symmetry, and (iii) because of its connections with other approaches to universal constructions (monads, etc., cf. [**122**], [**18**]).

In reading the statement and the proof of the Proposition, it can be helpful to keep in mind the example where V and F are respectively the underlying set functor $\mathbf{Group} \to \mathbf{Set}$, and the free group construction $\mathbf{Set} \to \mathbf{Group}$. In this case, i in (i) below corresponds to the bijection between homomorphisms from a free group into an arbitrary group, and set-maps from the free generating set into the underlying set of this group; and u_D in (ii) is the set-map taking each element of the given set D to the corresponding generator of the free group on D, there written R_D. In (iii) η is this same map, thought of as depending functorially on the set D. To describe the morphism ε of (iii), start with a group G and form the free group on its underlying set, writing $[g]$ for the free generator corresponding to each element g of G. Then $\varepsilon(G)\colon FV(G) \to G$ is the group homomorphism sending each group-theoretic word in these free generators $[g]$ to the value in G of the same word in the original elements g. The reader might check that the assertions of (iii) hold for the above example.

PROPOSITION 5.3. *Let* $\mathbf{C}$ *and* $\mathbf{D}$ *be categories. Then the following data are equivalent:*

(i) *A pair of functors* $V\colon \mathbf{C} \to \mathbf{D}$, $F\colon \mathbf{D} \to \mathbf{C}$, *and an isomorphism*

(5.4) $$i\colon\ \mathbf{C}(F(-), -) \cong \mathbf{D}(-, V(-))$$

as functors $\mathbf{D}^{\mathrm{op}} \times \mathbf{C} \to \mathbf{Set}$.

(ii) *A functor* $V\colon \mathbf{C} \to \mathbf{D}$, *and for every* $D \in \mathrm{Ob}(\mathbf{D})$, *an object* $R_D \in \mathrm{Ob}(\mathbf{C})$ *with a morphism* $u_D \in \mathbf{D}(D, V(R_D))$ *universal among morphisms* $D \to V(C)$ $(C \in \mathrm{Ob}(\mathbf{C}))$, *i.e., a representing object for the functor* $\mathbf{D}(D, V(-))\colon \mathbf{C} \to \mathbf{Set}$.

(ii*) *A functor* $F\colon \mathbf{D} \to \mathbf{C}$, *and for every* $C \in \mathrm{Ob}(\mathbf{C})$, *an object* $S_C \in \mathrm{Ob}(\mathbf{D})$ *and a morphism* $v_C \in \mathbf{C}(F(S_C), C)$ *universal among morphisms* $F(D) \to C$ $(D \in \mathrm{Ob}(\mathbf{D}))$, *i.e., a representing object for the contravariant functor* $\mathbf{C}(F(-), C)\colon \mathbf{D} \to \mathbf{Set}$.

(iii) *A pair of functors* $V\colon \mathbf{C} \to \mathbf{D}$, $F\colon \mathbf{D} \to \mathbf{C}$, *and a pair of morphisms of functors*

$$\eta\colon \mathrm{Id}_{\mathbf{D}} \to VF, \qquad \varepsilon\colon FV \to \mathrm{Id}_{\mathbf{C}},$$

such that the two composite morphisms

(5.5) $$V = \mathrm{Id}_{\mathbf{D}}\, V \xrightarrow{\eta V} VFV \xrightarrow{V\varepsilon} V\,\mathrm{Id}_{\mathbf{C}} = V,$$

(5.6) $$F = F\,\mathrm{Id}_{\mathbf{D}} \xrightarrow{F\eta} FVF \xrightarrow{\varepsilon F} \mathrm{Id}_{\mathbf{C}}\, F = F$$

are the identity morphisms of the functors V *and* F *respectively.*

SKETCH OF PROOF. To get (ii) from (i), put in a fixed D for the first argument in (5.4). The resulting isomorphism of functors $\mathbf{C} \to \mathbf{Set}$ says that $F(D)$ represents the functor referred to in the last phrase of (ii). The passage in the opposite direction is also easy: Given (ii), choose for each D a corresponding object R_D. The functors $\mathbf{D}(D, V(-))$ which these objects represent have morphisms among them induced by the various morphisms of $\mathbf{D}$, so by Yoneda's Lemma, the representing objects inherit such morphisms as well, and the system of these representing objects constitutes a functor F. (There is a reversal of variance in going from objects of $\mathbf{D}$ to represented covariant functors, and another in going from these back to objects of $\mathbf{C}$, with the result that F is, like V, covariant.) The system of one-variable isomorphisms of functors saying that each $F(D)$ represents $\mathbf{D}(D, V(-))$ now together give the desired isomorphism (5.4) of 2-variable functors. The equivalence of (i) and (ii*) holds by the dual arguments.

Given (ii) one obtains η for (iii) as the construction associating to each D the universal map $u_D: D \to VF(D)$, while to get $\varepsilon(C)$ one looks at the case $D = V(C)$ of (ii), and applies the universal property of $R_D = FV(C)$ to the identity map of $V(C)$.

One verifies that these ε and η are indeed morphisms of functors, and notes that the commutative diagram associated with the case of the universal property of R_D that gave ε is precisely the statement that the composite shown in (5.5) is 1_V. To obtain the corresponding result for (5.6), one could show that the morphisms ε and η arising from (ii*) agree with those obtained from (ii), and call on the dual of the above argument using (ii*) in place of (ii); however, let us see how what we want can be proved directly from (ii). The uniqueness part of the universal property of R_D just mentioned tells us that a morphism f in $\mathbf{C}$ from an object $F(D)$ to any other object is uniquely determined by the morphism that one gets on applying V to f, and then composing with $u_D = \eta(D)$ on the right. Thus, (5.6) will give the identity morphism of F if, on applying the functor V, and then putting in an $\eta: \mathrm{Id}_{\mathbf{D}} \to VF$ on the left side of the diagram, the resulting composite

$$\mathrm{Id}_{\mathbf{D}} \xrightarrow{\eta} VF = VF\,\mathrm{Id}_{\mathbf{D}} \xrightarrow{VF\eta} VFVF \xrightarrow{V\varepsilon F} VF$$

is just η. Now if we regard the leftmost arrow of the above diagram as $\eta\,\mathrm{Id}_{\mathbf{D}}: \mathrm{Id}_{\mathbf{D}}\,\mathrm{Id}_{\mathbf{D}} \to VF\,\mathrm{Id}_{\mathbf{D}}$, and apply the "interchange law" for morphisms of composite functors ([**122**, p. 43, (3)]) to the first two arrows of this diagram, and finally drop factors $\mathrm{Id}_{\mathbf{D}}$ from the new left-hand arrow, the diagram becomes

$$\mathrm{Id}_{\mathbf{D}} \xrightarrow{\eta} VF = \mathrm{Id}_{\mathbf{D}} \circ VF \xrightarrow{\eta VF} VFVF \xrightarrow{V\varepsilon F} VF.$$

We then see that the last two arrows of this diagram are simply the diagram (5.5), composed on the right with F. Applying the result already proved, that the composite of the maps of (5.5) is the identity morphism, we get the desired result.

Finally, assuming (iii), one wants to get an isomorphism i as in (i). Given $f: F(D) \to C$, one forms $V(f): VF(D) \to V(C)$, and composes on the right with $\eta(D): D \to VF(D)$ to get $i(f): D \to V(C)$. A similar construction transforms morphisms $g: D \to V(C)$ to morphisms $j(g): F(D) \to C$. The verification that these constructions are morphisms of functors is straightforward, and (5.5) and (5.6) supply the conditions needed to prove that they are inverse to one another. □

Definition 5.7. *The data described by Proposition 5.3* (i) *or* (iii) *is called an* adjunction *between* $V\colon \mathbf{C} \to \mathbf{D}$ *and* $F\colon \mathbf{D} \to \mathbf{C}$. *The functor* V *is called the* right *adjoint of* F, *and* F *the* left *adjoint of* V. *The morphisms* η *and* ε *are called respectively the* unit *and* counit *of the adjunction.*

The term "adjoint" was suggested by the analogy with *adjoint operators* in analysis, which satisfy $(Ax, y) = (x, By)$. "Left" and "right" point to the positions of the functors within the hom-sets in (5.4).

5.8. *Digression.* The concepts of left and right adjoint are mutually dual, in the sense that, if $\mathbf{C} \underset{F}{\overset{V}{\rightleftarrows}} \mathbf{D}$ are adjoints, and we turn these into functors between the opposite categories $\mathbf{C}^{\mathrm{op}} \underset{F}{\overset{V}{\rightleftarrows}} \mathbf{D}^{\mathrm{op}}$, then they remain adjoints, but with the roles of F and V (and hence of $\mathbf{C}$ and $\mathbf{D}$) interchanged. From this point of view, Proposition 5.3 is self-dual. Note that the functors V and F in that Proposition are both *covariant*. Suppose, however, that for two categories $\mathbf{C}$ and $\mathbf{D}$ one considers an adjoint pair $\mathbf{C}^{\mathrm{op}} \underset{F}{\overset{V}{\rightleftarrows}} \mathbf{D}$, or $\mathbf{C} \underset{F}{\overset{V}{\rightleftarrows}} \mathbf{D}^{\mathrm{op}}$. If we regard each of these adjunctions as a pair of contravariant functors between $\mathbf{C}$ and $\mathbf{D}$, we find that they define two *distinct* situations, which are dual to one another, but in each of which V and F (and hence $\mathbf{C}$ and $\mathbf{D}$) play equivalent roles. These are called respectively "a pair of mutually left adjoint contravariant functors" and "a pair of mutually right adjoint contravariant functors". We shall say a little about these phenomena in point 9.4 below.

Returning to covariant adjunctions, we note the very convenient result:

Lemma 5.9. *Suppose that* $\mathbf{C} \underset{F}{\overset{V}{\rightleftarrows}} \mathbf{D} \underset{G}{\overset{W}{\rightleftarrows}} \mathbf{E}$ *are adjoint pairs of functors* (F *left adjoint to* V, G *left adjoint to* W). *Then the composite functors* $\mathbf{C} \underset{FG}{\overset{WV}{\rightleftarrows}} \mathbf{E}$ *are also adjoint* (FG *left adjoint to* WV).

Sketch of Proof. $\mathbf{C}(FG(-), -) \cong \mathbf{D}(G(-), V(-)) \cong \mathbf{E}(-, WV(-))$. □

Example 5.10. Consider the diagram of "forgetful functors"

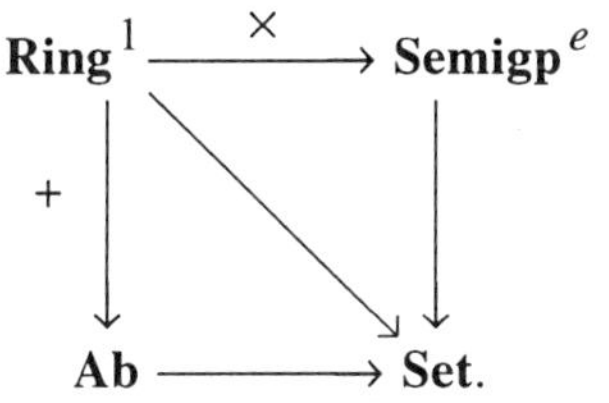

(The categories at the top of the diagram are those of associative rings with multiplicative neutral element 1, and semigroups with neutral element e respectively, with morphisms required to respect the neutral element as well as the other operations in each case. **Ab** is the category of abelian groups.)

Each of these functors has a left adjoint: The adjoints of the three functors to **Set** are the free ring, free abelian group, and free (nonabelian) semigroup constructions; those of the other two functors from $\mathbf{Ring}^1$ are the tensor ring construction, and the semigroup ring construction. Since the above diagram commutes, and composites of adjoints are adjoints of the composites, the corresponding diagram of left adjoints,

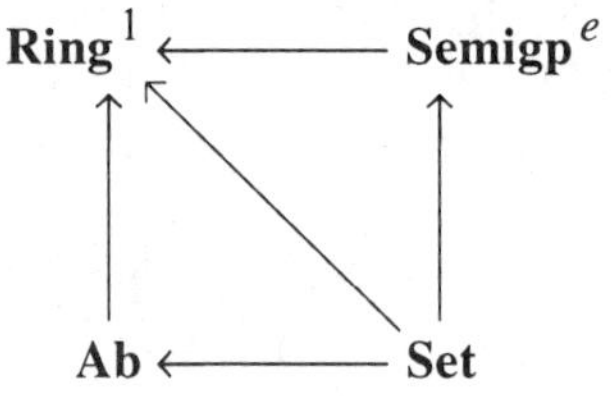

also commutes (up to natural isomorphism). These commutativity conditions say that a free ring (diagonal arrow) can be constructed stepwise, either as a semigroup ring on a free semigroup (upper-right path), or as a tensor ring on a free abelian group (lower-left path). These are well-known facts, but it is instructive to see the principle behind them.

A system of objects and arrows in a category **C** satisfying some specified composition-relations (e.g., a commuting square) can be interpreted as a functor from an appropriately defined "diagram category" into **C**. The **D** in the next definition should be thought of as a diagram category.

DEFINITION 5.11. *Let* $A: \mathbf{D} \to \mathbf{C}$ *be a functor. Then*

(i) *A* limit *of* A *means an object* $L \in \mathrm{Ob}(\mathbf{C})$ *given with morphisms* $p_D: L \to A(D)$ *for all* $D \in \mathrm{Ob}(\mathbf{D})$, *such that for every morphism* $f: D_1 \to D_2$ *of* $\mathbf{D}$, *the triangle formed by* p_{D_1}, p_{D_2} *and* $A(f)$ *commutes, and which moreover is* universal *for this property, in the sense that given any object* M *of* $\mathbf{C}$, *every system of morphisms* $r_D: M \to A(D)$ *forming commuting triangles with the morphisms* $A(f)$ *is induced by the system of morphisms* p_D, *via composition with a unique morphism of objects* $M \to L$. *In this situation we write* $L = \varprojlim A$, *and call* p_D *the* projection *of* L *to* $A(D)$.

(ii) *A* colimit *of* A *means an object* $L' \in \mathrm{Ob}(\mathbf{C})$ *given with morphisms* $q_D: A(D) \to L'$ *for all* $D \in \mathrm{Ob}(\mathbf{D})$, *such that for every morphism* $f: D_1 \to D_2$ *of* $\mathbf{D}$, *the triangle formed by* $A(f)$, q_{D_1}, *and* q_{D_2} *commutes, and universal for this property, in the sense that every system of morphisms of the* $A(D)$ *into a fixed object,* $r_D: A(D) \to M$, *forming commuting triangles with the morphisms* $A(f)$ *is induced by the* q_D, *via a unique morphism* $L' \to M$. *In this situation we write* $L' = \varinjlim A$, *and call* q_D *the* coprojection *of* $A(D)$ *to* L'.

Remark. Limits of systems of mathematical objects over diagrams of a sort exemplified by the picture $\ldots\to\cdot\to\cdot\to\cdot\to\cdot$ are commonly called *inverse limits* or *projective limits*, and *colimits* over diagrams of a form exemplified by the picture $\cdot\to\cdot\to\cdot\to\cdot\to\ldots$ are called *direct* or *inductive limits*. In these cases, it is clear why the constructed object is seen as a "limiting" case of the given chain of objects.

The term *limit* in Definition 5.11, and by dualization, the term *colimit*, arise by generalization of those concepts, though in fact, with other sorts of diagram-categories $\mathbf{D}$, these definitions include important constructions that have little to suggest "limiting" behavior: pushouts, pullbacks, fixed-point-sets, etc.; as well as products and coproducts, to be discussed in the paragraph after next. But the use of the terms "limit" and "colimit" has become standard.

Note that if *every* $A \in \mathbf{C}^{\mathbf{D}}$ has a limit (respectively, a colimit), then the construction $A \mapsto \underleftarrow{\mathrm{Lim}}\, A$, respectively $A \mapsto \underrightarrow{\mathrm{Lim}}\, A$, is a functor $\mathbf{C}^{\mathbf{D}} \to \mathbf{C}$ right adjoint (respectively left adjoint) to the *constant* or *diagonal* functor $\Delta: \mathbf{C} \to \mathbf{C}^{\mathbf{D}}$. (For each object C of $\mathbf{C}$, $\Delta(C) \in \mathbf{C}^{\mathbf{D}}$ is defined to take every object of $\mathbf{D}$ to C, and every morphism of $\mathbf{D}$ to 1_C.) Without the assumption that *every* member of $\mathbf{C}^{\mathbf{D}}$ has a (co)limit, we can still translate the concept of a limit or colimit for a particular $A \in \mathbf{C}^{\mathbf{D}}$ to that of an object of $\mathbf{C}$ representing the functor $\mathbf{C}^{\mathbf{D}}(\Delta(-), A)$, respectively $\mathbf{C}^{\mathbf{D}}(A, \Delta(-))$. For $M \in \mathrm{Ob}(\mathbf{C})$, elements of $\mathbf{C}^{\mathbf{D}}(A, \Delta(M))$, are often called *cones* from the diagram A to the object M, and elements of $\mathbf{C}^{\mathbf{D}}(\Delta(M), A)$ cones from M to A [**122**].

An important degenerate case of these concepts is that in which $\mathbf{D}$ is a *discrete* category, i.e., a set I of objects with no morphisms but identity maps. Then functors $A \in \mathbf{C}^{\mathbf{D}}$ may be regarded as I-tuples of objects of $\mathbf{C}$, $(A_i)_{i \in I}$, and the limit (when it exists) of such a system is familiar under the name of the *product* $\prod_I A_i$ of the A_i in $\mathbf{C}$, while the colimit is called the *coproduct* $\coprod_I A_i$. A still more degenerate case arises when $\mathbf{D}$ is the *empty* category $\mathbf{0}$ (no objects, no morphisms): the universal property characterizing the *colimit* of the unique object of $\mathbf{C}^{\mathbf{0}}$, i.e., the *coproduct* of the *empty* family of objects of $\mathbf{C}$, is the definition of an *initial object* of $\mathbf{C}$. A *limit* of the same functor, a product of the empty family, is a *final object*.

(All the universal constructions we have been considering can in fact be characterized in terms of these very degenerate ones; for an object representing an arbitrary set-valued functor $V: \mathbf{C} \to \mathbf{Set}$ can be described as an *initial object* in the category of pairs (C, x) with $C \in \mathrm{Ob}(\mathbf{C})$, $x \in V(C)$. E.g., the free group on n generators is an initial object in the category of groups with distinguished n-tuples of elements. Hence Lang [**115**], for instance, uses the concepts of initial and final object, under the names *universal repelling* and *universal attracting* objects, as the basic universal constructions in terms of which to develop all others. Final objects are also sometimes called *terminal* objects.)

It is easy to verify that the following is simply a restatement of the universal properties defining products and coproducts:

LEMMA 5.12. *Let* (A_i) *be a family of objects in a category* $\mathbf{C}$. *Then*

(i) *An object* P *given with a family of morphisms* $p_i: P \to A_i$ *is a product of the* A_i*'s in* $\mathbf{C}$ *if and only if for every object* X *of* $\mathbf{C}$, *the set* $\mathbf{C}(X, P)$, *with the induced maps to the sets* $\mathbf{C}(X, A_i)$, *is (up to isomorphism) the set-theoretic direct product of the latter sets, with its projection maps.*

(ii) *An object* Q *given with a family of morphisms* $q_i: A_i \to Q$ *is a coproduct of this family if and only if for every object* X *of* $\mathbf{C}$, *the set* $\mathbf{C}(Q, X)$, *with the induced maps to the sets* $\mathbf{C}(A_i, X)$, *is likewise (up to isomorphism) the set-theoretic direct product of those sets, with its projection maps.* □

Note that because the hom-sets of (i) above are covariant in P, while those of (ii) are contravariant in Q, both product and coproduct constructions get turned into set-theoretic direct products.

EXERCISE 5.13 (mostly easy). (i) *Prove a generalization of Lemma 5.12 which characterizes limits and colimits of arbitrary diagrams.*

(ii) *Suppose* $\mathbf{C} \underset{F}{\overset{V}{\rightleftarrows}} \mathbf{D}$ *are adjoint functors, and* $W: \mathbf{D} \to \mathbf{Set}$ *is a representable functor, with representing object* R. *Show that* $WV: \mathbf{C} \to \mathbf{Set}$ *is representable, with representing object* $F(R)$.

(iii) *Formulate precisely, and prove, a result to the effect that if* $\mathbf{C} \underset{F}{\overset{V}{\rightleftarrows}} \mathbf{D}$ *are adjoints, then* F *respects colimits, and* V *respects limits.*

(iv) *Verify that in the category of k-bimodules (k any associative unital ring) coproducts are direct sums; and that the forgetful functor from unital k-rings (Definition 10.3 below) to k-bimodules has as left adjoint the tensor ring construction* ((2.1) *above*). *Deduce that the coproduct of the tensor rings on a family of k-bimodules is the tensor ring on the direct sum of these bimodules.*

We end this section with some concepts that are less central to our work, but will occur from time to time.

DEFINITION 5.14 ([**122**, pp.88-90]). *A subcategory* $\mathbf{D}$ *of a category* $\mathbf{C}$ *is said to be* reflective *(N.B.: not "reflexive") if the inclusion of* $\mathbf{D}$ *in* $\mathbf{C}$ *has a left adjoint. Dually,* $\mathbf{D}$ *is said to be* coreflective *if that inclusion has a right adjoint.*

For example, **Ab** is a reflective subcategory of **Group**; we shall soon see a very general result, Corollary 8.18, of which this is a particular case. (Cf. also §9.2.) Coreflectivity is not so ubiquitous in algebra, but we will encounter some important examples in §14.

It is not hard to see that the left adjoint of the inclusion of a reflective full subcategory $\mathbf{D}$ in a category $\mathbf{C}$ is, up to isomorphism, a *left inverse* to that inclusion (e.g., the abelianization functor, which is the left adjoint to the inclusion of **Ab** in **Group**, when restricted to abelian groups is isomorphic to the identity functor). Indeed, adjunctions $\mathbf{C} \underset{F}{\overset{V}{\rightleftarrows}} \mathbf{D}$ such that the right adjoint is right inverse to the left adjoint, equivalently, such that the canonical map $\varepsilon: FV \to \mathrm{Id}_{\mathbf{C}}$ is an isomorphism, correspond, up to isomorphism, to instances of reflective full subcategories [**122**, Exercise IV.4.4]. Likewise, adjunctions where the right adjoint is *left* inverse to the left adjoint, i.e., where $\eta: \mathrm{Id}_{\mathbf{D}} \to VF$ is an isomorphism, correspond to instances of coreflective full subcategories.

It is not hard to verify that a full (co)reflective subcategory $\mathbf{D}$ of a category $\mathbf{C}$ is also (co)reflective in any intermediate full subcategory.

We recall next

DEFINITION 5.15. *Let* $f\colon A \to B$ *be a morphism in a category* **C**. *Then* f *is called a* monomorphism *if for every object* C *of* **C**, *and pair of morphisms* g_1, $g_2\colon C \rightrightarrows A$, *we have* $fg_1 = fg_2 \Rightarrow g_1 = g_2$. *Dually,* f *is called an* epimorphism *if for every* C *and pair of morphisms* e_1, $e_2\colon B \rightrightarrows C$, *we have* $e_1 f = e_2 f \Rightarrow e_1 = e_2$; *equivalently, if and only if as a morphism in* $\mathbf{C}^{\mathrm{op}}$, f *is a monomorphism.*

In most of the concrete categories commonly considered in mathematics (in particular, in any *variety of algebras*, definition recalled in the next section, and more generally in any concrete category whose underlying-set functor has a left adjoint), the concept of "monomorphism" translates to "map which is one-to-one on underlying sets". The relation between being an epimorphism and surjectivity is less predictable. The latter condition implies the former in any concrete category. In many categories, such as **Set** and **Group**, they are equivalent; but in many others such as $\mathbf{Semigp}^e$ and $\mathbf{Ring}^1$ there are epimorphisms which are not surjective. For instance, an extension of a semigroup or ring obtained by adjoining an inverse of a previously noninvertible element will be an epimorphism, but not surjective. Cf. [**122**], [**92-96**]. If A is an object of a category, then an object B given with an epimorphism $f\colon A \to B$ is called an *epimorph* of A. (In [**122**] epimorphisms and monomorphisms are renamed "epic" and "monic" morphisms, in an attempt to restore the longer words to their pre-category-theoretic meanings of surjective and injective homomorphism; we will not follow that terminology here.)

Finally, we shall at times refer to the concept of a *small category*, that is, a category in which the class of all objects and the class of all morphisms are both "small". Classically, this would mean that they are *sets* rather than proper classes. For example, the diagram category $\cdot\!\to\!\cdot\!\to\!\cdot\!\to\!\cdot\!\to\!\ldots$ mentioned previously, which has object-set indexed by the natural numbers, and at most one morphism in each hom-set, is small; the category of all groups is not. (Even the category of all finite groups is not, though it has a small *skeleton*, i.e., a full small subcategory such that every object of the large category is isomorphic to an object of the subcategory.) There are now alternative ways of formulating what one wants from this concept. For instance, suppose we define a *universe* as a *set* closed under all the operations of conventional set-theory, and suppose we assume a set-theory in which every set belongs to a universe. (Foundation theorists tell us this means assuming that there are enough inaccessible cardinals in our set-theory.) One may then define a set to be small or large *with respect to* a universe U, if it is a member, respectively a subset, of U. (A set not contained in U will be neither.) Now categories "large" with respect to one universe (i.e., having a large set of objects and a large set of morphisms between any two objects) will be "small" with respect to a bigger universe, and constructions like "the category of all categories" (i.e., all categories large with respect to a given universe) can be treated within a theory developed for ordinary (i.e., merely large), or even small categories. Cf. [**46**, p.6], [**24**, §6.4] where the above approach is used, or [**122**], [**121**] which use a weaker set-theoretic assumption; for discussion see [**122**, §I.6], [**121**], [**63**]. We leave it to the reader to choose which version of the concepts of "large" and "small" categories he or she prefers, since the difference will not be significant for the present work. Between the extremes of "small" and "large" categories is the concept of a *legitimate* category, one whose object-class is large, and where for any two objects X and Y, $\mathbf{C}(X, Y)$ is small. So the large categories of all small groups, sets, etc. (and most categories commonly encountered) are legitimate.

§6. Basic definitions and results of universal algebra.

In this section we shall bypass formal statements, since the main definitions and results are natural generalizations of cases with which the reader will be familiar. For details, cf. [**46**], [**73**], [**131**], [**24**] or any other general text on universal algebra.

An *algebra* $A = (|A|, (\alpha_A)_{\alpha \in I})$ in the sense of universal algebra means a set $|A|$ given with a family of operations $\alpha_A : |A|^{n(\alpha)} \to |A|$ $(\alpha \in I,\ 0 \leq n(\alpha) < \infty)$. For each α the integer $n(\alpha)$ is called the *arity* of α, and α is called an $n(\alpha)$*-ary* operation. Note that a zeroary operation is essentially a specification of a distinguished element of $|A|$. Two algebras whose operations are indexed by the same set I, with corresponding operations having the same arity, are said to be *of the same type*. For simplicity we have required all operations to have finite arities; however, we do allow algebras with infinitely *many* operations (e.g., modules over an infinite ring R). If A and B are algebras of the same type, a *homomorphism* $f: A \to B$ means a map of underlying sets which respects corresponding operations.

In the class of all algebras of a given type, a subclass consisting of all algebras satisfying a specified family of identities is called a *variety*; familiar examples are the variety of all groups, the variety of all associative rings, the variety of all lattices, the variety of all R-modules for a fixed ring R, and many well-known subvarieties (e.g., abelian groups) and over-varieties (e.g., not-necessarily-associative rings) of these.

By Birkhoff's famous theorem, a class of algebras of the same type is a variety if and only if it is closed in the class of all algebras of that type under taking *subalgebras, direct product algebras,* and *homomorphic images* [**46**], [**24**, Theorem 8.6.1].

We shall consider every *variety* of algebras as a *category*, with the homomorphisms (defined above) for morphisms. The category $\mathbf{C}$ of all algebras of a given type admits the construction of limits of functors $A: \mathbf{D} \to \mathbf{C}$ for all *small* categories $\mathbf{D}$; namely, $\varprojlim A$ may be obtained as the subalgebra of $\prod_{D \in \mathrm{Ob}(\mathbf{D})} A(D)$ whose underlying set is

$$\{(x_D) \in \textstyle\prod_{D \in \mathrm{Ob}(\mathbf{D})} |A(D)| \mid (\forall\, D_1, D_2 \in \mathrm{Ob}(\mathbf{D}), f \in \mathbf{D}(D_1, D_2)),\ A(f)(x_{D_1}) = x_{D_2}\}.$$

Since any variety $\mathbf{V} \subseteq \mathbf{C}$ is closed under products and subalgebras, it will also be closed under limits, and it is immediate that the limit in $\mathbf{C}$ of a system of objects and morphisms of $\mathbf{V}$ will be a limit of this system in $\mathbf{V}$ as well.

In addition to being closed under the three operations mentioned in Birkhoff's Theorem, any variety $\mathbf{V}$ admits the construction of free objects on any set (a key step in the proof of that Theorem), and more generally, of objects presented by generators and relations; hence of colimits of all functors from small categories. (The colimit of $A: \mathbf{D} \to \mathbf{V}$ may be presented by taking for a system of generators a disjoint union of the underlying sets of all the algebras $A(D)$, and imposing the relations specifying the internal operations of these algebras, and also relations identifying elements of different algebras that are mapped to one another under the morphisms $A(f)$, $f \in \mathrm{Ar}(\mathbf{D})$.)

While we have noted that *limits* are the same in the category of all algebras of a given type and in any subvariety thereof, the *colimit* of a diagram will in general depend on the variety in which it is taken. For instance, if A and B are abelian groups, their coproduct in the category of abelian groups is their direct sum, their coproduct in the category of all groups is what group-theorists call their "free product", which is in general noncommutative, and their coproduct in the category of all algebras of the same type as groups is an object with no particular name, which is not in general a group.

A *derived operation* of an algebra (or of a variety of algebras) means an operation formed by "composition" from the "primitive" operations. E.g., in the variety of groups, the derived unary operation of squaring, $x \mapsto x^2$, and the derived binary commutator operation, $(x, y) \mapsto xyx^{-1}y^{-1}$, are examples which happen to have names.

Although we have restricted our *primitive* operations to be finitary, we shall allow ourselves to speak of formally infinitary *derived* operations. This is so that for any set X, we can speak of the set of "all derived operations in an X-tuple of variables". But each such operation will, of course, in fact depend nontrivially on only finitely many members of this set of variables, since it is constructed from our finitary primitive operations.

§7. Some conventions followed throughout this work.

(These supplement the basic category-theoretic conventions noted in the initial paragraphs of §5.)

The words *algebra* and *variety of algebras* will be used in the sense of universal algebra, recalled in the preceding section. If $\mathbf{V}$ is a variety of algebras, an object of $\mathbf{V}$ will sometimes be called a $\mathbf{V}$*-algebra*. However, for k a commutative ring, the term *k-algebra* will be used in the ring-theoretic sense, with conditions such as *associative, commutative, Lie*, etc. specified in the context. There is little danger of confusion, since we will not use similar symbols for varieties $\mathbf{V}$ and rings k; moreover, we will deal less with k-algebras than with the more general concept of *k-ring*, defined in §10 below.

If A, B are two objects with coproduct $A \amalg B$ in a category $\mathbf{C}$, and $f\colon A \to C$, $g\colon B \to C$ are morphisms into a third object, then the morphism $A \amalg B \to C$ determined by these under the universal property of $A \amalg B$ will be written (f, g). On the other hand, given four objects and two morphisms, $a\colon A \to A'$, $b\colon B \to B'$ the morphism induced by the "functoriality" of the coproduct operation will be denoted $a \amalg b\colon A \amalg B \to A' \amalg B'$. (We put "functoriality" in quotes because to speak of $\amalg$ as a functor on $\mathbf{C} \times \mathbf{C}$, we need coproducts to be defined on all pairs of objects of $\mathbf{C}$, but for the above construction to make sense only $A \amalg B$ and $A' \amalg B'$ need exist. In fact, in terms of the preceding convention, one has $a \amalg b = (q_{A'}\, a,\ q_{B'}\, b)$, where $q_{A'}\colon A' \to A' \amalg B'$, $q_{B'}\colon B' \to A' \amalg B'$, are the coprojection morphisms.) Obvious extensions and analogs of these conventions apply to coproducts of more than two objects, and to products. (Hence, given morphisms $f, g\colon A \rightrightarrows B$ in a category, the symbol

(f, g) may, depending on context, denote the induced morphism $A \amalg A \to B$, the induced morphism $A \to B \times B$, or simply the ordered pair of maps.)

When we say that a category $\mathbf{C}$ *has finite products* or *has finite coproducts*, this will mean that every finite family of objects of $\mathbf{C}$ has a product, respectively a coproduct. These conditions will be understood to include the existence of a product of the *empty* family, i.e., a final object, respectively a coproduct of the empty family, an initial object.

Any variety of algebras with no zeroary operations includes an *empty algebra*, which is its initial object. (This would hardly need stating, except that some universal algebraists specifically require algebras to be nonempty. Under their definition, varieties may not have initial objects, nor some other very basic constructions, e.g., difference kernels.)

§8. Algebra and coalgebra objects in a category, and representable algebra-valued functors.

Though *coalgebra* objects in varieties of algebras are to be the main topic of this work, it will be heuristically most convenient to begin with the concept of an *algebra object* in a category, and arrive at coalgebras by duality.

Until the contrary is stated, $\mathbf{C}$ *will be a category with finite products.*

Note that for any category $\mathbf{D}$, $\mathbf{C}^{\mathbf{D}}$ will also have finite products: If $(A_i)_{i \in I}$ is a finite family of functors $\mathbf{D} \to \mathbf{C}$, we may construct their product $\prod A_i$ as the functor with $(\prod A_i)(X) = \prod A_i(X)$ $(X \in \mathrm{Ob}(\mathbf{D}))$.

For any object S of $\mathbf{C}$ and any nonnegative integer n, S^n will denote the product of n copies of S in $\mathbf{C}$, indexed by $\{1, \ldots, n\}$.

By an *n-ary operation* on an object S of $\mathbf{C}$, we will mean a morphism $\alpha = \alpha_S : S^n \to S$. Such an operation induces for each object A of $\mathbf{C}$ an n-ary operation $\alpha_{\mathbf{C}(A, S)}$ on the set $\mathbf{C}(A, S)$: An n-tuple of elements $\xi_1, \ldots, \xi_n \in \mathbf{C}(A, S)$ induces a single morphism in $\mathbf{C}(A, S^n)$, and composing this with $\alpha: S^n \to S$ gives the element of $\mathbf{C}(A, S)$ which we will call $\alpha_{\mathbf{C}(A, S)}(\xi_1, \ldots, \xi_n)$. This is just the category-theoretic version of the familiar idea of making the set of functions from a space A to an algebra S an algebra under *pointwise* operations.

In fact, we have

LEMMA 8.1. *Let n be a nonnegative integer and S an object of* $\mathbf{C}$. *Then the following structures are equivalent, via the construction described above:*

(i) *An n-ary operation* $\alpha_S : S^n \to S$.

(ii) *A morphism* $\alpha_{\mathbf{C}(-, S)} : \mathbf{C}(-, S)^n \to \mathbf{C}(-, S)$ *as functors* $\mathbf{C}^{\mathrm{op}} \to \mathbf{Set}$, *i.e., as contravariant set-valued functors on* $\mathbf{C}$.

(iii) *A way of defining on all sets* $\mathbf{C}(A, S)$ $(A \in \mathrm{Ob}(\mathbf{C}))$ *n-ary operations* $\alpha_{\mathbf{C}(A, S)}$: $\mathbf{C}(A, S)^n \to \mathbf{C}(A, S)$ *so that for every morphism* $f \in \mathbf{C}(A, B)$ *the induced map* $\mathbf{C}(B, S) \to \mathbf{C}(A, S)$ *respects the operations assigned to these sets.*

PROOF. (ii) and (iii) are clearly equivalent. To get the equivalence between (i) and (ii), note that by Yoneda's Lemma, a morphism $S^n \to S$ is equivalent to a morphism of represented functors, $\mathbf{C}(-, S^n) \to \mathbf{C}(-, S)$, and that by Lemma 5.12(i), $\mathbf{C}(-, S^n)$ can be written $\mathbf{C}(-, S)^n$. It is straightforward to verify that the resulting correspondence between operations on S and operations on $\mathbf{C}(-, S)$ is the one described in the preceding discussion. □

DEFINITION 8.2. *An* algebra object S *in the category* $\mathbf{C}$ *(or a* $\mathbf{C}$-based algebra) *will mean an object* $|S| \in \mathrm{Ob}(\mathbf{C})$ *given with a family* $(\alpha_S)_{\alpha \in I}$ *of operations of nonnegative integer arities,*

$$\alpha_S\colon\ |S|^{n(\alpha)} \to |S|.$$

Two algebra objects in the same or different *categories* $\mathbf{C}$, $\mathbf{C}'$ *are said to be of the* same type *if their operations are indexed by the same set* I, *and corresponding operations have the same arities* $n(\alpha)$. *A morphism between algebra objects of the same type in the same category* $\mathbf{C}$ *will mean a morphism between underlying objects in* $\mathbf{C}$ *which forms commuting squares with the operations.*

If S *is an algebra object of* $\mathbf{C}$, *and* A *any object of* $\mathbf{C}$, *then* $\mathbf{C}(A, S)$ *will denote the ordinary (i.e., set-based) algebra with underlying set* $\mathbf{C}(A, |S|)$, *and operations induced by those of* S *as described in Lemma 8.1 and the preceding discussion.*

Note that the $|\ |$-notation is relative; e.g., if $\mathbf{C}$ is a category of algebras, and S a $\mathbf{C}$-based algebra, then $|S|$ denotes the underlying $\mathbf{C}$-object of S, and if R is this object, $|R|$ denotes its underlying *set*.

The idea of defining algebra objects in a general category seems to have been introduced by Eckmann and Hilton [**61**, Part I]. In the present work, the term "algebra" will continue to mean "set-based algebra" except when the contrary is indicated, e.g. by writing "algebra object", "$\mathbf{C}$-based algebra", "algebra in $\mathbf{C}$", etc.. Occasionally, when referring to set-based algebras, we will add the words "set-based" for emphasis.

Yoneda's Lemma easily yields:

LEMMA 8.3. *Let* S *and* T *be algebra objects of the same type in* $\mathbf{C}$. *Then the following data are equivalent*:

(i) *A morphism of* $\mathbf{C}$*-based algebras* $S \to T$.

(ii) *A morphism* $f \in \mathbf{C}(|S|, |T|)$ *such that for every object* A *of* $\mathbf{C}$, *the induced set map* $\mathbf{C}(A, |S|) \to \mathbf{C}(A, |T|)$ *is a homomorphism of algebras* $\mathbf{C}(A, S) \to \mathbf{C}(A, T)$.

(iii) *A morphism* $\mathbf{C}(-, S) \to \mathbf{C}(-, T)$ *of functors from* $\mathbf{C}$ *to the category of set-based algebras of the given type.* □

Our next goal will be to define the *derived operations* of an algebra object S of $\mathbf{C}$, corresponding to the various derived operations of set-based algebras of the same type. This will allow us to say what it means for an *identity* (such as associativity of a binary operation) to hold in S, namely that the derived operations represented by the two sides of the identity are equal. It is not at first obvious how to go about this, since when $\mathbf{C}$ is not $\mathbf{Set}$, one cannot describe a derived operation by giving a formula for its value on a tuple of "elements of S". One standard approach is to express operations and identities by diagrams. For example, observe that if m is a binary operation on a *set* $|S|$, the condition that m be associative means that the diagram

$$\begin{array}{ccc} |S| \times |S| \times |S| & \xrightarrow{m \times 1_{|S|}} & |S| \times |S| \\ \Big\downarrow{\scriptstyle 1_{|S|} \times m} & & \Big\downarrow{\scriptstyle m} \\ |S| \times |S| & \xrightarrow{\quad m \quad} & |S| \end{array}$$

commutes, the path through the upper right-hand corner representing the ternary derived operation $(x, y, z) \mapsto m(m(x, y), z)$, the one through the lower left-hand corner the operation $(x, y, z) \mapsto m(x, m(y, z))$. Now given an object $|S|$ of our general category-with-products $\mathbf{C}$, and a binary operation $m\colon |S| \times |S| \to |S|$, the same diagram can be used to define two ternary "derived operations" of $S = (|S|, m)$, and their equality can be made the definition of associativity. This approach is followed in [**122**, pp. 2-3].

Another approach, which is equivalent to the above, but avoids the dependence on diagrams, is based on considering the set-based algebra $\mathbf{C}(A, S)$ for an appropriate *universal* choice of A. If we want to consider derived operations in n variables, let us look at $\mathbf{C}(|S|^n, S)$. Since this is a set-based algebra, we can construct derived n-ary operations there from the primitive operations in the "usual" way. Applying such a derived operation u to the n projections $p_1, \ldots, p_n\colon |S|^n \to |S|$, we get an element $u(p_1, \ldots, p_n) \in \mathbf{C}(|S|^n, |S|)$, which we define to be the corresponding derived n-ary operation u_S of the $\mathbf{C}$-based algebra S.

LEMMA 8.4. *Let* $\mathbf{C}$ *be a category with finite products, and* S *an algebra object of* $\mathbf{C}$. *Let* u, v *be two derived operations in* n *variables, for ordinary (i.e., set-based) algebras of the same type as* S. *Then the following conditions are equivalent:*

(i) *The "diagrammatic translation" of the identity* $u = v$ *is satisfied by the algebra-object* S.

(ii) *In the algebra* $\mathbf{C}(|S|^n, S)$, *one has* $u(p_1, \ldots, p_n) = v(p_1, \ldots, p_n)$, *where* $p_1, \ldots, p_n\colon |S|^n \to |S|$ *are the projection maps.*

(iii) *For all* $A \in \mathrm{Ob}(\mathbf{C})$, *the algebra* $\mathbf{C}(A, S)$ *satisfies the identity* $u = v$.

PROOF. (i) and (ii) are equivalent because the "diagrammatic" identity is simply a visual display of the condition (ii). Clearly, (ii) is a special case of (iii); to show conversely that (ii) $\Rightarrow$ (iii), consider any object A of $\mathbf{C}$ and n elements $f_1, \ldots, f_n \in \mathbf{C}(A, |S|)$. These determine a morphism $(f_1, \ldots, f_n)\colon A \to |S|^n$, and this induces an algebra-homomorphism $\mathbf{C}(|S|^n, S) \to \mathbf{C}(A, S)$ carrying $p_1, \ldots, p_n$ to $f_1, \ldots, f_n$. Hence, any equation satisfied by the former n-tuple is also satisfied by the latter. □

DEFINITION 8.5. *An algebra object* S *of* $\mathbf{C}$ *will be said to* satisfy *an identity* $u = v$ *if the equivalent conditions of the above Lemma are satisfied.*

If $\mathbf{V}$ *is a variety of algebras, then a* $\mathbf{V}$-object *of* $\mathbf{C}$ *will mean an algebra object of* **C** *of the type of* **V** *which satisfies in this sense the identities defining* **V**.

The preceding three Lemmas have shown us what is involved in putting **V**-algebra structures on the values of a representable set-valued functor. Now note that the concept of "a representable set-valued functor given with operations that make it **V**-valued" can also be looked at as "a **V**-valued functor whose composite with the forgetful functor $\mathbf{V} \to \mathbf{Set}$ is representable". This observation is used in the second half of the next definition, where we extend the term "representable functor" to include these algebra-valued constructions.

DEFINITION 8.6. *A functor* $\mathbf{C}^{\mathrm{op}} \to \mathbf{V}$ *will be called* representable *if it is isomorphic to a functor of the form* $\mathbf{C}(-, S)$, *for* S *a* **V**-*object of* **C**, *equivalently (by Lemmas 8.1 and 8.4) if its composite with the underlying-set functor* $\mathbf{V} \to \mathbf{Set}$ *is representable in the sense of Definition 5.2.*

(In a category *not necessarily* having finite products, the equivalent data of Lemma 8.1(ii) and (iii) give a reasonable substitute for the concept of an n-ary operation on an object, and the condition of Lemma 8.4(iii) can still be used as the definition of an identity holding among such operations. This approach is sometimes used to generalize to arbitrary categories **C** the concepts of a **V**-object of **C** and of a representable **V**-valued functor on **C**.)

We are now going to dualize, and get the concept of a *covariant* representable functor from **C** to **V**. We drop here the assumption that **C** has finite products; the dual hypothesis will be stated where needed. Before discussing algebra-valued functors, however, let us set down a characterization, noted informally in §1, of covariant representable *set*-valued functors whose domains are varieties of algebras, since it is functors on varieties that will ultimately interest us, and every algebra-valued functor has an underlying set-valued functor. Given any object R of a variety **C**, if we take a presentation $R = \langle X \mid Y \rangle$, then for any algebra A in **C**, homomorphisms $R \to A$ correspond to X-tuples of elements of A which satisfy the family of relations Y. Writing each such relation "$u(X) = v(X)$", where u, v are X-ary derived operations of **C**, we see that we have

LEMMA 8.7. *Let* **C** *be a variety of algebras, and* $F\colon \mathbf{C} \to \mathbf{Set}$ *a functor. Then the following conditions are equivalent:*

(i) F *is* representable *in the sense of Definition 5.2, i.e., there exists an object* R *of* **C** *such that* F *is isomorphic to the functor* $\mathbf{C}(R, -)$.

(ii) *There exists a set* X, *and a family*

$$(u_j(X) = v_j(X))_{j \in J}$$

of formal relations in the operations of **C** *and the elements of* X, *(i.e., each* u_j *and each* v_j *is a derived operation of the variety* **C** *in the set* X *of indeterminates), such that* F *is isomorphic to the functor associating to every object* A *of* **C** *the set of solutions in* A *to the above equations:*

$$\{(\xi) \in |A|^X \mid u_j(\xi) = v_j(\xi) \ (j \in J)\}. \ \square$$

Hence, for $R = \langle X \mid Y\rangle$ as above, we will often denote an element of $\mathbf{C}(R, A)$ by a letter such as ξ, and call the elements $\xi(x) \in |A|$ $(x \in X)$ the "coordinates" of ξ.

Let us now give the duals of the concepts of the first half of this section.

DEFINITION 8.8. *Let* $\mathbf{C}$ *be a category with finite* coproducts. *Then an n-ary* co-operation *on an object* $|R|$ *of* $\mathbf{C}$ *will mean a morphism of* $|R|$ *into the coproduct of* n *copies of* $|R|$*; in other words, an n-ary operation on* $|R|$ *in* $\mathbf{C}^{\mathrm{op}}$. *An object* $|R|$ *given with a family* $(\alpha_R)_{\alpha\in I}$ *of co-operations will be called a* coalgebra object R *in* $\mathbf{C}$, *and for any coalgebra object* R, *and object* A *of* $\mathbf{C}$, *the set* $\mathbf{C}(|R|, A)$, *made an* algebra *under the induced operations taking tuples* $(\xi_1, \dots, \xi_{n(\alpha)}) \in \mathbf{C}(|R|, A)^{n(\alpha)}$ $(\alpha \in I)$ *to the composite morphisms*

$$|R| \xrightarrow{\alpha_R} |R| \amalg \dots \amalg |R| \xrightarrow{(\xi_1, \dots, \xi_{n(\alpha)})} A$$

will be written $\mathbf{C}(R, A)$.

Having noted in point (ii) of the preceding Lemma the form that a representable **Set**-valued functor takes when the domain category $\mathbf{C}$ is a variety of algebras, let us now do the same for an operation on such a functor. We know that if our functor is represented by an object $|R|$, then an n-ary operation on this functor is induced by a co-operation

$$\alpha_R \colon |R| \to |R| \amalg \dots \amalg |R|.$$

If X is a generating set for $|R|$, then this co-operation α_R will be determined by its values at the elements $x \in X \subseteq |R|$, and these values may be written, using the operations of the variety $\mathbf{C}$, as expressions w_x in the elements of the n copies of X generating the n-fold coproduct $|R| \amalg \dots \amalg |R|$. If we now look at the composite morphism displayed in the preceding Definition, we find that the image of each $x \in X$ under this morphism will be gotten by applying this same derived operation w_x to the coordinates of the n maps $\xi_1, \dots, \xi_n$. Thus, each w_x provides a general formula for the xth coordinate of $\alpha(\xi_1, \dots, \xi_n) \in \mathbf{C}(|R|, A)$ in terms of the coordinates of the n X-tuples $\xi_1, \dots, \xi_n$. This gives the first half of the following Lemma; the second half notes the conditions under which, conversely, a family of such derived operations determines a co-operation α.

LEMMA 8.9. *Let* $\mathbf{C}$ *be a variety of algebras,* $|R|$ *an object of* $\mathbf{C}$, *and* $\langle X \mid Y\rangle$ *a presentation of* $|R|$ *by generators and relations, and let us, following Lemma 8.7*(ii), *regard elements of morphism-sets* $\mathbf{C}(|R|, A)$ *as determined by X-tuples of elements of* A *satisfying the relations given by* Y.

Then if $\alpha_R \colon |R| \to |R| \amalg \dots \amalg |R|$ *is an n-ary co-operation on* $|R|$, *there exists an X-tuple of derived* $X \times n$*-ary operations* $(w_x)_{x\in X}$ *of* $\mathbf{C}$ *such that for every object* A *of* $\mathbf{C}$, *every n-tuple of elements* $\xi_1, \dots, \xi_n \in \mathbf{C}(|R|, A)$, *and every* $x \in X$, *the* x*th coordinate of* $\alpha(\xi_1, \dots, \xi_n)$ *is computed from the coordinates of* $\xi_1, \dots, \xi_n$ *by the operation* w_x.

Conversely, given an X-tuple of derived $X \times n$*-ary operations of* $\mathbf{C}$, *if the identities of* $\mathbf{C}$ *guarantee that when applied to* n *X-tuples satisfying the relations*

Y, the w_x give an X-tuple also satisfying Y, then this family determines as above a morphism of functors $\mathbf{C}(|R|, -)^n \to \mathbf{C}(|R|, -)$, equivalently, a co-operation α_R: $|R| \to |R| \amalg \ldots \amalg |R|$. □

The reader not familiar with the above ideas might think through what the operations w_x are when $\mathbf{C}$ and $\mathbf{V}$ are the varieties of rings and groups respectively, X and Y are the generators and relations given in (1.2) for the ring R representing the group-valued functor GL_n, and α is the group multiplication.

Let us again return to coalgebras in a general category $\mathbf{C}$.

Definition 8.10. *Let $\mathbf{C}$ be a category with finite coproducts, and $\mathbf{V}$ a variety of algebras, consisting of all algebras of a specified type which satisfy a specified system of identities Σ. Then a* co-$\mathbf{V}$ object *of $\mathbf{C}$ will mean a coalgebra R in $\mathbf{C}$ of the type of $\mathbf{V}$, satisfying the following equivalent conditions:*

(i) *R satisfies the (dualized) diagrammatic conditions ("coidentities") corresponding to the identities of Σ.*

(ii) *For each identity $u = v$ in Σ, say in n variables, if we form the n-fold coproduct $|R| \amalg \ldots \amalg |R|$ with its canonical coprojections $q_1, \ldots, q_n$, then in the algebra $\mathbf{C}(R, |R| \amalg \ldots \amalg |R|)$, one has $u(q_1, \ldots, q_n) = v(q_1, \ldots, q_n)$.*

(iii) *For all objects A of $\mathbf{C}$, the algebra $\mathbf{C}(R, A)$ lies in $\mathbf{V}$.*

(iv) *Interpreted as an* algebra *object of $\mathbf{C}^{\mathrm{op}}$, R is a $\mathbf{V}$-object.*

Definition 8.11. *A covariant functor $\mathbf{C} \to \mathbf{V}$ will be called* representable *if it is isomorphic to a functor of the form $\mathbf{C}(R, -)$, for R a co-$\mathbf{V}$ object of $\mathbf{C}$; equivalently, if its composite with the forgetful functor $\mathbf{V} \to \mathbf{Set}$ is representable in the sense of Definition 5.2*(i)*, equivalently, in the case where $\mathbf{C}$ is a variety of algebras, if this composite satisfies Lemma 8.7*(ii).

The full subcategory of $\mathbf{V}^{\mathbf{C}}$ consisting of the representable covariant functors will be denoted $\mathbf{Rep}(\mathbf{C}, \mathbf{V})$.

Note that *algebra* objects represent *contra*variant functors, while *co*variant functors are represented by *coalgebra* objects. This is a consequence of the fact that the hom-set bifunctor on a category $\mathbf{C}$,

(8.12) $$\mathbf{C}(-, -)\colon\ \mathbf{C}^{\mathrm{op}} \times \mathbf{C} \to \mathbf{Set}$$

is covariant in one variable and contravariant in the other. Thus, if we put a "structured" object R in one position, obtaining a functor in the other variable, we will get contravariance *either* in the relation between the structure on R and the induced structure on the output sets, *or* in the relation between input object and output object, but not both.

For the same reason, morphisms among *covariant* representable functors correspond *contravariantly* to morphisms among the representing coalgebras:

Corollary 8.13 (to Lemma 8.3). *If $\mathbf{C}$ is a category with finite coproducts, and $\mathbf{V}$ a variety of algebras, then the full subcategory $\mathbf{Rep}(\mathbf{C}, \mathbf{V}) \subseteq \mathbf{V}^{\mathbf{C}}$ of covariant representable functors is equivalent to the opposite of the category of co-$\mathbf{V}$ objects* *of* **C**, *where a morphism of co-*$\mathbf{V}$ *objects means a morphism of underlying objects which respects all co-operations.* □

Some examples of covariant representable functors among varieties of algebras were sketched in §§1-2. We will see many more in subsequent chapters.

Turning again to the example GL_n, the reader will not find it hard to verify that for any group G, one can construct by generators and relations a ring R with a *universal* homomorphism $G \to \mathrm{GL}_n(R)$. This construction gives a left adjoint to GL_n. The next Theorem gives the general result (not as well known as it should be!) of which this observation is an instance. The construction of the left adjoint in the proof of this Theorem is essentially the same as in the above example, except that the generators-and-relations presentation is replaced by a colimit, allowing us to get our conclusion without requiring the domain of V to be a variety. We have already discussed the equivalence of conditions (ii) and (iii) below; we include both for completeness.

THEOREM 8.14 (Freyd [**69**]). *Let* $\mathbf{C}$ *be a category with small colimits (i.e., such that every functor from a small category into* $\mathbf{C}$ *has a colimit),* $\mathbf{V}$ *a variety of algebras, and* $V\colon \mathbf{C} \to \mathbf{V}$ *a (covariant) functor. Then the following conditions are equivalent*:

(i) V *has a left adjoint* $G\colon \mathbf{V} \to \mathbf{C}$.

(ii) $V\colon \mathbf{C} \to \mathbf{V}$ *is represented by a co-*$\mathbf{V}$ *object of* $\mathbf{C}$ *(Definition 8.10).*

(iii) *The composite of the functor* $V\colon \mathbf{C} \to \mathbf{V}$ *with the underlying set functor* $\mathbf{V} \to$ **Set** *is represented by an object of* $\mathbf{C}$ *(Definition 5.2).*

PROOF. As noted, (ii) ⇔ (iii) by Lemmas 8.1 and 8.4, applied to $\mathbf{C}^{\mathrm{op}}$.

(i) ⇒ (iii): Let $U\colon \mathbf{V} \to \mathbf{Set}$ denote the underlying set functor of $\mathbf{V}$, let $F\colon \mathbf{Set} \to \mathbf{V}$ be its left adjoint, the free algebra construction of $\mathbf{V}$ (§6), and let $1 \in \mathrm{Ob}(\mathbf{Set})$ be a 1-element set. Then $UV(-) \cong \mathbf{Set}(1, UV(-)) \cong \mathbf{C}(GF(1), -)$ (Lemma 5.9). So UV is represented by the object $GF(1)$ of $\mathbf{C}$.

We will complete the proof by showing (ii) ⇒ (i). Assume (ii). To prove (i), it will suffice (cf. Proposition 5.3(ii)) to show that for each $A \in \mathrm{Ob}(\mathbf{V})$, there exists $G(A) \in \mathrm{Ob}(\mathbf{C})$ such that $\mathbf{C}(G(A), -) \cong \mathbf{V}(A, V(-))$.

Let us take any presentation of A as an algebra in $\mathbf{V}$ by a generating set S, and a set of relations

$$u_j = v_j \quad (j \in J), \tag{8.15}$$

where the u_j and v_j are expressions in the elements of S and the operations of $\mathbf{V}$. Thus, $\mathbf{V}(A, V(-))$ can be described as associating to each $B \in \mathrm{Ob}(\mathbf{C})$ the set of all S-tuples of members of $V(B)$ that satisfy the relations (8.15). (This is the idea of Lemma 8.7(ii), but applied to the object A, rather than to the object representing V.)

Now let R be the co-$\mathbf{V}$ object of $\mathbf{C}$ representing the functor V. Let us form a coproduct $\coprod_{s \in S} |R|^{(s)} \in \mathrm{Ob}(\mathbf{C})$ of an S-tuple of copies, $|R|^{(s)}$ $(s \in S)$, of $|R|$. Then for any object B of $\mathbf{C}$, $\mathbf{C}(\coprod_S |R|^{(s)}, B)$ can be naturally identified with $\mathbf{C}(|R|, B)^S \cong |V(B)|^S$, the set of *all* S-tuples of elements of $V(B)$. To obtain the subset of these S-tuples satisfying the relations (8.15), we shall formally "impose the relations $u_j = v_j$ on $\coprod_S |R|^{(s)}$". Precisely, for each $j \in J$ let us form the two morphisms $|R| \rightrightarrows \coprod_S |R|^{(s)}$ corresponding to u_j and v_j (cf. Definition 8.10 (ii), though that is formulated for identities rather than relations), and define $G(A)$ to be the colimit of the diagram formed out of all these pairs of arrows:

$$\begin{array}{l} \vdots \\ |R| \rightrightarrows \\ |R| \rightrightarrows \quad \coprod_S |R|^{(s)} \longrightarrow G(A). \\ |R| \rightrightarrows \\ \vdots \end{array}$$

This object $G(A)$ is easily seen to have the desired universal property $\mathbf{C}(G(A), -) \cong \mathbf{V}(A, V(-))$. $\square$

As we have noted, if $\mathbf{C}$ is itself a variety of algebras, condition (iii) above also has the restatement given in Lemma 8.7(ii), and the operations on the resulting sets have the form described in Lemma 8.9.

An alternative arrangement of the above proof may be gotten by replacing the argument showing (ii)$\Rightarrow$(i) by the following proof that (iii)$\Rightarrow$(i): Let $\mathbf{P}$ denote the class of those objects A of $\mathbf{V}$ such that the functor $\mathbf{V}(A, V(-)) \colon \mathbf{C} \to \mathbf{Set}$ is representable, and $G_{\mathbf{P}}$ the construction associating to each object A in $\mathbf{P}$ the corresponding representing object; thus $G_{\mathbf{P}}$ is the maximal "partial adjoint" to V. One shows that (a) the free object $F(1)$ on one generator in $\mathbf{V}$ belongs to $\mathbf{P}$ (the required representability condition is exactly condition (iii)), and (b) $\mathbf{P}$ is closed under small colimits ($G_{\mathbf{P}}$ will respect colimits). One then combines these observations with the observation (c) every object of a variety $\mathbf{V}$ can be obtained from the free object on one generator by iterated small colimits. (Coproducts of $F(1)$ give all free algebras $F(X)$, and an arbitrary object A is the colimit of a two-object two-arrow diagram of free objects, $F(Y) \rightrightarrows F(X)$.) In this form, the proof is curiously similar to the proofs of both Freyd's Adjoint Functor Theorem [**122**, §V.6] and Birkhoff's Theorem [**46**, Theorem IV.3.1].

Combining the above Theorem with Lemma 5.9, we get

COROLLARY 8.16. *A composite of representable functors among varieties is representable.* $\square$

It is instructive to examine how the object representing a composite of representable functors, $\mathbf{U} \to \mathbf{V} \to \mathbf{W}$, is "put together" from the objects representing the given functors. The construction is essentially that of the proof of Theorem 8.14, (ii) $\Rightarrow$ (i): one takes a family of copies of the object of $\mathbf{U}$ that represents the functor $\mathbf{U} \to \mathbf{V}$, and glues them together according to a presentation of the object of $\mathbf{V}$ that represents the functor $\mathbf{V} \to \mathbf{W}$. The added complication is describing a co-$\mathbf{W}$ structure on the new object. Not unexpectedly, this arises from the co-$\mathbf{W}$ structure on the object representing the functor $\mathbf{V} \to \mathbf{W}$.

Freyd [**69**] calls the coalgebra object representing the composite of two representable functors the "tensor product" of the coalgebra objects representing these functors. And indeed, the construction of tensor products of bimodules is a special case of this concept: For any rings A and B, a representable functor from the category ${}_A\mathbf{Mod}$ of left A-modules to the category ${}_B\mathbf{Mod}$ of left B-modules has the form $({}_A\mathbf{Mod})({}_AM_B, -)$, where the subscripts on the M mean that it is an (A, B)-bimodule, and where the right B-module structure on M induces the left B-module structure on the sets $({}_A\mathbf{Mod})({}_AM_B, {}_AN)$. The left adjoint of this functor is ${}_AM_B \otimes_B -: {}_B\mathbf{Mod} \to {}_A\mathbf{Mod}$. Given two such adjoint pairs,

$$ {}_A\mathbf{Mod} \underset{{}_AM_B \otimes_B -}{\overset{({}_A\mathbf{Mod})({}_AM_B,\, -)}{\rightleftarrows}} {}_B\mathbf{Mod} \underset{{}_BN_C \otimes_C -}{\overset{({}_B\mathbf{Mod})({}_BN_C,\, -)}{\rightleftarrows}} {}_C\mathbf{Mod}, $$

the adjoint pair gotten by composing them corresponds to the (A, C)-bimodule ${}_AM_B \otimes_B {}_BN_C$. (This is most easily seen by composing the lower arrows.) So, as claimed, Freyd's "tensor product" operation on representing coalgebras is in this case the usual tensor product operation on bimodules. One might generalize the notation used for bimodules, and write a $\mathbf{V}$-coalgebra object R in a variety $\mathbf{U}$ as "${}_{\mathbf{U}}R_{\mathbf{V}}$", and what Freyd calls the tensor product of two such coalgebras as "${}_{\mathbf{U}}R_{\mathbf{V}} \otimes_{\mathbf{V}} {}_{\mathbf{V}}S_{\mathbf{W}}$". We shall not use this language or notation here; but see [**201**] and [**202**] for some development of this idea. This construction on representing objects will come up in §§63-64 below.

Incidentally, it is only for brevity that we assumed all three categories in Corollary 8.16 to be varieties: the same argument clearly works for a composite of representable functors $\mathbf{C} \to \mathbf{V} \to \mathbf{W}$ where $\mathbf{V}$ and $\mathbf{W}$ are varieties, and $\mathbf{C}$ any category with small colimits; but we shall not need this generality.

The following is an important case of Theorem 8.14 (though it is unfortunate that it is more widely known than the Theorem itself, and believed by many to be the "last word" on the subject).

COROLLARY 8.17. *Any functor between varieties of algebras,* $W: \mathbf{V} \to \mathbf{W}$, *which respects underlying sets has a left adjoint.*

PROOF. By Theorem 8.14 (iii)⇒(i), to show W has a left adjoint it suffices to show that there exist a set X, and a set of relations Y in indeterminates from X, such that for any S in $\mathbf{V}$, we can describe the functor $|W(-)|$ as taking each S to the set of X-tuples of members of S satisfying the relations Y. But the statement that W respects underlying sets means that $|W(S)|$ can, in fact, be described as the set of all 1-tuples of members of S satisfying the empty set of relations. □

This Corollary is applicable to such constructions as (i) the underlying-set functor $U_{\mathbf{V}}: \mathbf{V} \to \mathbf{Set}$ for any variety $\mathbf{V}$, the left adjoint of which, as we have already noted, is the *free algebra* construction, (ii) the "commutator brackets" functor from associative k-algebras to Lie algebras over k, whose left adjoint is the *universal enveloping algebra* construction, (iii) the functor taking associative k-algebras A to their underlying k-vector-spaces, whose left adjoint is the *tensor algebra* construction, and (iv) the functor taking associative k-algebras to their underlying multiplicative semigroups, whose left adjoint is the *semigroup-algebra* construction (cf. Example 5.10). On the other hand, some familiar functors which are shown to have left adjoints by Theorem 8.14, but are not covered by

Corollary 8.17, are the construction GL_n from rings to groups (and, on commutative rings, similarly SL_n, O_n, etc.), the formal power series ring construction, $A \mapsto A[[t]]$ (represented by the free ring on countably many generators), and the group-of-invertible-elements construction on semigroups with neutral element.

From the above Corollary, we see in particular

COROLLARY 8.18. *Every subvariety* $\mathbf{V}$ *of a variety* $\mathbf{W}$ *is a* reflective *subcategory of* $\mathbf{W}$ *(Definition 5.14).* □

Combining this with Corollary 8.16, we get

COROLLARY 8.19. *If a functor* $W: \mathbf{V} \to \mathbf{W}$ *between varieties of algebras is representable, so is its restriction to any subvariety of* $\mathbf{U} \subseteq \mathbf{V}$. □

For instance, having observed that GL_n is a representable functor on associative rings, we know automatically that it gives a representable functor on commutative associative rings. (What is the relation between the representing objects for these two functors?)

The dual to Theorem 8.14 applies to *contravariant* functors from a category with small *limits* to a variety $\mathbf{V}$. We shall not state it formally, since it is not relevant to our main results, but we will discuss the phenomenon briefly in point 9.4 below.

We remark that when a functor between varieties of algebras $W: \mathbf{V} \to \mathbf{W}$ is representable, this representability is usually easy to see and to prove – the construction of the underlying set of $W(A)$ is easily expressed in the form described in Lemma 8.7(ii). But when we seek to prove that a functor is *not* representable, this criterion is clearly not so helpful. A very useful criterion in this case is that of Exercise 5.13(iii): functors having left adjoints respect limits. Arbitrary limits can be constructed from difference kernels and products, so in seeking to show by this criterion that a functor is *not* representable, it suffices to look for failure to respect one of these two constructions. In the case of products, one may examine separately products of two objects, the product of the empty class of objects, and finally, if these are both respected, products of infinite families of objects.

EXERCISE 8.20. *Verify that none of the following covariant functors from abelian groups to abelian groups is representable:* (i) $F(A) = A \otimes A$, (ii) $G(A) =$ *the torsion subgroup of* A *(the group of elements of finite order),* (iii) $H(A) = A/nA$ *(*n *a fixed integer),* (iv) $J(A) = nA$ *(*n *a fixed integer).*

The condition of respecting limits is necessary for representability but not sufficient; so there *are* nonrepresentable functors which are not detected by this criterion, though these tend to be pathological. A strengthening of the above criterion to one that is necessary and sufficient is given by Freyd's "Adjoint Functor Theorem" [**122**, §V.6], [**24**, Theorem 7.9.4]. Some examples showing the insufficiency of the condition of respecting limits alone are given in [**24**, Exercises 7.9:5-7.9:6 and 9.4:4].

§9. Digressions on representable functors.

The observations in this section will not be used in the main results of this work.

9.1. *Generalizing "S" and "×".* The concepts and results developed in the preceding section can be generalized in various ways. On the one hand, note that if R is a $\mathbf{V}$-object of $\mathbf{C}$, then the construction by which a structure of $\mathbf{V}$-algebra is induced on every set $\mathbf{C}(S, |R|)$ $(S \in \mathrm{Ob}(\mathbf{C}))$ generalizes to say that for every functor $P\colon \mathbf{C} \to \mathbf{Set}$ which respects finite products, the set $P(|R|)$ becomes a $\mathbf{V}$-algebra $P(R)$.

On the other hand, one can define "generalized algebra objects" in a category $\mathbf{C}$ by replacing the operations $R^n \to R$ by maps $R * \ldots * R \to R$ for another bifunctor $*\colon \mathbf{C} \times \mathbf{C} \to \mathbf{C}$. Depending on how "product-like" $*$ is, these systems will behave to a greater or lesser degree like algebras.

A notable case of this sort (for coalgebras rather than algebras) is the concept which Hopf algebraists call a "coalgebra". These are k-vector-spaces V with a "comultiplication" $\Delta\colon V \to V \otimes_k V$. A k-vector-space having both a structure of associative k-algebra and a structure of coalgebra in the above sense, satisfying some compatibility conditions, is called a *bialgebra* over k. In the category of commutative k-algebras, the coproduct of two objects has for underlying k-module the tensor product of their underlying modules, and one finds that bialgebras with commutative multiplication *are* coalgebra objects in this category, in our sense. Noncommutative bialgebras, however, cannot be interpreted in that way. In §43 we will look at these concepts, using the term "$\otimes_k$-coalgebra" to avoid confusion with the concept of coalgebra used in this work; cf. also [**183**], [**21**]. In §§60-62 we note some analogous situations in other categories.

9.2. *Generalizing* $\mathbf{V}$. Varieties are not the most general classes $\mathbf{V}$ of algebras for which the concept of a "co-$\mathbf{V}$ object" of a category $\mathbf{C}$ can be developed.

Consider the class of *torsion-free groups,* i.e., groups satisfying the family of conditions

$$(9.3) \qquad (\forall\, x)\ \ x^n = e \ \Rightarrow\ x = e$$

for $n = 1, 2, \ldots$. The conditions (9.3) belong to a class of predicates known as *universal Horn sentences,* wider than the class of identities. Essentially, a universal Horn sentence means a condition of the form

$$(\forall\ x, y, \ldots)\ (\textit{finite conjunction of equations}) \Rightarrow (\textit{equation}).$$

A class of algebras defined by a family of universal Horn sentences is called a *quasivariety*. Quasivarieties have many of the good properties of varieties, including closure under taking subobjects and products within the class of all algebras of the given type (though *not*, in general, under taking homomorphic images), and the existence of free objects, small limits and colimits, and objects defined within the quasivariety by arbitrary systems of generators and relations. This last condition is equivalent to saying that any quasivariety, as a full subcategory of the category of algebras of the given type, is reflective.

The analog of Theorem 8.14 holds for functors to quasivarieties $\mathbf{V}$ if, in condition (ii) thereof, we interpret a "co-**V**-object of **C**" to mean a coalgebra of the desired type which satisfies condition (iii) of Definition 8.10 for the given quasivariety **V**.

We would, of course, also like to have equivalent conditions analogous to (i) and (ii) of that Definition, showing what properties of the co-operations insure that the Horn sentences defining **V** will be satisfied by the values of the functor represented by R. For brevity, let us illustrate the form these conditions take in the case of the particular Horn sentence (9.3). Suppose **C** is a category with finite colimits, n a positive integer, and R a *cogroup* object of **C**. If we form the two unary co-operations, i.e., morphisms $|R| \to |R|$, corresponding to the two unary derived group operations $x \mapsto x^n$ and $x \mapsto e$, then the difference cokernel (colimit) of the resulting diagram $|R| \rightrightarrows |R|$ will be an object S_n with a *universal* element $s \in \mathbf{C}(R, S_n)$ satisfying $s^n = e$. Thus, the condition that the values of the functor represented by R satisfy (9.3) is equivalent to the condition that the pair of maps $|R| \rightrightarrows S_n$ corresponding to s and e be equal. (Because of particular properties of the relation (9.3), this can also be expressed as saying that every group $\mathbf{C}(R, A)$ has exactly one solution to the equation $x^n = e$, hence that $\mathbf{C}(S_n, A)$ is a one-element set, hence that S_n is the initial object of **C**.) Thus, the functor $\mathbf{C}(R, -)\colon \mathbf{C} \to \mathbf{Group}$ will be torsion-free-group-valued (and so R may be called a "co-torsion-free" cogroup object of **C**) if and only if this condition holds for every positive integer n.

We shall make some further observations on representable functors among quasivarieties, and more general categories of algebras called "prevarieties", in §§59 and 65.

One can also "represent" functors to categories of objects whose structure involves *relations* as well as operations. One defines an n-ary *co-relation* on an object R of a category **C** to mean an epimorphism (Definition 5.15) $\rho\colon |R| \amalg \ldots \amalg |R| \to S$ for some object S. (To see that this is reasonable, note that in **Set**, an n-ary relation on an object X corresponds to a monomorphism $r\colon Y \to X^n$, where Y is the graph of the relation.) Such an epimorphism will correspond to a certain n-tuple of elements $x_1, \ldots, x_n \in \mathbf{C}(|R|, S)$, which will serve as a *universal* n-tuple satisfying this relation; an arbitrary n-tuple $(a_1, \ldots, a_n)$ of elements of a set $\mathbf{C}(|R|, A)$ is defined to satisfy the relation if and only if there exists a (necessarily unique) morphism $f\colon S \to A$ such that $a_i = fx_i$.

For example, let **C** be the category whose objects are pairs (X, r) where X is a set and r a symmetric binary relation on X, and where morphisms are set-maps respecting these relations. Consider the functor V from **Group** to **C** taking a group G to the pair $(|G|, \{(g, h) \in |G| \times |G| \mid gh = hg\})$. The construction taking G to the first component of this pair, $|G|$, is represented by the free group on one generator, $\langle x \rangle$. The binary *relation* on this set given by the second component is represented by the epimorphism of groups,

$$\langle x, y \rangle \;\to\; \langle x, y \mid xy = yx \rangle.$$

The reader can easily describe the left adjoint to this functor, carrying sets with such binary relations back to groups, and see that the construction is a natural generalization of that described in the proof of Theorem 8.14 (cf. [**18**]). As another example, the binary relation "$<$" on the underlying set of a lattice is induced by a certain epimorph of the free lattice on two generators, and this makes it possible to construct lattices universal for given systems of generators and order relations.

But in contrast with Lemma 8.1, it is not true that every functorial way of putting an n-ary relation on the values of a representable set-valued functor is "representable" in this sense. Some unary relations on underlying sets of groups G which respect group homomorphisms, but do not arise in this way from epimorphs of the free group on one generator, are (i) "g has exponent 2 or 3", (ii) "g has finite order", and (iii) "g is a square in G"; for there is no *universal* instance of any of these. (This is intuitively clear in the first two cases. In the last case, the element $x^2 \in <x>$ has a weak universal property, but the uniqueness condition that would be needed for a full universal property fails, because square roots in groups are nonunique, i.e., because the map $<y> \to <x>$ taking y to x^2 is not an epimorphism.) One can, in fact, prove a generalization of Theorem 8.14, saying that a functor to sets with operations and relations has a left adjoint if and only if as a functor to sets, it is representable in the usual sense, *and* the relational structure on the constructed sets is representable using epimorphisms, in the sense sketched above.

9.4. *Contravariant functors.* Applying duality to Theorem 8.14, we can conclude that a *contravariant* functor from a category $\mathbf{C}$ with small *limits* to a variety $\mathbf{V}$, say $V\colon \mathbf{C}^{\mathrm{op}} \to \mathbf{V}$, belongs to a *mutually right adjoint pair* of contravariant functors

$$V\colon \mathbf{C}^{\mathrm{op}} \to \mathbf{V}, \qquad W\colon \mathbf{V}^{\mathrm{op}} \to \mathbf{C}$$

(see 5.8 above) if and only if V is represented by a $\mathbf{V}$-*algebra* object R of $\mathbf{C}$. Suppose now that we also assume the category $\mathbf{C}$ to be a variety of algebras $\mathbf{W}$. A $\mathbf{V}$-algebra object R of $\mathbf{W}$ then turns out to be a set with two systems of operations, one making it a $\mathbf{V}$-object and the other making it a $\mathbf{W}$-object, which respect one another, in the sense that for every m-ary operation α of $\mathbf{V}$, n-ary operation β of $\mathbf{W}$, and mn elements x_{ij} of the underlying set of R, one has the *commutativity* identity

$$\begin{aligned}(9.5)\qquad &\alpha(\beta(x_{11},\dots,x_{1n}),\ \dots,\ \beta(x_{m1},\dots,x_{mn}))\\ &= \beta(\alpha(x_{11},\dots,x_{m1}),\ \dots,\ \alpha(x_{1n},\dots,x_{mn})).\end{aligned}$$

If R is such an object, then for any $A \in \mathrm{Ob}(\mathbf{W})$, the set of morphisms $\mathbf{W}(A, R)$ is closed under pointwise application of the $\mathbf{V}$-operations of R, and so becomes a $\mathbf{V}$-algebra. Not surprisingly, in view of the symmetric nature of contravariant adjunctions, this same object R, looked at as a $\mathbf{V}$-object of $\mathbf{W}$, is a representing object for the adjoint functor $W\colon \mathbf{V}^{\mathrm{op}} \to \mathbf{W}$. This general situation is examined in [**197**], [**53**] and [**97**].

One important way this situation can arise is if we have, in some third category $\mathbf{C}$, both a co-$\mathbf{V}$ object A and a $\mathbf{W}$-object B. Then the set $\mathbf{C}(|A|, |B|)$ acquires commuting structures of $\mathbf{V}$-object and of $\mathbf{W}$-object. For an interesting case of this phenomenon, let us begin by noting that it is easy to show that the only way a set can have a pair of group structures which commute with one another in the sense of

(9.5) is if the two group structures are equal, and are abelian [**106**]. Now the circle S^1 is a cogroup in the category **C** of topological spaces and homotopy classes of maps, representing the fundamental-group functor π_1. Suppose G is a group object in this same category, e.g., a Lie group. Then $\pi_1(G) = \mathbf{C}(S^1, G)$ has two commuting group structures; hence both are abelian. This yields the well-known result that the fundamental group of a Lie group is abelian.

Though the rest of this work will concern representable *covariant* functors, let us note here a few interesting examples of mutually right adjoint pairs of representable contravariant functors $\mathbf{V}(-, R)$, $\mathbf{W}(-, R)$ among categories of algebras:

(i) Let $\mathbf{V} = \mathbf{Set}$ (with the empty family of operations!), $\mathbf{W} = \mathbf{Bool}^1$, the variety of Boolean rings, and let $R = 2$, i.e., $\{0, 1\}$, which may be considered on the one hand as a set, and on the other as a Boolean ring. Then $\mathbf{Set}(-, 2)$, made a Boolean-ring-valued functor using this structure on 2, can be described as taking each set A to its Boolean ring of subsets, while its set-valued right adjoint, $\mathbf{Bool}^1(-, 2)$, can be described as taking each Boolean ring to its prime spectrum.

(ii) Let k be a field, let $\mathbf{W} = \mathbf{V} = \mathbf{Mod}_k$, the variety of vector spaces over k, and let $R = k$ regarded as a 1-dimensional k-vector-space. The $\mathbf{Mod}_k$ operations on k all commute with one another, so k is a $\mathbf{Mod}_k$ object of $\mathbf{Mod}_k$. Here $\mathbf{W}(-, k)$ and $\mathbf{V}(-, k)$ are the same construction, that of the *dual vector space*.

(iii) If we extend the concept of contravariant representable functor to structures involving relations as well as operations, using the approach sketched for covariant representable functors in §9.1 above, we can let $\mathbf{V}$ be the category of *partially ordered sets* and isotone ($\leq$-preserving) maps, and $\mathbf{W}$ the variety of *distributive lattices* with least element 0 and greatest element 1. We find that the structures of partially ordered set and of lattice with least and greatest element on $2 = \{0, 1\}$ "respect" one another, so again we get a pair of right adjoint contravariant functors, $\mathbf{POSet}(-, 2)$ and $\mathbf{DistLat}^{0,1}(-, 2)$, connecting the two categories.

In each of the above three examples, it happens that the adjoint pair of contravariant functors, when restricted to the full subcategories of *finitely generated* objects, which we may write $\mathbf{V}_{\mathrm{fg}}$ and $\mathbf{W}_{\mathrm{fg}}$, give a *duality*, i.e., a contravariant *equivalence* between these subcategories. The functors between the whole categories $\mathbf{V}$ and $\mathbf{W}$ are not equivalences, but if one modifies either category (but not both) by enlarging the set of operations to include all infinitary operations which respect the operations and relations of R as a member of the other variety (a *large set* of operations, in the sense of §5), these new operations turn out to be equivalent to a topological structure, and one then finds that one gets a full duality between the objects of one category and appropriately topologized objects of the other — Stone duality between Boolean rings and totally disconnected compact Hausdorff spaces, Lefschetz's duality between vector spaces and linearly compact vector spaces (reviewed in §24 below), etc.. For general results on such dualities see [**53**], [**104**, Ch.VI]. Note also the early paper [**9**], which develops systematically a wider class of examples than many subsequent authors seem to have been aware of.

We have described no examples of contravariant adjoint functors which are mutually *left adjoint*. In fact, it is shown in [**22**] that all such pairs of functors between varieties of algebras are quite degenerate, though interesting examples of such adjoint pairs of functors between full *sub*categories of varieties are noted.

9.6. *Coalgebras in* **Set**. Given a variety **V**, one finds, generally speaking, that the "richer" the structure of operations in **V**, the richer is the class of *covariant* representable functors on **V** (represented by coalgebras in **V**), and the scarcer are the *contravariant* representable functors (represented by algebra objects). This is because covariant constructions define their operations *using* derived operations of **V**, while the algebra objects representing contravariant functors must have operations that *commute* with those of **V**.

An extreme case in one direction is the variety **Set**, whose *algebra* objects form the subject-matter of all classical algebra, but which is virtually a desert in terms of coalgebra objects. For instance it is easy to verify that a binary co-operation on a nonempty set can never be cocommutative.

But on close examination, even this desert turns out to support some wildlife worthy of attention. Observe that for S a set, the n-fold coproduct $S \amalg \dots \amalg S$ can be written $S \times \{1, \dots, n\}$. Hence an n-ary co-operation on S can be described as a rule associating to every element of S a new element of S together with a member of $\{1, \dots, n\}$. This is a formalization of the concept of an *automaton*, with S as its set of states, and $\{1, \dots, n\}$ its set of outputs. An automaton which also takes an *input* in a set $\{1, \dots, m\}$ can be described as a set given with an m-tuple of n-ary co-operations. (However, in most category-theoretic developments of automata theory, $S \times \{1, \dots, n\}$ is interpreted more straightforwardly as a product of S with a fixed object, rather than as a coproduct of n copies of S. See [**8**], [**7**], [**54**], [**129**], [**130**] for some work in this area.)

9.7. *Coalgebras as bookkeeping devices.* In the remaining parts of this work we will be seeking to describe, completely or partially, the (covariant) representable functors from one variety of algebras **V** to another variety **W**. In the midst of the elaborate analyses of the possible structures of the representing coalgebras, the reader may sometimes wonder, "What is this all about?". We suggest keeping in mind that what we are studying is simply the possible ways of constructing objects of **W** from objects of **V** that have the elementary form indicated on page 1 above.

The concept of the representing *coalgebra* R may be regarded as an elegant bookkeeping device for such constructions, which frees us from particular choices of coordinates for the constructed objects. (A choice of coordinates corresponds to a choice of generating set for the **V**-object $|R|$.) Like many a good mathematical bookkeeping device, however, this one turns out to have a life of its own, and the study of its structure translates to a powerful technique for the study of the functors for which it "keeps book".

CHAPTER III.

Representable functors from rings to abelian groups.

In this chapter we shall obtain the key result of this work, the characterization of representable functors from k-rings (defined below) to abelian groups.

It turns out that the inverse operation in the definition of an abelian group does not come into the proof – we get it for free. Hence we shall develop the result first for the formally more general case of functors from k-rings to abelian semigroups with neutral element, then note that every functor obtained is actually group-valued, and deduce the corresponding statement for functors to abelian groups.

§10. k-Rings.

Let us begin by fixing some terminology.

We shall say "semigroup with neutral element" where followers of Bourbaki say "monoid". (When the semigroup is written additively we will generally write the neutral element 0, but we do not call it a "zero element", because to semigroup theorists, a "zero element" in a semigroup means an element z such that $a*z = z = z*a$ for all a.)

The two categories named in the first paragraph of the next Definition will be of primary importance in this chapter. Those named in the second paragraph will come up occasionally in examples, but will get most attention in Chapters IV, VII and VIII.

DEFINITION 10.1. *The category of abelian groups will be denoted* **Ab**. *The category of abelian semigroups with neutral element (and homomorphisms respecting neutral elements) will be denoted* $\mathbf{AbSemigp}^e$.

The category of all groups, and the category of all semigroups with neutral element, will be denoted **Group** *and* $\mathbf{Semigp}^e$ *respectively, and the categories of abelian and arbitrary semigroups with no assumption as to existence of neutral elements will be written* **AbSemigp** *and* **Semigp**.

DEFINITION 10.2. *"Ring" will mean "associative ring" unless the contrary is indicated.*

Throughout the remainder of this work, k will be a fixed unital ring (i.e., ring with multiplicative neutral element 1). It will not *be assumed commutative, except when so specified.*

In referring to rings, we will mention associativity explicitly only when nonassociative rings are also discussed (e.g., in Chapter V), or for emphasis, as in the first sentence of the next definition. Modules and bimodules will only be considered over unital rings, and will always be unital (i.e., the element 1 will act as the identity map).

For the reader who finds k-rings (about to be defined) and bimodules a bit exotic, we remark that the proof of our main Theorem uses no esoteric results about these classes of objects, so that the reader will not miss any important points if he or she reads it assuming k commutative, and replacing the categories of k-bimodules and k-rings by those of k-modules and k-algebras respectively. (More about this in §14. One could even assume k a field, but for that case there are shorter routes to our results than the one given here.) However, as we shall see in §23, it is useful to have results for general k-rings even if our ultimate interest is in algebras over fields.

DEFINITION 10.3. *A* unital k-ring R *means an associative unital ring* R *given with a unital homomorphism* $k \to R$. *A* homomorphism *of unital* k*-rings,* $R \to S$, *means a ring homomorphism* $R \to S$ *forming a commutative triangle with the maps from* k. *The variety of unital* rings *will be denoted* $\mathbf{Ring}^1$, *that of unital* k*-rings will be denoted* $k\text{-}\mathbf{Ring}^1$.

In regarding $k\text{-}\mathbf{Ring}^1$ as a variety, one of course considers the image of each element $\alpha \in k$ as given by a zeroary operation.

DEFINITION 10.4. *A* nonunital k-ring *will mean a* k*-bimodule* R *given with a nonunital associative ring structure such that the additive group structures as ring and* k*-bimodule are the same, and the ring and bimodule multiplications are related by the identities*

$$(\alpha x)y = \alpha(xy), \qquad (x\alpha)y = x(\alpha y), \qquad (xy)\alpha = x(y\alpha), \qquad (x, y \in R,\ \alpha \in k).$$

A homomorphism of nonunital k*-rings will mean a map of underlying sets which is a homomorphism both of rings and of* k*-bimodules. The variety of nonunital rings will be denoted* **Ring**, *and that of nonunital* k*-rings will be denoted* $k\text{-}\mathbf{Ring}$.

Note that in this definition, "nonunital" means "not *necessarily* unital". I.e., an object of this variety is not given with a zeroary operation specifying a unit element, and homomorphisms between objects that happen to have units are not required to respect these.

In considering this category a variety, we cannot associate a *zeroary* operation to each element $\alpha \in k$; rather, we have two *unary* operations, of left and right multiplication by α, which come from the k-bimodule structure. It is not hard to see that unital k-rings, as defined in Definition 10.3, can be naturally identified with nonunital k-rings which have multiplicative neutral elements, the morphisms of unital k-rings being those morphisms of nonunital k-rings which respect these neutral elements. We could have made this the definition of unital k-rings, but Definition 10.3 seemed more natural.

Clearly, there is a "forgetful" functor $k\text{-}\mathbf{Ring}^1 \to k\text{-}\mathbf{Ring}$. There is another fundamental construction connecting these two categories; to describe it we need:

DEFINITION 10.5. *If* $\mathbf{C}$ *is a category with* initial *object* I, *then an* augmented *object of* $\mathbf{C}$ *means a pair* (X, ε), *where* X *is an object of* $\mathbf{C}$ *and* ε *is a morphism* $\varepsilon: X \to I$. *The category of augmented objects of* $\mathbf{C}$, *where the morphisms are* $\mathbf{C}$*-morphisms of the first components making commuting triangles with* I, *will be written* $\mathbf{C}^{\mathrm{aug}}$.

Note that the initial object of k-$\mathbf{Ring}^1$ is k. We have

LEMMA 10.6. *There is an equivalence of categories,*

$$(k\text{-}\mathbf{Ring}^1)^{\mathrm{aug}} \simeq k\text{-}\mathbf{Ring},$$

where one goes to the right by taking the kernel of the augmentation, and to the left by a construction of "adjoining a unit" (described below).

SKETCH OF PROOF. If $(R, \varepsilon) \in (k\text{-}\mathbf{Ring}^1)^{\mathrm{aug}}$, then $\ker \varepsilon$ is an ideal of R, and as such it inherits a structure of nonunital k-ring.

If $R \in k$-$\mathbf{Ring}$, we define the ring R^1 to have underlying additive group $k \oplus R$ (where we think of (α, x), for $\alpha \in k$, $x \in R$, as representing the sum of x, and α times the adjoined unit element), and multiplication defined by $(\alpha, x)(\beta, y) = (\alpha\beta, \alpha y + x\beta + xy)$. We map k in by $\alpha \mapsto (\alpha, 0)$, and define the augmentation by $(\alpha, x) \mapsto \alpha$.

It is easy to verify that these two functors yield an equivalence $(k\text{-}\mathbf{Ring}^1)^{\mathrm{aug}} \simeq k\text{-}\mathbf{Ring}$. □

EXERCISE 10.7. *Suppose* $R \in k$-$\mathbf{Ring}$, *and let* (S, ε) *be the corresponding augmented unital k-ring. Show that the following conditions are equivalent:* (i) R *has a multiplicative neutral element.* (ii) S *can be written as a direct product* $k \times T$ *in* k-$\mathbf{Ring}^1$, *in such a way that the projection onto the first component is* ε. *(Warning: the direct sum* $R^1 = k \oplus R$ *in the proof of Lemma 10.6 is not a ring-theoretic direct product.)*

EXERCISE 10.8. *We have been considering unital, augmented unital, and nonunital k-rings, which, up to equivalence, comprise two categories,* k-$\mathbf{Ring}^1$ *and* $(k\text{-}\mathbf{Ring}^1)^{\mathrm{aug}} \simeq k\text{-}\mathbf{Ring}$. *Between these categories we have two forgetful functors*

$$(k\text{-}\mathbf{Ring}^1)^{\mathrm{aug}} \to k\text{-}\mathbf{Ring}^1 \quad \textit{and} \quad k\text{-}\mathbf{Ring}^1 \to k\text{-}\mathbf{Ring}.$$

Prove an adjointness relation between these functors. (First decide whether to take $(k\text{-}\mathbf{Ring}^1)^{\mathrm{aug}}$ *or* k-$\mathbf{Ring}$ *as representative of that pair of equivalent categories, and, depending on your choice, adjust appropriately the description of the functor that involves the other of these.)*

Though ultimately we will be most interested in functors on varieties of unital rings, it will be convenient to derive our main results *first* for functors on the variety k-$\mathbf{Ring}$ of nonunital k-rings. Let us see why.

Suppose $(R, \mathbf{a}, \varepsilon)$ is a co-$\mathbf{AbSemigp}^e$ object in k-$\mathbf{Ring}^1$. Here the coaddition is a morphism $\mathbf{a}: R \to R \amalg R$, while the co-neutral-element ε is a morphism from R to the coproduct of the empty family in k-$\mathbf{Ring}^1$, namely the initial object, k. Thus ε is an augmentation on R.

This augmentation ε on R induces (by the universal property of the coproduct) an augmentation $(\varepsilon, \varepsilon)$: $R \amalg R \to k$. The resulting augmented k-ring $(R \amalg R, (\varepsilon, \varepsilon))$ is easily shown to be the coproduct $(R, \varepsilon) \amalg (R, \varepsilon)$ in $(k\text{-}\mathbf{Ring}^1)^{\text{aug}}$. We claim that our coaddition map $\mathbf{a}$: $R \to R \amalg R$ respects the augmentations ε and $(\varepsilon, \varepsilon)$. Indeed, the reader should verify that this condition is the translation of the identity $0 + 0 = 0$ holding in $\mathbf{AbSemigp}^e$.

One finds that $((R, \varepsilon), \mathbf{a}, \varepsilon)$ is a co-$\mathbf{AbSemigp}^e$ object of $(k\text{-}\mathbf{Ring}^1)^{\text{aug}}$. Indeed, if V: $k\text{-}\mathbf{Ring}^1 \to \mathbf{AbSemigp}^e$ is the functor represented by $(R, \mathbf{a}, \varepsilon)$, then the functor $(k\text{-}\mathbf{Ring}^1)^{\text{aug}} \to \mathbf{AbSemigp}^e$ represented by $((R, \varepsilon), \mathbf{a}, \varepsilon)$ carries each augmented k-ring (S, ε_S) to the kernel of the abelian semigroup homomorphism

$$V(S) \xrightarrow{V(\varepsilon_S)} V(k).$$

Since $(k\text{-}\mathbf{Ring}^1)^{\text{aug}} \simeq k\text{-}\mathbf{Ring}$, we get a corresponding co-$\mathbf{AbSemigp}^e$ object $(R^0, \mathbf{a}^0, 0)$ in $k\text{-}\mathbf{Ring}$, where $R^0 = \ker \varepsilon$, $\mathbf{a}^0$: $R^0 \to R^0 \amalg R^0$ is the morphism corresponding to $\mathbf{a}$, and 0: $R^0 \to \{0\}$ denotes the unique morphism from R^0 to the zero ring, the initial-final object of $k\text{-}\mathbf{Ring}$.

Our original coalgebra $(R, \mathbf{a}, \varepsilon)$ in $k\text{-}\mathbf{Ring}^1$ may be recovered up to natural isomorphism from $(R^0, \mathbf{a}^0, 0)$ by reversing the above procedure: apply the adjunction-of-unit functor described in Lemma 10.6, and then forget the augmentation as a component of the structure of the underlying object. The functor represented by $(R, \mathbf{a}, \varepsilon)$ may also be recovered directly from the functor represented by $(R^0, \mathbf{a}^0, 0)$ by composing this with the forgetful functor $k\text{-}\mathbf{Ring}^1 \to k\text{-}\mathbf{Ring}$.

Hence the study of co-$\mathbf{AbSemigp}^e$ objects in $k\text{-}\mathbf{Ring}^1$ is equivalent to the study of such objects in $k\text{-}\mathbf{Ring}$. Using single symbols now for objects and morphisms of $k\text{-}\mathbf{Ring}$ (rather than expressing these in terms of kernels of augmentations on objects of $k\text{-}\mathbf{Ring}^1$), we may write any such co-$\mathbf{AbSemigp}^e$ object as $(R, \mathbf{a}, 0)$, since the only morphism from R to the initial object $\{0\}$ of $k\text{-}\mathbf{Ring}$ is the zero map.

Note that for such a co-$\mathbf{AbSemigp}^e$ object $(R, \mathbf{a}, 0)$, the pair $(R, \mathbf{a})$ is a co-$\mathbf{AbSemigp}$ object. Now if $(R, \mathbf{a})$ is *any* co-$\mathbf{AbSemigp}$ object of $k\text{-}\mathbf{Ring}$, the map 0: $R \to \{0\}$ is the unique zeroary co-operation which the k-ring R admits, and this co-operation will be coidempotent under the coaddition (otherwise we could get from it another distinct zeroary co-operation, contradicting uniqueness). We shall be studying the case where it also satisfies the co-neutral laws. For brevity, we will introduce a slight abuse of terminology, and use the term "co-$\mathbf{AbSemigp}^e$ object of $k\text{-}\mathbf{Ring}$" for a co-$\mathbf{AbSemigp}$ object $(R, \mathbf{a})$ such that the unique map 0: $R \to \{0\}$ satisfies the co-neutral laws; i.e., such that $(R, \mathbf{a}, 0)$ is a co-$\mathbf{AbSemigp}^e$ object in the strict sense.

The above reduction of co-$\mathbf{AbSemigp}^e$ objects in the category of unital k-rings to the corresponding objects in the category of nonunital k-rings simplifies the data we need to study, by replacing an arbitrary co-neutral-element morphism with the zero map. More important, it will allow us to take advantage of a convenient description of coproducts in $k\text{-}\mathbf{Ring}$ ((12.2) et seq. below) which has no analog in $k\text{-}\mathbf{Ring}^1$ except when an augmentation (or some generalization thereof, as in

[**19**, §8]) is given.

In the foundational material of Chapter II, we were careful to distinguish notationally between algebras and their underlying sets, etc.; but we shall be less formal now, so that if, say, A is a ring, we may use this same letter for its underlying set, its underlying additive group, additive semigroup, multiplicative semigroup, etc., as long as the context makes it clear what is meant. We shall also generally write $X \in \mathbf{C}$ where we have previously written $X \in \mathrm{Ob}(\mathbf{C})$.

§11. Representable functors and pointed categories.

The considerations by which we reduced the study of cosemigroups in k-$\mathbf{Ring}^1$ to the same question in $(k\text{-}\mathbf{Ring}^1)^{\mathrm{aug}}$ (though *not* the further identification of this category with k-$\mathbf{Ring}$) are special cases of some general results, which we will set down in this section. Proofs are mostly straightforward, and will be omitted or merely sketched. The reader who wishes may skip this section and refer back to it later, at the occasional places where we call on some of these general results.

DEFINITION 11.1. *A* pointed category *means a category having an object which is both initial and final.*

LEMMA 11.2. *A variety* $\mathbf{V}$ *of algebras is pointed if and only if it has a zeroary operation* e *whose value is respected by all other operations of* $\mathbf{V}$ *(i.e., such that in every object of* $\mathbf{V}$, $\{e\}$ *forms a one-element subalgebra); equivalently, if and only if the clone of derived operations of* $\mathbf{V}$ *contains exactly one zeroary derived operation. In such varieties, the initial-final object is the one-element algebra.* □

(Note that two zeroary operations "respect" one another if and only if they are equal.)

Thus, the varieties **Group**, $\mathbf{Semigp}^e$, and k-**Ring** are pointed. On the other hand, **Semigp** is not (final object: the 1-element semigroup; initial object: the empty semigroup) because there are no zeroary operations, and k-$\mathbf{Ring}^1$ is not (if $k \neq \{0\}$) because there is more than one zeroary operation. (Final object: $\{0\}$; initial object: k.)

For general categories, we note

LEMMA 11.3. *If* $\mathbf{C}$ *is any category with initial object* I, *then* $\mathbf{C}^{\mathrm{aug}}$ *(see Definition 10.5) is a pointed category. If* $(X^{\alpha}, \varepsilon_{\alpha})$ $(\alpha \in A)$ *is a family of objects of* $\mathbf{C}^{\mathrm{aug}}$, *and the coproduct* $\coprod X^{\alpha}$ *exists in* $\mathbf{C}$, *then writing* $(\varepsilon_{\alpha}) : \coprod X^{\alpha} \to I$ *for the map induced by the* ε_{α}*'s, the pair* $(\coprod X^{\alpha}, (\varepsilon_{\alpha}))$ *is the coproduct of the objects* $(X^{\alpha}, \varepsilon_{\alpha})$ *in* $\mathbf{C}^{\mathrm{aug}}$.

Dually, if $\mathbf{C}$ *is a category with a final object* J, *and we define* $\mathbf{C}^{\mathrm{pt}}$ *to have as objects all pairs* (X, η) *with* X *an object of* $\mathbf{C}$ *and* $\eta \in \mathbf{C}(J, X)$, *and to have as morphisms the morphisms of first components making commuting triangles with morphisms from* J, *then* $\mathbf{C}^{\mathrm{pt}}$ *is a pointed category, and products in* $\mathbf{C}$ *induce products in* $\mathbf{C}^{\mathrm{pt}}$.

If $\mathbf{C}$ *is itself pointed, then the forgetful functors give natural isomorphisms* $\mathbf{C}^{\mathrm{pt}} \cong \mathbf{C}$ *and* $\mathbf{C}^{\mathrm{aug}} \cong \mathbf{C}$. □

The term "pointed category" probably comes from the case of topological spaces with *base point*, an example of the $(\ )^{\mathrm{pt}}$ construction. (Caveat: the term "pointed" is often used, in phrases like "pointed group", to mean "given with a distinguished element". Pointed algebras of a variety $\mathbf{V}$ in that sense do form a variety, but it is not equivalent to $\mathbf{V}^{\mathrm{pt}}$ unless all operations of $\mathbf{V}$ satisfy the idempotent identity $s(x, \ldots, x) = x$. Thus "pointed object", like "epimorphism", is a term whose category-theoretic sense must be distinguished from the concrete concept that motivated it.)

In general, when $\mathbf{C}$ has both an initial and a final object, but is *not* pointed, $\mathbf{C}^{\mathrm{aug}}$ and $\mathbf{C}^{\mathrm{pt}}$ will not be equivalent. Thus, $(k\text{-}\mathbf{Ring}^1)^{\mathrm{aug}}$ has been characterized in the preceding section, but $(k\text{-}\mathbf{Ring}^1)^{\mathrm{pt}}$ consists only of the k-ring $\{0\}$ (with its identity map for η), since a unital ring admitting a homomorphism from the zero ring must satisfy $1 = 0$. $\mathbf{Semigp}^{\mathrm{aug}}$ has only the empty object, while $\mathbf{Semigp}^{\mathrm{pt}}$ can be identified with the variety of all semigroups given with a distinguished idempotent element.

The description of the degenerate category $(k\text{-}\mathbf{Ring}^1)^{\mathrm{pt}}$ in the above paragraph generalizes to the observation that whenever $\mathbf{V}$ is a variety with at least one zeroary operation, $\mathbf{V}^{\mathrm{pt}}$ may be identified with the subvariety of $\mathbf{V}$ determined by the family of identities saying that all derived zeroary operations are equal. The description of $\mathbf{Semigp}^{\mathrm{aug}}$ generalizes to say that if $\mathbf{V}$ is a variety without zeroary operations, $\mathbf{V}^{\mathrm{aug}}$ has only the empty algebra (with its identity morphism as augmentation).

The following easy consequence of Lemmas 11.2 and 11.3 generalizes the considerations of the preceding section relating co-$\mathbf{AbSemigp}^e$ objects in $k\text{-}\mathbf{Ring}^1$ and $(k\text{-}\mathbf{Ring}^1)^{\mathrm{aug}}$:

LEMMA 11.4. *Let* $\mathbf{C}$ *be a category with finite coproducts, and* $\mathbf{V}$ *any variety of algebras. Then, recalling that* $\mathbf{Rep}(\mathbf{C}, \mathbf{V})$ *denotes the category of all representable functors from* $\mathbf{C}$ *to* $\mathbf{V}$, *we have*

$$\mathbf{Rep}(\mathbf{C}, \mathbf{V}^{\mathrm{pt}}) \simeq \mathbf{Rep}(\mathbf{C}^{\mathrm{aug}}, \mathbf{V}^{\mathrm{pt}}) \simeq \mathbf{Rep}(\mathbf{C}^{\mathrm{aug}}, \mathbf{V}).$$

If either $\mathbf{C}$ *or* $\mathbf{V}$ *is pointed, then the above categories are also naturally equivalent to* $\mathbf{Rep}(\mathbf{C}, \mathbf{V})$. *If* $\mathbf{V}$ *is pointed (i.e., satisfies the equivalent conditions of Lemma 11.2) then any co-*$\mathbf{V}$ *object of* $\mathbf{C}^{\mathrm{aug}}$ *is uniquely determined by its underlying* $\mathbf{C}$*-object and its non-zeroary co-operations.* □

EXERCISE 11.5. *Consider the following conditions on a category* $\mathbf{C}$:

(i) $\mathbf{C}$ *is pointed.*

(ii) *It is possible to choose a morphism* $0_{X,Y}$ *in each hom-set* $\mathbf{C}(X, Y)$ *so that for all* $X, Y, Z \in \mathbf{C}$ *and* $f \in \mathbf{C}(X, Y)$, $g \in \mathbf{C}(Y, Z)$, *one has* $0_{Y,Z} f = 0_{X,Z}$ *and* $g\, 0_{X,Y} = 0_{X,Z}$.

(iii) *It is possible to choose a morphism* $0_{X,Y}$ *in each hom-set* $\mathbf{C}(X, Y)$ *so that for all* $X, Y, Z \in \mathbf{C}$ *one has* $0_{Y,Z}\, 0_{X,Y} = 0_{X,Z}$.

(iv) *For all* $X, Y \in \mathrm{Ob}(\mathbf{C})$, $\mathbf{C}(X, Y) \neq \varnothing$.

Verify that (i) $\Rightarrow$ (ii) $\Rightarrow$ (iii) $\Rightarrow$ (iv).

If **C** *has an initial object and a final object, show that* (iv) $\Rightarrow$ (i), *so that all four conditions are equivalent.*

If **C** *has an initial object* or *a final object, show that* (ii) $\Rightarrow$ (i). *Are both equations of* (ii) *needed to get this implication?*

Condition (ii) of the preceding Exercise is taken by some as the definition of a pointed category, avoiding the necessity of assuming the existence of an initial or final object.

Digressing from the subject of pointed categories, observe that if **C** is a variety *without* zeroary operations, it is hard to see how one can construct representable functors to any variety **V** *with* zeroary operations. (Cf. the characterization in Lemma 8.9 of the operations of an algebra given by a representable functor.) In fact

LEMMA 11.6. *If* **V** *is a variety without zeroary operations, and* **W** *a variety with at least one zeroary operation, then* **Rep(V, W)** *has (up to isomorphism) only one object, the functor carrying every object of* **V** *to the* 1*-element algebra of* **W**. *This functor is represented by the empty algebra of* **V**.

PROOF. If a coalgebra R in **V** represents a functor to **W**, then applying this functor to the empty algebra $\varnothing \in \mathbf{V}$, we see that $\mathbf{V}(|R|, \varnothing)$ has a **W**-structure, hence is *nonempty*, hence $|R| = \varnothing$. □

§12. Plans and preparations.

Recall that k is a fixed unital associative ring. In the following Definition the parenthesized convention will not be used until late in §13.

DEFINITION 12.1. *The variety of k-bimodules will be denoted* k-**Bimod**. *For two such bimodules* M *and* N, *their tensor product over* k *will be written* $M \otimes_k N \in k$-**Bimod**.

(If $U \subseteq M$, $V \subseteq N$ *are subbimodules, then in discussing* $M \otimes_k N$, *we shall write* $U \otimes V$ *without a subscript* k *for the subbimodule of* $M \otimes_k N$ *spanned by* $\{u \otimes v \mid u \in U,\ v \in V\}$, *in contrast with the tensor product bimodule* $U \otimes_k V$, *of which it will be an image, but to which it may not be isomorphic.)*

If $(R^{\alpha})_{\alpha \in A}$ is a family of objects of k-**Ring**, then it is not hard to show (cf. [**179**, §3.2], [**19**, Corollary 8.1]) that their coproduct in k-**Ring** has the k-bimodule structure

$$\coprod_{\alpha \in A} R^{\alpha} = \bigoplus_{(\alpha_1, \dots, \alpha_h)} R^{\alpha_1} \otimes_k \dots \otimes_k R^{\alpha_h}, \tag{12.2}$$

where the direct sum is over all strings

$$(\alpha_1, \dots, \alpha_h) \text{ with } h \geq 1, \text{ all } \alpha_i \in A, \text{ and } \alpha_i \neq \alpha_{i+1} \ (1 \leq i \leq h-1). \tag{12.3}$$

The multiplication of this ring (12.2) on a pair of summands

$$(R^{\alpha_1} \otimes_k \dots \otimes_k R^{\alpha_h}) \times (R^{\alpha'_1} \otimes_k \dots \otimes_k R^{\alpha'_{h'}})$$

is given, when $\alpha_h \neq \alpha'_1$, by tensor multiplication into

$$R^{\alpha_1} \otimes_k \ldots \otimes_k R^{\alpha_h} \otimes_k R^{\alpha'_1} \otimes_k \ldots \otimes_k R^{\alpha'_{h'}},$$

while when $\alpha_h = \alpha'_1 = \beta$, it is the map into

$$R^{\alpha_1} \otimes_k \ldots \otimes_k R^{\beta} \otimes_k \ldots \otimes_k R^{\alpha'_{h'}}$$

induced by the multiplication map $R^{\alpha_h} \times R^{\alpha'_1} = R^{\beta} \times R^{\beta} \to R^{\beta}$. The coprojections

$$\text{(12.4)} \qquad i^{\beta}\colon R^{\beta} \to \coprod R^{\alpha} \qquad (\beta \in A)$$

are, of course, the inclusion maps of the summands with $h = 1$ in the sum (12.2).

DEFINITION 12.5. *For any string* $\sigma = (\alpha_1, \ldots, \alpha_h)$ *as in* (12.3), h *will be called the* length *of* σ, *and the sub-k-bimodule* $R^{\alpha_1} \otimes_k \ldots \otimes_k R^{\alpha_h}$ *of the coproduct ring* (12.2) *will be abbreviated* R^{σ} *or* $R^{(\alpha_1, \ldots, \alpha_h)}$. *For any* $h \geq 0$, *the sum of all* R^{σ} *with* σ *of length* $\leq h$ *will be denoted* $R^{|h|}$. *For an element* x *of the coproduct ring, the least nonnegative integer* h *such that* $x \in R^{|h|}$ *will be called the* height *of* x, $\mathrm{ht}(x)$.

The coproduct k-ring is clearly *filtered* by height, i.e., each $R^{|h|}$ is a k-subbimodule, and $R^{|h|}\, R^{|i|} \subseteq R^{|h+i|}$ for all h and i.

Suppose now that $(R, \mathbf{a})$ is a co-$\mathbf{AbSemigp}^e$ object of k-**Ring**.

To study the coaddition map $\mathbf{a}\colon R \to R \amalg R$, we will want to distinguish the two copies of R in this coproduct (that is, the images of R under the two coprojection maps). We shall denote these R^{λ} and R^{ρ}, thinking of them as corresponding to the "left" and "right" arguments of the semigroup operation. We modify (12.4) trivially by writing the coprojection maps as

$$\text{(12.6)} \qquad i^{\lambda}\colon R \to R^{\lambda} \amalg R^{\rho}, \qquad i^{\rho}\colon R \to R^{\lambda} \amalg R^{\rho},$$

i.e., though we distinguish R^{λ} and R^{ρ} within the coproduct ring, we do not introduce two external copies of R. For every $x \in R$ we make the abbreviations

$$\text{(12.7)} \qquad i^{\lambda}(x) = x^{\lambda}, \qquad i^{\rho}(x) = x^{\rho}.$$

Since our index-set $\{\lambda, \rho\}$ has only two elements, the only strings (12.3) will be those of the forms $(\lambda, \rho, \lambda, \rho, \ldots)$ and $(\rho, \lambda, \rho, \lambda, \ldots)$. We will abbreviate these to $[h]$ and $[h]'$ respectively, where h is the length of the string. So,

$$R^{\lambda} = R^{[1]} \qquad R^{\rho} = R^{[1]'}$$
$$R^{\lambda} \otimes_k R^{\rho} = R^{[2]} \qquad R^{\rho} \otimes_k R^{\lambda} = R^{[2]'}$$
$$\ldots$$

and for all h,

$$R^{|h|} = \bigoplus_{1 \leq i \leq h} R^{[i]} \oplus R^{[i]'}.$$

We now introduce an invariant of elements of R which depends on the coaddition:

DEFINITION 12.8. *For* $x \in R$, *we define*

$$\deg(x) \;=\; \mathrm{ht}(\mathbf{a}(x)).$$

For $h \geq 0$ *we define*

$$R_h \;=\; \{x \in R \mid \deg(x) \leq h\} \;=\; \{x \in R \mid \mathbf{a}(x) \in R^{|h|}\}.$$

Let us consider now the interpretation of the simplest identities of $\mathbf{AbSemigp}^e$, the neutral-element laws, in our cosemigroup $(R, \mathbf{a})$. Recall that the co-neutral-element of this object is the unique map $0\colon R \to \{0\}$. The coidentity corresponding to the right neutral law,

$$u + 0 \;=\; u,$$

says that the composite map

$$R \xrightarrow{\;\mathbf{a}\;} R^{\lambda} \amalg R^{\rho} \xrightarrow{\;(1_R,\,0)\;} R$$

is the identity map of R. Concretely, this means that for $x \in R$, $\mathbf{a}(x) = x^{\lambda}$ + terms in summands R^{σ} such that σ involves at least one ρ. Similarly, the left coneutral law translates to the condition that $\mathbf{a}(x) = x^{\rho}$ + terms in summands involving at least one λ. Together these conditions say

(12.9) $$\mathbf{a}(x) \;=\; x^{\rho} + x^{\lambda} + \text{terms in summands } R^{\sigma} \text{ where } \sigma \text{ has length } \geq 2,$$

equivalently:

(12.10) $$\mathbf{a} \;=\; i^{\rho} + i^{\lambda} + \mathbf{b}, \text{ where } \mathbf{b} \text{ is a bimodule map of } R \text{ into } \textstyle\bigoplus_{h \geq 2} (R^{[h]} \oplus R^{[h]'}).$$

In particular, (12.9) tells us that

(12.11) $$x \in R_1 \;\Leftrightarrow\; \mathbf{a}(x) = x^{\rho} + x^{\lambda};$$

in other words, that for $\mathbf{b}$ the map defined by (12.10), we have

(12.12) $$R_1 \;=\; \ker \mathbf{b}.$$

We shall complete this section of ''plans and preparations'' by (a) constructing a certain class of co-$\mathbf{AbSemigp}^e$ objects of k-**Ring**, (b) indicating what will be involved in proving that *all* co-$\mathbf{AbSemigp}^e$ objects of k-**Ring** are isomorphic to members of this class, and then (c) preparing for the proof by again considering an arbitrary co-$\mathbf{AbSemigp}^e$ object of k-**Ring**, and setting up the notation needed to work with the coassociative and cocommutative laws.

If M is any k-bimodule, let $k\langle M\rangle$ denote, as usual, the *unital* tensor ring

(12.13) $$k\langle M\rangle \;=\; k \oplus M \oplus (M \otimes_k M) \oplus \dots .$$

This tensor-ring construction is the left adjoint to the forgetful functor k-$\mathbf{Ring}^1 \to$ k-**Bimod**. The nonunital k-ring with the corresponding universal property is the kernel of the obvious augmentation on $k\langle M\rangle$, the *nonunital tensor k-ring on* M, which we shall write

(12.14) $$[k]\langle M\rangle \;=\; M \oplus (M \otimes_k M) \oplus \dots .$$

Having given warning that we would be a bit loose notationally with forgetful functors when there is no danger of ambiguity, we write the universal property of this object as

(12.15) $$k\text{-}\mathbf{Ring}([k]\langle M\rangle, -) \;\cong\; k\text{-}\mathbf{Bimod}(M, -).$$

The values of the functor on the right have natural abelian group structures, so the object $[k]\langle M\rangle$ must have a structure of co-**Ab** object, and in particular, of co-$\mathbf{AbSemigp}^e$ object, in the category k-**Ring**. To determine the coaddition, we take the coproduct

$$[k]\langle M\rangle^{\lambda} \amalg [k]\langle M\rangle^{\rho} \;=\; [k]\langle M^{\lambda} \oplus M^{\rho}\rangle$$

(cf. Exercise 5.13(iv)), and form the sum of the two coprojections

$$i^{\lambda},\; i^{\rho}\colon\; [k]\langle M\rangle \;\to\; [k]\langle M^{\lambda} \oplus M^{\rho}\rangle$$

under the operation corresponding to adding bimodule maps on the right-hand-side of (12.15). This sum, the coaddition map of our cosemigroup, is the map

(12.16) $$\mathbf{a}_M\colon\; [k]\langle M\rangle \;\to\; [k]\langle M^{\lambda} \oplus M^{\rho}\rangle$$

induced by the diagonal map $M \to M^{\lambda} \oplus M^{\rho}$, i.e., defined by

(12.17) $$\mathbf{a}_M(x) \;=\; x^{\lambda} + x^{\rho} \qquad \text{for } x \in M.$$

Thus, for instance, $\mathbf{a}_M$ will carry an element $x \otimes y \in M \otimes_k M \subseteq [k]\langle M\rangle$ to $(x^{\lambda}+x^{\rho})(y^{\lambda}+y^{\rho}) = x^{\lambda} \otimes y^{\lambda} + x^{\lambda} \otimes y^{\rho} + x^{\rho} \otimes y^{\lambda} + x^{\rho} \otimes y^{\rho}$. More generally, an element of the h-fold tensor product,

$$r \in M \otimes_k \dots \otimes_k M,$$

is mapped to the sum of its images in the 2^h mutually isomorphic summands $M^{\alpha_1} \otimes_k \dots \otimes_k M^{\alpha_h}$ $(\alpha_i \in \{\lambda, \rho\})$ of $[k]\langle M^{\lambda} \oplus M^{\rho}\rangle$. Note that of these summands, exactly two have height h as defined in Definition 12.5: the summands with $(\alpha_1, \dots, \alpha_h) = [h]$ and $[h]'$. (All other summands involve repeated tensor-factors, e.g., $\dots \otimes_k M^{\rho} \otimes_k M^{\rho} \otimes_k \dots$. In such cases, since two successive factors come from the same coproduct-constituent, $[k]\langle M\rangle^{\lambda}$ or $[k]\langle M\rangle^{\rho}$, the term belongs to a summand $[k]\langle M\rangle^{\sigma}$ with σ of length $<h$.) From the presence of height-h terms, we can see that $\mathbf{a}_M(r)$ has height h, and hence that the *degree function* on $[k]\langle M\rangle$ (Definition 12.8) relative to the coaddition $\mathbf{a}_M$ coincides with the usual degree function on $[k]\langle M\rangle$ as a tensor ring. (However, as the preceding parenthetical observations showed, the *height* function on $[k]\langle M\rangle^{\lambda} \amalg [k]\langle M\rangle^{\rho}$ does not coincide with, but is, rather, $\leq$ the degree function on the tensor ring $[k]\langle M^{\lambda} \oplus M^{\rho}\rangle$.)

In particular, the subbimodule $[k]\langle M\rangle_1$ (Definition 12.8) of this cosemigroup object is precisely the generating subbimodule M.

What is our program for characterizing arbitrary co-**AbSemigp** objects $(R, \mathbf{a})$ in k-**Ring**? The major step will be to show that for every such $(R, \mathbf{a})$,

(12.18) R is generated as a ring by the k-subbimodule R_1.

This condition (12.18), together with the observation (12.11), clearly determines the coaddition map $\mathbf{a}$ on R. In fact, it means that if we let $M = R_1$, then the co-$\mathbf{AbSemigp}^e$ object $(R, \mathbf{a})$ is the image of the co-$\mathbf{AbSemigp}^e$ object $([k]\langle M\rangle, \mathbf{a}_M)$ ($\mathbf{a}_M$ defined in (12.17)) under a homomorphism of co-$\mathbf{AbSemigp}^e$ objects which is one-to-one on M. We shall then show that such a homomorphism must be an isomorphism, so that $(R, \mathbf{a})$ actually has the form $([k]\langle M\rangle, \mathbf{a}_M)$, as desired. All morphisms between such cosemigroups,

$$f\colon ([k]\langle M\rangle, \mathbf{a}_M) \to ([k]\langle N\rangle, \mathbf{a}_N)$$

are easily shown to be induced by bimodule homomorphisms $M \to N$, so we will have a complete description of the category of co-$\mathbf{AbSemigp}^e$ objects in k-$\mathbf{Ring}$, and hence of the category of representable functors k-$\mathbf{Ring} \to \mathbf{AbSemigp}^e$.

To prove the results sketched, we will have to understand the coassociativity and cocommutativity conditions on the coaddition $\mathbf{a}$. (We have already interpreted the co-neutral laws as (12.9).) Coassociativity says that the diagram

$$\begin{array}{ccc} R & \xrightarrow{\ \mathbf{a}\ } & R \amalg R \\ \Big\downarrow \mathbf{a} & & \Big\downarrow \mathbf{a} \amalg 1_R \\ R \amalg R & \xrightarrow{\ 1_R \amalg \mathbf{a}\ } & R \amalg R \amalg R \end{array} \tag{12.19}$$

commutes. We need symbols for the *three* copies of R in the final coproduct; we shall use R^λ, R^μ, and R^ρ, corresponding to the "left", "middle" and "right" variables in the associative law $(u + v) + w = u + (v + w)$. The two k-ring homomorphisms

$$\mathbf{a} \amalg 1_R,\ 1_R \amalg \mathbf{a} : R^\lambda \amalg R^\rho \rightrightarrows R^\lambda \amalg R^\mu \amalg R^\rho$$

will be called $\mathbf{a}_\lambda$ and $\mathbf{a}_\rho$ respectively. Thus, $\mathbf{a}_\lambda$ takes each element x^ρ $(x \in R)$ to itself, and carries x^λ to the element $\mathbf{a}(x)$ but with all superscripts $^\rho$ changed to $^\mu$; and likewise $\mathbf{a}_\rho$ takes x^λ to itself, and x^ρ to $\mathbf{a}(x)$ with all superscripts $^\lambda$ changed to $^\mu$. The coassociative law may now be written

$$\mathbf{a}_\lambda \mathbf{a} = \mathbf{a}_\rho \mathbf{a}. \tag{12.20}$$

To study this condition, we need some further notation.

DEFINITION 12.21. *An* index-string *will mean a finite nonempty sequence* $(\alpha_1, \ldots, \alpha_h)$ $(h \geq 1)$ *of symbols from the set* $\{\lambda, \mu, \rho\}$, *in which no two successive symbols are the same.*

If $\sigma = (\alpha_1, \ldots, \alpha_h)$ *is an index-string, then* $(\mathbf{aa})^\sigma$ *will denote the composite bimodule homomorphism*

$$R \xrightarrow{\ \mathbf{a}_\lambda \mathbf{a} = \mathbf{a}_\rho \mathbf{a}\ } R^\lambda \amalg R^\mu \amalg R^\rho \longrightarrow R^\sigma,$$

where the second map is the projection of the coproduct as a bimodule onto its direct summand $R^\sigma = R^{\alpha_1} \otimes_k \ldots \otimes_k R^{\alpha_h}$.

If σ *is an index-string consisting of* λ*'s and* ρ*'s only, then* $\mathbf{a}^\sigma$ *(N.B.,* $\mathbf{a}$ *not doubled) will denote the composite map* $R \xrightarrow{\mathbf{a}} R^\lambda \amalg R^\rho \longrightarrow R^\sigma$ *(where the last map is again the obvious bimodule projection). If* σ *consists of* λ*'s and* μ*'s only, or of* μ*'s and* ρ*'s only, then* $\mathbf{a}^\sigma$ *will denote the composite of* $\mathbf{a}$ *with the operation of changing all superscripts* ρ *(respectively* λ*) to* μ*, followed by projection onto* R^σ.

For any σ involving only two indices, it is easy to deduce from (12.9) that $\mathbf{a}^\sigma = (\mathbf{aa})^\sigma$. (These observations correspond to the semigroup identities $u+v = 0+u+v = u+0+v = u+v+0$.) Thus we might have consolidated our notation by writing $\mathbf{a}^\sigma$ rather than $(\mathbf{aa})^\sigma$ in the case of arbitrary index-strings σ; but we judged the present notation more suggestive. We do, however, need to note that $\mathbf{a}^{(\lambda)}$, $\mathbf{a}^{(\mu)}$, and $\mathbf{a}^{(\rho)}$ are well-defined, independent of whether we interpret (λ) as a (length-one) string of terms from $\{\lambda, \mu\}$ or from $\{\lambda, \rho\}$, (μ) as a string of terms from $\{\lambda, \mu\}$ or from $\{\rho, \mu\}$, etc.. In fact, (12.9) gives us

(12.22) $$\mathbf{a}^{(\lambda)} = i^\lambda, \qquad \mathbf{a}^{(\mu)} = i^\mu, \qquad \mathbf{a}^{(\rho)} = i^\rho,$$

under both interpretations of each term.

Finally, the *cocommutative* law says that $\mathbf{a}$ remains unchanged if we follow it by the automorphism of $R^\lambda \amalg R^\rho$ which takes x^λ to x^ρ and vice versa for all $x \in R$. We could write this condition down by introducing a symbol for that automorphism, but there is another way of expressing cocommutativity that we shall find more useful. We need

DEFINITION 12.23. *If* $\sigma = (\alpha_1, \ldots, \alpha_h)$ *is any index-string, then the "forgetful" isomorphism*

$$R^\sigma = R^{\alpha_1} \otimes_k \ldots \otimes_k R^{\alpha_h} \longrightarrow R \otimes_k \ldots \otimes_k R,$$

which acts by "dropping superscripts", will be denoted f.

Recalling that $[h]$ and $[h]'$ denote the index-strings of length h, $(\lambda, \rho, \ldots)$ and $(\rho, \lambda, \ldots)$ respectively, we find that cocommutativity of $\mathbf{a}$ means that for every $h \geq 1$,

(12.24) $$f\mathbf{a}^{[h]} = f\mathbf{a}^{[h]'}.$$

(Here the f on the left drops superscripts in $R^{[h]}$, while the f on the right drops them in $R^{[h]'}$. We could have distinguished our forgetful isomorphisms f with indices, but clearly their meaning is determined by the object to which they are applied. In the other direction, we might have built the forgetful maps we are calling f into the definitions of the linear maps $\mathbf{a}^\sigma$ and $(\mathbf{aa})^\sigma$, since each such map has for codomain only a single bimodule R^σ, so that the labeling of this bimodule is unnecessary. Indeed, the maps f and $\mathbf{a}^\sigma$ (respectively $(\mathbf{aa})^\sigma$) will virtually always appear in combination in the remainder of our proof. This would lead to a more compact notation, but we felt it preferable to show things explicitly.)

§13. Proof of the structure theorem for co-AbSemigp^e objects.

$(R, \mathbf{a})$ will continue to denote a fixed co-**AbSemigp**^{e} object in k-**Ring**.

We shall now use the direct sum decomposition (12.2) of the coproduct $R^\lambda \amalg R^\mu \amalg R^\rho$ to "dissect" the coassociative law (12.20).

DEFINITION 13.1. *If* $\sigma = (\alpha_1, \dots, \alpha_h)$ *is an index-string, then the* λ-decomposition *of* σ *will mean the finite sequence, generally of shorter length,* $\sigma_\lambda = (\dots, \rho, \theta_i, \rho, \theta_{i+1}, \dots)$ *obtained by breaking* σ *into substrings* θ_i *of consecutive* λ*'s and* μ*'s, alternating with single* ρ*'s. The* θ_i *will be called the* $\{\lambda, \mu\}$*-segments of* σ.

The λ-reduct *of* σ *will mean the index-string* $\bar{\sigma}_\lambda$ *obtained from* σ *by replacing each* $\{\lambda, \mu\}$*-segment by the symbol* λ.

The ρ-decomposition σ_ρ, *into* $\{\rho, \mu\}$*-segments and terms* λ, *and the* ρ-reduct $\bar{\sigma}_\rho$ *of* σ, *are defined analogously.*

Note that in writing out σ_λ in the above definition, we avoided specifying whether it began with θ_1 or with ρ. This, of course, depends on σ.

We now isolate the σ-component of the coassociative law (12.20), for an arbitrary index-string σ.

LEMMA 13.2. *Let* $\sigma = (\alpha_1, \dots, \alpha_h)$ *be an index-string, with* λ- *and* ρ*-decompositions*

$$\sigma_\lambda = (\dots, \rho, \theta_i, \rho, \theta_{i+1}, \dots), \qquad \sigma_\rho = (\dots, \lambda, \varphi_j, \lambda, \varphi_{j+1}, \dots).$$

Then $(\mathbf{aa})^\sigma$ *(Definition 12.21) may be expressed both as the composite map*

$$R \xrightarrow{\mathbf{a}^{\bar{\sigma}_\lambda}} R^{\bar{\sigma}_\lambda} \xrightarrow{f} \dots \otimes_k R \otimes_k R \otimes_k \dots \xrightarrow{\dots \otimes i^\rho \otimes \mathbf{a}^{\theta_i} \otimes i^\rho \otimes \mathbf{a}^{\theta_{i+1}} \otimes \dots} R^\sigma$$

and as the composite

$$R \xrightarrow{\mathbf{a}^{\bar{\sigma}_\rho}} R^{\bar{\sigma}_\rho} \xrightarrow{f} \dots \otimes_k R \otimes_k R \otimes_k \dots \xrightarrow{\dots \otimes i^\lambda \otimes \mathbf{a}^{\varphi_j} \otimes i^\lambda \otimes \mathbf{a}^{\varphi_{j+1}} \otimes \dots} R^\sigma. \quad \square$$

EXAMPLE 13.3. Say $\sigma = (\rho, \mu, \lambda, \mu, \lambda)$. Then

$$\sigma_\lambda = (\rho, (\mu, \lambda, \mu, \lambda)), \qquad \sigma_\rho = ((\rho, \mu), \lambda, (\mu), \lambda),$$

$$\bar{\sigma}_\lambda = (\rho, \quad \lambda \quad), \qquad \bar{\sigma}_\rho = (\quad \rho \quad, \lambda, \ \rho\ , \lambda).$$

For this σ, the calculations of the σ-components of the two sides of the coassociativity equation (12.20) as described in the above Lemma are:

48   III. REPRESENTABLE FUNCTORS FROM RINGS TO ABELIAN GROUPS

$$
\begin{array}{c}
R \\
\downarrow \mathbf{a}^{(\rho,\lambda)} \\
\overbrace{R^\rho \otimes_k R^\lambda} \\
\| \S \quad\quad \| \S \\
R \quad\quad\quad R \\
\downarrow i^\rho \quad\quad \downarrow \mathbf{a}^{(\mu,\lambda,\mu,\lambda)} \\
\underbrace{R^\rho \otimes_k R^\mu \otimes_k R^\lambda \otimes_k R^\mu \otimes_k R^\lambda} = \underbrace{R^\rho \otimes_k R^\mu \otimes_k R^\lambda \otimes_k R^\mu \otimes_k R^\lambda}
\end{array}
$$

$$
\begin{array}{c}
R \\
\downarrow \mathbf{a}^{(\rho,\lambda,\rho,\lambda)} \\
\overbrace{R^\rho \otimes_k R^\lambda \otimes_k R^\rho \otimes_k R^\lambda} \\
\| \S \quad \| \S \quad \| \S \quad \| \S \\
R \quad R \quad R \quad R \\
\downarrow \mathbf{a}^{(\rho,\mu)} \quad \downarrow i^\lambda \quad \downarrow \mathbf{a}^{(\mu)} \quad \downarrow i^\lambda
\end{array}
$$

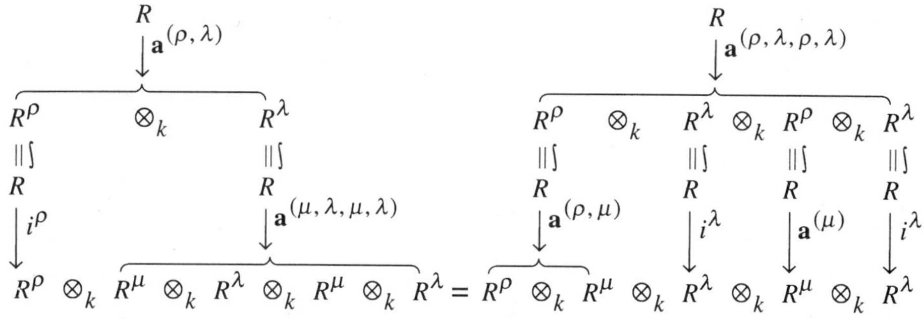

Note that in this example, $\bar\sigma_\lambda$ and $\bar\sigma_\rho$ have unequal lengths, 2 and 4 respectively. Now suppose we apply $(\mathbf{aa})^{(\rho,\mu,\lambda,\mu,\lambda)}$ to an element $r \in R$ of degree <4, i.e., an element such that $\mathbf{a}(r) \in R^\lambda \amalg R^\rho$ has height <4. We can see using the right-hand decomposition shown above for $(\mathbf{aa})^{(\rho,\mu,\lambda,\mu,\lambda)}$ that $(\mathbf{aa})^{(\rho,\mu,\lambda,\mu,\lambda)}(r)$ will be zero (for r is killed off by $\mathbf{a}^{(\rho,\lambda,\rho,\lambda)}$, since this map looks for a height 4 component in $\mathbf{a}(r)$), but there is no apparent reason for it to be killed by the left-hand side. Thus, the fact that the right-hand side is zero gives a nontrivial statement about the left-hand expression. We shall exploit this phenomenon systematically below.

Let us fix a positive integer h. For any index-string σ, an equation such as $(\mathbf{aa})^\sigma(R_h) = 0$, thought of as a condition on the operator $(\mathbf{aa})^\sigma$, will be abbreviated $(\mathbf{aa})^\sigma =_h 0$. From Lemma 13.2 we clearly have:

COROLLARY 13.4. *If σ is an index-string, and either $\bar\sigma_\lambda$ or $\bar\sigma_\rho$ has length $>h$, then $(\mathbf{aa})^\sigma =_h 0$.* □

We shall apply this result to a family of index-strings which is general enough to capture the strength of the coassociative and cocommutative laws, but simple enough to calculate with easily. We construct these strings σ as follows: Start with an index-string of λ's and ρ's, of length h (i.e., either $[h]$ or $[h]'$). For some single $i \in \{1, \ldots, h\}$, replace the ith term of this string, if it is λ, by an index-string θ of λ's and μ's; if it is ρ, by an index-string φ of ρ's and μ's. We shall do our calculations in the first of these two cases, so that

$$\sigma_\lambda = (\ldots, \rho, (\lambda), \rho, \theta, \rho, (\lambda), \rho, \ldots),$$

but we shall keep in mind that for each statement we make, the $\{\lambda,\rho\}$-dual statement holds as well.

If θ involves more than one λ, then $\bar\sigma_\rho$ will have length $>h$, so by Corollary 13.4,

(13.5) $\quad\quad\quad\quad\quad\quad\quad (\mathbf{aa})^\sigma =_h 0.$

The remaining case, where θ does not involve more than one λ, comprises five subcases:

(13.6) $\quad\quad\quad\quad \theta = (\mu),\ (\lambda),\ (\mu,\lambda),\ (\lambda,\mu),\ (\mu,\lambda,\mu).$

The coassociative law alone is insufficient to imply (13.5) in any of these cases, but we shall obtain (13.5) in the last three cases by bringing the cocommutative law into play as well. Applying that law, (12.24), to a single tensor-factor "$\ldots \otimes \mathbf{a}^{\theta_i} \otimes \ldots$" or "$\ldots \otimes \mathbf{a}^{\varphi_j} \otimes \ldots$" of either of the long arrows in Lemma 13.2 gives the general result

LEMMA 13.7. *Let σ be any index-string, and τ an index-string obtained from σ by interchanging the symbols λ and μ throughout some $\{\lambda, \mu\}$-segment of σ, or by interchanging the symbols ρ and μ throughout some $\{\rho, \mu\}$-segment of σ (Definition 13.1). Then $f(\mathbf{aa})^{\sigma} = f(\mathbf{aa})^{\tau}$.* □

Now if σ is the string obtained from $[h]$ by replacing some λ by $\theta = (\mu, \lambda, \mu)$, and we let τ denote the string formed from σ by replacing this θ by $\theta' = (\lambda, \mu, \lambda)$, then $(\mathbf{aa})^{\tau} =_h 0$ because θ' involves more than one λ, hence as $f(\mathbf{aa})^{\sigma} = f(\mathbf{aa})^{\tau}$, we see that $(\mathbf{aa})^{\sigma} =_h 0$, as claimed. (Recall that the maps f are isomorphisms).

The cases $\theta = (\mu, \lambda)$, (λ, μ) are trickier. Note that for $\theta = (\mu, \lambda)$ there is one case where $\bar{\sigma}_{\rho}$ in fact has length $> h$, namely when θ occurs at the beginning of σ, so that $\sigma = (\mu, \lambda, \rho, \lambda, \ldots)$, which has ρ-decomposition $((\mu), \lambda, (\rho), \lambda, \ldots)$. Starting with this case, let us apply Lemma 13.7 repeatedly, alternately to $\{\lambda, \mu\}$-segments and $\{\rho, \mu\}$-segments. We get

$$(13.8) \qquad 0 =_h f(\mathbf{aa})^{(\mu, \lambda, \rho, \lambda, \rho, \ldots)} = f(\mathbf{aa})^{(\lambda, \mu, \rho, \lambda, \rho, \ldots)} = f(\mathbf{aa})^{(\lambda, \rho, \mu, \lambda, \rho, \ldots)} = f(\mathbf{aa})^{(\lambda, \rho, \lambda, \mu, \rho, \ldots)} = \ldots .$$

We see that in this way we go through precisely all strings obtainable from $[h]$ by substituting $\theta = (\lambda, \mu)$ or (μ, λ) for one term λ, or $\varphi = (\rho, \mu)$ or (μ, ρ) for one term ρ. The same results for strings formed starting from $[h]'$ follow by the $\{\lambda, \rho\}$-dual considerations. So (13.5) holds whenever θ has length > 1.

We have obtained these results by noting that our string σ, or a related string, has ρ-reduct of length greater than h, and evaluating $(\mathbf{aa})^{\sigma}$ by the second display of Lemma 13.2. On the other hand, by construction σ always has λ-reduct $\bar{\sigma}_{\lambda} = [h]$ or $[h]'$. Let us now evaluate $(\mathbf{aa})^{\sigma}$ using the first display of Lemma 13.2. However, let us trim our superscripts somewhat. We note that all but one of the $\{\lambda, \mu\}$-segments of σ are (λ), and that $\mathbf{a}^{(\lambda)} = i^{\lambda}$, $\mathbf{a}^{(\rho)} = i^{\rho}$. Hence all but one of the tensor factors of the "long" arrow in our composite map will be an i^{λ} or an i^{ρ}. These merely restore the superscripts that f has "forgotten". Let us leave these superscripts "forgotten", replacing these i^{λ}'s and i^{ρ}'s by the identity map of R. Also, the leftmost map $\mathbf{a}^{\bar{\sigma}_{\lambda}}$ may be $\mathbf{a}^{[h]}$ or $\mathbf{a}^{[h]'}$, but by cocommutativity these have the same composite with f, so we may as well write the composite map as $f\mathbf{a}^{[h]}$. Our result now says that the following composite map annihilates $R_h \subseteq R$,

$$R \xrightarrow{f\mathbf{a}^{[h]}} R \otimes_k \ldots \otimes_k R \xrightarrow{1_R \otimes \ldots \otimes \mathbf{a}^{\theta} \otimes \ldots \otimes 1_R} R \otimes_k \ldots \otimes_k R^{\theta} \otimes_k \ldots \otimes_k R,$$

for any index-string θ of λ's and μ's of length strictly greater than 1, and any choice of position for the lone "$\otimes \mathbf{a}^{\theta} \otimes$".

But if we look at the definition of $\mathbf{a}^{\theta}$ (Definition 12.21, last sentence), for θ consisting entirely of λ's and μ's we see that it is essentially one component of $\mathbf{a}$, with all ρ's changed to μ's in the output. Since the above annihilation statement is purely module-theoretic, we could equally well use an index-string θ of λ's and ρ's. Now the maps $\mathbf{a}^{\theta}$ where θ is an index-string of λ's and ρ's of length strictly greater than 1 are just the components of the map $\mathbf{b}\colon R \to R^{\lambda} \amalg R^{\rho}$ in (12.10). Hence the above arguments give:

LEMMA 13.9. *For any* $h > 0$, *and any choice of location for the one* "$\mathbf{b}$", *the following composite map is zero*:

$$R_h \hookrightarrow R \xrightarrow{f\mathbf{a}^{[h]}} R \otimes_k \dots \otimes_k R$$

$$\xrightarrow{1_R \otimes \dots \otimes \mathbf{b} \otimes \dots \otimes 1_R} R \otimes_k \dots \otimes_k (R^{\lambda} \amalg R^{\rho}) \otimes_k \dots \otimes_k R. \quad \square$$

The idea of this result is that if an element of R yields only terms of heights $\leq h$ after one application of $\mathbf{a}$ (i.e., if it lies in R_h) it will not yield terms of greater height on repeated applications of $\mathbf{a}$. (We have only proved here that *certain* higher components, which we shall use in our proof, are zero. In fact, one can deduce much more by these methods, namely that for all index-strings σ of length h, $f(\mathbf{aa})^{\sigma} = f\mathbf{a}^{[h]}$. This will be done, though in slightly different language, in §§17-18; cf. also §60.)

The above innocuous-looking Lemma is the watershed of our proof. Before beginning the headlong descent, let us take our bearings, and see where we want to go.

To prove that R is generated by R_1 (12.18), we must show that given any $x \in R$, the coaddition map $\mathbf{a}$ can somehow be made use of to produce an expression for x in terms of elements of R_1. For guidance, let us write down a sample element of the subring of R generated by R_1, and see how it behaves under $\mathbf{a}$:

EXAMPLE 13.10. Suppose

$$x = rst + uv, \quad \text{where } r, s, t, u, v \in R_1.$$

The relations $\mathbf{a}(r) = r^{\lambda} + r^{\rho}$ etc. allow us to expand $\mathbf{a}(x)$. Let us write out this expansion with terms grouped according to the summand $R^{\sigma} \subseteq R^{\lambda} \amalg R^{\rho}$ in which they lie:

$$\begin{aligned}\mathbf{a}(x) = \; & r^{\lambda} \otimes s^{\rho} \otimes t^{\lambda} && + r^{\rho} \otimes s^{\lambda} \otimes t^{\rho} \\ & + (r^{\lambda} \otimes (st)^{\rho} + (rs)^{\lambda} \otimes t^{\rho} + u^{\lambda} \otimes v^{\rho}) && + (r^{\rho} \otimes (st)^{\lambda} + (rs)^{\rho} \otimes t^{\lambda} + u^{\rho} \otimes v^{\lambda}) \\ & + (\, rst \, + \, uv \,)^{\lambda} && + (\, rst \, + \, uv \,)^{\rho}.\end{aligned}$$

Looking at the (λ, ρ, λ)-component of $\mathbf{a}(x)$, we see that it displays in tensorial form a decomposition of the leading term rst of x in terms of elements of R_1. This suggests that we should try to prove an analogous statement for all elements of R; that is, first, that for every $h > 1$,

$$(13.11) \qquad f\mathbf{a}^{[h]}(R_h) \subseteq R_1 \otimes \dots \otimes R_1$$

(as subbimodule of the h-fold tensor product $R \otimes_k \dots \otimes_k R$; here for the first time we are using the convention on tensor-symbols without subscripts k, made parenthetically in Definition 12.1); and secondly, that if we define m: $R \otimes_k \ldots \otimes_k R \to R$ by

$$m(x_1 \otimes \ldots \otimes x_h) = x_1 \ldots x_h,$$

then for $x \in R_h$, $m f \mathbf{a}^{[h]}(x)$ gives a "leading term" of x in the sense that

$$\deg(x - m f \mathbf{a}^{[h]}(x)) < h \tag{13.12}$$

(for deg as defined in Definition 12.8). If (13.11) and (13.12) are true, they will imply, by an induction on $\deg(x)$, that R is generated by R_1.

Now Lemma 13.9 says that $f\mathbf{a}^{[h]}(R_h) \subseteq R \otimes_k \ldots \otimes_k R$ is annihilated on applying $\mathbf{b}$ to any of the h tensor factors of $R \otimes_k \ldots \otimes_k R$. This is certainly an approximation to the desired result (13.11) that $f\mathbf{a}^{[h]}(R_h)$ lies in the product of the submodules $R_1 = \ker \mathbf{b}$. If k were a division ring, the two statements would be equivalent by a splitting argument, but for general k they are not. We shall in fact first prove (13.12), and then get (13.11) using this.

We begin by noting the following immediate consequence of Lemma 13.9. If $h \geq 1$, and $u_1, \ldots, u_h$ are any bimodule maps of R into $R^\lambda \amalg R^\rho$, such that *at least one* u_i is equal to $\mathbf{b}$, then the tensor product map $u_1 \otimes \ldots \otimes u_h$: $R \otimes_k \ldots \otimes_k R \to (R^\lambda \amalg R^\rho) \otimes_k \ldots \otimes_k (R^\lambda \amalg R^\rho)$ will annihilate $f \mathbf{a}^{[h]}(R_h)$. For the tensor product map can be factored

$$R \otimes_k \ldots \otimes_k R \xrightarrow{1_R \otimes \ldots \otimes u_i \otimes \ldots \otimes 1_R} R \otimes_k \ldots \otimes_k (R^\lambda \amalg R^\rho) \otimes_k \ldots \otimes_k R$$
$$\xrightarrow{u_1 \otimes \ldots \otimes 1_{(R^\lambda \amalg R^\rho)} \otimes \ldots \otimes u_h} (R^\lambda \amalg R^\rho) \otimes_k \ldots \otimes_k (R^\lambda \amalg R^\rho)$$

where the lone factor u_i of the first arrow is the one which equals $\mathbf{b}$; the first arrow therefore already kills $f \mathbf{a}^{[h]}(R_h)$.

Now consider the diagram

$$\begin{array}{ccccc} R_h & \xrightarrow{f\mathbf{a}^{[h]}} & R \otimes_k \ldots \otimes_k R & \xrightarrow{\mathbf{a} \otimes \ldots \otimes \mathbf{a}} & (R^\lambda \amalg R^\rho) \otimes_k \ldots \otimes_k (R^\lambda \amalg R^\rho) \\ & & \downarrow m & & \downarrow m \\ & & R & \xrightarrow{\mathbf{a}} & R^\lambda \amalg R^\rho, \end{array} \tag{13.13}$$

where the downward maps "m" are induced by multiplication in the rings R and $R^\lambda \amalg R^\rho$ respectively. The square commutes because the coaddition $\mathbf{a}$ is a ring homomorphism. If we write each factor $\mathbf{a}$ in $\mathbf{a} \otimes \ldots \otimes \mathbf{a}$ as a sum of two maps, $(i^\lambda + i^\rho) + \mathbf{b}$, and use this to expand that tensor product of maps into a sum of 2^h terms, we see that each of these will involve a factor $\mathbf{b}$ except for the one term $(i^\lambda + i^\rho) \otimes \ldots \otimes (i^\lambda + i^\rho)$. Hence when we restrict to $f \mathbf{a}^{[h]}(R_h)$ (see upper left arrow of (13.13)) all these other terms become 0. So the top arrow can be replaced, in considering behavior on elements of this submodule, by $(i^\lambda + i^\rho) \otimes \ldots \otimes (i^\lambda + i^\rho)$.

But when we apply $m((i^\lambda + i^\rho) \otimes \ldots \otimes (i^\lambda + i^\rho))f$ to a decomposable element $y = x_1^\lambda x_2^\rho \ldots \in R^{[h]}$, the resulting element of $R^\lambda \amalg R^\rho$ is $(x_1^\lambda + x_1^\rho)(x_2^\lambda + x_2^\rho) \ldots,$

whose components of greatest height are $x_1^{\lambda} x_2^{\rho} \ldots \in R^{[h]}$ and $x_1^{\rho} x_2^{\lambda} \ldots \in R^{[h]'}$. Thus, this element has height $\leq h$, and its $[h]$-component reproduces the element y we started with. Hence the same will be true for an arbitrary element $y \in R^{[h]}$, i.e., a sum of such products; hence switching to the lower branch of (13.13) we deduce that for any $x \in R_h$, $\mathbf{a}\, mf\, \mathbf{a}^{[h]}(x)$ has height $\leq h$ and $[h]$-component precisely $\mathbf{a}^{[h]}(x)$. Thus it has the same $[h]$-component as $\mathbf{a}(x)$, and by cocommutativity of $\mathbf{a}$, it also has the same $[h]'$-component as $\mathbf{a}(x)$. Hence $\mathbf{a}(x - mf\,\mathbf{a}^{[h]}(x))$ has $[h]$- and $[h]'$-components zero, hence has height $< h$. This means that $x - mf\,\mathbf{a}^{[h]}(x)$ has degree $< h$, giving (13.12).

(The idea of the above proof is that, though Lemma 13.9 does not exactly tell us that for every $x \in R_h$, $f\,\mathbf{a}^{[h]}(x)$ lies in $R_1 \otimes \ldots \otimes R_1$, it tells us that it behaves so nearly as though it did that we can mimic in abstract form the considerations of Example 13.10 that showed us that $mf\,\mathbf{a}^{[h]}(x)$ should give the "leading term" of x.)

From (13.12) we will now prove (13.11). Let $x \in R_h$, let $y = f\,\mathbf{a}^{[h]}(x) \in R \otimes_k \ldots \otimes_k R$, and choose $g_1, \ldots, g_h \geq 1$ large enough so that

$$y \in R_{g_1} \otimes \ldots \otimes R_{g_h}.$$

If $g_1 = \ldots = g_h = 1$, we are done. If not, say $g_i > 1$. Then $\mathbf{a}^{[g_i]}$ is a component of $\mathbf{b}$, so by Lemma 13.9, $1_R \otimes \ldots \otimes \mathbf{a}^{[g_i]} \otimes \ldots \otimes 1_R$ annihilates y. Composing with

$$1_R \otimes \ldots \otimes (mf) \otimes \ldots \otimes 1_R,$$

we get

$$0 = (1_R \otimes \ldots \otimes (mf\mathbf{a}^{[g_i]}) \otimes \ldots \otimes 1_R)(y),$$

hence

$$y = (1_R \otimes \ldots \otimes (1_R - mf\mathbf{a}^{[g_i]}) \otimes \ldots \otimes 1_R)(y).$$

But by (13.12), $1_R - mf\mathbf{a}^{[g_i]}$ carries R_{g_i} into $R_{g_i - 1}$, hence the above equation tells us that $y \in R_{g_1} \otimes \ldots \otimes R_{g_i - 1} \otimes \ldots \otimes R_{g_h}$. By induction on Σg_i we clearly get (13.11). As we have noted, this and (13.12) tell us that R is generated by R_1, which is the desired result (12.18).

Thus, if we write

$$R_1 = M,$$

and let T denote the tensor ring on this k-bimodule:

$$T = [k]\langle M \rangle,$$

the homomorphism

$$q \colon T \to R$$

which acts as the identity on $T_1 = M = R_1$ will be *surjective*. If we make T a cosemigroup as described previously, using the coaddition

$$\mathbf{a}_M\colon\ T \to T^\lambda \amalg T^\rho$$

induced by the diagonal bimodule map $M \to M^\lambda \oplus M^\rho$ (12.16-17), then comparing (12.11) with (12.17), we see that our map q respects cosemigroup structure, and hence also respects the filtration by degree (i.e., satisfies $q(T_h) \subseteq R_h$ for all h).

We want to prove that q is in fact an isomorphism. Note that as a tensor ring, T has a natural *grading* (12.14), and that the subbimodule T_h (cf. Definition 12.8) will be the sum of the homogeneous components of all degrees $\leq h$ under this grading. Now for any $h \geq 1$, consider the commuting diagram

$$\begin{array}{ccccccc} T_h \hookrightarrow T & \xrightarrow{f\mathbf{a}_M^{[h]}} & T \otimes_k \dots \otimes_k T & \xrightarrow{m} & T \\ \downarrow q & & \downarrow q\otimes \dots \otimes q & & \downarrow q \\ R & \xrightarrow{f\mathbf{a}^{[h]}} & R \otimes_k \dots \otimes_k R & \xrightarrow{m} & R. \end{array}$$

From the explicit descriptions of T and $\mathbf{a}_M$ it is easy to verify that the composite of the top arrows is the projection of T_h onto its homogeneous component of degree h. From the commutativity of the diagram, we deduce that $\ker q$ is a *homogeneous* ideal of T (i.e., if an element lies in this kernel, so do its homogeneous components of all degrees h). It follows that the grading on T induces a grading on R. Let $T_{(h)}$, $R_{(h)}$ denote the homogeneous components of degree h in these two k-rings. By (13.11), we can refine the left-hand square of the above diagram to

(13.14)
$$\begin{array}{ccc} T_{(h)} & \xrightarrow{f\mathbf{a}_M^{[h]}} & T_1 \otimes \dots \otimes T_1 \subseteq T \otimes_k \dots \otimes_k T \\ \downarrow q & & \downarrow q\otimes \dots \otimes q \\ R_{(h)} & \xrightarrow{f\mathbf{a}^{[h]}} & R_1 \otimes \dots \otimes R_1 \subseteq R \otimes_k \dots \otimes_k R \end{array}$$

(again following the convention on tensor product signs without subscripts at the end of Definition 12.1). Because T and R are graded, T_1 and R_1 are k-bimodule *direct summands* in these rings. Hence the tensor product subbimodules in the above diagram can be identified with the abstract tensor products $T_1 \otimes_k \dots \otimes_k T_1 = M \otimes_k \dots \otimes_k M$ and $R_1 \otimes_k \dots \otimes_k R_1 = M \otimes_k \dots \otimes_k M$. So the right hand downward map ($q \otimes \dots \otimes q$ restricted to the indicated domain and codomain) is an isomorphism. By the structure of our cosemigroup T, the top map is also an isomorphism, hence the left-hand downward map must be one-to-one. Hence q is one-to-one, hence an isomorphism.

We have obtained the desired description of our cosemigroup object: $(R, \mathbf{a}) \cong ([k]\langle M\rangle, \mathbf{a}_M)$!

Remarks: If k were a division ring, it would be immediate by a splitting argument that the submodule $R_1 \otimes \dots \otimes R_1 \subseteq R \otimes_k \dots \otimes_k R$ was naturally isomorphic to $R_1 \otimes_k \dots \otimes_k R_1$, and a slight simplification of the above proof could have been achieved, bypassing the argument to show that R could be graded. As in the proof of (13.11), an extra application of the properties of the coaddition **a** has somehow made up for the non-left-exactness of tensor products over general rings.

Note that when we showed that R was a *homomorphic image* of $T = [k]\langle M\rangle$, this meant that the set-valued functor represented by R can be described as taking every k-ring S to a *subset* of k-**Bimod**(M, S) determined by a set of *k-ring equations* in the "coordinates"; here the coordinates are the images in S of the elements of M, and the k-ring equations in question correspond to the elements of the kernel of the homomorphism $q\colon [k]\langle M\rangle \to R$. The fact that R was an image of $[k]\langle M\rangle$ as a *cosemigroup* meant that these equations in the coordinates have the property that for every k-ring S, the set of their solutions is an additive subsemigroup of k-**Bimod**(M, S). Now this system of equations can contain no nontrivial equations of degree 1 in the elements of M (for these would correspond to bimodule-relations other than those defining M, which would mean that our M was the wrong bimodule to start with). What we have succeeded in deducing is that there are no equations of higher degrees either. Thus, the proof that the surjective map q is one-to-one is essentially a verification of the intuitive idea that a system of nonlinear equations which entails no linear equation cannot be expected to have a solution-set closed under addition. (In *commutative* ring theory, this principle is false in positive characteristic, with consequences that we will note in Chapter VIII.)

Recall that by Corollary 8.13, the category **Rep**(k-**Ring**, **AbSemigp**^{e}) of all representable functors k-**Ring** $\to$ **AbSemigp**^{e} is equivalent to the *opposite* of the category of co-**AbSemigp**^{e} objects in k-**Ring**. Now if M and N are k-bimodules, a k-ring homomorphism $g\colon [k]\langle M\rangle \to [k]\langle N\rangle$ which respects the coaddition maps $\mathbf{a}_M$ and $\mathbf{a}_N$ must carry $[k]\langle M\rangle_1 = M$ into $[k]\langle N\rangle_1 = N$, hence must be induced by a bimodule homomorphism $M \to N$. It takes just a few more observations to deduce:

THEOREM 13.15. *Let k be an associative unital ring. Then the functors*

$$k\text{-}\mathbf{Bimod}^{\mathrm{op}} \xrightarrow{\alpha} \mathbf{Rep}(k\text{-}\mathbf{Ring}, \mathbf{Ab}) \xrightarrow{\beta} \mathbf{Rep}(k\text{-}\mathbf{Ring}, \mathbf{AbSemigp}^e)$$

*are equivalences, where α takes a bimodule M to the functor k-**Bimod**$(M, -)$: k-**Ring** $\to$ **Ab**, and β denotes composition with the forgetful functor **Ab** $\to$ **AbSemigp**^{e}. For M a k-bimodule, the representing objects for $\alpha(M)$ and $\beta\alpha(M)$ have underlying object $[k]\langle M\rangle$. The coaddition on this object in both cases is $\mathbf{a}_M$ (see (12.16-17)), the co-neutral element is the map carrying all elements of M to 0; and the coinverse in the co-**Ab** structure representing $\alpha(M)$ is given by the automorphism of $[k]\langle M\rangle$ taking each $x \in M$ to $-x$.*

*The corresponding result holds with k-**Ring**^{1} in place of k-**Ring**, and $k\langle M\rangle$ in place of $[k]\langle M\rangle$.*

REMAINDER OF PROOF. What we have proved so far is that $\beta\alpha$ is an equivalence. Because the forgetful functor **Ab** $\to$ **AbSemigp**^{e} is full and faithful, β is also full and faithful, and it follows that α and β must each be equivalences. The observations made at the end of §10 allow us to deduce the corresponding results for functors on k-**Ring**^{1}. □

We recall

DEFINITION 13.16. *An* **Ab**-*category means a category* **A** *given with a structure of abelian group on each hom-set* $\mathbf{A}(X, Y)$, *such that for all* $X, Y, Z \in \mathrm{Ob}(\mathbf{A})$ *the composition map* $\mathbf{A}(Y, Z) \times \mathbf{A}(X, Y) \to \mathbf{A}(X, Z)$ *is bilinear.*

If **C** is an arbitrary category and **A** an **Ab**-category, then the category of all functors from **C** to **A** has a natural structure of **Ab**-category, and hence so does any full subcategory thereof. In particular, since **Ab** is an **Ab**-category, **Rep**(k-**Ring**^{1}, **Ab**) has a natural structure of **Ab**-category. Now k-**Bimod** is also an **Ab**-category in a natural way (this was used implicitly in defining α above), and it is straightforward to verify

COROLLARY 13.17. *The equivalence* α *of Theorem 13.15 respects* **Ab**-*structures.* □

We could define **AbSemigp**^{e}-categories analogously and record the corresponding property for $\beta\alpha$; but this is implied by the above Corollary anyway.

§14. Some immediate consequences.

DEFINITION 14.1. *If* k *is a* commutative *associative ring with* 1, *then by*

$$\mathbf{Mod}_k \subseteq k\text{-}\mathbf{Bimod}, \qquad \mathbf{Ring}_k \subseteq k\text{-}\mathbf{Ring}, \qquad \mathbf{Ring}_k^1 \subseteq k\text{-}\mathbf{Ring}^1$$

we shall understand, respectively, the category of k-modules, *considered as* k-*bimodules with the same action on the left and the right, and the categories of nonunital and unital associative* k-algebras, *identified similarly with subvarieties of the varieties of nonunital and unital* k-*rings.*

Now, characterizations of representable **Ab**- and **AbSemigp**^{e}-valued functors on the varieties $\mathbf{Ring}_k$ and $\mathbf{Ring}_k^1$ for k a commutative ring cannot be obtained simply as cases of the characterizations of such functors on k-**Ring** and k-**Ring**^{1}, since $\mathbf{Ring}_k$ and $\mathbf{Ring}_k^1$ are in general proper subcategories of k-**Ring** and k-**Ring**^{1}, and a subcategory **D** of a category **C** may admit representable functors which do not extend to **C**. (E.g., if $\mathbf{D} \subseteq \mathbf{C}$ are varieties of algebras, the identity functor $\mathbf{D} \to \mathbf{D}$ is representable, but does not in general extend to a representable functor $\mathbf{C} \to \mathbf{D}$.) However, for certain pairs of categories, such an extension result can be proved, and we shall see that this is true of the subvarieties of the above Definition, and indeed, of a wide class of subvarieties of ring and module categories, of which these are a special case. We start with the following general observation.

LEMMA 14.2. *Let* **C** *be a category with finite coproducts, and* $\mathbf{D} \subseteq \mathbf{C}$ *a full subcategory. Suppose that*

(a) *All finite coproducts in* **C** *of objects of* **D** *again lie in* **D**.

Then the coproduct in $\mathbf{C}$ *of any finite family of objects of* $\mathbf{D}$ *is also their coproduct in* $\mathbf{D}$ (*so that* $\mathbf{D}$ *also has finite coproducts*), *and for any variety* $\mathbf{V}$, *the co-*$\mathbf{V}$ *objects in* $\mathbf{D}$ *are precisely the co-*$\mathbf{V}$ *objects of* $\mathbf{C}$ *whose underlying* $\mathbf{C}$*-objects lie in* $\mathbf{D}$. *That is, the representable functors* $\mathbf{D} \to \mathbf{V}$ *are the restrictions to* $\mathbf{D}$ *of the representable functors* $\mathbf{C} \to \mathbf{V}$ *whose representing objects lie in* $\mathbf{D}$.

If $\mathbf{C}$ *is a* variety of algebras, *then a sufficient condition for* (a) *to hold is*

(b) *Every algebra* X *in* $\mathbf{C}$ *which is generated by homomorphic images of finitely many algebras in the subcategory* $\mathbf{D}$ *again lies in* $\mathbf{D}$. □

Now if k is a commutative associative ring with 1, then the elements of any k-ring R (unital or nonunital) which *centralize* k form a k-subring. It follows easily that the pairs of varieties $\mathbf{Ring}_k \subseteq k\text{-}\mathbf{Ring}$ and $\mathbf{Ring}^1_k \subseteq k\text{-}\mathbf{Ring}^1$ each satisfy condition (b) of the above Lemma. From the same observations we see that the (unital or nonunital) tensor k-ring on a k-bimodule M will be a *k-algebra* if and only if $M \in \mathbf{Mod}_k$. Combining the above Lemma and Theorem 13.15, we conclude that tensor *algebras* on *k-modules*, with co-operations defined as in that Theorem, give all co-$\mathbf{AbSemigp}^e$ objects and co-$\mathbf{Ab}$ objects in $\mathbf{Ring}_k$ and $\mathbf{Ring}^1_k$.

More generally, if K is any unital associative ring, and k a subring of the center of K, then the elements of any K-ring R which centralize k form a K-subring of R. So let us make

DEFINITION 14.3. *If* K *is an associative ring with* 1, *and* k *a unital subring of the center of* K, *then*

$$K\text{-}\mathbf{Bimod}_k, \qquad K\text{-}\mathbf{Ring}_k, \qquad K\text{-}\mathbf{Ring}^1_k$$

will denote the subvarieties of K-$\mathbf{Bimod}$, K-$\mathbf{Ring}$ *and* K-$\mathbf{Ring}^1$ *consisting of those objects* X *which for all* $\alpha \in k$ *satisfy the identity*

$$\alpha x = x \alpha.$$

(*Thus when* $k = K$, *these are* $\mathbf{Mod}_k$, $\mathbf{Ring}_k$, $\mathbf{Ring}^1_k$ *respectively, while when* k *is the image of* $\mathbf{Z}$ *in* K, *they are the whole varieties* K-$\mathbf{Bimod}$, K-$\mathbf{Ring}$ *and* K-$\mathbf{Ring}^1$.)

Let us apply Lemma 14.2 to these subvarieties. The next Corollary begins with the formulation of the results for the "classical" concepts of k-algebra and k-module, then gives them for the more general categories just discussed.

COROLLARY 14.4. *The statement of Theorem 13.15 remains true if the categories* k-$\mathbf{Ring}$, k-$\mathbf{Ring}^1$ *and* k-$\mathbf{Bimod}$ *are respectively replaced by* $\mathbf{Ring}_k$, $\mathbf{Ring}^1_k$ *and* $\mathbf{Mod}_k$ (*k commutative*), *or more generally, by* K-$\mathbf{Ring}_k$, K-$\mathbf{Ring}^1_k$ *and* K-$\mathbf{Bimod}_k$ (*K a unital ring and k a subring of its center*). □

14.5. *Remarks on some conditions.* Let us look again at conditions (a) and (b) of Lemma 14.2 on a full subcategory $\mathbf{D}$ of a category $\mathbf{C}$, and let us write (a′) and (b′) for the stronger conditions obtained by assuming $\mathbf{C}$ has arbitrary coproducts, and deleting the finiteness restriction from (a) and (b). Recalling from Definition 5.14 the meaning of a *coreflective* subcategory, it is not hard to verify the following implications among properties of a full subcategory **D** of a variety of algebras **C**:

(14.6) (coreflective subvariety) $\Rightarrow$ (b′) $\Rightarrow$ (coreflective full subcategory) $\Rightarrow$ (a′).

The subvarieties of Definition 14.3 actually satisfy the strongest (leftmost) of these conditions: indeed, right adjoints to the inclusions into the larger varieties are given by the "centralizer of k" construction. Here are examples of various subcategories of varieties, $\mathbf{D} \subseteq \mathbf{C}$, with weaker properties. For $\mathbf{C} = \mathbf{Semigp}^e$, the subcategory **D** of "groups", in the loose sense of semigroups all of whose elements are invertible (i.e., the image of the variety **Group** under the forgetful functor into $\mathbf{Semigp}^e$) is clearly not a subvariety, but does satisfy (b′). The image of $\mathbf{Mod_Q}$ in **Ab** under the forgetful functor fails to satisfy (b′) or even (b), since $\mathbf{Q}/\mathbf{Z}$ is a homomorphic image of an object of this subvariety, but does not lie in it. However, it is a coreflective full subcategory, for if we regard the forgetful functor $\mathbf{Mod_Q} \to \mathbf{Ab} = \mathbf{Mod_Z}$ as the operation of tensoring over **Q** with the $(\mathbf{Z}, \mathbf{Q})$-bimodule **Q**, we see that it has the right adjoint $\mathbf{Ab}(\mathbf{Q}, -)\colon \mathbf{Ab} \to \mathbf{Mod_Q}$ (cf. discussion following Corollary 8.16). Next, the full subcategory of *free algebras* in any variety is generally not coreflective, but always satisfies (a′). Finally, in any nontrivial variety **C**, the full subcategory of *finitely generated* algebras satisfies (a) and (b), but not (a′) or (b′).

Most subvarieties of varieties do not even satisfy (a); in particular, the subvariety of *commutative* rings in the variety of *all* associative rings does not. And indeed we shall see in §40 that there are more representable functors from commutative rings to **Ab** (and still more to $\mathbf{AbSemigp}^e$) than those that are restrictions of representable functors on associative rings.

14.7. It is of interest to note that if **D** is a full coreflective subcategory of **C** (i.e., if the third condition of (14.6) is satisfied), then for any representable functor V from **D** to a variety **V**, the functor $\mathbf{C} \to \mathbf{V}$ represented by the same coalgebra can be described as the composite $\mathbf{C} \to \mathbf{D} \to \mathbf{V}$ of the right adjoint to the inclusion of **D** in **C** (the "coreflection") with the given functor V. E.g., if **C** is K-**Ring** and **D** is K-$\mathbf{Ring}_k$, and we are given a representable functor V on the latter subcategory, then the corresponding functor on the larger category is obtained by applying V to the centralizers of k in general K-rings. As a consequence, we see that the functor on the larger category **C** takes on as values exactly the same set of objects and morphisms as the functor on the subcategory **D**.

Returning to **Ab**-valued functors on varieties of rings, let us note a special case of Corollary 14.4. If k is a field, then a k-module M has a *basis* X. Hence $\mathbf{Mod}_k(M, -) \cong \mathbf{Set}(X, -) = (-)^X$. We deduce

Corollary 14.8. *If k is a field, then every representable functor F from $\mathbf{Ring}^1_k$ or $\mathbf{Ring}_k$ to* **Ab** *is isomorphic to a power of the underlying-additive-group functor; i.e., $F(S) = (S^{\mathrm{add}})^X$ for a fixed set X.* □

However, this description of F is noncanonical, because the choice of a basis of a vector space is nonunique. In contrast, the description given by Theorem 13.15

is canonical, as may be seen both from its proof, where $M \subseteq R$ is obtained without making arbitrary choices, and from the functoriality of the statement. We shall obtain a different canonical description of these same functors F under the hypotheses of the above Corollary in Theorem 25.1(i).

Let us consider one more variation on our result. If k is commutative, one may note that the values of a functor $\mathbf{Mod}_k(M, -)\colon \mathbf{Ring}_k \to \mathbf{Ab}$ have natural structures not only of abelian group, but of k-module, and one may ask whether all representable functors from k-algebras to k-modules arise in this way.

The answer is no. The right context in which to look at the question is: Given an arbitrary associative unital ring k, and another such ring L, what are the representable functors from k-**Ring** to the category of left L-modules? Now a left L-module is an abelian group given with a homomorphism of L into its endomorphism ring. We can see from Theorem 13.15 and Corollary 13.17 that the endomorphism ring of the functor k-**Bimod**$(M, -)\colon k$-**Ring** $\to$ **Ab** is the opposite of the endomorphism ring of the k-bimodule M. We deduce

COROLLARY 14.9. *Every representable functor from k-**Ring** or k-**Ring**^{1} to the category of left L-modules has the form k-**Bimod**$(M, -)$, where M is a k-bimodule given with a unital homomorphism $L \to k$-**Bimod**$(M, M)^{\mathrm{op}}$, which induces the L-module structure on the values of the functor.* □

If we consider k-bimodules as left $k \otimes_{\mathbf{Z}} k^{\mathrm{op}}$-modules, then such objects M correspond to left $k \otimes_{\mathbf{Z}} k^{\mathrm{op}} \otimes_{\mathbf{Z}} L^{\mathrm{op}}$-modules.

If k is commutative, then every k-module has, it is true, a *canonical* homomorphism of k into its endomorphism ring, giving the canonical k-module structure referred to; but this is generally not the *unique* such homomorphism, hence the negative answer to our question.

We could proliferate Corollaries of the above sort: We could characterize functors from k-rings (or k-algebras, or K-rings$_k$) to (L, L')-bimodules, etc., and in subsequent sections when we talk about functors from k-**Ring**^{1} to rings, we could state what is required to get L-module or bimodule structures on the underlying additive groups of those rings, and the compatibility conditions that are then needed to make them into L-rings, L-algebras, etc.. We shall, however, refrain from doing so. The approach is clear, and the reader who wishes to can easily write out the details, but we prefer to leave these considerations aside in the present development, so as not to dilute matters of greater interest.

(While on the subject of various sorts of module structures, we feel obliged to mention that the notation ${}_A\mathbf{Mod}$ for the category of left A-modules that we introduced in the discussion of composition of representable functors following Corollary 8.16 does not form a coherent system with the notations K-**Bimod**, $\mathbf{Mod}_k$ and K-$\mathbf{Bimod}_k$ introduced in this chapter (Definitions 12.1, 14.1 and 14.3). If we were going to be dealing further with left and right modules, and with bimodules with possibly different rings acting on the two sides, we might, for any two unital rings (respectively, k-algebras) A and B define (A, B)-**Bimod** (respectively, (A, B)-$\mathbf{Bimod}_k$) to be the category of all (respectively, of all

k-centralizing) (A, B)-bimodules. Then for any unital ring A, we could write the categories of left and right A-modules as $(A, \mathbf{Z})$-**Bimod** and $(\mathbf{Z}, A)$-**Bimod**, while what we have been writing K-**Bimod** and K-$\mathbf{Bimod}_k$ would become (K, K)-**Bimod** and (K, K)-$\mathbf{Bimod}_k$. However, since the categories named in Definitions 12.1, 14.1 and 14.3 will be the ones of importance for most of this work, we will stick with the more limited notation introduced in this chapter. Note, incidentally, that if A is a k-algebra, then (using for the moment the more versatile notation just sketched), $(A, \mathbf{Z})$-**Bimod** can be naturally identified with (A, k)-$\mathbf{Bimod}_k$, and similarly for right modules. In particular, $(k, \mathbf{Z})$-**Bimod** $\simeq$ (k, k)-$\mathbf{Bimod}_k \simeq (\mathbf{Z}, k)$-**Bimod**; i.e., for commutative k, left k-modules, right k-modules, and ''central'' k-modules can all be identified.)

CHAPTER IV.

Digressions on semigroups, etc.

The material in this chapter falls into two parts.

In §§15-18 we continue to look at a ring R with a cocommutative binary co-operation **a**, but we examine what can be said if we drop the assumption of coassociativity, or of existence of a co-neutral-element; in other words, what one can say about representable functors from rings to sets with a commutative binary operation with neutral element (§15), or to commutative semigroups without neutral element (§§16-18). In the process, in §17, we develop some curious results about symmetric elements in coproducts of rings, to which we add some further observations and questions in §19. (Later, in Chapter VII, we shall discuss representable functors from rings to *noncommutative* groups and semigroups.)

In §§20-21 we examine representable functors *among* categories of semigroups, mostly noncommutative. It might appear more natural to put this material with the later discussion of functors from rings to noncommutative groups and semigroups; we have put it here, however, because the approach is reminiscent of that of the preceding chapter. Results on representable functors between these and some other related varieties, in particular the variety of *heaps* and that of *Mal'cev algebras*, are sketched in §22.

§15. Representable functors to $\mathbf{AbBinar}^e$.

In this section (though not in the remaining sections of this chapter) we shall for simplicity only consider functors on categories of k-algebras, where k is a field.

Let us denote by $\mathbf{AbBinar}^e$ the variety whose objects are sets given with a commutative, not necessarily associative binary operation, and a two-sided neutral element e for this operation. (There is no generally accepted word for a set with a not necessarily associative binary operation. The word *groupoid* is used by many universal algebraists, but workers in category theory and related areas object strongly to this usage because they use the same word to mean "category in which all morphisms are invertible". The term *magma* was used by Serre in [**174**]. The symbol **Binar** which we are using here was suggested by the Russian term *unar* for a set with a single unary operation [**176**].)

In studying representable functors from categories of k-algebras to $\mathbf{AbBinar}^e$, we can, for the same reasons as before (cf. §10 or 11), restrict attention to functors on categories of nonunital algebras.

Representable functors from $\mathbf{Ring}_k$ to $\mathbf{AbBinar}^e$ are in fact very abundant. E.g., to obtain the general commutative binary derived operation with 0 as neutral element on the underlying set functor of $\mathbf{Ring}_k$, one chooses from the free associative algebra in two indeterminates $[k]\langle x, y\rangle$ any element $b(x, y)$ which is symmetric in x and y, and has the property that every monomial occurring in b with nonzero coefficient involves at least one x (and hence, by symmetry, such that every such monomial involves at least one y); for example, $xy + yx$, or $x^2y + y^2x$. Then for every $S \in \mathbf{Ring}_k$, the operation $\xi * \eta = \xi + \eta + b(\xi, \eta)$ on $|S|$ is commutative and has 0 as neutral element. Let us denote by V_b the underlying-set-preserving functor $\mathbf{Ring}_k \to \mathbf{AbBinar}^e$ obtained in this way from the polynomial b.

We claim that V_b will not be $\mathbf{AbSemigp}^e$-valued (i.e., that the above operation $*$ will not be associative) unless $b = 0$. For if V_b is $\mathbf{AbSemigp}^e$-valued, then by Theorem 13.15, there must be an isomorphism of the representing algebra $R = [k]\langle x\rangle$ with an algebra $[k]\langle M\rangle$ $(M \in \mathbf{Mod}_k)$, which transforms the coaddition of R, given by

$$(15.1) \qquad \mathbf{a}(x) = x^{\lambda} + x^{\rho} + b(x^{\lambda}, x^{\rho}),$$

into the coaddition $\mathbf{a}_M$ satisfying $\mathbf{a}_M(r) = r^{\lambda} + r^{\rho}$ for all $r \in M$. Now if $[k]\langle M\rangle \cong R = [k]\langle x\rangle$, then looking at R/R^2 we see that M is isomorphic to the free module of rank 1 over k. So the isomorphism of R with $k\langle M\rangle$ can be viewed as an *automorphism* of $R = [k]\langle x\rangle$ such that the conjugate of the coaddition (15.1) by this automorphism is the co-operation carrying x to $x^{\lambda} + x^{\rho}$. However, the only automorphisms of $[k]\langle x\rangle$ are those carrying x to αx $(\alpha \in k - \{0\})$, and these cannot bring a co-operation (15.1) with $b \neq 0$ to the one with $b = 0$.

Examples of functors to $\mathbf{AbBinar}^e$ represented by free algebras in more than one indeterminate are likewise easy to write down. There is not such a simple criterion for associativity, i.e., for the existence of an isomorphism with a coalgebra of the form $([k]\langle M\rangle, \mathbf{a}_M)$, because free algebras in more than one indeterminate have rather complicated automorphism groups, but, of course, "most" examples are nonassociative-valued.

Curiously, we do not know whether there exist representable functors $\mathbf{Ring}_k \to \mathbf{AbBinar}^e$ for which the representing object is not a free algebra. The next Lemma shows that this question is equivalent to the open problem of whether every retract of a free k-algebra is free. (See [**72**, Problem 1], where the question is asked for unital algebras. The unital and nonunital cases are equivalent.) Recall that a "retract" of an object X of a category means an object which can be mapped to X by a left invertible morphism. In a variety of algebras, this means an object isomorphic to the image of an idempotent endomorphism of X.

LEMMA 15.2. *Let k be a field and R an object of* $\mathbf{Ring}_k$. *Then the following conditions are equivalent:*

(i) *R is a retract of a tensor algebra $[k]\langle M\rangle$ (M a vector space over k), equivalently, of a free algebra $[k]\langle X\rangle$ (X a set).*

(ii) *R admits a co-*$\mathbf{AbBinar}^e$ *structure in* $\mathbf{Ring}_k$.

PROOF. (i) $\Rightarrow$ (ii): Say $R \subseteq [k]\langle M\rangle$, with $r\colon [k]\langle M\rangle \to R$ a left inverse to the inclusion map. Letting $\mathbf{a}_M$ denote the standard coaddition on $[k]\langle M\rangle$

(12.16-17), it is easy to verify that the map

$$R \hookrightarrow [k]\langle M\rangle \xrightarrow{\mathbf{a}_M} [k]\langle M\rangle^{\lambda} \amalg [k]\langle M\rangle^{\rho} \xrightarrow{r \amalg r} R^{\lambda} \amalg R^{\rho}$$

is cocommutative, and has $0\colon R \to \{0\}$ as co-neutral-element.

(ii) $\Rightarrow$ (i): Suppose $\mathbf{a}\colon R \to R^{\lambda} \amalg R^{\rho}$ is a cocommutative map for which the map $0\colon R \to \{0\}$ is a co-neutral-element. Cocommutativity says that $\mathbf{a}$ carries R into the subalgebra $S \subseteq R^{\lambda} \amalg R^{\rho}$ of elements invariant under the automorphism interchanging the indices λ and ρ. Now by [**29**], this algebra S of symmetric elements has the structure of the tensor algebra on the vector space of all elements $x^{\lambda} + x^{\rho}$ $(x \in R)$. (To get this, apply [**29**, Theorem 1] with $X = \{\lambda, \rho\}$, $G =$ the permutation group on X, and $Q(X)$ ordered so that $\lambda > \rho$. Then the Γ of that Theorem is simply $\{\lambda\}$. Cf. [**29**, (17)].) The coneutral law for $\mathbf{a}$ says that the maps

$$R \xrightarrow{\mathbf{a}} S \subseteq R^{\lambda} \amalg R^{\rho} \xrightarrow{(1_R, 0)} R$$

compose to the identity, giving a retraction of the tensor algebra S onto R, as desired. □

This phenomenon of "weak" identities, such as commutativity and the neutral element law, putting strong restrictions on the structures of objects which admit operations satisfying them has also been encountered in topology; see [**190**], [**191**].

(The open question mentioned above on retracts of free algebras is known to have a positive answer in the 2-generator case [**47**, p.243, Ex.6]. The result of [**13**, Prop. 2.1] on the fixed ring of an action of a group on a ring with 2-term weak algorithm might be of use in approaching the general case.)

An object of **Binar** with two additional binary operations, x/y and $x\backslash y$, satisfying the identities $(x\cdot y)/y = (x/y)\cdot y = x$ and $x\backslash(x\cdot y) = x\cdot(x\backslash y) = y$ is called a *quasigroup*, and a quasigroup with neutral element is called a *loop*. It might be interesting to try

PROBLEM 15.3. *Investigate representable functors from* $\mathbf{Ring}_k$ *or* $\mathbf{Ring}^1_k$ *to the variety of quasigroups and/or loops, or to any of the subvarieties thereof described in* [**41**].

§16. Representable functors to abelian semigroups without neutral element – easy results.

Once again k is an arbitrary ring with 1.

Recall that **AbSemigp** denotes the category of commutative semigroups not necessarily having neutral elements. In this section we shall deduce partial information about representable functors from categories of k-rings to **AbSemigp** from what we have already proved about functors from these categories to $\mathbf{AbSemigp}^e$. In §§17-18, we shall improve these results with the help of new ring-theoretic arguments.

There is an important difference, from the start, between the study of **AbSemigp**-valued functors and our earlier investigations. When we considered representable functors to $\mathbf{AbSemigp}^e$, and even to $\mathbf{AbBinar}^e$, the presence of the unique zeroary operation giving the neutral element made the consideration of such functors on categories of *unital* k-rings and on the corresponding categories of *nonunital* k-rings equivalent. For semigroups without neutral element, this equivalence no longer holds, though we shall see that there are still important implications relating results for the two cases. Let us start with the unital case.

PROPOSITION 16.1. *Suppose* $V: k\text{-}\mathbf{Ring}^1 \to \mathbf{AbSemigp}$ *is a representable functor. If* $S \in k\text{-}\mathbf{Ring}^1$ *and* $V(S)$ *contains an idempotent, i.e., an element* i *satisfying*:

$$i + i = i,$$

then the subsemigroup

$$i + V(S) = \{i + a \mid a \in V(S)\} = \{b \in V(S) \mid b = i + b\}$$

is an abelian group, *with neutral element* i. *Further,* $V(S)$ *can have at most one idempotent element.*

PROOF. Let S be as above, and consider the composite functor

$$(16.2) \qquad S\text{-}\mathbf{Ring}^1 \xrightarrow{\text{forget}} k\text{-}\mathbf{Ring}^1 \xrightarrow{V} \mathbf{AbSemigp}.$$

Being a composite of representable functors, this is representable (Corollary 8.16). (If R is the k-ring representing V, we can explicitly describe the S-ring representing (16.2) as $S \amalg R$, the coproduct being taken in $k\text{-}\mathbf{Ring}^1$.) Further, the functor (16.2) admits a zeroary operation i, associating to every S-ring T the image $i_T \in V(T)$ of $i \in V(S)$, and the elements i_T are idempotent because $i \in V(S)$ is. Hence if we let $\mathbf{AbSemigp}^i$ denote the category of abelian semigroups given with one distinguished idempotent element, (16.2) can be strengthened to a representable functor $S\text{-}\mathbf{Ring}^1 \to \mathbf{AbSemigp}^i$. (Incidentally, as noted in the discussion following Lemma 11.3, the variety $\mathbf{AbSemigp}^i$ can be identified with $\mathbf{AbSemigp}^{\mathrm{pt}}$.)

Suppose we now compose this functor $S\text{-}\mathbf{Ring}^1 \to \mathbf{AbSemigp}^i$ with the functor

$$(16.3) \qquad \mathbf{AbSemigp}^i \to \mathbf{AbSemigp}^e$$

which takes an abelian semigroup A with distinguished idempotent i to $i + A = \{i + a\} = \{b \mid b = i + b\}$. Note that $i + A$ is indeed a subsemigroup, and has i as neutral element. This functor (16.3) is represented by the object of $\mathbf{AbSemigp}^i$ presented by one generator x and one relation $i + x = x$ (isomorphic to the additive semigroup of natural numbers, with $0 = i$, $1 = x$). The result of composing (16.2) (enhanced to an $\mathbf{AbSemigp}^i$-valued functor) with (16.3):

$$S\text{-}\mathbf{Ring}^1 \to \mathbf{AbSemigp}^i \to \mathbf{AbSemigp}^e$$

is a representable functor to which we can apply Theorem 13.15! That Theorem tells us that it will be *group*-valued; moreover its value on S is $i + V(S)$. This proves our main assertion.

To see that $V(S)$ contains at most one idempotent element, observe that if

$i, j \in V(S)$ are idempotent, then $i+j$ is an idempotent element of $i+V(S)$. But the only idempotent element in a group is the neutral element, so $i+j = i$. Similarly, $i+j = j$, hence $i = j$. $\square$

Note that for brevity we are saying that a semigroup S "is a group" when it is of the form $U(G)$, where U is the forgetful functor from groups to semigroups, and G is a group in the strict sense; we will use this shorthand frequently in this chapter. Also, though in the above proof we distinguished notationally between the idempotent $i \in V(S)$ and its images $i_T \in V(T)$, in corresponding situations below we will generally call that image i as well, on the principle that it is a zeroary operation, and corresponding operations on different objects of a variety are denoted by the same symbol unless there is danger of ambiguity.

For functors on categories of nonunital rings, we have a similar result, but we get the idempotent element automatically:

PROPOSITION 16.4. *Suppose that* V: k-**Ring** $\to$ **AbSemigp** *is a representable functor, with representing k-ring* R. *Then the map* $0\colon R \to \{0\}$ *gives a zeroary operation* i *on the semigroups* $V(S)$, *whose value in each* $V(S)$ *is idempotent, thus allowing us to consider* V *an* **AbSemigp**^{i}*-valued functor. The* **AbSemigp**^{e}*-valued functor* $S \mapsto i+V(S)$ *is representable, hence has a structure as described in Theorem 13.15; in particular, it is group valued. Again,* i *is the unique idempotent element in each of the semigroups* $V(S)$.

PROOF. All but the last sentence is straightforward. In getting this last assertion we cannot as before use *symmetry* to conclude that any other idempotent j in $V(S)$ also has the property that $j+V(S)$ is a group, because such a j has not arisen in the same way as i. Instead, let us reduce this question to the unital case. Let V' denote the composite functor k-**Ring**$^1 \xrightarrow{\text{forget}} k$-**Ring** $\xrightarrow{V}$ **AbSemigp**. By Proposition 16.1 the values of V' are semigroups with at most one idempotent. Forming the unital k-ring $S^1 = k \oplus S$, we find that $V(S) \subseteq V'(S^1)$, so $V(S)$ can have at most one idempotent. $\square$

We will now show by examples that the differences between the properties established above for **AbSemigp**-valued functors, and the stronger properties of **AbSemigp**^{e}-valued functors proved in the previous chapter, are real; namely, that in the situation of the above Proposition, the group $i+V(S)$ may be a proper subsemigroup of $V(S)$, and that in the unital case, there may be no idempotents in $V(S)$. In these examples we will let k be a commutative ring, and describe representable functors on k-algebras. By the observations of §14.7, these representable functors assume as their values the same classes of semigroups as the functors on the larger categories k-**Ring** and k-**Ring**^{1} with the same representing objects, so these examples also illustrate properties of functors on the latter varieties.

16.5. *Examples with* $i+V(S) \neq V(S)$. For our first example, let $f \in [k]\langle x, x', y, y'\rangle$ be any element of this free algebra which is invariant under the automorphism that simultaneously interchanges x and y and interchanges x' and

y', and such that every monomial appearing with nonzero coefficient in f contains at least one factor x (whence every such monomial also contains at least one factor y); e.g., $f = xy^2y' + yx^2x'$. For any $S \in \mathbf{Ring}_k$, we define on pairs of elements of S the commutative operation $(\xi, \xi') + (\eta, \eta') = (0, f(\xi, \xi', \eta, \eta'))$. It is easy to check that any 3-fold sum $(\xi, \xi') + ((\eta, \eta') + (\zeta, \zeta'))$ or $((\xi, \xi') + (\eta, \eta')) + (\zeta, \zeta')$ gives $(0,0)$, so we have "associativity by default". The unique idempotent i in this semigroup is $(0,0)$; the group $i + V(S)$ is $\{i\}$; in particular, it is properly contained in $V(S)$.

The above functor is represented by $[k]<x, x'>$, but in contrast to the situation of the preceding section (functors to $\mathbf{AbBinar}^e$), we can modify the example easily to get functors with nonfree representing objects. For the above discussion goes over unchanged if we restrict our semigroups to consist of those pairs (ξ, ξ') that satisfy some further equations, as long as these do not exclude any pairs of the form $(0, \xi')$. E.g., we may impose the restrictions $\xi^3 - \xi^2 - \xi = 0$ and $\xi\xi' = \xi'^2\xi$, getting a functor represented by the k-algebra $[k]<x, x' \mid x^3 - x^2 - x = 0, xx' = x'^2x>$.

We can also modify these examples to make the subgroups $i + V(S)$ nontrivial, by taking the direct product of one of these functors with a nontrivial functor to $\mathbf{AbSemigp}^e$, as characterized in Theorem 13.15. From such a direct product functor we can in turn obtain subfunctors which are not themselves products by again imposing equations that do not exclude any tuples which arise under the operation. For instance, on 3-tuples of elements of S, subject to any equations which are satisfied by all 3-tuples of the form $(0, \xi', \xi'')$, we can define an associative operation

$$(16.6) \qquad (\xi, \xi', \xi'') + (\eta, \eta', \eta'') = (0,\ f(\xi, \xi', \xi'', \eta, \eta', \eta''),\ \xi'' + \eta'').$$

for every $f \in k<x, x', x'', y, y', y''>$ which is invariant under simultaneously interchanging x, x', x'' with y, y', y'' respectively, and which has the property that every monomial appearing in f contains at least one x.

16.7. *Examples with no idempotents.* To get functors on unital algebras which can give semigroups that have no idempotents, let $k = \mathbf{Z}$, the ring of integers, and let R be either the polynomial ring $\mathbf{Q}[x]$ over the rational numbers, or the polynomial ring $(\mathbf{Z}/p\mathbf{Z})[x]$ over the field of p elements for some prime p. Observe that R represents the functor which associates to a unital ring S its underlying set if the $\mathbf{Z}$-algebra structure on S extends to a $\mathbf{Q}$-algebra structure, respectively a $(\mathbf{Z}/p\mathbf{Z})$-algebra structure, and otherwise gives the empty set. The empty set has a unique structure of semigroup, and the unique map from the empty semigroup to any other semigroup is a homomorphism; hence if we make the above set-valued functors semigroup-valued by putting on the underlying set of S its additive semigroup structure, and on the empty set its unique semigroup structure, we get functors to $\mathbf{AbSemigp}$. Although when nonempty these semigroups have idempotent elements (indeed, neutral elements), in the empty case they do not.

The relation between these peculiar representable functors on $\mathbf{Ring}^1_{\mathbf{Z}}$, and the underlying-additive-semigroup functors on $\mathbf{Ring}^1_{\mathbf{Q}}$, respectively $\mathbf{Ring}^1_{\mathbf{Z}/p\mathbf{Z}}$, is explained by the observation that each of the latter varieties can be identified with a full subcategory of $\mathbf{Ring}^1_{\mathbf{Z}}$ which satisfies a slight weakening of conditions (a) and (b) of Lemma 14.2, namely the conditions gotten by replacing references to *finite* families by *finite nonempty* families in each statement. These weakened conditions are sufficient to show that $\mathbf{V}$-coalgebras in the subcategory $\mathbf{D}$ are also $\mathbf{V}$-coalgebras in the larger category $\mathbf{C}$ for varieties $\mathbf{V}$ with no zeroary operations, e.g., **AbSemigp**, though not for varieties with zeroary operations, e.g., $\mathbf{AbSemigp}^e$.

Observe that in the first class of examples above, the ''associativity by default'' trick was based on an addition operation that ''died'' when iterated, with the result that any threefold sum landed in the subgroup $i+V(S)$; and that in the second class of examples, the only way we avoided getting idempotent elements was by having semigroups with no elements at all. We shall prove in the next two sections that these two restrictions hold in general, i.e., that if V is any representable **AbSemigp**-valued functor on k-$\mathbf{Ring}^1$ and S any object such that the semigroup $V(S)$ is *nonempty*, then this semigroup contains an idempotent i, and every sum of three elements of $V(S)$ belongs to $i+V(S)$. (It will follow that the last assertion is also true for representable **AbSemigp**-valued functors on the category of nonunital k-rings, where the hypothesis and the first conclusion are automatic.) We shall also show that **AbSemigp**-valued functors on unital k-rings which are empty-valued on some objects arise in ''essentially the same way'' as in Example 16.7.

Let us end this section by recording a trivial consequence of Propositions 16.1 and 16.4. We must first define the kinds of triviality in question.

DEFINITION 16.8. *An algebra-valued functor V on a category $\mathbf{C}$ will be called* trivial *if for all objects Y of $\mathbf{C}$, $V(Y)$ is the 1-element algebra, while V will be called* quasitrivial *if for all Y, $V(Y)$ is* either *the 1-element algebra or the* empty *algebra.*

An object X of a category $\mathbf{C}$ will be called quasi-initial *if for all objects Y of $\mathbf{C}$, $\mathbf{C}(X, Y)$ has* at most *one element.*

Thus, a representable functor is *trivial* if and only if it is represented by an initial object of $\mathbf{C}$, *quasitrivial* if and only if it is represented by a quasi-initial object. If a category has an initial object I, then the quasi-initial objects are precisely the *epimorphs* of I (the objects for which the unique morphism from I is an epimorphism). An initial object has a unique structure of co-$\mathbf{V}$ object for *every* variety $\mathbf{V}$, a quasi-initial object has a unique co-$\mathbf{V}$ structure for every variety $\mathbf{V}$ without zeroary operations.

Now recall that a *semilattice* is an abelian semigroup in which all elements are idempotent. In particular, both the meet and the join operation of a lattice are semilattice operations. From the results on unicity of idempotents in Propositions 16.1 and 16.4 we see:

COROLLARY 16.9. *Any representable functor from k-**Ring** to semilattices or lattices is trivial. Any representable functor from k-$\mathbf{Ring}^1$ to semilattices or lattices is quasitrivial.* □

§17. Symmetry conditions, and cocommutativity.

So far, our investigation of co-**AbSemigp** objects in categories of k-rings has lived off the ring-theoretic work we did in Chapter III. To go further we need some new ring-theoretic arguments. In Chapter III we emphasized the coassociative law, with cocommutativity playing a secondary role. Below we shall look more closely at the latter condition.

A co-operation $R \to R^{\rho} \amalg R^{\lambda}$ is cocommutative if and only if it takes R into the subring of elements of $R^{\rho} \amalg R^{\lambda}$ that are *symmetric* in λ and ρ. In setting up notation for classes of symmetric elements, let us use general indices α, β etc., for which we will later be able to substitute any of λ, μ, ρ.

Throughout this section, "k-ring" will mean unital *k-ring.*

DEFINITION 17.1. *In a coproduct $R^{\alpha} \amalg R^{\beta}$ of two copies of a k-ring R, an element will be called* symmetric *if it is invariant under the automorphism which interchanges x^{α} and x^{β} for all $x \in R$. The k-ring of symmetric elements of $R^{\alpha} \amalg R^{\beta}$ will be denoted* $\mathrm{Sym}(R^{\alpha} \amalg R^{\beta})$.

In a threefold coproduct there are several kinds of symmetry to consider:

DEFINITION 17.2. *In a coproduct of k-rings $R^{\alpha} \amalg R^{\beta} \amalg R^{\gamma}$, where R^{α} and R^{β} are two copies of the same k-ring R (but R^{γ} is for the moment arbitrary), we shall call an element symmetric in α and β if it is invariant under the automorphism interchanging x^{α} and x^{β} for all $x \in R$ and fixing all elements of R^{γ}. The k-ring of all elements symmetric in α and β will be denoted*

$$\mathrm{Sym}_{\{\alpha,\beta\}}(R^{\alpha} \amalg R^{\beta} \amalg R^{\gamma}).$$

On the other hand, an element of $R^{\alpha} \amalg R^{\beta} \amalg R^{\gamma}$ will be called strongly symmetric *in α and β if it belongs to the subring generated by the images of* $\mathrm{Sym}(R^{\alpha} \amalg R^{\beta})$ *and R^{γ}. This subring will be denoted*

$$\mathrm{SSym}_{\{\alpha,\beta\}}(R^{\alpha} \amalg R^{\beta} \amalg R^{\gamma}).$$

If R^{α}, R^{β} and R^{γ} are all copies of the same k-ring R, then we shall write

$$\mathrm{Sym}(R^{\alpha} \amalg R^{\beta} \amalg R^{\gamma}) = \mathrm{Sym}_{\{\alpha,\beta\}}(R^{\alpha} \amalg R^{\beta} \amalg R^{\gamma}) \cap \mathrm{Sym}_{\{\beta,\gamma\}}(R^{\alpha} \amalg R^{\beta} \amalg R^{\gamma})$$

for the subring of symmetric *elements, i.e., elements invariant under the natural action of the full permutation group on α, β and γ. Likewise we shall write*

$$\mathrm{SSym}(R^{\alpha} \amalg R^{\beta} \amalg R^{\gamma}) = \mathrm{SSym}_{\{\alpha,\beta\}}(R^{\alpha} \amalg R^{\beta} \amalg R^{\gamma}) \cap \mathrm{Sym}(R^{\alpha} \amalg R^{\beta} \amalg R^{\gamma})$$

and call elements of this subring strongly symmetric (*in all three indices*).

Note that an element symmetric in α and β and also in β and γ is indeed invariant under the full permutation group on α, β and γ, since this group is generated by the two transpositions (α, β) and (β, γ).

Likewise, though the definition of $\mathrm{SSym}(R^{\alpha} \amalg R^{\beta} \amalg R^{\gamma})$ stated above involves the pair of indices $\{\alpha, \beta\}$ in a distinguished way, we note that because an element r of this subring is required to lie in $\mathrm{Sym}(R^{\alpha} \amalg R^{\beta} \amalg R^{\gamma})$, any property that r satisfies with respect to one pair of indices will also hold with respect to every pair.

So, for instance, it is easy to verify the alternative characterization

(17.3)
$$\begin{aligned}&\mathrm{SSym}(R^\alpha \amalg R^\beta \amalg R^\gamma)\\ &= \mathrm{SSym}_{\{\alpha,\beta\}}(R^\alpha \amalg R^\beta \amalg R^\gamma) \cap \mathrm{SSym}_{\{\beta,\gamma\}}(R^\alpha \amalg R^\beta \amalg R^\gamma).\end{aligned}$$

17.4. *Examples.* Let us again look at algebras over a field k, recalling that their coproducts as k-rings and as k-algebras are the same. Suppose R^α, R^β, R^γ are all copies of the free algebra $k\langle x, y, z\rangle$. The element $x^\alpha y^\beta z^\gamma + x^\beta y^\alpha z^\gamma$ is strongly symmetric in α and β, because it is the product of $x^\alpha y^\beta + x^\beta y^\alpha \in \mathrm{Sym}(R^\alpha \amalg R^\beta)$ with $z^\gamma \in R^\gamma$. On the other hand, $x^\alpha y^\gamma z^\beta + x^\beta y^\gamma z^\alpha$ is symmetric in α and β but not strongly symmetric, though we don't yet have an easy way of proving the latter fact. (We shall soon.) However $x^\alpha y^\gamma z^\alpha + x^\alpha y^\gamma z^\beta + x^\beta y^\gamma z^\alpha + x^\beta y^\gamma z^\beta$ is again strongly symmetric in α and β, being expressible as $(x^\alpha + x^\beta)\, y^\gamma\, (z^\alpha + z^\beta)$.

An example of an element of the same algebra that is strongly symmetric in all three indices is the sum of all terms $x^{\alpha_1} y^{\alpha_2} z^{\alpha_3}$ where $(\alpha_1, \alpha_2, \alpha_3)$ ranges over the 12 strings of length three in α, β and γ satisfying $\alpha_1 \neq \alpha_2$, $\alpha_2 \neq \alpha_3$. We could verify this explicitly now, but it will follow easily from a general criterion we shall obtain. On the other hand, the same criterion will show that the sum of the six terms $x^{\alpha_1} y^{\alpha_2} z^{\alpha_3}$ with α_1, α_2, α_3 all distinct, though it is obviously symmetric, is not strongly so.

To develop the desired criterion for an element to be strongly symmetric, we first need a description of a coproduct $\coprod_{\alpha \in A} R^\alpha$ of unital k-rings, and to get this we must assume that each of the k-rings R^α is given with an augmentation $\varepsilon^\alpha\colon R^\alpha \to k$ (though we shall not restrict attention to ring homomorphisms respecting augmentations). This means that each R^α can be described as the result of adjoining a unit to a nonunital k-ring $B^\alpha = \ker(\varepsilon^\alpha)$ (Lemma 10.6):

(17.5)
$$R^\alpha = k \oplus B^\alpha.$$

We can now describe coproducts almost as before. Given a family of augmented k-rings $(R^\alpha, \varepsilon^\alpha)_{\alpha \in A}$, their coproduct as unital k-rings has the k-bimodule structure

(17.6)
$$\coprod_{\alpha \in A} R^\alpha = \bigoplus B^\sigma = \bigoplus B^{\alpha_1} \otimes_k \dots \otimes_k B^{\alpha_h}$$

where the direct sum is taken over index-strings $\sigma = (\alpha_1, \dots, \alpha_h)$ $(h \geq 0)$ such that $\alpha_i \neq \alpha_{i+1}$ $(1 \leq i < h)$. Note that we are now including a summand indexed by the empty string, of length $h = 0$, understanding the tensor product of the empty sequence of k-bimodules to mean the bimodule k.

Let us denote the projection of the coproduct (17.6) onto the summand associated with an index-string σ by

$$p_\sigma\colon \coprod_{\alpha \in A} R^\alpha \to R^\sigma.$$

(In Chapter III we invoked the projection onto such a component by putting a superscript σ on our map **a** or **aa**. But here we are studying symmetric elements in the abstract before passing to consideration of a co-operation **a**, so we need the symbol p_σ.) In cases where some of the R^α are assumed to be copies of the same k-ring, distinguished only by superscripts, we shall again use f to denote maps that "forget superscripts", i.e., the isomorphisms from the various tensor product bimodules B^{σ} in (17.6) to the corresponding tensor products with these superscripts omitted.

In the case where our family of augmented k-rings has just two members, $(R^{\alpha}, \varepsilon^{\alpha})$ and $(R^{\beta}, \varepsilon^{\beta})$, which are copies of a single augmented k-ring (R, ε), it is clear that an element x of the coproduct ring is symmetric in these two indices if and only if we have

(17.7) $fp_{\sigma}(x) = fp_{\tau}(x)$ whenever σ is an index-string, and τ the string formed from σ by interchanging α and β throughout.

We can deduce that $\mathrm{Sym}(R^{\alpha} \amalg R^{\beta})$ is a bimodule direct summand in $R^{\alpha} \amalg R^{\beta}$: a bimodule projection of the latter ring onto the former is the map taking every element y to the element x which has the same σ-component as y for each string σ beginning with α, and also for σ the empty string, while the τ-components for strings τ beginning with β are determined by (17.7) from the σ-components for strings σ beginning with α.

Let S for the moment denote the kernel of the natural augmentation on $\mathrm{Sym}(R^{\alpha} \amalg R^{\beta})$, i.e., the set of elements of $R^{\alpha} \amalg R^{\beta}$ satisfying (17.7) and having zero component in k, and let T denote the kernel of the bimodule projection $R^{\alpha} \amalg R^{\beta} \to \mathrm{Sym}(R^{\alpha} \amalg R^{\beta})$ described above. Thus

(17.8) $$\mathrm{Sym}(R^{\alpha} \amalg R^{\beta}) = k \oplus S, \qquad R^{\alpha} \amalg R^{\beta} = k \oplus S \oplus T.$$

These splittings will now help us study symmetric elements of threefold coproducts. If σ is an index-string formed from the indices α, β, γ, let us define the $\{\alpha, \beta\}$-segments of σ to be the maximal consecutive substrings consisting of α's and β's.

LEMMA 17.9. *Let* $(R^{\alpha}, \varepsilon^{\alpha})$, $(R^{\beta}, \varepsilon^{\beta})$, $(R^{\gamma}, \varepsilon^{\gamma})$ *be augmented k-rings, of which the first two are copies of a common augmented k-ring* (R, ε). *Let* $R^{\alpha} \amalg R^{\beta} \amalg R^{\gamma}$ *be written as a bimodule direct sum* (17.6). *Then*

(i) *An element* $x \in R^{\alpha} \amalg R^{\beta} \amalg R^{\gamma}$ *is symmetric in* α *and* β *if and only if whenever* σ *and* τ *are index-strings such that* τ *is obtained from* σ *by interchanging* α *and* β *everywhere, one has* $fp_{\sigma}(x) = fp_{\tau}(x)$.

(ii) *An element* $x \in R^{\alpha} \amalg R^{\beta} \amalg R^{\gamma}$ *is* strongly *symmetric in* α *and* β *if and only if whenever* σ *and* τ *are index-strings such that* τ *is obtained from* σ *by interchanging* α *and* β *throughout a* single $\{\alpha, \beta\}$*-segment, one has* $fp_{\sigma}(x) = fp_{\tau}(x)$.

PROOF. (i) is clear. To show (ii), let us regard $R^{\alpha} \amalg R^{\beta} \amalg R^{\gamma}$ as $(R^{\alpha} \amalg R^{\beta}) \amalg R^{\gamma}$, and write this coproduct as a bimodule direct sum (17.6), using the given augmentation on R^{γ} with kernel B^{γ}, and the augmentation $(\varepsilon^{\alpha}, \varepsilon^{\beta})$: $R^{\alpha} \amalg R^{\beta} \to k$, which has kernel $S \oplus T$ (17.8). Thus, the summands in this expansion have the form

(17.10) $$\ldots \otimes_k (S \oplus T) \otimes_k B^{\gamma} \otimes_k (S \oplus T) \otimes_k B^{\gamma} \otimes_k \ldots .$$

Clearly the subring generated by $\mathrm{Sym}(R^{\alpha} \amalg R^{\beta}) = k \oplus S$ and $R^{\gamma} = k \oplus B^{\gamma}$ will have components

(17.11) $$\dots \otimes S \otimes B^{\gamma} \otimes S \otimes B^{\gamma} \otimes \dots .$$

Recalling that S consists of the elements of $S \oplus T$ which satisfy (17.7), we see that an element of (17.10) belongs to the sum of the bimodules (17.11) if and only if it satisfies the condition described in (ii). □

As a result of the splittings (17.8), we also have

COROLLARY 17.12. *Let* R^{α}, R^{β}, R^{γ} *be k-rings such that the first two are copies of a common k-ring* R, *and such that all three k-rings admit augmentations. Then the natural surjective homomorphism*

$$\mathrm{Sym}(R^{\alpha} \amalg R^{\beta}) \amalg R^{\gamma} \to \mathrm{SSym}_{\{\alpha,\beta\}}(R^{\alpha} \amalg R^{\beta} \amalg R^{\gamma})$$

is an isomorphism. (I.e., the natural map $\mathrm{Sym}(R^{\alpha} \amalg R^{\beta}) \amalg R^{\gamma} \to R^{\alpha} \amalg R^{\beta} \amalg R^{\gamma}$ *is one-to-one.)*

PROOF. The components (17.11) of $\mathrm{SSym}_{\{\alpha,\beta\}}(R^{\alpha} \amalg R^{\beta} \amalg R^{\gamma})$ are precisely the components in the decomposition of $\mathrm{Sym}(R^{\alpha} \amalg R^{\beta}) \amalg R^{\gamma}$ as in (17.6). □

We now turn to the conditions of symmetry in all three indices.

LEMMA 17.13. *Let* $(R^{\alpha}, \varepsilon^{\alpha})$, $(R^{\beta}, \varepsilon^{\beta})$, $(R^{\gamma}, \varepsilon^{\gamma})$ *be three copies of an augmented k-ring* (R, ε), *and let* $R^{\alpha} \amalg R^{\beta} \amalg R^{\gamma}$ *be written as a direct sum* (17.6) *using these augmentations. Then*

(i) *An element* $x \in R^{\alpha} \amalg R^{\beta} \amalg R^{\gamma}$ *is symmetric (with respect to all three indices) if and only if whenever* σ *and* τ *are index-strings such that* τ *is obtained from* σ *by applying a fixed permutation of the indices* α, β *and* γ, *one has* $fp_{\sigma}(x) = fp_{\tau}(x)$.

(ii) *An element* $x \in R^{\alpha} \amalg R^{\beta} \amalg R^{\gamma}$ *is strongly symmetric (with respect to all three indices) if and only if for every pair of index-strings* σ *and* τ *of the same length, one has* $fp_{\sigma}(x) = fp_{\tau}(x)$.

PROOF. Again, (i) is clear. In (ii), "if" is true because an element x with the indicated property will satisfy the condition of the preceding Lemma for strong symmetry with respect to α and β on the one hand, and with respect to β and γ on the other. "Only if" will follow by a repeated application of that same Lemma if we can show that, given any two index-strings σ and τ of the same length h, we can get from one to the other by successive operations of interchanging α and β throughout a single $\{\alpha,\beta\}$-segment, interchanging β and γ throughout a single $\{\beta,\gamma\}$-segment, and interchanging γ and α throughout a single $\{\gamma,\alpha\}$-segment.

So let index-strings σ and τ of the same length h be given, and assume inductively that we can go by such steps from σ to a string σ' which agrees with τ in the first i terms, for some i with $0 \leq i < h$. If the $i{+}1$st terms of σ' and τ do not agree, we may assume without loss of generality that they are α and β respectively. Note that this implies that if $i > 0$, the common ith term of σ' and τ is γ; it follows that (whether $i > 0$ or $i = 0$), the $i{+}1$st term α of σ' begins an $\{\alpha,\beta\}$-segment. So by interchanging α and β throughout that $\{\alpha,\beta\}$-segment of σ', we get a string σ'' agreeing with τ in at least $i+1$ terms. Hence by induction, we can get from σ to τ, as desired. □

(Using the criteria of Lemmas 17.9 and 17.13, it is now easy to verify the assertions in Example 17.4 that certain expressions are, and others are not, strongly symmetric in various indices.)

Let us note a surprising consequence of the above Lemma. Let $(R^\alpha, \varepsilon^\alpha)$, $(R^\beta, \varepsilon^\beta)$, $(R^\gamma, \varepsilon^\gamma)$ be as in that result. Then in an element of either $\mathrm{SSym}(R^\alpha \amalg R^\beta \amalg R^\gamma)$ or $\mathrm{Sym}(R^\alpha \amalg R^\beta)$, all components of every degree are determined by any one component of that degree, hence as k-bimodules, each of these k-rings is isomorphic in a natural way to

(17.14) $$k \oplus B \oplus (B \otimes_k B) \oplus \dots .$$

Though this bimodule is the same as the underlying k-bimodule of the familiar tensor k-ring $k\langle B\rangle$, the ring structure on this object is not the standard one. (We shall describe it in §19.) However, we claim that the ring structures induced on (17.14) by the identifications with $\mathrm{SSym}(R^\alpha \amalg R^\beta \amalg R^\gamma)$ and with $\mathrm{Sym}(R^\alpha \amalg R^\beta)$ are the same. To see this, note that the k-ring homomorphism $(i^\alpha, i^\beta, \varepsilon^\gamma)\colon R^\alpha \amalg R^\beta \amalg R^\gamma \to R^\alpha \amalg R^\beta$ can be described bimodule-theoretically as simply throwing away all components indexed by strings involving γ. We can see from this that the restriction of this homomorphism to a map $\mathrm{SSym}(R^\alpha \amalg R^\beta \amalg R^\gamma) \to \mathrm{Sym}(R^\alpha \amalg R^\beta)$ respects the bimodule isomorphisms with (17.14), hence, being a homomorphism of k-rings, it is an isomorphism as k-rings. We record this as

LEMMA 17.15. *Let R^α, R^β, R^γ be three copies of a unital k-ring R. Then for any augmentation $\varepsilon\colon R \to k$, the homomorphism $(i^\alpha, i^\beta, \varepsilon^\gamma)\colon R^\alpha \amalg R^\beta \amalg R^\gamma \to R^\alpha \amalg R^\beta$ carries* $\mathrm{SSym}(R^\alpha \amalg R^\beta \amalg R^\gamma)$ *isomorphically onto* $\mathrm{Sym}(R^\alpha \amalg R^\beta)$. □

We are now ready to apply these concepts to co-operations on k-rings. Let us first remark that the results we have obtained so far about symmetric elements in unital k-rings with augmentation are equivalent to statements about symmetric elements in nonunital k-rings; and if we were to stop here, the latter formulations would have the advantage of simplicity. But it is of key importance in what will follow that the *co-operations* we shall consider will not be assumed to respect the given augmentations. Indeed, in considering a cosemigroup $(R, \mathbf{a})$ representing a functor $V\colon k\text{-}\mathbf{Ring}^1 \to \mathbf{AbSemigp}$, the existence of an augmentation ε *respected* by the coaddition amounts to the existence of an idempotent element in $V(k)$, and this is what we want to *prove*, from the weaker assumption that $V(k)$ is nonempty, i.e., that R admits *some* augmentation ε.

Recall that a morphism $R \to R^{\alpha_1} \amalg \dots \amalg R^{\alpha_n}$ of k-rings corresponds to a morphism $V_1 \times \dots \times V_n \to V$ among representable functors. We shall here speak of such a morphism as an "n-ary operation", even when we do not assume that all the functors $V_1, \dots, V_n$ and V are the same. Given representable functors V, $V'\colon k\text{-}\mathbf{Ring}^1 \to \mathbf{Set}$, a binary operation $a\colon V \times V \to V'$ is commutative (satisfies the identity $a(x, y) = a(y, x)$ for all $x, y \in V(S)$ and all k-rings S) if and only if the corresponding ring homomorphism $\mathbf{a}\colon R' \to R^{\alpha} \amalg R^{\beta}$ (where R^{α} and R^{β} are copies of the ring R representing V) has range in $\mathrm{Sym}(R^{\alpha} \amalg R^{\beta})$. Likewise, a ternary operation $a\colon V \times V \times V' \to V''$ is commutative in its first two variables if and only if the corresponding ring-map $\mathbf{a}$ has range in $\mathrm{Sym}_{\{\alpha,\beta\}}(R^{\alpha} \amalg R^{\beta} \amalg R^{\gamma})$; and in the case where $V' = V$, it will be commutative in all three variables if and only if $\mathbf{a}$ has range in $\mathrm{Sym}(R^{\alpha} \amalg R^{\beta} \amalg R^{\gamma})$.

The significance of *strong symmetry* in the first two variables is described in

LEMMA 17.16. *Let* $a\colon V \times V \times V' \to V''$ *be an operation on representable functors* k-**Ring**$^1 \to$ **Set** *such that* $V(k)$ *and* $V'(k)$ *(and hence also* $V''(k)$*) are nonempty; thus, the operation* a *corresponds to a k-ring homomorphism*

$$\mathbf{a}\colon\ R'' \to R^{\alpha} \amalg R^{\beta} \amalg R^{\gamma}$$

where R^{α} *and* R^{β} *are copies of a common k-ring, and where all these k-rings admit augmentations. Then the following conditions are equivalent:*

(i) *The co-operation* $\mathbf{a}$ *has range in* $\mathrm{SSym}_{\{\alpha,\beta\}}(R^{\alpha} \amalg R^{\beta} \amalg R^{\gamma})$.

(ii) *The operation* a *can be factored*

$$a(x, y, z) = b(c(x, y), z) \quad \text{with } c \text{ commutative,} \tag{17.17}$$

that is, it can be written as a composite

$$V \times V \times V' \xrightarrow{(c,\,1_{V'})} W \times V' \xrightarrow{\ b\ } V''$$

where W *is another representable functor* k-**Ring**$^1 \to$ **Set** *and* $c\colon V \times V \to W$ *is a commutative operation.*

PROOF. Assuming (i), Corollary 17.12 tells us that we can factor $\mathbf{a}$ through the embedding of $\mathrm{Sym}(R^{\alpha} \amalg R^{\beta}) \amalg R^{\gamma}$ in $R^{\alpha} \amalg R^{\beta} \amalg R^{\gamma}$. (It is to use that Corollary that we need to assume in this Lemma the existence of augmentations on the given k-rings.) Letting W denote the functor represented by $\mathrm{Sym}(R^{\alpha} \amalg R^{\beta})$, we see that the inclusion of $\mathrm{Sym}(R^{\alpha} \amalg R^{\beta})$ in $R^{\alpha} \amalg R^{\beta}$ is a cocommutative binary co-operation, corresponding to a commutative operation $c\colon V \times V \to W$, which gives the desired factorization.

Conversely, assuming (ii) let us write T for the k-ring representing the (now arbitrary) W, so that we have a factorization of the co-operation $\mathbf{a}$ as

$$R'' \xrightarrow{\ \mathbf{b}\ } T \amalg R^{\gamma} \xrightarrow{\mathbf{c} \amalg 1_{R^{\gamma}}} R^{\alpha} \amalg R^{\beta} \amalg R^{\gamma}$$

with $\mathbf{c}\colon T \to R^{\alpha} \amalg R^{\beta}$ cocommutative. Thus $\mathbf{c}$ has range in $\mathrm{Sym}(R^{\alpha} \amalg R^{\beta})$, hence $\mathbf{c} \amalg 1_{R^{\gamma}}$ has range in $\mathrm{SSym}_{\{\alpha,\beta\}}(R^{\alpha} \amalg R^{\beta} \amalg R^{\gamma})$, proving (i). $\square$

Hence let us call a ternary operation $a\colon V \times V \times V' \to V''$ among representable functors k-**Ring**$^1 \to$ **Set** *strongly commutative* in its first two variables if it admits a factorization (17.17). We shall likewise call a ternary operation $a\colon V \times V \times V \to V'$ *strongly commutative* (in all three variables) if it is strongly commutative in each pair of variables.

(One can use the corresponding factorization properties to define a condition of "strong commutativity" on operations among objects of any category **C** which has finite products – for instance, **C** = **Set** – in place of the functor-category **C** = **Rep**(k-**Ring**^{1}, **Set**) of the present discussion. We remark, however, that in **Set** itself, strong commutativity is easily shown equivalent to ordinary commutativity.)

Finally, let us translate our "unexpected result", Lemma 17.15, into a statement about co-operations. The conclusion of that Lemma is equivalent to the statement that for any k-ring R', homomorphisms from R' into $R^{\alpha} \amalg R^{\beta}$ which take values in the subring of symmetric elements correspond bijectively to homomorphisms from R' into $R^{\alpha} \amalg R^{\beta} \amalg R^{\gamma}$ which take values in the subring of strongly symmetric elements, the bijection being given by composition of the latter maps with the map $(i^{\alpha}, i^{\beta}, \varepsilon^{\gamma})$. Writing V and W for the functors represented by the given k-ring R and the arbitrary k-ring R' respectively, and q for the element of $V(k)$ corresponding to the augmentation ε, this says

LEMMA 17.18. *Let* V *and* W *be representable functors* k-**Ring**$^1 \to$ **Set**, *and* q *any element of* $V(k)$, *equivalently, any zeroary operation admitted by* V. *Then for every commutative binary operation* $a\colon V \times V \to W$, *there exists a* unique strongly commutative *ternary operation* $b\colon V \times V \times V \to W$ *satisfying the identity*

$$a(x, y) = b(x, y, q)$$

for all $x, y \in V(S)$, $S \in k$-**Ring**^{1}. □

We are now ready to bring in associativity –

§18. Application to AbSemigp-valued functors.

Suppose that $+$ is a commutative associative binary operation on a representable functor $V\colon k$-**Ring**$^1 \to$ **Set**. Then the ternary operation $(x + y) + z = x + (y + z)$ is clearly strongly commutative in the first and the last pair of variables, hence in all three variables. Also, for each $q \in V(k)$ the preceding Lemma, applied to the commutative binary operation $+$, gives us another strongly commutative ternary operation b_q. The next result shows that in its relation to $+$, this operation looks like "$x + y + z - q$". It is this "subtraction-like" quality that will allow us to deduce that $V(k) + V(k) + V(k)$ forms a group.

LEMMA 18.1. *Suppose* $V\colon k$-**Ring**$^1 \to$ **AbSemigp** *is a representable functor. Then for each* $q \in V(k)$, *the strongly commutative ternary operation* b_q *on* V *satisfying*

$$x + y = b_q(x, y, q) \tag{18.2}$$

given by Lemma 17.18 also satisfies the identity

$$x + y + z = b_q(x, y, z) + q. \tag{18.3}$$

Consequently,

$$V(k) + V(k) + V(k) = \textstyle\bigcap_{q \in V(k)} (V(k) + q). \tag{18.4}$$

PROOF. By the uniqueness condition of Lemma 17.18, any strongly commutative ternary operation is determined by the commutative binary operation one gets on substituting q for one variable. Hence to prove (18.3) it will suffice to verify that both sides are strongly commutative, and that they agree after this substitution. We have observed that the left hand side is strongly commutative; that the right hand side is so follows from the fact that b_q is. If we substitute q for z in (18.3), the equation we get holds as a consequence of (18.2), hence (18.3) holds for all x, y and z.

The assertion (18.4) is vacuous if $V(k)$ is empty (the intersection being understood as taken within $V(k)$). Assuming the contrary, "$\supseteq$" is immediate because the left hand side contains the term of the right-hand intersection corresponding to any $q \in V(k)+V(k)$. To get "$\subseteq$" we need to see that the left hand side is contained in every term of that intersection; but this follows from (18.3). □

We shall now deduce from (18.4) that $V(k)+V(k)+V(k)$ is a group if $V(k)$ is nonempty. The semigroup-theoretic argument we shall use does not require commutativity, so we will set it down in multiplicative notation.

LEMMA 18.5. *Let* S *be a semigroup. Then the following conditions are equivalent:* (i) $\bigcap_{s,s' \in S} sSs'$ *is nonempty,* (ii) $(\bigcap_{s \in S} sS) \cap (\bigcap_{s \in S} Ss)$ *is nonempty,* (iii) $\bigcap_{s \in S} sS$ *and* $\bigcap_{s \in S} Ss$ *are both nonempty.*

When these equivalent conditions hold, then (iv) $\bigcap sSs' = \bigcap sS = \bigcap Ss$, (v) *this set forms a subgroup* G *of the semigroup* S, (vi) *for all* $s \in S$, $sG = G = Gs$, *and* (vii) *there exists a unique retraction* $S \to G$ *as semigroups, namely,* $s \mapsto is$, *where* i *is the neutral element of* G.

PROOF. (i)⇒(ii)⇒(iii) is clear, and given (iii), we can obtain a member of $\bigcap sSs'$ by multiplying a member of $\bigcap sS$ by a member of $\bigcap Ss$, proving (i). Assuming the equivalent conditions (i)-(iii), let

$$G = \bigcap_{s,s' \in S} sSs'. \tag{18.6}$$

We shall now prove (v) and (vi) for this definition of G, and then deduce (iv) and (vii).

First, it is easy to see from the definition (18.6) that

$$GSG \cup GG \subseteq G. \tag{18.7}$$

In particular, G is closed under multiplication.

Now given any $g \in G$ and $s \in S$, we can by the definition of G write $g = (g)t(gs)$ for some $t \in S$. Hence

$$g = hs \tag{18.8}$$

with $h = gtg$, and the latter element belongs to G by (18.7). This says that we can "extract" an arbitrary element $s \in S$ from the right side of an arbitrary element $g \in G$ and still leave a member of G. By symmetry, we can do the same on the left. In particular, every member of G is right and left divisible within G by every other; and this condition on a nonempty semigroup G is easily shown to imply that G is a group. (Outline: take $x \in G$. Then we can write $x = ix$ with

$i \in G$. Since every $y \in G$ is a right multiple of x, we also have $y = iy$, so i is a left neutral element for G. Similarly, G has a right neutral element, hence it has a neutral element i. Since i is a left and right multiple of every element of G, every such element is invertible.) This gives (v).

Given any $s \in S$, $g \in G$, if we take $h \in G$ so that (18.8) holds, then we have $G = Gg = Ghs = Gs$, and by symmetry we also have the other half of (vi).

Note that for any $g \in G$, $\bigcap_{s \in S} sS \subseteq gS \subseteq GS = G$ (last equality by (vi)), and also $\bigcap_{s \in S} sS \supseteq \bigcap sG = G$, hence $\bigcap sS = G$. By symmetry $\bigcap Ss = G$ as well, establishing (iv).

Finally, writing i for the neutral element of the group G, note that for $s, t \in S$, we have $(is)(it) = ((is)i)t = (is)t = i(st)$ (the second equality holding because $is \in G$); thus, $s \mapsto is$ is a semigroup homomorphism $S \to G$. It is clear that this is a retraction. Conversely, if r is any retraction of S to G, then for all $s \in S$ we have $r(s) = i\,r(s) = r(i)\,r(s) = r(is) = is$, so the retraction r is unique, proving (vii). □

Now suppose V as in Lemma 18.1, and that $V(k)$ is nonempty. Then the left hand side of (18.4) is nonempty, hence so is the right hand side, i.e., $V(k)$ satisfies condition (iii) of Lemma 18.5. Hence by assertions (v) and (vii) of that Lemma, the common value of the two sides of (18.4) is a group, and if we write i for the neutral element of this group, then the whole group is equal to $i + V(k)$.

More generally, suppose we merely know for some $S \in k\text{-}\mathbf{Ring}^1$ that $V(S)$ is nonempty. Then the composite

$$S\text{-}\mathbf{Ring}^1 \xrightarrow{\text{forget}} k\text{-}\mathbf{Ring}^1 \xrightarrow{\ V\ } \mathbf{AbSemigp}$$

is a representable functor which takes the initial object S of $S\text{-}\mathbf{Ring}^1$ to a nonempty semigroup; so by the above observations, with S in place of k, we have the promised result:

Proposition 18.9. *Let* $V\colon k\text{-}\mathbf{Ring}^1 \to \mathbf{AbSemigp}$ *be a representable functor and* S *an object of* $k\text{-}\mathbf{Ring}^1_k$. *Then if the semigroup* $V(S)$ *is nonempty, it contains an idempotent* i, *and the group* $i + V(S)$ *is precisely* $V(S)+V(S)+V(S)$. □

For functors on varieties of *nonunital* k-rings, we deduce

Corollary 18.10. *Let* $V\colon k\text{-}\mathbf{Ring} \to \mathbf{AbSemigp}$ *be a representable functor, and* i *the unique zeroary operation admitted by* V *(which we know is idempotent-valued). Then for every* k*-ring* S, *the group* $i + V(S)$ *is precisely* $V(S)+V(S)+V(S)$. □

Remark: One may prefer a briefer, though more ad hoc proof of the $S = k$ case of Proposition 18.9 (from which we have seen that the general case follows) which, though still based on Lemma 18.1, avoids the digression through Lemma 18.5. Given any $w = x+y+z \in V(k)+V(k)+V(k)$, apply (18.3) with $w+w$ for q. Thus we have

(18.11) $$w \;=\; b_{w+w}(x,y,z) + (w+w).$$

Adding $b_{w+w}(x,y,z)$ to both sides, we see that the element $i = b_{w+w}(x,y,z)+w$ is idempotent. By Proposition 16.1 this is the unique idempotent in $V(k)$, and by (18.11), our arbitrary element w of $V(k)+V(k)+V(k)$ belongs to the group $i+V(k)$, as desired.

We shall see in §§39 and 42 that for most choices of k (including $\mathbf{Z}$), the analog of Proposition 18.9 fails for representable functors from k-rings to *not necessarily abelian* semigroups on the one hand, and from *commutative* k-algebras to abelian *or* general semigroups on the other; but that for certain k, including all fields and some, but not all, commutative local rings, all these statements do hold.

The next Lemma shows that semigroup-valued functors V on varieties of unital k-rings which give *empty* semigroups on some objects arise essentially as in Example 16.7. (We could have given this Lemma in §16, as a further result obtained from those of Chapter III, but it would have had to refer to conditions for a semigroup $V(S)$ to have an idempotent, rather than for it to be nonempty – Proposition 18.9, which we just proved, is needed to make these equivalent.)

LEMMA 18.12. *Let* $V\colon k\text{-}\mathbf{Ring}^1 \to \mathbf{AbSemigp}$ *be a representable functor, with representing k-ring* R. *Then:*

(i) *The k-ring structure of* R *can be factored* $k \to K \to R$, *where the map* $k \to K$ *is an epimorphism, and* $V(K) \neq \varnothing$.

(ii) *The epimorph* K *of* k *with this property is unique up to unique isomorphism of k-rings.*

This k-ring K *has the further properties:*

(iii) *For any k-ring* S, $V(S)$ *is nonempty if and only the k-ring structure of* S *extends (necessarily uniquely) to a K-ring structure.*

(iv) *Coproducts of nonempty families of K-rings coincide with the coproducts of the same families as k-rings; hence the co-***AbSemigp** *structure on* R *in* k-$\mathbf{Ring}^1$ *is equivalent to such a structure on* R *in* K-$\mathbf{Ring}^1$.

*Thus, by our previous result, this co-***AbSemigp** *structure on* R *in* K-$\mathbf{Ring}^1$ *may be strengthened (in a unique way) to a co-*$\mathbf{AbSemigp}^i$ *structure, which in turn corresponds to a co-***AbSemigp** *structure in* K-**Ring** *on the kernel* B *of the co-idempotent-element map* $\mathbf{i}\colon R \to K$.

PROOF. Let K be the k-ring such that $V(K)$ has a *universal* idempotent element i; this may be constructed by applying the left adjoint of V to the 1-element semigroup. (For an explicit description, define the *co-squaring* co-operation as the composite $\mathbf{s}\colon R \xrightarrow{\mathbf{a}} R^{\lambda} \amalg R^{\rho} \xrightarrow{(1_R,\,1_R)} R$, and let K be the factor ring of R by the ideal generated by $\{x - \mathbf{s}(x) \mid x \in R\}$. Then the quotient map into this factor-ring, $i\colon R \to K$, is an idempotent element of $V(K) = k\text{-}\mathbf{Ring}^1(R, K)$ with the required universal property.)

Combining the universal property of $i \in V(K)$ with the fact that for any k-ring S, $V(S)$ can have at most one idempotent, we see that in k-$\mathbf{Ring}^1$, K has at most one map into any object S, i.e., that in $\mathbf{Ring}^1$, $k \to K$ is an epimorphism. (In the language of Definition 16.8, K is a quasi-initial object of k-$\mathbf{Ring}^1$.) Since we have proved that a semigroup $V(S)$ is nonempty if and only if it has an idempotent (and this idempotent is then unique), the universal property of S yields (iii). Since $V(R)$ is certainly nonempty, the case of (iii) with $S = R$ gives the factorization of (i), completing the proof of (i). If K_1 and K_2 both have the property (i), then each admits a k-ring homomorphism to the other (because each admits maps into and out of R), and these will be isomorphisms and unique because both objects are epimorphs of k, proving (ii).

The first assertion of (iv) is a case of the general fact that in any category, pushouts over an epimorph K of an object k coincide with the corresponding pushouts over k. The second assertion follows.

Finally, since $V(K) \neq \varnothing$, the resulting co-**AbSemigp** structure extends, as in the proof of Proposition 16.1, to a co-$\mathbf{AbSemigp}^i$ structure, and by Lemma 10.6 and the results of §11, this is equivalent to a co-**AbSemigp** structure (or to a co-$\mathbf{AbSemigp}^i$ structure) on the kernel of $\mathbf{i}$, regarded as a nonunital K-ring. □

Thus, to define a representable functor from k-$\mathbf{Ring}^1$ to **AbSemigp** is equivalent to choosing (a) an epimorph K of k, and then (b) a representable functor from K-$\mathbf{Ring}^1$ to $\mathbf{AbSemigp}^i$, equivalently, a representable functor from K-**Ring** to **AbSemigp**.

General references on epimorphisms in $\mathbf{Ring}^1$ are [**175**] and [**181**]. Incidentally, if k is a field (or more generally, a simple von Neumann regular ring), it has no proper epimorphs other than the trivial ring $\{0\}$; hence for such k, every representable functor V: k-$\mathbf{Ring}^1 \to \mathbf{AbSemigp}$ except the one represented by the k-ring $\{0\}$ satisfies $V(k) \neq \varnothing$.

§19. Some observations and questions on rings of symmetric elements.

This section is only tangentially relevant to representable functors. Readers who find they are not interested in the points considered here may prefer to skip it.

We have noted that if R is a unital k-ring with an augmentation ε, then the isomorphic k-rings $\mathrm{Sym}(R^\alpha \amalg R^\beta) \cong \mathrm{SSym}(R^\alpha \amalg R^\beta \amalg R^\gamma)$ (where R^α, R^β, R^γ are copies of R) both have the k-bimodule structure

$$(19.1) \qquad k \oplus B \oplus (B \otimes_k B) \oplus \ldots,$$

where $B = \ker(\varepsilon)$. The bimodule isomorphism between (19.1) and $\mathrm{Sym}(R^\alpha \amalg R^\beta)$ is given by

$$(19.2) \qquad x_1 \otimes x_2 \otimes x_3 \otimes \ldots \otimes x_h \mapsto (x_1^\alpha x_2^\beta x_3^\alpha \ldots x_h^\iota) + (x_1^\beta x_2^\alpha x_3^\beta \ldots x_h^{\iota'})$$

where ι is α or β depending on whether h is odd or even, and ι' denotes the reverse choice.

We can use (19.2) to figure out the multiplication on (19.1) corresponding to that of $\mathrm{Sym}(R^\alpha \amalg R^\beta)$. We find that on multiplying a member of the h-fold tensor product of B with a member of the i-fold tensor product, we get not only the element of the $(h+i)$-fold tensor product which is their product in the tensor k-ring $k\langle B\rangle$, but also a term in the $(h+i-1)$-fold tensor product, in which the last term of the former factor is contracted with the first term of the latter factor using the (nonunital) ring structure of B:

$$\begin{aligned}
& (x_1 \otimes \ldots \otimes x_h)(x_{h+1} \otimes \ldots \otimes x_{h+i}) \\
(19.3) \quad & = (x_1 \otimes \ldots \otimes x_h \otimes x_{h+1} \otimes \ldots \otimes x_{h+i}) + (x_1 \otimes \ldots \otimes (x_h x_{h+1}) \otimes \ldots \otimes x_{h+i}) \\
& = x_1 \otimes \ldots \otimes x_{h-1} \otimes (x_h \otimes x_{h+1} + x_h x_{h+1}) \otimes x_{h+2} \otimes \ldots x_{h+i}.
\end{aligned}$$

However, this "deformation" of $k<B>$ is *isomorphic* to $k<B>$! For the universal property of $k<B>$ gives a k-ring homomorphism from $k<B>$ to the ring with multiplication (19.3), which is the identity on B; and this is seen to preserve highest-degree components, hence is bijective.

Let us compare the kinds of symmetry introduced here with those studied in [**29**]. That paper considered a coproduct $\coprod_A R^\alpha$ of copies of one ring R (in the category of algebras over a field; but this condition can be weakened), and determined the structure of the fixed ring of an arbitrary group G acting on the index set A with finite orbits none of which were singletons. The fixed ring was shown always to be a tensor algebra; the isomorphism $\mathrm{Sym}(R^\alpha \amalg R^\beta) \cong k<B>$ noted above corresponds to the result of that paper in the simplest case, $A = \{\alpha, \beta\}$. (We cited this result in the proof of Lemma 15.2.)

Such subrings were not sufficient for our present purposes, and we have introduced here further subrings obtained from these by taking intersections and subrings-generated, i.e., constructions in the *lattice* of subrings of a coproduct generated by fixed subrings of groups acting on the coproducts of various subfamilies of the given family. For instance, $\mathrm{SSym}(R^\alpha \amalg R^\beta \amalg R^\gamma)$ might be written lattice-theoretically, in an ad hoc but suggestive notation, as

$$(\mathrm{Sym}(R^\alpha,\, R^\beta) \vee R^\gamma) \wedge (R^\alpha \vee \mathrm{Sym}(R^\beta,\, R^\gamma)).$$

For a given family of k-rings, any member of this lattice will correspond to a condition on operations among representable functors, and results on the subrings in question should be applicable to the study of such operations.

In particular, one can define, for an operation among any number of representable functors, the concept of *strong commutativity* in some family of variables that are copies of a common functor. One can then prove a generalization of Lemma 17.18, obtaining for any representable functors V and W on k-$\mathbf{Ring}^1$, any element $q \in V(k)$, and any $n>0$, a bijection between the n-ary operations $V^n \to W$ that are strongly commutative in all variables, and the commutative binary operations $V^2 \to W$.

The kinds of symmetry conditions become more numerous as one goes to higher arities. For instance, suppose we regard the system of strongly commutative ternary operations $b_q(x, y, z)$ of Lemma 19.2 as defining a quaternary operation $b(q, x, y, z)$. (This involves a little juggling, because although the definition of $b_q(x, y, z)$ allows x, y and z to come from $V(S)$ for an arbitrary k-ring S, q was assumed to lie in $V(k)$. However, given $q \in V(S)$ one can perform the same construction in S-$\mathbf{Ring}^1$; and one can verify functoriality in q of the resulting operation.) So far as the authors can see, this quaternary operation need not be "strongly commutative in x and y" in the sense of being expressible as $c(q, d(x, y), z)$ with d commutative; but the fact that for each q it gives a ternary operation in x, y and z strongly commutative in x and y has the consequence that it *can* be expressed as $c(d(q,x,y),z)$ where d is commutative in its last two arguments. (Since $b(q,x,y,z)$ is symmetric in x, y and z, we also get such expressions with the variables x, y and z permuted arbitrarily.)

On the other hand, let us record one case where two subrings (and hence two sorts of commutativity) that a priori might be distinct are equal.

EXERCISE 19.4. *Suppose* R^α *and* R^β *are two copies of one k-ring,* R^γ, R^δ *are two copies of a possibly different k-ring, and all these k-rings admit augmentations. Verify that within* $R^\alpha \amalg R^\beta \amalg R^\gamma \amalg R^\delta$, *we have*

$$(\mathrm{Sym}(R^\alpha, R^\beta) \vee R^\gamma \vee R^\delta) \wedge (R^\alpha \vee R^\beta \vee \mathrm{Sym}(R^\gamma, R^\delta))$$
$$= \mathrm{Sym}(R^\alpha, R^\beta) \vee \mathrm{Sym}(R^\gamma, R^\delta).$$

(*Here only the inclusion* $\supseteq$ *follows directly from the definition.*)

These considerations lead to

PROBLEM 19.5. (i) *Suppose* R *is a unital k-ring admitting an augmentation, and* $C = R^{\alpha_1} \amalg \ldots \amalg R^{\alpha_n}$ *a coproduct of finitely many copies of* R. *For each subset* $\Sigma \subseteq \{1, \ldots, n\}$, *and each subgroup* G *of the permutation group on* Σ, *form the fixed ring of the action of* G *on the subring of* C *generated by the family of copies of* R *corresponding to* Σ, *and let* L *denote the lattice of subrings of* C *generated by all these fixed rings. To what extent can one describe* L? *(In this investigation one can again simplify things by replacing "unital k-rings admitting augmentation" by "nonunital k-rings".)*

(ii) *Study the analogous question for other classes of algebras, e.g., unital k-rings not necessarily admitting augmentation, commutative algebras over a field, etc..*

The condition that a k-ring R admit an augmentation says that the map $\eta\colon k \to R$ which defines the k-ring structure (a) has a left inverse $\varepsilon\colon R \to k$ as a ring homomorphism. Some obvious weakenings of this assumption which one might consider are that this map (b) has a left inverse as a k-bimodule homomorphism, or (c) has a left inverse as a left k-module homomorphism, or (d) does not factor through any proper epimorphism $k \to K$, or (d′) is one-to-one; one can also, of course, assume (d″) no condition at all. It is not hard to prove that (a) $\Rightarrow$ (b) $\Rightarrow$ (c) $\Rightarrow$ (d) $\Rightarrow$ (d′) $\Rightarrow$ (d″).

For any ring k (or more generally, any object k of any variety of algebras), one can show that the isomorphism classes of epimorphs of k form a complete lattice (which, if one takes one representative from each such isomorphism class, has small underlying set), with k as least element, and that for a given object R, the class of epimorphs of k through which the structure map $k \to R$ factors has a greatest element K_R. In our ring-theoretic context, we can then regard R as a K_R-ring, and by looking at our rings over K_R rather than over k (or if several R's are involved, by passing to K-rings, where K is the join of the various K_R's in our complete lattice), the study of coproducts and their subalgebras in situation (d″) can be essentially reduced to that of situation (d), which is why we used related symbols for (d), (d′), (d″). Lemma 18.12(i) shows that for k-rings admitting an

AbSemigp structure one has (d)$\Rightarrow$(a), though this is certainly not true for arbitrary k-rings.

PROBLEM 19.6. *Which, if any, of* (b)-(d) *above* (*or similar conditions*) *can replace the augmentability hypothesis in Corollary 17.12* (*the result that the map*

$$\mathrm{Sym}(R^{\alpha} \amalg R^{\beta}) \amalg R^{\gamma} \to R^{\alpha} \amalg R^{\beta} \amalg R^{\gamma} \tag{19.7}$$

is one-to-one)?

Let us note a counterexample to the analog of that Corollary, with the augmentability hypothesis entirely deleted, in the variety $\mathbf{Semigp}^{e,s}$ of semigroups given with both a neutral element e and a distinguished element s. Let R^{α} and R^{β} be two copies of the object R of $\mathbf{Semigp}^{e,s}$ presented by a single right inverse x to s, and R^{γ} the object presented by a single left inverse y to s. One finds that $\mathrm{Sym}(R^{\alpha} \amalg R^{\beta})$ is simply the semigroup with neutral element generated by s, i.e., the initial object of the category, so $\mathrm{Sym}(R^{\alpha} \amalg R^{\beta}) \amalg R^{\gamma} \cong R^{\gamma}$. But within $R^{\alpha} \amalg R^{\beta} \amalg R^{\gamma}$, general properties of inverses tell us that x^{α}, x^{β} and y must fall together, so there y becomes a *two-sided* inverse to s, and so (19.7) is not one-to-one.

Can we get an analogous ring-theoretic example, by passing to semigroup algebras of the above semigroups? Let k_0 be a fixed commutative ring, let $k = k_0[s]$, i.e., the semigroup algebra over k_0 on the initial object of $\mathbf{Semigp}^{e,s}$, and let us now write R, R^{α}, R^{β}, R^{γ} for the semigroup algebras over k_0 of the semigroups that were denoted by those same symbols in the preceding paragraph, regarded as $k_0[s]$-rings. (E.g., $R = k_0\langle s, x \mid sx = 1\rangle$, regarded as a $k_0[s]$-ring.) It is not immediately obvious whether for these rings, (19.7) also fails to be one-to-one, because there are in general more fixed elements in a semigroup algebra over a semigroup on which a group acts than those arising from fixed elements in the semigroup alone. In fact, if 2 is invertible in k_0, we see that $\frac{1}{2}(x^{\alpha} + x^{\beta})$ is a right inverse to s which lies in $\mathrm{Sym}(R^{\alpha} \amalg R^{\beta})$, hence the promotion of $y \in R^{\gamma}$ to a two-sided inverse of x occurs already in $\mathrm{Sym}(R^{\alpha} \amalg R^{\beta}) \amalg R^{\gamma}$, and one can show that (19.7) is indeed one-to-one. However, if k_0 has characteristic 2, it can be shown that (19.7) is not one-to-one (giving a partial negative answer to Problem 19.6), though we shall not go into the details here.

Note that in this example, the ring R^{γ} satisfies condition (c) above, while $R^{\alpha} \cong R^{\beta}$ satisfies the right-left dual thereof. One may still hope that a positive result holds when all rings satisfy this condition on the same side. It is also plausible that such a result may be true whenever 2 is invertible in k, since we are taking symmetric elements with respect to a group of order 2. Finally, we remark that we know no analog of the above counterexamples in any pointed variety.

In a variety of algebras, such as $\mathbf{Semigp}^{e,s}$ or $(\mathbf{Z}/2\mathbf{Z})[s]$-$\mathbf{Ring}^1$, where the analog of Corollary 17.12 fails, we may note that there are two candidates for the definition of a co-operation $R \to R^{\alpha} \amalg R^{\beta} \amalg R^{\gamma}$ being "strongly cocommutative in α and β", namely that it factor through $\mathrm{Sym}(R^{\alpha} \amalg R^{\beta}) \amalg R^{\gamma}$, and the weaker condition that it land in the image thereof in $R^{\alpha} \amalg R^{\beta} \amalg R^{\gamma}$, namely $\mathrm{SSym}_{\{\alpha,\beta\}}(R^{\alpha} \amalg R^{\beta} \amalg R^{\gamma})$. (There is even a third possible definition: As these examples show, coproducts in a general variety need not be faithful, hence we can distinguish between the image of $\mathrm{Sym}(R^\alpha \amalg R^\beta)$ in $R^\alpha \amalg R^\beta \amalg R^\gamma$, and the subalgebra of $R^\alpha \amalg R^\beta \amalg R^\gamma$ consisting of elements of the image of $R^\alpha \amalg R^\beta$ invariant under interchange of the indices α and β. The subring of $R^\alpha \amalg R^\beta \amalg R^\gamma$ generated by the latter subring and R^γ gives a variant of the definition of $\mathrm{SSym}_{\{\alpha,\beta\}}(R^\alpha \amalg R^\beta \amalg R^\gamma)$, and hence of strong cocommutativity. However, the map $R^\alpha \amalg R^\beta \to R^\alpha \amalg R^\beta \amalg R^\gamma$ is easily shown to be one-to-one whenever R^γ admits a homomorphism to $R^\alpha \amalg R^\beta$, in particular if $R^\alpha \cong R^\beta \cong R^\gamma$, so this distinction is not relevant to operations on three copies of a common functor.) We do not know which of these definitions is likely to be most useful in general.

§20. Representable functors from $\mathbf{Semigp}^e$ to $\mathbf{Semigp}^e$.

In the remaining sections of this chapter, we put ring theory aside, and examine representable functors among varieties of semigroups, and some related varieties. In the present section we shall characterize all representable functors from the category $\mathbf{Semigp}^e$ to itself. In the next we will obtain, using this result and some easier arguments, descriptions of all representable functors among this and a number of closely related varieties: **Group**, **Semigp**, $\mathbf{AbSemigp}^e$, etc.. In the final section, we bring in some less familiar sorts of objects.

Let us recall the description of a coproduct

(20.1) $$\coprod_{\alpha\in I} R^\alpha$$

in $\mathbf{Semigp}^e$. This has the same form as the coproduct construction in **Group**, classically called the "free product" of groups; if we assume for notational convenience that the R^α are disjoint, then each element of (20.1) can be written uniquely as a product

(20.2) $$r_1 r_2 \dots r_h, \quad \text{with } h \geq 0, \text{ each } r_i \text{ in some } R^{\alpha_i} - \{e\}, \text{ and } \alpha_i \neq \alpha_{i+1} \text{ for } 1 \leq i < h$$

([**24**, §§3.6 and 3.10]). An arbitrary product of elements of the R^α's is reduced to one of the above form by combining any successive factors that belong to the same R^α, dropping factors e if these occur, and repeating these steps as long as possible. The neutral element e of the coproduct is understood to be the case $h = 0$ of (20.2).

Now let $(R, \mathbf{m}, \mathbf{e})$ be a co-$\mathbf{Semigp}^e$ object in $\mathbf{Semigp}^e$, with $\mathbf{m}$ the comultiplication and $\mathbf{e}$ the co-neutral-element. Our analysis of this situation will in many ways resemble our earlier analysis of co-$\mathbf{AbSemigp}^e$ objects in $\mathbf{Ring}_k$. The zeroary co-operation $\mathbf{e}$ is uniquely determined because the domain category $\mathbf{Semigp}^e$ is pointed; it is the unique homomorphism of R into the trivial semigroup $\{e\}$, and we shall suppress it, writing our coalgebra as

$$(R, \mathbf{m}).$$

We will write the codomain of the morphism $\mathbf{m}$ as $R^\lambda \amalg R^\rho$, where $R^\lambda = \{x^\lambda \mid x \in R\}$ and $R^\rho = \{x^\rho \mid x \in R\}$ are copies of R; in expressing the coassociative law we will also introduce a "middle" copy R^{μ}. We again define an index-string to mean a sequence of terms from $\{\lambda, \mu, \rho\}$ with no two successive terms equal, but this time we include the empty string. As before, for $h > 0$ we define $[h]$ and $[h]'$ to be the strings of alternating λ's and ρ's of length h that begin with λ and with ρ respectively.

For every index-string $\sigma = (\alpha_1, \ldots, \alpha_h)$, we define

$$R^{\sigma} = (R^{\alpha_1} - \{e\}) \ldots (R^{\alpha_h} - \{e\}).$$

Each of the semigroups $R^{\lambda} \amalg R^{\rho}$ and $R^{\lambda} \amalg R^{\mu} \amalg R^{\rho}$ is now the *disjoint union* of its subsets R^{σ}. We define the *height* of $s \in R^{\lambda} \amalg R^{\rho}$ as the length of the unique σ such that $s \in R^{\sigma}$, and the *degree* of $x \in R$ as the height of $\mathbf{m}(x)$.

When we write out the coneutral laws for $\mathbf{m}$, we begin to lose the parallelism with the ring-theoretic case. We see that these take the form

(20.3) If $\mathbf{m}(x) = \ldots y_i^{\lambda} z_i^{\rho} y_{i+1}^{\lambda} z_{i+1}^{\rho} \ldots$, then $x = \ldots y_i y_{i+1} \ldots = \ldots z_i z_{i+1} \ldots$.

(Cf. (12.9). As usual we leave it open whether $\mathbf{m}(x)$ begins and/or ends with an element of R^{λ} or of R^{ρ}.)

Note that (20.3) implies

(20.4) If $x \neq e$, then $\deg(x) \geq 2$.

On the two sorts of elements of degree exactly 2, (20.3) precisely determines the action of $\mathbf{m}$:

$$(20.5)\qquad \begin{cases} \text{If } \mathbf{m}(x) \in R^{[2]}, \text{ then } \mathbf{m}(x) = x^{\lambda} x^{\rho}. \\ \text{If } \mathbf{m}(x) \in R^{[2]'}, \text{ then } \mathbf{m}(x) = x^{\rho} x^{\lambda}. \end{cases}$$

Let us also record what (20.3) tells us about the degree 3 case:

$$(20.6)\qquad \begin{cases} \text{If } \mathbf{m}(x) \in R^{[3]}, \text{ then } \mathbf{m}(x) = y_1^{\lambda} x^{\rho} y_2^{\lambda} \text{ where } y_1 y_2 = x. \\ \text{If } \mathbf{m}(x) \in R^{[3]'}, \text{ then } \mathbf{m}(x) = z_1^{\rho} x^{\lambda} z_2^{\rho} \text{ where } z_1 z_2 = x. \end{cases}$$

We now take an arbitrary element $x \in R - \{e\}$, and consider the coassociative law:

$$(20.7)\qquad \mathbf{m}_{\lambda}\, \mathbf{m}(x) = \mathbf{m}_{\rho}\, \mathbf{m}(x)$$

in $R^{\lambda} \amalg R^{\mu} \amalg R^{\rho}$ (notation analogous to (12.20)). Say $\mathbf{m}(x)$ belongs to R^{σ}, and the common value of both sides of (20.7) belongs to R^{τ}. Note that each "λ" in σ yields a single λ in τ on evaluating the right-hand side of (20.7), since a factor y_i^{λ} of $\mathbf{m}(x)$ is unaffected by $\mathbf{m}_{\rho}$; but looking at the left hand side of (20.7) it gives *at least* one λ in τ, in view of (20.4). Since the two sides of (20.7) are equal, all of the "at least one"'s in this statement must be exactly one. For this to happen, the elements y_i in the expansion (20.3) must all have degree ≤ 3. By a symmetric argument (comparing terms "ρ" in σ and τ) we get the same conclusion for the elements z_i. Note also that if τ begins with μ, then the left hand side of (20.7) tells us σ must begin with a λ, but the right-hand side says it must begin with a ρ, a contradiction. Hence τ can only begin with a λ or a ρ. In the former case, σ must begin with a λ which expands to $\lambda\mu$ on the left-hand side of (20.7) (so as not to yield more than one λ); in the latter case it must begin with a ρ which expands to $\rho\mu$ on the right-hand side. In either case, we conclude that the first factor in the expansion of $\mathbf{m}(x)$ must have degree 2. The same arguments apply to the last factor. In summary:

(20.8) All terms y_i and z_i in (20.3) have degree ≤ 3; hence by (20.3), every element of R is a product of elements of degree ≤ 3. Moreover, the terms giving the *first* and *last* factors of $\mathbf{m}(x)$ have degree 2.

But the observation about first and last factors, applied to the rightmost equations in each line of (20.6), gives

(20.9) Every element of R of degree 3 is a product of two elements of degree 2.

(20.8) and (20.9) together allow one to express every element of R as a product of elements of degree 2, showing that R is generated by these elements. We can prove still more:

LEMMA 20.10. *Let* $(R, \mathbf{m})$ *be a co-*$\mathbf{Semigp}^e$ *object in* $\mathbf{Semigp}^e$. *Then every element* $x \in R$ *has an expression as a product*

$$x_1 \ldots x_h \quad (h \geq 0),$$

where all x_i *are of degree* 2, *and this expression is* unique *subject only to the condition that there be no two successive terms* x_i, x_{i+1} *such that one of* $\mathbf{m}(x_i)$, $\mathbf{m}(x_{i+1})$ *belongs to* $R^{[2]}$ *and the other to* $R^{[2]'}$, *and such that* $x_i x_{i+1} = e$.

PROOF. Since R is generated by elements of degree 2, and since any expression involving two successive terms whose product is e can be simplified to a shorter expression, we can clearly express every element in the indicated form subject to the conditions noted. To show that this form is unique, it suffices to prove that given an element and such an expression for it,

$$x = x_1 \ldots x_h,$$

we can recover the factors x_i from x. We claim in fact that if we apply $\mathbf{mm}$ (that is, the map $\mathbf{m}_\lambda \mathbf{m} = \mathbf{m}_\rho \mathbf{m}$) to x and write the result as a reduced product of elements of R^λ, R^μ and R^ρ, i.e., as in (20.2), then the sequence of factors belonging to R^μ will be precisely $x_1^\mu, \ldots, x_h^\mu$, recovering the x_i, as required.

Indeed, it follows from (20.5) that $\mathbf{mm}$ carries each x such that $\mathbf{m}(x) \in R^{[2]}$ to $x^\lambda x^\mu x^\rho$, and each x such that $\mathbf{m}(x) \in R^{[2]'}$ to $x^\rho x^\mu x^\lambda$. Thus, when we expand $\mathbf{mm}(x_1 \ldots x_h)$, the μ-terms comprise, *initially*, the sequence claimed. They will continue to do so after we reduce this product to the form (20.2), unless, in the process of reduction, the λ- and/or ρ-terms separating some pair of successive μ-terms cancel, so that these μ-terms end up in a single μ-term of $\mathbf{mm}(x)$. Now if $\mathbf{m}(x_i)$ and $\mathbf{m}(x_{i+1})$ both belong to $R^{[2]}$ or both belong to $R^{[2]'}$, then between x_i^μ and x_{i+1}^μ we will have exactly one λ-term and one ρ-term, and these cannot cancel. In the case where one belongs to $R^{[2]}$ and the other to $R^{[2]'}$, we get adjacent factors in the same set R^ρ or R^λ; these will combine into one term, but they will cancel only if $x_i x_{i+1} = e$ in R, which is the case excluded by our hypothesis. □

In the above Lemma, we could clearly have asserted that every element can be reduced to a product of the indicated form in which no two successive terms *whatever* have product e. However, we have proved uniqueness subject to a weaker condition than this, so we have gotten a stronger uniqueness result. Indeed, this result clearly implies (as the weaker uniqueness statement would not):

(20.11) All semigroup relations satisfied by elements of degree 2 in R are consequences of relations of the form $x_1 x_2 = e$, such that one of $\mathbf{m}(x_1)$, $\mathbf{m}(x_2)$ belongs to $R^{[2]}$, and the other belongs to $R^{[2]'}$.

Thus, the next step in studying our cosemigroup should evidently be to examine the properties of the relation $x_1 x_2 = e$ on elements of degree 2 in R. So let us make

DEFINITION 20.12. *If* $(R, \mathbf{m})$ *is a co-***Semigp**^{e} *object in* **Semigp**^{e}, *then* $P(R, \mathbf{m})$ *will denote the 4-tuple* (u, X^+, X^-, E), *where*

$$u = e, \text{ the neutral element of } R,$$

$$X^+ = \{x \in R \mid \mathbf{m}(x) = x^\lambda x^\rho\} = \{x \in R \mid \mathbf{m}(x) \in R^{[2]}\} \cup \{u\},$$

$$X^- = \{x \in R \mid \mathbf{m}(x) = x^\rho x^\lambda\} = \{x \in R \mid \mathbf{m}(x) \in R^{[2]'}\} \cup \{u\},$$

and

$$E = \{(x_1, x_2) \mid \deg(x_1), \deg(x_2) \le 2,\ x_1 x_2 = e\} \subseteq (X^+ \times X^-) \cup (X^- \times X^+).$$

Note a property of this binary relation E: if both (x_1, x_2) and (x_2, x_3) belong to it, then since x_2 has x_1 as a left inverse and has x_3 as a right inverse, x_1 must equal x_3.

We now formalize the type of combinatorial object we have obtained.

DEFINITION 20.13. *An E-*system *will mean a 4-tuple* (u, X^+, X^-, E) *where* u *is an element,* X^+ *and* X^- *are sets such that*

$$X^+ \cap X^- = \{u\},$$

and

$$E \subseteq (X^+ \times X^-) \cup (X^- \times X^+)$$

is a relation such that

$$u \, E \, u \tag{20.14}$$

and

$$x_1 E x_2,\ x_2 E x_3 \ \Rightarrow\ x_1 = x_3. \tag{20.15}$$

A morphism *of E-systems* $(u, X^+, X^-, E) \to (u', X'^+, X'^-, E')$ *will mean a map* $X^+ \cup X^- \to X'^+ \cup X'^-$ *carrying* u *to* u', X^+ *into* X'^+, X^- *into* X'^-, *and the relation* E *into the relation* E'.

The next result shows that these systems correspond precisely to the cosemigroups we have been considering:

THEOREM 20.16. *The construction* P *of Definition 20.12 yields an equivalence between the category of co-***Semigp**^{e} *objects in* **Semigp**^{e} *and the category of E-systems. Its inverse (up to isomorphism of functors) is the functor* Q *taking an E-system* (u, X^+, X^-, E) *to the cosemigroup object whose underlying semigroup has the presentation*

$$(20.17) \qquad < X^+ \cup X^- - \{u\} \mid x_1 x_2 = e \text{ whenever } x_1 E x_2 >,$$

(equivalently, the same presentation with u *included among the generators, and with the additional relation* $u = e$*), and whose co-***Semigp**^{e} *structure is defined by*

$$(20.18) \qquad \mathbf{m}(x) = \begin{cases} x^\lambda x^\rho & \text{if } x \in X^+, \\ x^\rho x^\lambda & \text{if } x \in X^-. \end{cases}$$

SKETCH OF PROOF. Given an E-system (u, X^+, X^-, E), let R be the semigroup with presentation (20.17). To see that (20.18) defines a homomorphism $R \to R^\lambda \amalg R^\rho$, one checks compatibility with the defining relations of (20.17), a quick calculation. One verifies coassociativity and the coneutral laws by checking these on the generators, again immediate. It is easy to see that the constructions P and Q define functors. (For the purpose of seeing the functoriality of Q, the presentation which includes the generator u is most convenient, since a morphism of E-systems may take elements $x \neq u$ to u.)

Lemma 20.10 and its consequence (20.11) say that QP is isomorphic to the identity functor on the category of co-**Semigp**^{e} objects of **Semigp**^{e}.

In verifying that PQ is isomorphic to the identity on the category of E-systems, the first step is to show that if we start with an arbitrary E-system, then the semigroup R with presentation (20.17) will have a normal form given by all products $x_1 \dots x_h$ $(x_i \in X^+ \cup X^- - \{u\})$ in which no pair of successive factors $x_i x_{i+1}$ satisfies $x_i E x_{i+1}$; i.e., that the formally distinct products of this form give distinct elements of the semigroup. This is easily shown using the Diamond Lemma [**19**]. In particular, the images of the elements of $X^+ \cup X^-$ are distinct, and the only relations $x_1 x_2 = e$ holding among these elements are those determined by E; hence for any E-system T, the natural map from T to $PQ(T)$ is an embedding. One also verifies easily that the only elements of $Q(T)$ having degree 2 under the co-**Semigp**^{e} structure are the elements from the generating set $X^+ \cup X^- - \{u\}$, so this embedding of E-systems is an isomorphism. □

In terms of representable functors, we get:

COROLLARY 20.19. *Every representable functor* V *from* **Semigp**^{e} *to* **Semigp**^{e} *is a subfunctor of a direct product of copies of the* identity *functor and of the* opposite-semigroup *functor. Precisely, the functor corresponding to the E-system* (u, X^+, X^-, E) *takes a semigroup* S *to the subsemigroup of*

$$S^{(X^+-\{u\})} \times (S^{\mathrm{op}})^{(X^--\{u\})}$$

consisting of those elements s *such that for each* $(x, y) \in E-\{(u, u)\}$, *the* x*th coordinate of* s *is a left inverse to the* y*th coordinate.*

All morphisms between these functors are induced (contravariantly) by morphisms of E-systems. □

For the purpose of describing these morphisms of representable functors, it is actually most convenient to treat the functor V: **Semigp**^{e} $\to$ **Semigp**^{e} represented by the E-system (u, X^+, X^-, E) as taking a semigroup S to a subsemigroup of

$$S^{(X^+-\{u\})} \times \{e\} \times (S^{\mathrm{op}})^{(X^--\{u\})};$$

i.e., to introduce an extra slot, indexed by the element u of the E-system, such that the coordinate of $V(S)$ in that slot is always required to be the neutral element $e \in S$. We can then say that if $\mathbf{f}$ is a morphism of E-systems, and f: $V' \to V$ the corresponding morphism of representable functors, then for a semigroup S and an element $\xi \in V'(S)$, the image $f(S)(\xi)$ has for xth coordinate the $\mathbf{f}(x)$th coordinate of ξ, whether $\mathbf{f}(x)$ happens to be u or a member of $X^+ \cup X^- - \{u\}$.

Let us give some simple examples of E-systems and the corresponding representable functors. We shall display an E-system by showing the elements of $X^+-\{u\}$ and $X^--\{u\}$ respectively as points in two boxes, 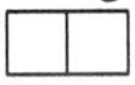, and indicating a condition $x_1 E x_2$ by an arrow from the point x_1 to the point x_2. (The element u will not be shown; it may be thought of as embedded in the dividing line between the boxes.)

20.20. *Examples of E-systems and the corresponding representable functors.*

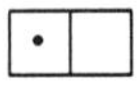 The representing cosemigroup R is given by the free semigroup on one generator x, with comultiplication such that $\mathbf{m}(x) = x^\lambda x^\rho$, hence the functor is the *identity* functor **Semigp**^{e} $\to$ **Semigp**^{e}.

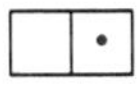 This similarly gives the *opposite semigroup* functor.

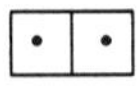 (the relation $E-\{(u, u)\}$ still being empty). This gives the direct product of the above two functors, i.e., the functor associating to every semigroup S the semigroup

$$\{(\alpha, \beta) \mid \alpha, \beta \in S\},$$

with multiplication

(20.21) $$(\alpha_1, \beta_1)\,(\alpha_2, \beta_2) = (\alpha_1 \alpha_2, \beta_2 \beta_1).$$

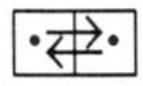 This corresponds to the subfunctor of the functor described in the preceding example determined by adding to the description of the underlying set the conditions

$$\alpha\beta = e = \beta\alpha.$$

Since under these conditions α uniquely determines β, the second coordinate provides no new information, and we can describe this functor as associating to S

its *group of invertible elements* α (regarded as a semigroup).

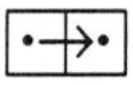 As above, except that only the condition $\alpha\beta = e$, and not $\beta\alpha = e$ is imposed. Right inverses are not generally unique, so we describe this functor as associating to S the semigroup of elements $\alpha \in S$ given with a *specified* right inverse β. The multiplication is again as in (20.21).

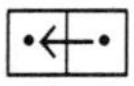 This associates to S the semigroup of elements α given with a specified *left inverse* β, multiplied as in (20.21). Set-theoretically, this construction is isomorphic to the preceding, via $(\alpha, \beta) \longleftrightarrow (\beta, \alpha)$, but the semigroup structures are opposite to one another. (We have indicated this in our paraphrases by naming, after the words "semigroup of", the elements which are multiplied as in S, those with the opposite multiplication being referred to as specified inverses of these elements.)

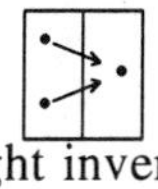 "The semigroup of *pairs* of elements of S with a specified *common* right inverse."

And so on. We note that for a general diagram such as

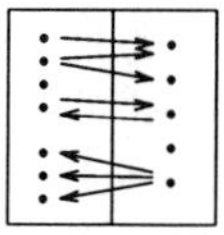

the associated functor is the direct product of the functors associated with the graph-theoretic "connected components" of the diagram. Each of these components, *except* those of the form 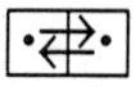 must have, by (20.15), the property that arrows, if any, all go in the same direction, i.e., from left to right or from right to left. Subject to this restriction, the arrows are independent.

It is interesting to note that, although for every *positive* cardinality r, the construction that associates to a semigroup S the semigroup of its right invertible elements given with a specified r-tuple of right inverses is a representable functor, this is false for $r = 0$:

EXERCISE 20.22. *Let* F: $\mathbf{Semigp}^e \to \mathbf{Semigp}^e$ *be the functor associating to a semigroup* S *its subsemigroup of right invertible elements* (*a subfunctor of the identity functor*).

(i) *Show that* F *is not representable.* (*Hint: any representable functor respects intersections of subalgebras.*)

(ii) *Show, however, that the composite functor* FF *is representable, and concisely describe this functor.*

(iii) *Show that* F *is a direct limit of representable functors.* (*Hint: write the empty set as the intersection of a downward directed family of nonempty sets.*)

Given a pair of functors which *are* representable, it is natural to ask how to describe the E-system associated with their composite in terms of the E-systems associated with the given functors. This is answered by

LEMMA 20.23. *Let* V, W: **Semigp**$^{e} \to$ **Semigp**^{e} *be representable functors.*

The task of describing the E-system corresponding to the functor WV *is reduced to the case where the E-systems corresponding to* V *and to* W *each have connected diagram, by the result*

(i) *If* $V = \prod_I V_i$ *and* $W = \prod_J W_j$, *then* $WV = \prod_{J,I} W_j V_i$.

The case where at least one of the two given functors has diagram with just one point is settled by the two results:

(ii) *Composition on either side with the identity functor (diagram* 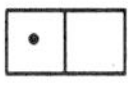*) leaves any functor unchanged.*

(iii) *Composition on the left with the opposite-semigroup functor (diagram* 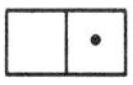*) corresponds to forming the "mirror image" of the diagram of a functor, while composition on the right with this construction corresponds to such a mirror reflection followed by reversal of the directions of all the arrows.*

The remaining case is covered by

(iv) *If* V *and* W *both have connected diagrams with more than one point, then* WV *is isomorphic to the semigroup-of-invertible-elements functor, i.e., the functor with diagram* 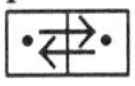.

SKETCH OF PROOF. Statement (i) follows from the fact that representable functors respect products (Exercise 5.13(iii) above, or [**122**, p.117, Theorem 2], or [**24**, Corollary 7.7.2 or Proposition 7.7.3]). Statements (ii) and (iii) are straightforward to verify.

Statement (iv) is easily deduced from three observations: First, if V is a representable functor whose diagram is connected and has more than one point, and we describe $V(S)$ as a semigroup of $X^{+} \cup X^{-} - \{u\}$-tuples of elements of S, as in Corollary 20.19, then each coordinate in these tuples either must always be filled by elements of S which are left invertible, or must always be filled by elements which are right invertible. Second, if s, t are two left invertible elements of S, or two right invertible elements, then the condition $st = e$ implies that they are both invertible. And third, if V is a representable functor whose diagram is connected, and some $\xi \in V(S)$ has an invertible element of S in at least one coordinate, then it has invertible elements in all coordinates, these are determined by that one coordinate, and the set of ξ with these properties forms a subsemigroup of $V(S)$ isomorphic to the group of invertible elements of S. □

It is also not hard to extend the above result with a description of how *morphisms* among representable functors behave under composition on the right and left with fixed representable functors; the interested reader can easily work out the details.

§21. Representable functors among varieties of groups and semigroups.

The category **Group** can be identified with a full subcategory of **Semigp**^{e}, via the forgetful functor F: **Group** $\to$ **Semigp**^{e}. This functor F, in addition to having a left adjoint, as any underlying-set-preserving functor among varieties must, has a right adjoint G, the functor taking a semigroup to its group of invertible elements. That is, the image of **Group** in **Semigp**^{e} is a coreflective subcategory;

hence, by the observations made in §14, representable functors from **Group** to any variety **C** can be identified with certain representable functors from $\mathbf{Semigp}^e$ to **C**, namely, those which are invariant under composition on the right with the retraction FG of $\mathbf{Semigp}^e$ to its subcategory equivalent to **Group**. Representable functors from any category with finite products **C** to **Group** can likewise be identified with those representable functors $\mathbf{C} \to \mathbf{Semigp}^e$ which are invariant under composition on the left with FG.

This allows us to deduce the following results, of which (i) recovers a classical result of Kan's.

Corollary 21.1. (i) (Kan [**106**]) *Every representable functor from* **Group** *to* $\mathbf{Semigp}^e$ *is a power of the forgetful functor* F.

(ii) *Every representable functor from* $\mathbf{Semigp}^e$ *to* **Group** *is a power of the group-of-invertible-elements functor* G.

(iii) *Every representable functor from* **Group** *to* **Group** *is a power of the identity functor.*

Proof. Let us take for **C** in the preceding discussion the category $\mathbf{Semigp}^e$. Now FG: $\mathbf{Semigp}^e \to \mathbf{Semigp}^e$ is the semigroup-of-invertible-elements construction, whose E-system has diagram 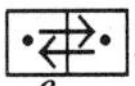. We see from Lemma 20.23 that a representable functor $\mathbf{Semigp}^e \to \mathbf{Semigp}^e$ will be invariant under composition on the right or left with this functor if and only if all connected components in its diagram have the form 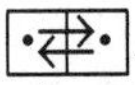. The functors $\mathbf{Group} \to \mathbf{Semigp}^e$ and $\mathbf{Semigp}^e \to$ **Group** corresponding in this way to the functor with diagram 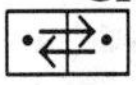 are respectively F and G; from this we easily deduce (i) and (ii). Representable functors $\mathbf{Group} \to \mathbf{Group}$ likewise correspond to representable functors $\mathbf{Semigp}^e \to \mathbf{Semigp}^e$ invariant under composition on both sides with FG, and this yields assertion (iii). □

(Remark: Since F: $\mathbf{Group} \to \mathbf{Semigp}^e$ has both a left and a right adjoint, both F and its right adjoint G are representable, hence so is their composite FG, which is why it is "reasonable" to consider the property that a representable functor be invariant under composition with this functor on one side or another, as in the above proof. If we denote by E: $\mathbf{Semigp}^e \to \mathbf{Group}$, the *left adjoint* of F, i.e., the "universal group" or "noncommutative Grothendieckization" functor, then the pair of functors E, F gives another retraction of $\mathbf{Semigp}^e$ to **Group**, and the composite FE: $\mathbf{Semigp}^e \to \mathbf{Semigp}^e$ plays the same role, in looking at left adjoint functors into and out of **Group**, that FG plays in looking at right adjoint, i.e., representable, functors.)

We now consider morphisms among the functors characterized in the above Corollary:

Corollary 21.2. *In each of the three cases of the preceding Corollary, the indicated category of representable functors is equivalent to the opposite of the category of pointed sets,* $(\mathbf{Set}^{\mathrm{pt}})^{\mathrm{op}}$.

Precisely, if we let V *denote* (i) *the forgetful functor from* **Group** *to*

Semigp^e, *or* (ii) *the group-of-invertible-elements functor on* **Semigp**^{e}, *or* (iii) *the identity functor of* **Group**, *then in each case, for sets* X *and* Y, *the morphisms* $f\colon V^X \to V^Y$ *correspond to the morphisms of* pointed sets $\mathbf{f}\colon Y \cup \{u\} \to X \cup \{u\}$, *where "u" denotes an adjoined basepoint. Namely, for* $\xi \in V^X(S)$, *the image* $f(\xi) \in V^Y(S)$ *has* y*th coordinate given by the* $\mathbf{f}(y)$*th coordinate of* ξ *if* $\mathbf{f}(y) \in X$, *or by* e *if* $\mathbf{f}(y) = u$.

PROOF. Let W denote the endofunctor of **Semigp**^{e} with diagram 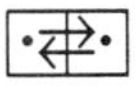. Then for any set X, the diagram corresponding to the functor W^X consists of the union of an X-tuple of copies of this diagram. This implies that the corresponding E-system is the union of an X-tuple of copies of the E-system of this diagram, with amalgamation of the element u. It is easy to see that a morphism from this E-system to the E-system similarly constructed as the union of a Y-tuple of copies of the system corresponding to W is determined by a morphism of pointed sets $X \cup \{u\} \to Y \cup \{u\}$. E-systems correspond covariantly to coalgebras, hence contravariantly to representable functors, yielding the "op" in the above results. The explicit description follows from the observations following Corollary 20.19. □

We shall now obtain, by different methods, some easy characterizations of representable functors among other varieties of groups and semigroups. These results and those obtained above will be collected in a chart at the end of this section.

In the category **Ab** of abelian groups (in which we will use additive notation), a normal form for elements of a coproduct $A \amalg B$ is $a + b$. (In other words, $A \amalg B$ is naturally isomorphic to $A \times B$.) Hence any binary co-operation on an abelian group R has the form

$$\mathbf{m}(a) = f(a)^{\lambda} + g(a)^{\rho}, \tag{21.3}$$

for some set-maps $f, g\colon R \to R$. These set-maps f and g must clearly be group homomorphisms for $\mathbf{m}$ to be one.

If $\mathbf{m}$ also satisfies the coneutral laws with respect to the unique co-zeroary operation, we see that f and g must each be the identity map. Thus, there is a unique possibility for (21.3), $\mathbf{m}(a) = a^{\lambda} + a^{\rho}$. It is easily verified that this co-operation is always coassociative, and that the induced structure on $\mathbf{Ab}(R, -)$ is the familiar addition of homomorphisms of abelian groups. We conclude that **Rep**(**Ab**, **Semigp**^{e}) is naturally equivalent to **Ab**^{op}. Since the functors obtained are actually **Ab**-valued, this characterization also describes the categories of representable functors from **Ab** to each of **AbSemigp**^{e}, **Group** and **Ab**. (We could also throw in **Binar**^{e} and **AbBinar**^{e}, but we will here stick to familiar categories.)

Coproducts in **AbSemigp**^{e} also have the normal form $a + b$, and we find as above that every object of **AbSemigp**^{e} has a unique co-**Semigp**^{e} structure, which is in fact a co-**AbSemigp**^{e} structure. These are not, however, in general co-**Group** structures. Rather, a *coinverse* $\mathbf{i}\colon R \to R$ on a co-**Semigp**^{e} object R (i.e., *any* object R) of this category turns out to be the same thing as an *inverse* operation. So those objects whose unique co-**Semigp**^{e} structures are actually co-**Group** (equivalently, co-**Ab**) structures comprise the subcategory **Ab** of **AbSemigp**^{e}.

The above arguments made use of the co-neutral law to pin down the maps f and g in (21.3). If we consider representable functors from **Ab** or $\mathbf{AbSemigp}^e$ to varieties of semigroups *without* neutral elements, the descriptions become more complicated. One calculates that the co-operation (21.3) determined by $f, g\colon R \to R$ is coassociative if and only if

$$(21.4) \qquad f^2 = f, \qquad g^2 = g, \qquad fg = gf.$$

Thus, the co-**Semigp** objects in **Ab** or $\mathbf{AbSemigp}^e$ correspond to arbitrary objects R of these categories, given with an action on R of the semigroup with neutral element presented by the relations (21.4), a 4-element semigroup with underlying set $\{1, f, g, fg\}$.

For $R \in \mathbf{Ab}$, such an action is equivalent to a decomposition of R as a direct sum of 4 subgroups,

$$R^{11} = \{a \mid f(a) = a,\ g(a) = a\}, \qquad R^{10} = \{a \mid f(a) = a,\ g(a) = 0\},$$

$$R^{01} = \{a \mid f(a) = 0,\ g(a) = a\}, \qquad R^{00} = \{a \mid f(a) = 0,\ g(a) = 0\}.$$

Given this decomposition, we have $\mathbf{Ab}(R, A) = \prod_{i,j \in \{0,1\}} \mathbf{Ab}(R^{ij}, A)$, and the semigroup structure on this set is given by

$$(21.5) \qquad (\alpha, \beta, \gamma, \delta) \cdot (\alpha', \beta', \gamma', \delta') = (\alpha + \alpha',\ \beta,\ \gamma',\ 0).$$

Conversely, for any four objects $R^{ij} \in \mathbf{Ab}$, this construction clearly defines a functor $\mathbf{Ab} \to \mathbf{Semigp}$. The semigroups so constructed will be *abelian* if and only R^{01} and R^{10} are trivial, equivalently, if and only if $f = g$ in (21.4).

Although for $R \in \mathbf{Ab}$ a pair of operators f, g satisfying (21.4) corresponds to a 4-fold direct sum decomposition of R as described, this is not so for $R \in \mathbf{AbSemigp}^e$. (To get a counterexample, start with an abelian semigroup that is a 2-fold direct sum, $S = A \oplus B$, let $f = g =$ projection onto A, and then take $R \subseteq S$ to be a subsemigroup *properly* containing A, and so in particular closed under f and g, but having trivial intersection with B.) So we must be content simply to describe a co-**Semigp** object in $\mathbf{AbSemigp}^e$ as equivalent to an abelian semigroup given with two commuting idempotent endomorphisms f and g, i.e., endomorphisms satisfying (21.4).

Consider now representable functors on the variety **AbSemigp** (no neutral element!). Since there can be no nontrivial representable functors from a variety without zeroary operations to one with such operations (Lemma 11.6), the only categories among those considered above to which there can be nontrivial representable functors are **Semigp** and **AbSemigp**. Now in a coproduct $A \amalg B$ in **AbSemigp**, the general element can have any of the three forms a, b, or $a + b$ $(a \in A,\ b \in B)$. This coproduct can be most conveniently described as $(A \cup \{0\}) \times (B \cup \{0\}) - \{(0, 0)\}$, where the 0's represent adjoined neutral elements. Using this description, we find that a co-**Semigp** structure on an object $R \in \mathbf{AbSemigp}$ corresponds to a pair of endomorphisms $f, g\colon R \cup \{0\} \to R \cup \{0\}$ satisfying (21.4) and the additional condition that the *intersection* of the kernels of f and g is zero (so that the combined map will not take anything nonzero to $(0, 0)$). Again, this cosemigroup will be coabelian if and only if $f = g$.

Another class of degenerate cases is that of representable functors from our varieties of nonabelian objects, **Group**, $\mathbf{Semigp}^e$ and **Semigp**, to varieties of abelian objects **Ab**, $\mathbf{AbSemigp}^e$ and **AbSemigp**. In each of the former categories, we find that the *symmetric* elements of a coproduct $R^\lambda \amalg R^\rho$ form the least subobject, $\{e\}$, $\{e\}$ and $\varnothing$ respectively. In the last of these cases, for R to admit a cocommutative co-operation, $R \to \varnothing \subseteq R^\lambda \amalg R^\rho$, we must have $R = \varnothing$. In the other two cases, we see that every R admits a unique cocommutative co-operation given by the trivial homomorphism $R \to \{e\} \subseteq R^\lambda \amalg R^\rho$; this corresponds to a functor which to every object S assigns the set $\mathbf{C}(R, S)$ with the multiplication under which all pairs of elements have the same product, namely the trivial homomorphism (in the notation of Exercise 11.5, $0_{R,S} \in \mathbf{C}(R, S)$). We shall call this the *zero multiplication* on $\mathbf{C}(R, S)$. Thus, every object of **Group** or $\mathbf{Semigp}^e$ can be made a co-**AbSemigp** object in a unique (but uninteresting) way. Since a nontrivial semigroup with a zero multiplication cannot have a neutral element, we get no nontrivial representable functors from **Group** or $\mathbf{Semigp}^e$ to $\mathbf{AbSemigp}^e$ or **Ab**.

We display the above conclusions in the chart below. The entry at the intersection of the row labeled **C** and the column labeled **D** shows, up to equivalence, the structure of the category of objects of **C** with co-**D** structures (the opposite of the category $\mathbf{Rep}(\mathbf{C}, \mathbf{D})$). Here **1** denotes the category with one object and one morphism (the "trivial functor only" case), and E-**Syst** the category of E-systems, as defined in Definition 20.13. The symbol z under the name of a *pointed* category means that the set of morphisms is given the zero multiplication, described above. The symbol l under the name of an arbitrary category refers to *left-zero* multiplication, defined by the identity $x \cdot y = x$, and the symbol r to *right-zero* multiplication, defined by $x \cdot y = y$ (these occur in the middle two coordinates in (21.5)). The symbol a denotes the natural additive group or semigroup structure on homomorphisms among abelian groups or semigroups, and the symbol p (for "power") under $\mathbf{Set}^{\mathrm{pt}}$ means that to a pointed set $X \cup \{u\}$, we associate the Xth power of some "basic" functor V, as described in Corollaries 21.1 and 21.2. The notation "$\simeq$ above" means a construction resembling the one in the top box in the same column, though not as elegantly describable, and discussed in the paragraphs beginning shortly after (21.5). Finally, the three **?**'s in the last column indicate cases where we do not know the structures of all representable functors.

(21.6)

	Ab	**AbSemigp**^{e}	**AbSemigp**	**Group**	**Semigp**^{e}	**Semigp**
Ab	**Ab** a	**Ab** a	**Ab** × **Ab** a z	**Ab** a	**Ab** a	**Ab** × **Ab** × **Ab** × **Ab** a l r z
AbSemigp^e	**Ab** a	**AbSemigp**^{e} a	≂ above	**Ab** a	**AbSemigp**^{e} a	≂ above
AbSemigp	**1**	**1**	≂ above	**1**	**1**	≂ above
Group	**1**	**1**	**Group** z	**Set**^{pt} p	**Set**^{pt} p	?
Semigp^e	**1**	**1**	**Semigp**^{e} z	**Set**^{pt} p	***E*-Syst**	?
Semigp	**1**	**1**	**1**	**1**	**1**	?

PROBLEM 21.7. *Investigate the three cases marked* **?** *above.*

In some of these cases there may well be no simple description. There are certainly many phenomena that such descriptions would have to cover. For instance, functors from **Semigp**^{e} to **Semigp** clearly include functors determined by E-systems and constructions based on zero, right-zero, and left-zero multiplications. Among constructions from **Group** to **Semigp** we find the last three phenomena and also examples of "associativity by default" such as we saw in representable functors from rings to nonunital semigroups (§16); e.g., consider the operation, on pairs (ξ, η) of elements of a group, defined by $(\xi, \eta)\cdot(\xi', \eta') = (e,\ \xi\xi'\xi^{-1}\xi'^{-1})$. One also gets this phenomenon in functors from **Semigp**^{e} to **Semigp**, e.g., by following the group-of-invertible-elements functor by the one just described.

One might bring into these investigations still other varieties, such as that of semigroups with zero-element, **Semigp**^{z}, semigroups with neutral element *and* zero-element, **Semigp**^{e,z}, S-semigroups (S a fixed semigroup with neutral element) defined analogously to k-rings, and groups or semigroups satisfying other identities than commutativity.

Let us briefly sketch results on one case that has been studied. A semigroup satisfying the rather degenerate identities $sut = st$ and $ss = s$ is called by specialists a *rectangular band*. Any direct product of two objects $X \times Y$ in a category **C** with finite direct products has a structure of rectangular band object, given by the direct product of the left-zero multiplication of X and the right-zero multiplication of Y. (E.g., if **C** = **Set** this structure is described by $(x, y)(x', y') = (x, y')$.) By duality, every coproduct of two objects in a category with finite coproducts has a canonical structure of co-rectangular-band object, and [**23**, §§2-3] looks at the question of when this construction defines an equivalence between **C** × **C** and the category of co-rectangular band objects of **C**. It is shown that this is so when the coproduct operation of **C** behaves "nicely" in certain ways; for instance, this holds if **C** is any of the 6 categories of our chart (21.6), or the full category of *nonzero* objects of $\mathbf{Ring}_k^1$ for k a field. Two examples of categories for which it fails are the full variety $\mathbf{Ring}_k^1$ (because information is lost when one takes the coproduct of a general k-algebra with the zero algebra), and the variety of rectangular bands itself (in which the coalgebra representing the identity functor of that category does not arise from a coproduct).

§22. Some related varieties: binars, heaps, and Mal'cev algebras.

Let us recall a result from the literature, reminiscent of Lemma 15.2.

THEOREM 22.1 (Eckmann-Hilton-Kneser [**60**, Theorem 1.4]). *If* $(R, \mathbf{m}, \mathbf{e})$ *is a co-*$\mathbf{Binar}^e$ *object of* **Group**, *then* R *is a free group.*

PROOF. The map $\mathbf{e}: R \to \{e\}$ is, of course, the trivial homomorphism. The coneutral laws for $\mathbf{m}$ say that $(1_R, \mathbf{e})\mathbf{m} = 1_R = (\mathbf{e}, 1_R)\mathbf{m}$; hence $\mathbf{m}$ embeds R in the group

$$\{r \in R^{\lambda} \amalg R^{\rho} \mid (1_R, \mathbf{e})(r) = (\mathbf{e}, 1_R)(r)\}$$

(cf. (20.3)). Now the above subgroup of $R^{\lambda} \amalg R^{\rho}$ has trivial intersection with all conjugates of R^{λ} and R^{ρ}. Hence by the Kurosh Subgroup Theorem [**113**, Theorem 6.3] [**124**, Corollary 4.9.1, p.243], it is free. (One can also verify directly that it is free on $\{x^{\lambda}x^{\rho} \mid x \in R - \{e\}\}$.) The assertion now follows from the fact that a subgroup of a free group is free. (Actually, by the argument used to prove Lemma 15.2, it suffices to know that a *retract* of a free group is free.) □

Note that this result does not pick out a distinguished free generating set for R. It would be interesting to try to establish formally the nonexistence of a canonical choice of free generating set, say by finding a co-$\mathbf{Binar}^e$ object $(R, \mathbf{m}, \mathbf{e})$ having automorphisms not respecting any free generating set for R. One may also ask whether objects of $\mathbf{Semigp}^e$ admitting co-$\mathbf{Binar}^e$ structures can be characterized; in particular, whether they coincide with the objects admitting co-$\mathbf{Semigp}^e$ structures (i.e., corresponding to E-systems).

A variety closely related to **Group** is **Heap**. A *heap* is a set given with a ternary operation t that satisfies the identities of the derived operation $xy^{-1}z$ on groups. A generating set for these is given by the three identities

$$t(x, x, y) = y = t(y, x, x) \tag{22.2}$$

$$t(v, w, t(x, y, z)) = t(t(v, w, x), y, z). \tag{22.3}$$

This concept has been rediscovered many times and given other *ad hoc* names. A brief presentation of basic results about this variety was given by one of the present authors in [**20**, pp.60-61], where, however, he called them *isogroups*, because neither he nor anyone he asked at the time knew the standard term.

Every nonempty heap in fact arises from a group by taking the derived operation $xy^{-1}z$, and this group is unique up to isomorphism [**20**, Proposition 19]; but the category of heaps has more *morphisms* than the group homomorphisms, since every left or right translation on a group is an automorphism as a heap. Intuitively, the heap structure of a group contains all the information about the group structure except the location of the neutral element. And indeed,

$$\textbf{Group} \cong \textbf{Heap}^{\mathrm{pt}} \tag{22.4}$$

[**20**, Proposition 19 (b)].

Given two objects X, Y of a category $\mathbf{C}$, the set of *isomorphisms* $i: X \to Y$ is easily seen to admit the ternary operation $(i, i', i'') \mapsto i\, i'^{-1} i''$, and to become a heap under this operation, which, if nonempty, is isomorphic to the heap of automorphisms of each object.

(The concept of heap was apparently first introduced by H. Prüfer [**153**, §3] under the name *Schar*. Prüfer was dealing with abelian groups so his *Scharen* are abelian heaps; Baer [**10**] appears to have silently removed the commutativity hypothesis. *Schar* (''crowd'' or ''flock'') was rendered loosely in Russian by груда (''heap'') in [**182**], and this term was later translated literally into English as *heap*, into French as *amas*, etc.. These structures were also studied in [**38**]. Because of the structural equivalence with groups noted above, there are not many works devoted to heaps themselves, though generalizations called ''semiheaps'' etc. receive some attention. For more information see [**170**] (the author of which, however, introduces in the English translation, in place of standard term ''heap'', the coinage ''groud''), or the *Mathematical Reviews* subject index under the heading **20N10**. The present authors are indebted to Boris Schein for most of these references. Incidentally, there is an unrelated notion with the name ''heap'' in the theory of data structures [**198**, p.72].)

Combining (22.4) with Lemma 11.4, we see that our preceding results yield characterizations of all representable functors from all the varieties of groups and of semigroups with *neutral element* in our chart (21.6) to **Heap**, while Lemma 11.6 shows that representable functors in the opposite direction are all trivial.

One can also prove triviality of representable functors from our two varieties of semigroups *without* neutral elements to **Heap**. This result, sketched in the next exercise, uses only the two identities (22.2) of **Heap**. This pair of identities is of independent interest, because Mal'cev ([**128**], [**131**, Theorem 4.141]) showed that a variety $\mathbf{V}$ ''has permuting congruences'' (i.e., relational composition on the set of congruences of every object A of $\mathbf{V}$ is commutative) if and only if $\mathbf{V}$ has some derived operation satisfying these identities; such an operation is now called a ''Mal'cev term''. Let us denote by **Mal'cev** the variety of algebras $S = (|S|, t)$ where t is a ternary operation satisfying (22.2).

EXERCISE 22.5. *Suppose* $(R, \mathbf{t})$ *is a co-Mal'cev object (e.g., a coheap) in* **Semigp** *or* **AbSemigp**.

(i) *Deduce from the first equation of* (22.2) *that for every* $r \in R$, *the element* $\mathbf{t}(r) \in R^{\lambda} \amalg R^{\mu} \amalg R^{\rho}$ *involves terms from* R^{ρ} *(i.e., does not lie in the subsemigroup generated by* R^{λ} *and* R^{μ}*).*

(ii) *Deduce from general properties of coproducts in* **Semigp** *and* **AbSemigp** *that on applying the map* $(1_{R^{\lambda}}, 1_{R^{\rho}}, 1_{R^{\rho}}): R^{\lambda} \amalg R^{\mu} \amalg R^{\rho} \to R^{\lambda} \amalg R^{\rho}$, $\mathbf{t}(r)$ *must be carried to an element still involving terms from* R^{ρ}.

(iii) *Obtain a contradiction between the conclusion of* (ii) *and the coidentity corresponding to the second equation of* (22.2), *and deduce that* R *must be the empty semigroup.*

(iv) *Generalize the above result, in the spirit of Lemma 11.6.*

The arguments sketched above suffice to fill in most of the new boxes one gets by adjoining a row and column labeled **Heap** to (21.6), and, with some minor additional arguments, a row and column labeled **AbHeap** as well (where an abelian heap means a heap which corresponds to an abelian group, equivalently, which satisfies the identity $t(x, y, z) = t(z, y, x)$). It would, of course, be of interest to fill in the remaining boxes.

We end this section by sketching how to characterize all co-Mal'cev objects in $\mathbf{Semigp}^e$ and **Group**.

EXERCISE 22.6. *Suppose* $\mathbf{t}\colon R \to R^\lambda \amalg R^\mu \amalg R^\rho$ *is a ternary co-operation on an object* R *of* $\mathbf{Semigp}^e$.

(i) *Show that* $\mathbf{t}$ *is a co-Mal'cev operation if and only if for every* $r \in R$,

(a) *each* (λ, μ)*-segment and each* (μ, ρ)*-segment of* $\mathbf{t}(r)$ *satisfies* $f(x) = e$, *where* f *is the superscript-forgetting map* $(1_R, 1_R, 1_R)\colon R^\lambda \amalg R^\mu \amalg R^\rho \to R$, *and*

(b) *the product of all terms* x^λ *occurring in* $\mathbf{t}(r)$ *is* r^λ, *and the product of all terms* x^ρ *occurring in* $\mathbf{t}(r)$ *is* r^ρ.

Now assume that the above conditions hold.

(ii) *Deduce from* (a) *above that every element* x^μ *occurring in the normal-form expression for* $\mathbf{t}(r) \in R^\lambda \amalg R^\mu \amalg R^\rho$ *is flanked on one side by a member of* R^λ *and on the other by a member of* R^ρ, *and is invertible in* R^μ. *Deduce using* (a) *and* (b) *together that* r *is invertible, and thus that* R *is a group.*

We will now want to take $\mathbf{t}(r)$, *which we have so far written in terms of elements* x^λ, x^ρ *and* x^μ, *and rewrite it in terms of elements* x^λ, x^ρ *and a new family of elements:*

$$x^* = x^\lambda (x^\mu)^{-1} x^\rho. \tag{22.7}$$

Before we do this translation, we want to

(iii) *Show that the subgroup of* $R^\lambda \amalg R^\mu \amalg R^\rho$ *generated by the elements* (22.7) *with* $x \in |R| - \{e\}$ *is a free group on these generators.*

Unfortunately, when we consider elements of R^λ *and* R^ρ *as well, we no longer have a simple normal form for expressions in these elements and elements* x^*. *E.g., whenever an equation* $xy = z$ *holds in* R, *we get a relation between* x^*, y^*, z^* *and the corresponding elements of* R^λ *and* R^ρ. *Nevertheless, for the elements* $\mathbf{t}(r)$ *there is a particularly nice choice of expression:*

(iv) *Suppose that in the expression for* $\mathbf{t}(r)$, *we write each term* x^μ *that is flanked on the left by a member of* R^λ *and on the right by a member of* R^ρ *as* $x^\lambda (x^{-1})^* x^\rho$, *and each term* x^μ *that is flanked on the left by a member of* R^ρ *and on the right by a member of* R^λ *as* $x^\rho (x^*)^{-1} x^\lambda$ *(so that no new* λ*- or* ρ*-terms are created). Show that the resulting expression reduces to a product of terms* x^* *and their inverses only. Deduce from* (iii) *that* R *is a free group.*

(v) *Conclude that the most general co-Mal'cev-operation* **t** *on a group* R *can be constructed as follows. Let* $F(|R|-\{e\})$ *denote the free group on a system of generators* $[r]$ $(r\in|R|-\{e\})$. *Let*

$$\varphi\colon\ F(|R|-\{e\}) \ \to\ R$$

be the homomorphism taking each $[r]$ *to* r, *and let*

$$\tau\colon\ F(|R|-\{e\}) \ \to\ R^{\lambda} \amalg R^{\mu} \amalg R^{\rho}$$

be the homomorphism taking each $[r]$ *to* r^*, *defined as in* (22.7). *Choose a homomorphism*

$$\psi\colon\ R \ \to\ F(|R|-\{e\})$$

right inverse to φ *(possible if and only if* R *is a free group!). Then the composite*

$$\mathbf{t} \ =\ \tau\psi\colon\ R \ \to\ R^{\lambda} \amalg R^{\mu} \amalg R^{\rho}$$

is the desired general co-Mal'cev-operation.

Combining this result with the last statement of (ii), *and using the results of §14, we see that the general co-Mal'cev object of* **Semigp**^{e} *has the same form, but with the free group* R *now regarded as a semigroup.*

Let us record a particular case of (v)*:*

(vi) *Taking for* R *the free group on one generator, deduce that every Mal'cev term for the variety of groups (i.e., every derived ternary operation* t *on groups which satisfies* (22.2)) *may be gotten by choosing arbitrary integers* $a_1, \dots, a_n$, $b_1, \dots, b_n$ *satisfying* $\Sigma\, a_i b_i = 1$, *and defining*

$$t(\xi,\eta,\zeta) \ =\ (\xi^{a_1}\eta^{-a_1}\zeta^{a_1})^{b_1} \dots (\xi^{a_n}\eta^{-a_n}\zeta^{a_n})^{b_n}.$$

Moreover, the a_i *and* b_i *become unique if we require that successive* a*'s be distinct, and that all* a_i *and* b_i *be nonzero.*

CHAPTER V.

Representable functors from algebras over a field to rings.

A ring, associative or nonassociative, can be described as an object $A \in \mathbf{Ab}$ given with a bilinear map $m: A \times A \to A$. In Chapter III we characterized representable functors from categories of associative k-rings to $\mathbf{Ab}$; in §23 below, we shall characterize "bilinear maps" among these abelian group valued functors. We will then use this characterization to study representable functors from associative k-rings to various sorts of associative and nonassociative rings.

In Chapter III, there were technical advantages in working with representable functors whose domains were categories of *nonunital* associative rings, owing to the convenient behavior of coproducts in those categories. However, now that we have the results of that chapter in hand, we can work in the more agreeable categories of *unital* k-rings. In fact there will now be a technical advantage in the latter choice, for a morphism $R \to S$ in k-$\mathbf{Ring}^1$ can be interpreted as an R-ring structure on S, i.e., an object of R-$\mathbf{Ring}^1$, allowing us to push structure up and down among these categories.

Hence, "k-ring" will henceforth mean object of k-$\mathbf{Ring}^1$, i.e., *unital associative* k-ring, unless the contrary is indicated. Results for functors on categories of *nonunital* rings will generally be discussed only when there are non-obvious differences from the unital case. As previously mentioned, when we refer to categories of *not necessarily associative* rings (as codomains of our functors; they will not occur as domains except in Chapter IX), this will be made clear by a qualifying adjective: "nonassociative", "Lie", etc..

After proving the general criterion for bilinearity in the next section, we shall restrict attention for the remainder of this chapter to the case where k is a field and our k-rings are k-algebras, so that we can take advantage of an elegant description of the opposite of the category of k-vector-spaces, recalled in §24. In Chapter VI we shall see how to extend the results so obtained to more general categories of k-rings.

§23. Bilinear maps.

Suppose F, G, H are representable functors k-$\mathbf{Ring}^1 \to \mathbf{Set}$, with representing objects R_1, R_2, R_3. Then the functor $F \times G$ is represented by $R_1 \amalg R_2$, hence morphisms of functors $F \times G \to H$ correspond to ring homomorphisms $R_3 \to R_1 \amalg R_2$. Suppose now that F, G and H have $\mathbf{Ab}$ structures, so that by Theorem 13.15 they may be written $R_1 = k\langle L\rangle$, $R_2 = k\langle M\rangle$, $R_3 = k\langle N\rangle$ (L, M, $N \in k$-$\mathbf{Bimod}$) with coaddition maps as described in that Theorem. Then $R_1 \amalg R_2$ has the form $k\langle L\rangle \amalg k\langle M\rangle \cong k\langle L \oplus M\rangle$. We want to characterize those k-ring homomorphisms

$$\mathbf{m}\colon\ k\langle N\rangle \to k\langle L \oplus M\rangle$$

such that the corresponding morphism of functors

$$m\colon\ F \times G \to H$$

is *bilinear*, in the sense that for each $T \in k\text{-}\mathbf{Ring}^1$, the map $m(T)$: $F(T) \times G(T) \to H(T)$ is a bilinear map of abelian groups.

Let us recall here a point that regularly escapes students in first-year graduate algebra courses. If A, B and C are abelian groups, a bilinear map $A \times B \to C$ is *not* a homomorphism from the *group* $A \times B$ into C. Rather, the group structure that one learns, earlier in such a course, to put on $A \times B$ is irrelevant to the definition of bilinearity; the relevant structure is that of a set given with two projection maps, $A \times B \to A$ and $A \times B \to B$, and abelian group structures on the *fibers* of these maps (the inverse images of elements of the codomains). Each fiber of the projection to A can be identified with B and given its group structure, and each fiber of the projection to B similarly inherits its group structure from A. A bilinear map is a map from the set $A \times B$ into C which acts as a group homomorphism on the fibers of both projections.

Let us see what it means to have algebraic structures on the fibers of a morphism of representable functors. Suppose

$$R \to R' \tag{23.1}$$

is a k-ring homomorphism, and

$$F' \to F \tag{23.2}$$

the induced morphism in $\mathbf{Rep}(k\text{-}\mathbf{Ring}^1, \mathbf{Set})$ (F represented by R, and F' by R'). Then to put $\mathbf{V}$-structures, for an arbitrary variety $\mathbf{V}$, on the fibers of the maps $F'(S) \to F(S)$ ($S \in k\text{-}\mathbf{Ring}^1$) in a functorial way is, we claim, equivalent to giving R', considered an R-ring via (23.1), a co-$\mathbf{V}$ structure in the category $R\text{-}\mathbf{Ring}^1$. Indeed, a single element $x \in F(S) = k\text{-}\mathbf{Ring}^1(R, S)$ corresponds to a structure of R-ring on S; if we call S with this R-ring structure S_x, then we see that the inverse image of x in $F'(S)$ is

$$R\text{-}\mathbf{Ring}^1(R', S_x). \tag{23.3}$$

To put $\mathbf{V}$-structures on the sets (23.3) for every R-ring S_x in a functorial manner is to make the functor $R\text{-}\mathbf{Ring}^1(R', -)$ $\mathbf{V}$-valued, i.e., to make R' a co-$\mathbf{V}$ object in $R\text{-}\mathbf{Ring}^1$, as claimed.

Taking $\mathbf{V} = \mathbf{Ab}$, we get:

LEMMA 23.4. *Let* (23.1) *be a map in* $k\text{-}\mathbf{Ring}^1$, *and* (23.2) *the induced map of set-valued representable functors. Then to put abelian group structures on the fibers of* (23.2) *in a functorial manner is equivalent to writing* R' *as a tensor algebra* $R' = R\langle B\rangle$ ($B \in R\text{-}\mathbf{Bimod}$). *If, further,* H *is any representable functor* $k\text{-}\mathbf{Ring}^1 \to \mathbf{Ab}$, *with representing object* $k\langle N\rangle$, *and we are given a k-ring homomorphism*

$$k\langle N\rangle \;\to\; R' = R\langle B\rangle, \tag{23.5}$$

then the induced maps $F'(S) \to H(S)$ $(S \in k\text{-}\mathbf{Ring}^1)$ *give group homomorphisms on the fibers of* $F'(S) \to F(S)$ *if and only if* (23.5) *carries* N *into* B.

Sketch of Proof. The first assertion follows from the preceding observations and Theorem 13.15. The condition that (23.5) yield maps that are homomorphisms on the fibers of $F' \to F$ means that it should "respect" the coaddition maps of $k\langle N\rangle$ and $R\langle B\rangle$. These maps are defined by

$$\mathbf{a}(x) \;=\; x^{\lambda} + x^{\mu} \tag{23.6}$$

for $x \in N,\ B$ respectively, and conversely these submodules comprise all elements satisfying (23.6). It follows that a map will respect these coaddition maps if and only if it carries N into B. □

Now consider the special case where the functor F' has the form $F \times G$, the morphism (23.2) is the projection $F \times G \to F$, and the **Ab** structure on the fibers comes from an **Ab** structure on the functor G. Thus G is represented by a tensor algebra $k\langle M\rangle$; we will continue to write R for the object representing F. Then we have $R' = R \amalg k\langle M\rangle$, and the tensor algebra structure referred to in Lemma 23.4 is easily seen to be $R' = R\langle M'\rangle$, where M' is the R-bimodule $R \otimes_k M \otimes_k R$.

Finally, suppose F and G are *both* **Ab**-valued, with representing objects

$$R_1 \;=\; k\langle L\rangle \qquad \text{and} \qquad R_2 \;=\; k\langle M\rangle.$$

Then $F \times G$ has **Ab** structures on the fibers of each projection, and these correspond to two tensor-decompositions of $R_1 \amalg R_2 = k\langle L \oplus M\rangle$, namely as

$$k\langle L \oplus M\rangle \;=\; R_1\langle M'\rangle \qquad \text{and} \qquad k\langle L \oplus M\rangle \;=\; R_2\langle L''\rangle$$

where

$$M' \;=\; R_1 \otimes_k M \otimes_k R_1 \qquad \text{and} \qquad L'' \;=\; R_2 \otimes_k L \otimes_k R_2.$$

If we write $k\langle L \oplus M\rangle$ as the direct sum of all tensor products of finite strings of the k-bimodules L and M, we see that M' consists of the sum of all such products in which M occurs exactly once, and L'' of the sum of those in which L occurs exactly once. Hence if a map $k\langle N\rangle \to k\langle L \oplus M\rangle$ is to induce maps linear on both sorts of fiber, it must carry N into the intersection of these two sums, i.e., the sum of all tensor products in which L and M each occur exactly once. There are just two such summands, $L \otimes_k M$ and $M \otimes_k L$, and we deduce

Proposition 23.7. *Let* $F,\ G,\ H \in \mathbf{Rep}(k\text{-}\mathbf{Ring}^1, \mathbf{Ab})$, *with representing objects* $k\langle L\rangle$, $k\langle M\rangle$, $k\langle N\rangle$ *and cogroup structures as in Theorem 13.15. Then the morphisms* $m\colon F \times G \to H$ *that are* bilinear *with respect to these structures correspond to the ring homomorphisms* $\mathbf{m}\colon k\langle N\rangle \to k\langle L \oplus M\rangle$ *which carry* N *into* $(L \otimes_k M) \oplus (M \otimes_k L)$. □

(Some other classes of maps among **Ab**-valued functors on k-$\mathbf{Ring}^1$ —

multilinear maps, quadratic maps, etc. — will receive similar characterizations in Chapter X.)

We confess that Proposition 23.7 could have been obtained without the discussion of fibers of projection maps, by a direct calculation, as the reader is encouraged to verify. For instance, the equation for right linearity, $m(x, y+z) = m(x, y) + m(x, z)$ translates to commutativity of the diagram of co-operations

$$\begin{array}{ccc} k\langle N\rangle & \xrightarrow{\quad\mathbf{m}\quad} & k\langle L\oplus M\rangle \\ \Big\downarrow{\scriptstyle \mathbf{a}_N} & & \Big\downarrow{\scriptstyle 1_L \amalg \mathbf{a}_M} \\ k\langle N^\lambda \oplus N^\rho\rangle & \xrightarrow{(\mathbf{m}^\lambda,\, \mathbf{m}^\rho)} & k\langle L\oplus M^\lambda \oplus M^\rho\rangle, \end{array}$$

where $\mathbf{m}^\lambda$ and $\mathbf{m}^\rho$ denote respectively the maps $k\langle N\rangle \to k\langle L\oplus M^\lambda\rangle \subseteq k\langle L\oplus M^\lambda \oplus M^\rho\rangle$ and $k\langle N\rangle \to k\langle L\oplus M^\rho\rangle \subseteq k\langle L\oplus M^\lambda \oplus M^\rho\rangle$ induced by $\mathbf{m}\colon k\langle N\rangle \to k\langle L\oplus M\rangle$. Again looking at $k\langle L\oplus M\rangle$ as a sum of tensor products of L and M, one finds that this diagram will commute if and only if $\mathbf{m}$ carries N into the subsum we called M' above. Left linearity gives the complementary condition, and Proposition 23.7 follows. However, we hope that the development we sketched for that Proposition was illuminating, despite a certain amount of handwaving.

Note that the $\mathbf{m}$ of Proposition 23.7 is determined by the induced bimodule map $N \to (L\otimes_k M) \oplus (M\otimes_k L)$. In the next Theorem we shall consider the case of that Proposition where $F = G = H$, also repeating for completeness the information given by Theorem 13.15. We first want

DEFINITION 23.8. **NARing** *will denote the category of nonunital not necessarily associative rings, i.e., abelian groups* A *given with arbitrary bilinear maps* $A\times A \to A$.

THEOREM 23.9. *Let* R *be a co-***NARing** *object in* k-**Ring**^{1}. *Then* R *has the form* $k\langle M\rangle$ *for some* $M\in k$-**Bimod**; *its coaddition* $\mathbf{a}$ *is induced by the diagonal map* $M\to M^\lambda\oplus M^\rho$, *and its comultiplication* $\mathbf{m}$ *by a bimodule homomorphism (which we shall denote by the same symbol):*

$$\mathbf{m}\colon M \to (M^\lambda\otimes_k M^\rho)\oplus(M^\rho\otimes_k M^\lambda). \qquad (23.10)$$

The category **Rep**$(k$-**Ring**^{1}, **NARing**$)$ *is in fact equivalent to the opposite of the category having for objects all pairs* $(M, \mathbf{m})$ *with* $M\in k$-**Bimod** *and* $\mathbf{m}$ *a bimodule map* (23.10), *and for morphisms all bimodule homomorphisms respecting the maps* $\mathbf{m}$. □

(Note that in the first statement of the above Theorem, we have described a co-**Ab** object by giving only its coaddition. We shall continue to use this abbreviated form of description throughout this chapter.)

As in §14, we have:

COROLLARY 23.11. *Theorem 23.9 continues to hold if* k-**Ring**^{1} *is replaced by* $\mathbf{Ring}_k^1$ *for* k *a commutative ring (or more generally, by* K-$\mathbf{Ring}_k^1$ *for* k *a commutative ring and* K *a* k*-algebra), and* k-**Bimod** *by* $\mathbf{Mod}_k$ *(respectively* K-$\mathbf{Bimod}_k$*); and/or if the above domain categories of unital* k*-rings are replaced by the corresponding categories of nonunital* k*-rings (unital tensor* k*-rings being replaced by nonunital tensor* k*-rings).* □

The simplest nontrivial case of these results is

EXAMPLE 23.12. Let k be a commutative ring and M the free k-module on one generator x. Thus the abelian group valued functor on $\mathbf{Ring}_k^1$ represented by $R = k\langle M\rangle = k[x]$ is the underlying additive group functor. The module $(M^\lambda \otimes_k M^\rho) \oplus (M^\rho \otimes_k M^\lambda)$ is free of rank 2 on the generators $x^\lambda x^\rho$ and $x^\rho x^\lambda$. (Here we have suppressed the $\otimes$ signs, thinking of these elements as products in the ring $k\langle M^\lambda \oplus M^\rho\rangle = k\langle x^\lambda, x^\rho\rangle$.)

What ring structures will we get on the underlying-abelian-group functor by choosing various maps $\mathbf{m}$ of M into this module? If we set $\mathbf{m}(x) = x^\lambda x^\rho$, then the induced multiplication m on $F(T) = T$ is just $m(\xi, \eta) = \xi\eta$, the original multiplication of T. The resulting functor F is essentially the identity (except that it forgets the k-module structure and multiplicative neutral element); in particular, F is associative-ring valued. If we send the generator x to $x^\rho x^\lambda$, we similarly get the functor taking T to the opposite ring, T^{op}, again associative. But in general, the comultiplication sending x to $a x^\lambda x^\rho + b x^\rho x^\lambda$ $(a, b \in k)$ gives the operation $m(\xi, \eta) = a\xi\eta + b\eta\xi$, which is not associative unless $ab = 0$. For instance, using $\mathbf{m}(x) = x^\lambda x^\rho - x^\rho x^\lambda$ we get the *commutator* operation, which does not in general make T an associative ring but a Lie ring; similarly, the map given by $\mathbf{m}(x) = x^\lambda x^\rho + x^\rho x^\lambda$ makes T a Jordan ring.

23.13. *Pedagogic digression on "the point that regularly escapes first-year graduate students".* When the first author found that some students in his beginning graduate algebra course, after learning the definition of a bilinear map $A \times B \to C$, believed that such a map was a homomorphism on the group $A \times B$, he questioned others in the class and found this misimpression almost universal! The approach he has used to avoid this misunderstanding since then is to first call such a function a bilinear map $f\colon (A, B) \to C$, rather than $f\colon A \times B \to C$.

Note that if we distinguished, as universal algebraists do, between a group A and its underlying set $|A|$, then a bilinear map would be a map on the set $|A| \times |B|$, not on the group $A \times B$. Yet since the definition of bilinearity uses the group structures of A and B, the notation "a bilinear map $f\colon |A| \times |B| \to |C|$" is also inadequate. The introduction of an object with underlying set $|A| \times |B|$ and group structures on the fibers of the projection maps to $|A|$ and $|B|$, though enlightening for the expert to see, is too sophisticated for an elementary course, and too cumbersome to impose for everyday use on ring-theorists who already know what they mean.

So as the least of several pedagogic evils in teaching beginning graduate algebra, the first author now starts by writing $f\colon (A, B) \to C$ as indicated above, then acknowledges the convenient standard notation $f\colon A \times B \to C$, but emphasizes that despite the suggestiveness of that notation, such maps are *not* homomorphisms on, or even characterized in terms of, the "product group" (cf. [**24**, §3.9]). A useful exercise to assign is to show that a map $f\colon |A| \times |B| \to |C|$ which is both bilinear *and* a homomorphism is 0.

§24. Review of linearly compact vector spaces.

A bimodule map $\mathbf{m}$ as in (23.10) is equivalent to a pair of maps from M to $M \otimes_k M$. Each of these maps may be thought of intuitively as a "multiplication" on the object corresponding to M in the opposite category $(k\text{-}\mathbf{Bimod})^{\mathrm{op}}$. A difficulty in working with such objects (or with the analogous structures that we get in considering representable functors on categories $\mathbf{Ring}_k^1$, $K\text{-}\mathbf{Ring}_k^1$, etc.) is the unfamiliarity of the opposite categories, $(k\text{-}\mathbf{Bimod})^{\mathrm{op}}$, $(\mathbf{Mod}_k)^{\mathrm{op}}$, etc.. In this section we shall review a useful description of an important class of cases: If k is a field, then the opposite of the category of k-vector-spaces, $(\mathbf{Mod}_k)^{\mathrm{op}}$, is equivalent to the category of *linearly compact* topological k-vector-spaces (Definition 24.5 below). This will allow us to put the description of representable **NARing**-valued functors on $\mathbf{Ring}_k^1$ for k a field into a form easy to conceptualize and calculate with, and in terms of which we shall obtain, in §§25-26, necessary and sufficient conditions for associativity and various other identities to hold for such a functor. Guided by these results, we shall see at the beginning of the next Chapter how to obtain analogous characterizations of the corresponding classes of functors on general k-rings.

CONVENTION 24.1. *Throughout the remainder of this chapter, k will be a field, with the discrete topology, and "vector space" will mean k-vector-space. A* topology *on a vector space will always be understood to mean a Hausdorff topology under which the vector space operations are continuous.*

In speaking of topological vector spaces, subspace *will continue to mean sub-*vector-*space.*

It is well-known that a topology on a vector space is determined by a neighborhood basis of 0. Let us now make

DEFINITION 24.2. *A topology on a vector space V is called* linear *if it has a neighborhood basis of 0 consisting of vector subspaces open in the topology. The category of linearly topologized vector spaces, with continuous linear maps for morphisms, will be denoted $\mathbf{LTopMod}_k$. The category $\mathbf{Mod}_k$ will be identified with the full subcategory of $\mathbf{LTopMod}_k$ having for objects the discrete vector spaces.*

We see that a linear topology on a vector space V is uniquely determined by choosing a separating filter of subspaces of V, and making these the open subspaces. (We recall that a *filter* of subspaces of a vector space means a nonempty family of subspaces closed under pairwise intersections and passage to larger subspaces. Such a filter is called *separating* if the intersection of all its members is $\{0\}$. The condition that the filter of open subspaces of a linear topology be separating corresponds to the Hausdorffness assumption of Convention 24.1.) All open subspaces are closed; the general closed subspaces are the intersections of possibly infinite families of open subspaces. A linear topology on a *finite-dimensional* space must be discrete, since the filter of open subspaces will have a least element, which must be $\{0\}$.

The category $\mathbf{LTopMod}_k$ has products, given by direct products of underlying vector spaces under the product topology. In this category, as in all $\mathbf{Ab}$-categories, finite products are also coproducts.

Given any topology (in the sense of Convention 24.1) on V, we recall that one has the concept of V being *complete* with respect to the topology, and if V is not necessarily complete, one can form its *completion*. (The general context in which the concepts of completeness and completion make sense is that of a set with a *uniform structure* [**34**, Ch. 2], [**91**]. In this case, the operation of translating open sets makes a vector-space topology equivalent to such a structure.) Among the criteria for V to be complete is:

(24.3) Suppose B is any neighborhood basis of 0 in V. Then V is complete if and only if every system of translates $(x_U + U)_{U\in B}$ of these subspaces which has the finite intersection property in fact has nonempty intersection.

Any discrete vector space is trivially complete. If V is a not necessarily complete linearly topologized vector space and B a neighborhood basis of 0 consisting of open subspaces, then it is easy to see that each point of V lies in a unique system of translates of these subspaces, and the completion of V can be constructed as the set of all systems $(x_U + U)_{U\in B}$ of translates of members of B with the finite intersection property; equivalently, the property $U \subseteq U' \Rightarrow x_U + U \subseteq x_{U'} + U'$.

We will give only sketches of proofs for results of this section which are well-known in the theory of topological vector spaces. We also remark that, as usual, results given in Exercises are *not* required for the sequel.

PROPOSITION 24.4. (After Lefschetz [**117**, Ch. II, 27.6 and 32.1], cf. [**56**]). *Let A be a k-vector-space with a linear topology. Then the following conditions are equivalent:*

(a) *Every filter of cosets of closed subspaces of A has nonempty intersection.*

(b) *A is complete, and all its open subspaces are of finite codimension.*

(c) *A is an inverse limit of an inversely directed system of discrete finite-dimensional vector spaces, with the inverse limit topology.*

(d) *A is isomorphic as a topological vector space to the dual $\hat{V} = \mathbf{Mod}_k(V, k)$ of a discrete vector space V, with the function topology. Equivalently, A is isomorphic to k^X, with the product topology, for some set X.*

Under interpretation (d), *the open subspaces of A are precisely the annihilators of the finite-dimensional subspaces of V, and the closed subspaces of A are the annihilators of arbitrary subspaces of V.*

SKETCH OF PROOF. (a)$\Rightarrow$(b). If we look at those cases of (a) where the subspaces whose cosets are being considered are open and form a neighborhood basis of 0, we have precisely the first condition of (b). Further, if an open subspace $B \subseteq A$ had infinite codimension, the discrete infinite-dimensional space A/B would not satisfy condition (a), and a counterexample to that condition could be lifted to a counterexample in A.

(b)⇒(c). Use the system of vector spaces A/B, B ranging over the open subspaces of A.

(c)⇒(d). Take for V the direct limit of the directed system of duals of the spaces in the given inverse system. The second form of (d) is equivalent to the first because if X is a basis of V, then $\mathbf{Mod}_k(V, k) \cong k^X$, and the indicated topologies on these spaces coincide.

From the first sentence of (d) one easily deduces the last sentence of the Proposition, and with the help of this one gets:

(d)⇒(a). As noted, a closed subspace $B \subseteq A$ is the annihilator of a subspace $U \subseteq V$; in this situation one finds that a coset of B corresponds to a linear functional on U. Now a linear functional specified in a consistent manner on all members of a directed system of subspaces of V (corresponding to a filter of closed subspaces of A) yields a linear functional on the union of this system, and can be extended from there to all of V. □

DEFINITION 24.5. *A topological vector space with the above equivalent properties is called* linearly compact. *The category of all linearly compact k-vector-spaces and continuous linear maps among them will be denoted* $\mathbf{LCpMod}_k$, *a full subcategory of* $\mathbf{LTopMod}_k$.

The dual $\mathbf{Mod}_k(V, k)$ *of a discrete vector space* V, *taken with the (linearly compact) function topology (Proposition 24.4*(d)), *will be denoted* $\hat{V}$.

(For topological modules over *arbitrary* rings k, where the conditions of Proposition 24.4 are not equivalent, the analog of condition (a) is generally taken as the definition of linear compactness.)

The most familiar example in algebra of a linearly compact topology on an infinite-dimensional k-vector-space is the valuation topology on the formal power series algebra, $k[[t]]$, i.e., the topology in which two power series are "close" if they agree up to high-degree terms. $k[[t]]$ may be identified as a linearly compact vector space with $\hat{V}$, where V is a discrete vector space with basis $\{c_0, c_1, \ldots\}$, by letting a formal power series $f = \Sigma\, \alpha_i t^i$ correspond to the linear functional on V taking $c_i \in V$ to $\alpha_i \in k$. Thus, for each i, the basis-element $c_i \in V$ is a "formal ith coefficient".

Observe that

$$\mathbf{LCpMod}_k \subseteq \mathbf{LTopMod}_k \supseteq \mathbf{Mod}_k,$$

and that these two subclasses, the linearly compact and the discrete vector spaces, intersect only in the finite-dimensional spaces.

EXERCISE 24.6. (i) *Show that a topology on a vector space* A *is* linear *if and only if* A *admits an embedding, as topological vector space, in a product of discrete spaces* $\prod A_i$.

(ii) *Given such an embedding, show that* A *is* complete *if and only if it is closed in* $\prod A_i$.
(iii) *If the equivalent conditions of* (ii) *hold, and we assume without loss of generality that the projection of* A *onto each* A_i *is surjective, show that* A *is* linearly compact *if and only if all* A_i *are finite-dimensional.*

EXERCISE 24.7. *Let* A *be a* complete *linear topological vector space. Show that the following conditions are equivalent:* (a) A *is linearly compact,* (b) *the given topology on* A *is minimal among linear topologies on* A, (c) A *cannot be written as the product in* $\mathbf{LTopMod}_k$ *of two objects, one of which is discrete and infinite-dimensional.*

(*Hint for* (a)⇒(b): *If* $A = \hat{V}$, *show that the linear functionals on* A *continuous in any properly weaker linear topology* (*one with a smaller class of open subspaces*) *correspond to the elements of a proper subspace* $U \subset V$. *Show that a nonzero linear functional* $a \in \hat{V} = A$ *annihilating* U *can't be separated from* 0 *in the weakened topology on* A.)

Can every linear topology on A *be weakened to a linearly compact topology? Can at least every* complete *linear topology on* A *be so weakened? Are conditions* (a)-(c) *above equivalent to minimality of the topology of* A *among linear topologies under which* A *is complete?*

The reader not familiar with the **Pro** (formal inverse limit) construction may ignore part (c) of the next result; however the equivalence of parts (a) and (b) is of fundamental importance for the results to come.

(The symbol **Pro** is based on the term "projective limit", a synonym for "inverse limit". For the construction of **Pro**, and of the dual formal direct limit construction **Ind**, whose symbol is based on the synonym "inductive limit" for "direct limit", see [**104**, VI.1]. We recall that an "inverse limit" means the limit, i.e., $\varprojlim$ in the sense of Definition 5.11, of a system indexed by an inversely directed partially ordered set, and that a "direct limit" means the colimit, i.e., $\varinjlim$, of a system indexed by a directed partially ordered set.)

PROPOSITION 24.8. *The following categories are equivalent* (*under functors to be described*):
(a) $\mathbf{LCpMod}_k$, *the category of linearly compact k-vector-spaces and continuous linear maps.*
(b) $(\mathbf{Mod}_k)^{\mathrm{op}}$, *the* opposite *of the category of* discrete *k-vector-spaces and linear maps.*
(c) $\mathbf{Pro}(\mathbf{FGMod}_k)$, *the category of formal inverse limits of* (*discrete*) *finite-dimensional k-vector-spaces, with morphisms constructed as equivalence classes of compatible systems of maps.*

DESCRIPTION OF EQUIVALENCES. We shall show (a) equivalent to each of (b) and (c).

We already have a functor $(\mathbf{Mod}_k)^{\mathrm{op}} \to \mathbf{LCpMod}_k$, taking a vector space V to the dual space $\hat{V} = \mathbf{Mod}_k(V, k)$, topologized in a natural way, as noted in Proposition 24.4(d). Inversely, for $A \in \mathbf{LCpMod}_k$ we may give the k-vector-space

$\mathbf{LCpMod}_k(A, k)$ the discrete topology, getting a functor the other way. These constructions are inverses up to natural isomorphism. (This is an example of the kind of duality alluded to in §9.4, arising, in this case, from "commuting" structures of k-vector-space and linearly compact k-vector-space on k. One may also note the parallel with the Pontrjagin duality of locally compact abelian groups, which in particular gives a contravariant equivalence between discrete and compact groups [**168**, p.28].)

To get a functor from $\mathbf{LCpMod}_k$ to $\mathbf{Pro}(\mathbf{FGMod}_k)$, associate to every linearly compact A the inverse system of its discrete quotient spaces (quotients A/B with B open). To go in the opposite direction, associate to an object of $\mathbf{Pro}(\mathbf{FGMod}_k)$ its inverse limit, with the inverse limit of the discrete topologies.

It is also instructive to describe directly the equivalence between (b) and (c): $\mathbf{Mod}_k$, like all varieties of (finitary) algebras, is equivalent to the category of formal direct limits on its subcategory of finitely presented algebras, which in this case coincide with the finitely generated algebras: $\mathbf{Mod}_k \simeq \mathbf{Ind}(\mathbf{FGMod}_k)$. Dualizing, we have $(\mathbf{Mod}_k)^{\mathrm{op}} \simeq \mathbf{Pro}(\mathbf{FGMod}_k{}^{\mathrm{op}}) \simeq \mathbf{Pro}(\mathbf{FGMod}_k)$, where the last step invokes the familiar contravariant equivalence $V \longleftrightarrow \hat{V}$ on finite-dimensional vector spaces. □

We turn now to *tensor products* of topological vector spaces. Observe that if A_1, A_2 are linear topological vector spaces, there will exist a strongest linear topology (one with most open sets) on $A_1 \otimes_k A_2$ making the tensor multiplication map $A_1 \times A_2 \to A_1 \otimes_k A_2$ continuous, namely the topology in which a subspace $U \subseteq A_1 \otimes_k A_2$ is open if and only if the inverse image in $A_1 \times A_2$ of every coset of the form

$$a_1 \otimes a_2 + U \tag{24.9}$$

contains a cylinder

$$(a_1+B_1) \times (a_2+B_2), \tag{24.10}$$

with B_1 and B_2 open subspaces of A_1 and A_2 respectively. (One must verify that the topology so described is Hausdorff, as required by Convention 24.1. Idea: Show that for every nonzero $x \in A_1 \otimes_k A_2$, there exist continuous linear functionals, f_1 on A_1 and f_2 on A_2, such that $(f_1 \otimes f_2)(x) \neq 0$. Now $\ker(f_1 \otimes f_2 \colon A_1 \otimes_k A_2 \to k)$ will not contain x, and will be open because the inverse image of an element $a_1 \otimes a_2$ contains $(a_1 + \ker(f_1)) \times (a_2 + \ker(f_2))$.)

Is there a convenient way to generate this topology? A first guess one might make, that a basis of open subspaces should be given by tensor products $B_1 \otimes B_2$ of open subspaces $B_1 \subseteq A_1$ and $B_2 \subseteq A_2$ turns out to be far from the mark: the tensor multiplication into $A_1 \otimes_k A_2$ with this topology is almost never continuous. For if B_2 is any proper open subspace of A_2, then under this topology $A_1 \otimes B_2$ is open, hence the map from $A_1 \otimes_k A_2$ with this topology to $A_1 \otimes_k (A_2/B_2)$ with the *discrete* topology is continuous. But the composite map $A_1 \times A_2 \to A_1 \otimes_k A_2 \to A_1 \otimes_k (A_2/B_2)$ cannot be continuous with respect to the discrete topology on the last space unless A_1 is discrete.

A next guess for a basis of this topology might be the system of subspaces

(24.11) $\quad (B_1 \otimes A_2) + (A_1 \otimes B_2) \quad (B_1 \subseteq A_1$ and $B_2 \subseteq A_2$ open subspaces).

This *does* define a topology under which tensor multiplication is continuous, but we have overshot in the opposite direction: it is not generally the strongest such topology. Consider, for instance, the field of fractions $k((t))$ of the formal power series algebra $k[[t]]$, under the valuation topology. As a topological vector space this is the product of the linearly compact space $k[[t]]$ and the discrete space spanned by $\{t^{-1}, t^{-2}, \dots\}$. Since multiplication is continuous in the valuation topology, the induced linear map

$$k((t)) \otimes_k k((t)) \to k((t)) \tag{24.12}$$

will be continuous if we put on the tensor product the strongest linear topology consistent with continuity of the tensor multiplication. But (24.12) is not continuous in the topology with basis (24.11): the inverse image of the open subspace $k[[t]]$ does not contain any member of that basis. A variant of this example has properties which will be particularly striking when compared with a result below. Let us replace the first factor in (24.12) by the subalgebra $k[[t]]$, and note that the multiplication will then carry the open subspace $k[[t]]$ of the second factor into the open subspace $k[[t]]$ of the codomain. Hence we get an induced continuous multiplication

$$k[[t]] \times (k((t))/k[[t]]) \to k((t))/k[[t]]. \tag{24.13}$$

Again the map does not factor continuously through the topologization of $k[[t]] \otimes_k (k((t))/k[[t]])$ given by (24.11) (the inverse image of the open subspace $\{0\}$ of the discrete codomain does not contain any product $B_1 \otimes A_2$ with B_1 open); though here, one factor is linearly compact, and the other discrete.

A correct description of the topology we have been discussing is given in

LEMMA 24.14. *Let A_1 and A_2 be linear topological vector spaces, and U a subspace of $A_1 \otimes_k A_2$. Then the following conditions are equivalent:*

(i) *U is open in the strongest topology on $A_1 \otimes_k A_2$ under which the tensor multiplication map $A_1 \times A_2 \to A_1 \otimes_k A_2$ is continuous.*

(ii) *U contains a tensor product $B_1 \otimes B_2$ of open subspaces $B_1 \subseteq A_1$, $B_2 \subseteq A_2$, and moreover, for every $a_1 \in A_1$ there exists an open subspace $B_2' \subseteq A_2$ such that $a_1 \otimes B_2' \subseteq U$, and for every $a_2 \in A_2$ there exists an open subspace $B_1' \subseteq A_1$ such that $B_1' \otimes a_2 \subseteq U$.*

(ii′) *U contains a tensor product $B_1 \otimes B_2$ of open subspaces $B_1 \subseteq A_1$, $B_2 \subseteq A_2$, and moreover, for every finite-dimensional subspace $C_1 \subseteq A_1$ there exists an open subspace $B_2' \subseteq A_2$ such that $C_1 \otimes B_2' \subseteq U$, and for every finite-dimensional subspace $C_2 \subseteq A_2$ there exists an open subspace $B_1' \subseteq A_1$ such that $B_1' \otimes C_2 \subseteq U$.*

(iii) *U contains a sum (over some index set I)*

(24.15) *$\Sigma_I B_1^{(i)} \otimes B_2^{(i)}$, where the $B_1^{(i)}$ are open subsets of A_1 and the $B_2^{(i)}$ are open subsets of A_2, with $\Sigma_I B_1^{(i)} = A_1$ and $\Sigma_I B_2^{(i)} = A_2$.*

Sketch of Proof. Condition (ii) is easily shown equivalent to our original characterization of the open sets in our desired topology in terms of "cylinders" (24.10), and hence to condition (i). To get the equivalence of (ii) and (ii′), note that (ii) is the case of (ii′) where the C_i are one-dimensional, and one can pass from the one-dimensional statement to the finite-dimensional statement using the fact that the class of open subspaces of A_1 (respectively A_2) is closed under finite intersections. Given (ii′), one gets (iii) by noting that U will contain the sum of the system of subspaces $(B_1 + C_1) \otimes (B_2 \cap B_2')$ and $(B_1 \cap B_1') \otimes (B_2 + C_2)$, where the spaces named are related as in (ii′); the reverse direction is straightforward. □

In fact, however, the topology on a tensor product of linearly topologized vector spaces that will be important to us here will not be the above universal one, but the weaker one given by the basis (24.11), because, as we shall see, completions of certain tensor products so topologized correspond to our representable functors. This topology also has a natural universal property, namely, it is easy to verify

Lemma 24.16. *If* A_1 *and* A_2 *are topological vector spaces, then the topology on* $A_1 \otimes A_2$ *with basis of open sets* (24.11) *is the strongest linear topology with respect to which the tensor multiplication map* $A_1 \times A_2 \to A_1 \otimes_k A_2$ *is* uniformly *continuous, i.e., the strongest linear topology such that the inverse image of every coset* (24.9) *of an open subset* U *contains a cylinder* (24.10) *with* B_1 *and* B_2 *depending only on* U. □

Here are some further properties of this topology.

Lemma 24.17. *The construction of tensor products of linearly topologized vector spaces with the topology universal for uniform continuity of tensor multiplication is functorial, and is commutative and associative up to natural isomorphism.*

If two linearly topologized vector spaces A_1 *and* A_2 *are either* both *discrete or* both *linearly compact, then this topology on* $A_1 \otimes_k A_2$ *agrees with the linear topology universal for (not necessarily uniform) continuity of tensor multiplication, characterized in Lemma 24.14.*

Sketch of Proof. Functoriality, and commutativity up to natural isomorphism, are straightforward. One gets associativity by noting that on a 3-fold tensor product $A_1 \otimes_k A_2 \otimes_k A_3$, the two topologies one obtains both have bases of open sets given by the subspaces

$$(B_1 \otimes A_2 \otimes A_3) + (A_1 \otimes B_2 \otimes A_3) + (A_1 \otimes A_2 \otimes B_3)$$

where B_1, B_2, B_3 range over the open subspaces of A_1, A_2, A_3 respectively. (This topology can also be characterized as the strongest topology on this space making 3-fold tensor multiplication uniformly continuous.)

To get the last assertion, let us first note that the strongest topology on a tensor product making tensor multiplication *uniformly continuous* in particular makes that operation *continuous*, hence is *contained in* the strongest topology with the latter property, so what we need to show is that (in the indicated situations) it also *contains* this topology. The case where both topologies are discrete is immediate:

the family of subspaces (24.11) includes $\{0\}\otimes A_2 + A_1\otimes\{0\} = \{0\}$, so this topology is discrete, and hence certainly majorizes the other topology.

So suppose A_1 and A_2 are both linearly compact. Let U be a subspace of $A_1\otimes_k A_2$ satisfying Lemma 24.14(ii′); we shall show that it contains a subspace of the form (24.11), as required. By assumption, U contains a tensor product $B_1\otimes B_2$ of subspaces open in A_1 and A_2 respectively. By Proposition 24.4(b), B_1 is of finite codimension in A_1; hence we can write $A_1 = B_1 + C_1$ with C_1 finite-dimensional. Now by assumption there exists an open subspace $B_2'\subseteq A_2$ such that $C_1\otimes B_2'\subseteq U$. Hence U contains $(B_1+C_1)\otimes(B_2\cap B_2')$, a term having the form of the second summand of (24.11). Similarly, U contains a term having the form of the first summand of (24.11), as required. □

We remark that the last assertion of the above Lemma does not go over to the case where *one* of the two spaces is discrete and the other is linearly compact. This may be seen from the continuity of the map (24.13). Since the codomain space is taken discrete, the kernel of the induced map $k[[t]]\otimes_k(k((t))/k[[t]])\to k((t))/k[[t]]$ must be open in the topology of Lemma 24.14; and indeed it satisfies condition (iii) of that Lemma, since it contains

$$(\Sigma_{i\geq 0}\, t^i k[[t]])\otimes_k(t^{-i}k[[t]]/k[[t]]).$$

But as we noted earlier, it contains no subspace of the form (24.11).

We now come to the construction for the sake of which we have been looking at tensor products of topological vector spaces:

Definition 24.18. *If A_1 and A_2 are linear topological vector spaces, then $A_1\hat{\otimes}A_2$ will denote the completion of $A_1\otimes_k A_2$ under the topology described in Lemma 24.16.*

(Note the somewhat confusing notation: On objects, the mark "$\wedge$" denotes dualization, while over the symbol $\otimes$, it denotes completion.)

The importance of the operation $\hat{\otimes}$ for us is that it makes it possible to describe covariant representable functors on vector spaces as completed tensor product operations ((24.21) below). Later we will drop the topology on the resulting spaces, but we include it in

Lemma 24.19. *If $S\in\mathbf{LTopMod}_k$ and $V\in\mathbf{Mod}_k$, let $\mathbf{Mod}_k(V,S)$ be given the function topology, equivalently, the topology having for a basis of open subspaces the spaces $(U{:}C)=\{a\mid a(C)\subseteq U\}$, where C ranges over the finite-dimensional subspaces of V and U over the open subspaces of S.*

Then the natural map

$$i\colon\ \hat{V}\otimes_k S\to\mathbf{Mod}_k(V,S), \tag{24.20}$$

taking a decomposable tensor $a\otimes s$ to the composite $V\xrightarrow{a}k\xrightarrow{s}S$, is a dense embedding of topological vector spaces. Taking any basis X of V, so that $\hat{V}$ can be identified with k^X and $\mathbf{Mod}_k(V,S)$ with S^X, each with the product topology, this map can be described as taking a decomposable tensor $(\alpha_x)_{x\in X}\otimes s$ in $k^X\otimes S$ to the element $(\alpha_x s)_{x\in X}$ of S^X; its full image therefore consists of *all elements* $(s_x)_{x\in X}\in S^X$ *such that the span of the components* s_x *is a finite-dimensional subspace of* S.

If S *is complete (possibly discrete) then* i *induces an isomorphism as topological vector spaces*

$$\hat{V}\hat{\otimes} S \cong \mathbf{Mod}_k(V, S). \tag{24.21}$$

Sketch of Proof. Take a basis X for V, and verify that (24.20) has the indicated description in terms of this basis, and in this form gives an embedding of topological vector spaces (i.e., that $\hat{V}\otimes_k S$ has precisely the topology of its image in S^X). The image of this map is clearly dense in S^X, which is complete if S is; hence if S is complete, S^X can be identified with the completion of the domain of (24.20), which is assertion (24.21). □

24.22. *Example.* Recall that the formal power series algebra $k[[t]]$ may be identified with the dual of a discrete space V having basis $\{c_0, c_1, \dots\}$, where c_i is a "formal ith coefficient". Now for R any k-algebra taken with the discrete topology (or indeed, taken with any complete linear topology), we claim that

$$k[[t]]\hat{\otimes} R \cong R[[t]].$$

Indeed, writing the left-hand side as $\hat{V}\hat{\otimes} R$, Lemma 24.19 allows us to identify this with $\mathbf{Mod}_k(V, R)\cong R[[t]]$, a decomposable element $(\Sigma\,\alpha_i t^i)\otimes r$ going to $\Sigma(\alpha_i r)\,t^i$. Within this object, the uncompleted tensor product $k[[t]]\otimes_k R$ corresponds to the set of those $\Sigma\, r_i t^i\in R[[t]]$ such that the coefficients r_i span a finite-dimensional k-subspace of R.

(Note that the *ring* structures of R and of these power series algebras have not yet come under discussion; at this point we are simply describing isomorphisms as k-vector spaces.)

We also note

Lemma 24.23. *In the context of Lemma 24.19,* $\mathbf{Mod}_k(V, S)$ *is functorially isomorphic to* $\varprojlim \mathbf{Mod}_k(V_0, S)$, *where* V_0 *ranges over the finite-dimensional subspaces of* V; *equivalently, to* $\varprojlim \hat{V}_0\otimes S$.

Hence if A *is a linearly compact vector space and* S *any complete linearly topologized vector space,* $A\hat{\otimes} S$ *is naturally isomorphic to* $\varprojlim A_0\otimes_k S$, *as* A_0 *ranges over the factor-spaces of* A *by open subspaces.* □

We take a last look at the stronger topology on a tensor product characterized in Lemma 24.14 in the next exercise, which compares the completion of the tensor product under that topology with the above completion $\hat{\otimes}$ in a particular case.

Exercise 24.24. *Let* $k[s]$ *denote the underlying discrete vector space of the polynomial algebra in one indeterminate* s *over* k, *and* $k[[t]]$, *as above, the linearly compact vector space of formal power series in an indeterminate* t. *Let us identify the tensor product of these two vector spaces with the underlying vector space of the polynomial algebra over the formal power series algebra:* $k[[t]][s]$. *(This is a case of the identification* $R\otimes_k k[s]\cong R[s]$, *valid for an arbitrary*

k-algebra R.)

(i) *Show that the linear topology of Lemma 24.14 on this space (the strongest linear topology making tensor multiplication $k[\![t]\!] \times k[s] \to k[\![t]\!][s]$ continuous) has a basis of open subspaces indexed by all sequences of nonnegative integers, such that the subspace corresponding to a sequence $(n_0, n_1, \dots, n_i, \dots)$ consists of those polynomials in which for every i the coefficient of s^i is divisible in $k[\![t]\!]$ by t^{n_i}.*

(ii) *Show that $k[\![t]\!][s]$ is already complete in this (nondiscrete) topology.*

Thus the completion of the above tensor product with respect to this universal topology may still be naturally identified as a vector space with $k[\![t]\!][s]$, while by Example 24.22 the completion with respect to the topology of uniformly continuous tensor multiplication looks like $k[s][\![t]\!]$. The latter ring is much larger than the former; for example

(iii) *Show that the formal power series $(1-st)^{-1}$ belongs to $k[s][\![t]\!]$, but not to $k[\![t]\!][s]$.*

The next result follows easily from Lemma 24.19.

COROLLARY 24.25. *If $U,\ V \in \mathbf{Mod}_k$, then the natural map $\hat{U} \otimes_k \hat{V} \to \widehat{U \otimes_k V}$ taking a decomposable tensor $a \otimes b$ $(a \in \hat{U},\ b \in \hat{V})$ to the functional $u \otimes v \mapsto a(u)\,b(v)$ extends to an isomorphism*

$$\hat{U} \mathbin{\hat{\otimes}} \hat{V} \cong \widehat{U \otimes_k V}. \quad \square$$

In other words, under the duality $\hat{\ }$ between $\mathbf{Mod}_k$ and $\mathbf{LCpMod}_k$, the bifunctor $\otimes_k$ on $\mathbf{Mod}_k$ corresponds to the bifunctor $\hat{\otimes}$ on $\mathbf{LCpMod}_k$. (This gives *some* justification to our double use of the symbol $\hat{\ }$.)

As an example, consider the completed tensor product $k[\![s]\!] \mathbin{\hat{\otimes}} k[\![t]\!]$ of two copies of the formal power series algebra in one indeterminate. By Example 24.22, this has the form $k[\![s]\!]\,[\![t]\!]$, or as this is commonly written, $k[\![s,\, t]\!]$, the formal power series algebra in two indeterminates. If we write $k[\![s]\!]$ and $k[\![t]\!]$ as duals of discrete spaces with bases of formal coefficients $\{c_0,\ c_1,\ \dots\}$ and $\{d_0,\ d_1,\ \dots\}$, the above Corollary tells us that their completed tensor product will be the dual of a vector space with basis $\{c_i \otimes d_j \mid i, j \geq 0\}$; and indeed, we can clearly describe $k[\![s,\, t]\!]$ in this way, regarding $c_i \otimes d_j$ as a formal coefficient of $s^i t^j$.

It is curious that in (24.21), because V is discrete, the underlying vector space of the right hand side does not depend on the topology on S, hence the same must be true of the left hand side, as long as the hypothesis that S is complete is satisfied. (This condition is needed; for if we take $V = k$, and an arbitrary linearly topologized vector space for S, the right-hand side gives S, but the left-hand side gives the completion of S.) Thus, in the above example, we would still have gotten $k[\![s,\, t]\!]$ for the underlying vector space of $k[\![s]\!] \mathbin{\hat{\otimes}} k[\![t]\!]$ if we had taken only *one* of $k[\![s]\!]$, $k[\![t]\!]$ to have the linearly compact topology, and made the other discrete, or given it any other complete linear topology. You can prove a converse of this observation, as the (c)$\Rightarrow$(a) part of:

EXERCISE 24.26. *Let* A *be a complete linear topological vector space. Show that the following conditions are equivalent:*

(a) A *is linearly compact.*

(b) *For every surjective homomorphism of complete linear topological vector spaces,* $f\colon B \to C$, *the homomorphism* $A \mathbin{\hat{\otimes}} f\colon A \mathbin{\hat{\otimes}} B \to A \mathbin{\hat{\otimes}} C$ *is also surjective.*

(c) *Same condition as* (b), *for the particular situation where* C *is a linearly compact vector space,* B *the same space with the discrete topology, and* f *the map which is the identity on underlying sets.*

(*Suggestion: recall Exercise 24.7,* (a)$\Leftrightarrow$(c).)

Though the spaces we have used in our examples were underlying spaces of topological rings, we have not yet dealt with the ring structures. Let us define a not necessarily associative *linear topological k-algebra* as a linear topological k-vector-space given with a continuous k-bilinear multiplication, $A \times A \to A$. By Lemma 24.17(iii), if A is linearly compact or discrete, this multiplication is equivalent to a continuous linear map $A \mathbin{\hat{\otimes}} A \to A$. For a general linear topological k-vector-space A, that equivalence no longer holds (cf. our earlier example of $k((t))$), but one direction clearly remains valid: a continuous linear map $A \mathbin{\hat{\otimes}} A \to A$ yields a structure of linear topological k-algebra on A.

If A and B are linear topological k-algebras whose multiplications are induced by maps $A \mathbin{\hat{\otimes}} A \to A$, $B \mathbin{\hat{\otimes}} B \to B$ (for instance, if each is linearly compact or discrete), then as a consequence of the associativity of $\hat{\otimes}$, $A \mathbin{\hat{\otimes}} B$ again becomes a linear topological algebra. Recall that elements of the non-completed tensor product, $A \otimes_k B$, are dense in $A \mathbin{\hat{\otimes}} B$. It follows that the ring-theoretic identities holding in $A \mathbin{\hat{\otimes}} B$ will be the same as those holding in $A \otimes_k B$. For instance, since a tensor product of associative k-algebras is associative, a *completed* tensor product of two linearly compact associative k-algebras, or of a linearly compact and a discrete associative k-algebra, is again associative.

The nondiscrete algebras of importance to us will be the linearly compact ones. We make

CONVENTION 24.27. *Categories of linearly compact topological k-algebras will be symbolized by prefixing the symbol* **LCp** *to the symbols for the corresponding categories of nontopological algebras. For example, the category of linearly compact associative unital k-algebras will be denoted* $\mathbf{LCpRing}^1_k$.

§25. Functors to associative rings, Lie rings, and Jordan rings.

We are now ready to apply the above ideas to the study of representable **Ab**-valued functors on categories of k-algebras, and bilinear operations on such functors. Let us begin by translating what we know about such functors and operations, Theorem 23.9, into descriptions in terms of linearly compact vector spaces in the case where k is a field and our k-rings are k-algebras. In addition to bilinear operations, we shall also note the description of the general zeroary operation (so as to be able to handle functors to varieties of rings with 1), and of the general unary linear operation (though we will not use these here).

We continue to assume k a field (Convention 24.27). In writing decomposable elements of objects $A \hat{\otimes} S$, we shall suppress the sign $\otimes$, writing as for $a \otimes s$. We will generally specify a linear or bilinear map on a completed tensor product $A \hat{\otimes} S$ by describing its action on such elements as, respectively on pairs of such elements, in a manner continuous in our topology.

In statement (ii) of the next Theorem, I denotes the *trivial* functor $\mathbf{Ring}^1_k \to \mathbf{Set}$, taking every k-algebra S to a 1-element set. This can be described as

$$S \mapsto |S|^0 \cong \mathbf{Set}(\varnothing, |S|) \cong \mathbf{Mod}_k(\{0\}, S) \cong \mathbf{Ring}^1_k(k, S).$$

We also write 1_S (which until now has denoted the identity morphism of an object S) for the multiplicative neutral element of the unital ring S. (We shall use this notation at a few places in subsequent chapters as well. Fortunately, we have no further need for 1_S as a symbol for identity morphisms.)

THEOREM 25.1. *Let k be a field, and $F\colon \mathbf{Ring}^1_k \to \mathbf{Ab}$ a representable functor. Then*

(i) *F can be written $S \mapsto A \hat{\otimes} S$ for a fixed linearly compact k-vector-space A.*

(ii) *Zeroary operations $z\colon I \to F$ correspond to elements $\zeta \in A$, by the formula*

$$z_S = \zeta 1_S \quad (\text{i.e., } \zeta \otimes 1_S \in A \hat{\otimes} S). \tag{25.2}$$

(iii) *Linear operations $f\colon F \to F$ correspond to continuous k-linear maps $\varphi\colon A \to A$, by the formula*

$$f(as) = \varphi(a)\, s. \tag{25.3}$$

(iv) *Bilinear operations $m\colon F \times F \to F$ correspond to* pairs *of continuous k-bilinear maps*

$$^*\colon A \times A \to A, \qquad {}_*\colon A \times A \to A \tag{25.4}$$

via the formula

$$m(as, bt) = (a^*b)(st) + (b_*a)(ts). \tag{25.5}$$

The same results hold for representable functors on $\mathbf{Ring}_k$ except for (ii)*: There are no zeroary operations on such an F other than the zero operation, $z_S = 0$.*

PROOF. By Theorem 13.15 and the observations of §14, F is represented by $k\langle V\rangle$ for some $V \in \mathbf{Mod}_k$. Let $A = \hat{V} \in \mathbf{LCpMod}_k$. Then (i) follows by (24.21).

To get (ii), suppose z is any zeroary operation on F, and consider its value at $S = k$. This element $z_k \in A \hat{\otimes} k \cong A$ may be written $\zeta 1_k$ for some $\zeta \in A$. By functoriality of F, this maps to $\zeta 1_S$ in each $F(S)$, as required.

(We can get the same result by going back to first principles: A zeroary co-operation on R corresponds to a homomorphism $\mathbf{z}\colon R = k\langle V\rangle \to k$, which is determined by its restriction to V, a functional $\zeta \in \hat{V} = A$. Following the identifications $F(S) = \mathbf{Ring}^1_k(k\langle V\rangle, S) \cong \mathbf{Mod}_k(V, S) \cong A \hat{\otimes} S$, we find that for each S the element of $A \hat{\otimes} S$ determined by $\mathbf{z}$ is $\zeta \otimes 1_S$, as desired.)

(iii) is straightforward by Proposition 24.8(a)⇔(b).

Finally, suppose m is a bilinear operation on F. By Theorem 23.9 and

Corollary 23.11, m corresponds to a vector-space map

$$\mathbf{m}\colon\ V \to (V^\lambda \otimes_k V^\rho) \oplus (V^\rho \otimes_k V^\lambda).$$

Let us write this as a sum of two maps, $\mathbf{m}'\colon V \to V^\lambda \otimes_k V^\rho$ and $\mathbf{m}''\colon V \to V^\rho \otimes_k V^\lambda$. These dualize to continuous linear maps $A^\lambda \hat{\otimes} A^\rho \to A$ and $A^\rho \hat{\otimes} A^\lambda \to A$ which are equivalent to continuous bilinear maps * and $_*$ as in (25.4). We shall now show that m is described by (25.5). Let us be fairly detailed in our derivation of this result, because of its key importance and its unfamiliar nature.

Given $a, b \in A$, and $s, t \in S$, we recall that as means $a \otimes s \in A \hat{\otimes} S$, which corresponds to the element of $\mathbf{Mod}_k(V, S)$ given by the composite map

$$V \xrightarrow{\ a\ } k \xrightarrow{\ 1 \mapsto s\ } S, \tag{25.6}$$

and thus to the element of $\mathbf{Ring}^1_k(k\langle V\rangle, S)$ which this induces; and bt is likewise given by

$$V \xrightarrow{\ b\ } k \xrightarrow{\ 1 \mapsto t\ } S. \tag{25.7}$$

The map $m(as, bt)\colon k\langle V\rangle \to S$ is gotten by composing $\mathbf{m}\colon k\langle V\rangle \to k\langle V^\lambda \oplus V^\rho\rangle$ with the map $k\langle V^\lambda \oplus V^\rho\rangle \to S$ induced by this pair of maps. To describe this composite it suffices to determine what it does on V. Since the first factor, $\mathbf{m}$, acts on V as a sum of two maps $\mathbf{m}'$, $\mathbf{m}''$, and the second factor acts on V^λ by (25.6) and on V^ρ by (25.7), the total action can be expressed as the sum of the upper and lower routes of the (noncommutative) diagram

$$V \begin{array}{l} \xrightarrow{\ \mathbf{m}'\ } V^\lambda \otimes_k V^\rho \\[2ex] \xrightarrow{\ \mathbf{m}''\ } V^\rho \otimes_k V^\lambda \end{array} \left\{\begin{array}{l} (25.6)\colon v^\lambda \mapsto a(v)s \\ (25.7)\colon v^\rho \mapsto b(v)t \end{array}\right\} \rightrightarrows S. \tag{25.8}$$

Now the top branch can be written

$$V \xrightarrow{\ \mathbf{m}'\ } V^\lambda \otimes_k V^\rho \xrightarrow{\ a \otimes b\ } k \xrightarrow{\ 1 \mapsto st\ } S, \tag{25.9}$$

and by our definition of * (as the multiplication on A corresponding to the operation $\mathbf{m}'$ on V), the first two arrows compose to a^*b, so the full composite is the element $(a^*b)(st)$. The lower branch similarly gives $(b_*a)(ts)$. We have thus established (25.5).

Everything we have done here except the proof of (ii) goes over essentially unchanged to functors on the variety $\mathbf{Ring}_k$ of nonunital k-algebras. (Note that in (25.{6, 7, 9}), k does not appear as a k-algebra, but as the 1-dimensional vector space in terms of which the duality between V and A is defined; i.e., "1" is used not as a multiplicative unit but as a canonical basis element.) The statement replacing (ii) in the nonunital case, namely that all zeroary operations are trivial, is immediate. (A representable functor from a pointed category $\mathbf{C}$ to $\mathbf{Set}$ clearly admits one and only one zeroary operation.) □

Our tools are now complete! Their first job will be the characterization of representable functors from $\mathbf{Ring}_k^1$ to the category $\mathbf{Ring}^1$ of *unital associative rings*.

By the above Theorem, such a functor F is determined by a linearly compact vector space A, an element of A, which we shall denote ε, giving the multiplicative neutral element, and two continuous bilinear maps ${}^*, {}_*: A \times A \to A$, determining the multiplication on F. Let us write out the *associative law* for the multiplication of $F(S) = A \hat{\otimes} S$ $(S \in \mathbf{Ring}_k^1)$ (which we shall symbolize by $\cdot$ rather than m). Since the law is multilinear, it will hold on all elements of $A \hat{\otimes} S$ if and only if it holds on decomposable elements; so it can be written

$$as\cdot(bt\cdot cu) \;=\; (as\cdot bt)\cdot cu \qquad (a,\ b,\ c \in A;\ \ s,\ t,\ u \in S).$$

Expanding by (25.5), we get

$$\begin{aligned} &a^*(b^*c)(stu) + a^*(c_*b)(sut) + (b^*c)_*a(tus) + (c_*b)_*a(uts) \\ (25.10)\qquad &= (a^*b)^*c(stu) + (b_*a)^*c(tsu) + c_*(a^*b)(ust) + c_*(b_*a)(uts). \end{aligned}$$

In particular, this must hold for $S =$ the free algebra $k\langle s, t, u\rangle$, hence we can equate coefficients for each product stu, tsu, etc. in this equation. We get six equations

$$(25.11)\qquad a^*(b^*c) = (a^*b)^*c, \qquad\qquad (c_*b)_*a = c_*(b_*a)$$

$$(25.12)\qquad a^*(c_*b) = 0, \qquad (b^*c)_*a = 0, \qquad 0 = (b_*a)^*c, \qquad 0 = c_*(a^*b).$$

In other words, * and ${}_*$ are each associative; and each annihilates, on both the right and the left, the range of the other.

We now bring in the multiplicative neutral element $\varepsilon 1$. Left-multiplying $as \in F(S)$ by this element, equating the result with as, and taking the coefficient of s, we get

$$(25.13)\qquad a \;=\; \varepsilon^*a + a_*\varepsilon.$$

This gives us a decomposition of a as a sum of an element annihilating ${}_*$ on both sides and one annihilating * on both sides (by (25.12)). Now putting ε^*a in place of a throughout (25.13), and using this annihilation property, we find that the linear operator ε^*- is idempotent, hence it induces a direct-sum decomposition

$$(25.14)\qquad A \;=\; A' \oplus A'',$$

which we see must coincide with the decomposition (25.13). In particular, $\varepsilon^*\varepsilon$ will lie in A', and be a left neutral element for this summand under *, and by a like argument, $\varepsilon_*\varepsilon$ lies in A'' and is a right neutral element for this summand under ${}_*$.

Although we have so far used only the left neutral law for $\varepsilon 1$, which is not left-right symmetric, these observations allow us to deduce a left-right symmetric characterization of the summands in (25.14); namely, A' is the range of *, and A'' the range of ${}_*$. Hence if we now write down the *right* neutral law, $a = a^*\varepsilon + \varepsilon_*a$, it will give the same decomposition; but this time we find that $\varepsilon^*\varepsilon$ is a multiplicative right neutral element for A' under * and $\varepsilon_*\varepsilon$ a multiplicative left neutral element for A'' under ${}_*$. Thus $\varepsilon^*\varepsilon$ and $\varepsilon_*\varepsilon$ are two-sided neutral elements for the appropriate spaces and multiplications.

Hence the structure we have on A is determined by a decomposition (25.14) as linearly compact vector spaces, a continuous associative algebra structure on A', given by * and having a neutral element $\varepsilon^*\varepsilon$, and a continuous associative algebra structure on A'', given by $_*$ and having a neutral element $\varepsilon_*\varepsilon$. In summary:

THEOREM 25.15. *Any representable functor* $\mathbf{Ring}_k^1 \to \mathbf{Ring}^1$ *has the form*

$$S \mapsto (A' \hat{\otimes} S) \times (A'' \hat{\otimes} S)^{\mathrm{op}}, \tag{25.16}$$

where A', A'' *are linearly compact associative unital k-algebras. This construction induces an equivalence of categories*

$$\mathbf{LCpRing}_k^1 \times \mathbf{LCpRing}_k^1 \xrightarrow{(A',A'') \mapsto (25.16)} \mathbf{Rep}(\mathbf{Ring}_k^1, \mathbf{Ring}^1). \quad \square \tag{25.17}$$

25.18. *Remark.* In (25.16), and in similar constructions below, when we refer to a (completed or ordinary) tensor product of k-algebras, we understand it to be initially made an algebra under the "ordinary" tensor product of the given operations. E.g., above, $(A'', {}_*) \hat{\otimes} S$ initially has the operation $(as)\cdot(bt) = (a_*b)(st)$. Other operations considered will be described in terms of these initial operations; e.g., in (25.16) we applied the operator $(\ )^{\mathrm{op}}$ to this multiplication on $A'' \hat{\otimes} S$.

Note also that we are suppressing the topologies and k-vector-space structures on these rings; that is, there is a tacit "underlying discrete ring" functor applied to each of the completed tensor products in (25.16), since the codomain of the representable functors we are considering is $\mathbf{Ring}^1$.

25.19. *Examples*: For each of A', A'' in (25.16) we might take such linearly compact algebras as $\{0\}$, k, $M_n(k)$ ($n \times n$ matrix ring), $\prod_{i \geq 1} M_i(k)$, $k[[t]]$, $k[[s,t]]$, $k\langle\langle s,t\rangle\rangle$ (formal power series in two noncommuting indeterminates), etc.. So a typical representable functor is $S \mapsto S[[s,t]] \times S\langle\langle s,t\rangle\rangle^{\mathrm{op}}$.

25.20. *Examples.* A *nonrepresentable* functor between these same categories is given by the polynomial ring construction $S \mapsto S[t]$. For if it admitted a representing object R, then the identity morphism of R would correspond to a universal polynomial in t over a k-algebra, and letting d be the degree of this polynomial, it would follow from its universal property that all polynomials in t over all k-algebras had degree $\leq d$, which is absurd. Alternatively, one can deduce that this functor is nonrepresentable by noting that it does not respect direct products — in an infinite direct product of polynomial rings, consider an element whose components have unbounded degrees.

However, in some ways this functor is "close to" representable. Note that we had to look at *infinite* products to get our counterexample. Observe also that a representing object, if it existed, would be a k-algebra R whose homomorphisms into any k-algebra S corresponded to all sequences of elements in S indexed by the nonnegative integers, *with only finitely many nonzero terms.* We cannot choose

R so as to get this finiteness condition, but if we drop that condition, we are left with the description of the functor $S \mapsto S[[t]]$, which is indeed representable. In fact, as we shall note in §58.3, if we look at $\mathbf{Ring}_k^1$ within the larger category $\mathbf{LTopRing}_k^1$, then the polynomial-ring construction on the former category can be expressed as the restriction of a representable functor on the latter category.

A less subtle example of a nonrepresentable ring-valued functor is $S \mapsto S \otimes_k S$; its nonrepresentability can be seen from the fact that it does not even respect finite direct products.

If F is a representable functor $\mathbf{Ring}_k^1 \to \mathbf{Ring}^1$, as characterized in the above Theorem, it is straightforward to characterize L-algebra or L-ring structures on F, whether for $L = k$ or another ring; cf. Theorem 25.1(ii). But as promised in the last paragraph of §14, we shall not go into this.

25.21. *Remark.* In the above two Theorems, we might have chosen to reverse the order in which the arguments are written in our second multiplication map on A, defining, say, $a \circ b = b_* a$, and expressing our structure in terms of * and $^\circ$. Then a and b would appear in constant order in (25.5), and the general representable functor to $\mathbf{Ring}_k^1$ would assume the form

$$S \mapsto (A' \hat{\otimes} S) \times (A'' \hat{\otimes} S^{\mathrm{op}}).$$

This looks neither better nor worse than (25.16); but on the basis of other considerations, we claim that the choice we made was the right one. If one generalizes Theorem 25.1 to situations where k is not central, this reversed map $^\circ$ is not bilinear with respect to the given bimodule structure on A, but with respect to the *reversed* bimodule structure. Likewise, Lemma 25.27 below is more naturally expressed in the present notation. Finally, the fact that the terms a, b, c etc. permute themselves in the same way as s, t, u etc. will prove convenient in general arguments concerning the forms of the identities we obtain.

A curious consequence of Theorem 25.15 is:

COROLLARY 25.22. *The category* $\mathbf{Rep}(\mathbf{Ring}_k^1, \mathbf{Ring}^1)$ *has for initial object the functor* $S \mapsto S \times S^{\mathrm{op}}$. □

It will be shown in [**26**] that for arbitrary varieties of algebras $\mathbf{C}$ and $\mathbf{D}$, the category $\mathbf{Rep}(\mathbf{C}, \mathbf{D})$ has an initial object. Surprisingly, this is not easy to establish. Examples will show that these initial objects can have quite varied forms.

Another immediate consequence of Theorem 25.15 is:

COROLLARY 25.23. *There is no nontrivial representable functor from* $\mathbf{Ring}_k^1$ *to the category* $\mathbf{Comm}^1$ *of* commutative *associative unital rings.* □

In the proof of Theorem 25.15, we got a great deal of mileage from the *left* neutral identity for $\varepsilon 1$, including the fundamental decomposition (25.14), and only had to call on the right neutral identity in showing that $\varepsilon^* \varepsilon$ and $\varepsilon_* \varepsilon$ were two-sided neutral elements for A' and A''. Without the latter identity, we get

COROLLARY 25.24 (to the proof of Theorem 25.15). *Every representable functor* F *from* $\mathbf{Ring}_k^1$ *to the variety of associative rings given with a multiplicative* left neutral *element has the form* $S \mapsto (A' \hat{\otimes} S) \times (A'' \hat{\otimes} S)^{\mathrm{op}}$, *where* A' *is a linearly compact associative algebra given with a left neutral element, and* A'' *is a linearly compact associative algebra given with a right neutral element.* □

In the absence of even a one-sided neutral element, we have a less elegant description of F, already mentioned informally:

THEOREM 25.25. *Every representable functor* $F\colon \mathbf{Ring}_k^1 \to \mathbf{Ring}$ *or* $\mathbf{Ring}_k \to \mathbf{Ring}$ *has the form* $S \mapsto A \hat{\otimes} S$, *where* A *is a linearly compact vector space given with two associative multiplications* * *and* $_*$, *such that the range of each lies in the two-sided annihilator ideal of the other, and where the multiplication of* $F(S)$ *is the sum of the multiplications of* $(A, {}^*) \hat{\otimes} S$ *and* $((A, {}_*) \hat{\otimes} S)^{\mathrm{op}}$. □

The annihilator ideals of * and $_*$ in the above construction need not be disjoint, nor need the images of these multiplications. As an example, if we take for * any cube-zero multiplication on A (any bilinear operation under which $A{}^*(A{}^*A) = (A{}^*A){}^*A = 0$), and for $_*$ the same operation or its negative, we get a *commutative*, respectively *anticommutative* operation on $A \hat{\otimes} S$ which has cube zero, hence is associative. For instance, letting $A = tk[t]/t^3k[t]$, a 2-dimensional k-algebra, with * its ordinary multiplication, we get functors F such that the additive group of $F(S)$ is $S \times S$, and the product $(a, b)\cdot(a', b')$ is given by $(0, aa' + a'a)$, respectively $(0, aa' - a'a)$. It is in fact easy to verify that all functors from $\mathbf{Ring}_k^1$ to the varieties of commutative, respectively anticommutative, associative rings arise in this way from cube-zero operations. These are further examples of "associativity by default", such as we saw in §§16 and 21.

A somewhat more general criterion than Corollary 25.24 for our functors to have descriptions of the form (25.16), and some negative examples that are not simply cases of "associativity by default", are developed in

EXERCISE 25.26. (i) *Suppose* $F\colon \mathbf{Ring}_k^1 \to \mathbf{Ring}$ *is a representable functor such that the ring* $F(k)$ (a) *is idempotent:* $F(k) = F(k)F(k)$ (*that is, every element of* $F(k)$ *is a sum of products*), *and* (b) *contains no nonzero element* x *such that* $xF(k) = \{0\} = F(k)x$. *Show that* F *can be written as in* (25.16), *for unique linearly compact algebras* A' *and* A'' *each satisfying conditions* (a) *and* (b).

The next two parts will show that if either condition (a) *or* (b) *above is removed, we may not have a decomposition as in* (25.16). (*The functors we will consider will correspond to objects* $(A, {}^*, {}_*)$ *which can be constructed in one case as subobjects and in the other as homomorphic images of objects which do admit such a decomposition.*)

(ii) *Let* G *be the functor associating to* $S \in \mathbf{Ring}_k^1$ *the ring* $[S][[t]] \times [S][[t]]^{\mathrm{op}}$. *Let* $F(S) \subseteq G(S)$ *denote the set of those pairs of formal power series* $(\Sigma\, \xi_i t^i, \Sigma\, \zeta_i t^i)$ *such that* $\xi_1 = \zeta_1$. *Show that* $F(S)$ *forms a subring of* $G(S)$, *and gives a representable functor, which satisfies condition* (b) *but not condition* (a) *of part* (i) *above, and which cannot be expressed as in* (25.16). (*More generally, for any positive integers* m *and* n, *if one replaces the two occurrences of* $[S][[t]]$ *in the above description by* $t^m S[[t]]$ *and* $t^n S[[t]]$ *respectively, and the condition of equality of the first coefficients by* any *family of* k*-linear relations among* $\xi_m, \dots, \xi_{2m-1}, \zeta_n, \dots, \zeta_{2n-1}$, *one can verify that one gets a subring-valued functor. E.g., letting* $m = n = 2$, *one finds that the set of pairs of formal power series both divisible by* t^2, *satisfying the unlikely conditions* $\xi_2 = \zeta_3$, $\xi_3 = \zeta_2$ *gives a ring, and the resulting representable functor is not of the form* (25.16).)

(iii) *Let* $A' = A'' =$ *the* 4*-dimensional subalgebra of the* 3×3 *matrix algebra* $M_3(k)$ *spanned by* $e_{12}, e_{13}, e_{22}, e_{23}$. *Let* H *be the functor associating to* $S \in \mathbf{Ring}^1_k$ *the ring* $A \otimes_k S \times (A \otimes_k S)^{\mathrm{op}}$. *Show that within* $H(S)$, *the subset consisting of all elements* $(\xi e_{13}, -\xi e_{13})$ $(\xi \in S)$ *forms an ideal* $I(S)$, *and that the functor defined by* $F(S) = H(S)/I(S)$ *is representable and satisfies condition* (a) *but not* (b) *of part* (i), *and cannot be expressed as in* (25.16).

(iv) *Show that there exists a functor which, like the* F *of part* (ii) *above, is representable and satisfies condition* (b) *but not* (a), *and does not admit a decomposition* (25.16), *and which moreover, like the* F *of part* (iii), *has* $F(k)$ *finite-dimensional.* (*Hint: the* 8*-dimensional algebra* $A' \subseteq M_4(k)$ *generated by the five elements* $e_{11}, e_{12}, e_{23}, e_{34}, e_{44}$ *has zero two-sided annihilator ideal, but no product of two elements of* A' *involves* e_{23}.)

Though commutative or anticommutative associative ring-valued functors on $\mathbf{Ring}^1_k$ must, as we have seen, be quite degenerate, this is not true of commutative or anticommutative *nonassociative* ring-valued functors. Let us record the easily verified criteria for the operation (25.5) to satisfy each of these conditions.

LEMMA 25.27. *The operation* m *described in* (25.5) *is commutative if and only if the operations* * *and* $_*$ *are equal, anticommutative if and only if they are negatives of one another.*

In particular, for such an m *the anticommutativity identity* $m(x, y) + m(y, x) = 0$ *is equivalent to the alternating identity* $m(x, x) = 0$, *even if* char $k = 2$. □

Starting with these observations, we can now obtain characterizations of representable functors from $\mathbf{Ring}^1_k$ to Lie and Jordan rings with surprising ease. The definitions of these two varieties of nonassociative rings will be recalled below; general references are Jacobson's books [**100**] and [**101**].

First, suppose F is as in Theorem 25.1, m as in (25.5), and for conformity with Lie notation let us write $m(x, y)$ as $[x, y]$. Assuming this operation is alternating, we have by the above Lemma and (25.5)

$$[as,\ bt] \;=\; a^*b\,(st) - b^*a\,(ts) \qquad (a, b \in A;\ s, t \in S). \tag{25.28}$$

To be a Lie operation, this bracket product must also satisfy the Jacobi identity:

$$[x,\ [y,\ z]] + [y,\ [z,\ x]] + [z,\ [x,\ y]] \;=\; 0 \qquad (x, y, z \in A \mathbin{\hat{\otimes}} S). \tag{25.29}$$

Let us substitute $x = as$, $y = bt$, $z = cu$ in this identity, evaluate using (25.28), and take the coefficient of each monomial in s, t, u. Since the left hand side of

(25.29) is alternating in the three variables, it actually suffices to look at the coefficient of the single monomial stu; this gives the equation

$$a^*(b^*c) + 0 - (a^*b)^*c = 0 \tag{25.30}$$

– the condition of associativity for $*$! So the data determining the functor F reduce to a linearly compact vector space A with a single continuous *associative* multiplication $*$, i.e., an object $(A, *) \in \mathbf{LCpRing}_k$.

The operation (25.28) is that of commutator brackets with respect to the multiplication $(as)\cdot(bt) = a^*b(st)$. So letting

$$L\colon \mathbf{Ring} \to \mathbf{Lie}$$

denote the functor carrying every associative ring to the Lie ring with the same additive group, and with commutator brackets for the Lie multiplication, we have

THEOREM 25.31. *Any representable functor* $\mathbf{Ring}_k^1 \to \mathbf{Lie}$ *can be written as a composite*

$$\mathbf{Ring}_k^1 \xrightarrow{A\,\hat{\otimes}\,-} \mathbf{Ring} \xrightarrow{L} \mathbf{Lie}, \tag{25.32}$$

where $A \in \mathbf{LCpRing}_k$ *and* L *is as defined above. This construction induces an equivalence of categories:*

$$\mathbf{LCpRing}_k \xrightarrow{A \mapsto (25.32)} \mathbf{Rep}(\mathbf{Ring}_k^1, \mathbf{Lie}). \tag{25.33}$$

The same statements hold with $\mathbf{Ring}_k$ *in place of* $\mathbf{Ring}_k^1$. □

From one point of view this is a dull result: it says that any representable functor constructing Lie rings from associative algebras reduces to the long-known construction L. From another point of view it is surprising. It says that every Lie structure on an object of the category $(\mathbf{Ring}_k^1)^{\mathrm{op}}$ is induced by an associative-ring structure on the same object, which is much stronger than the nearest result holding for set-based Lie algebras over a field, namely the consequence of the Poincaré-Birkhoff-Witt Theorem that every Lie algebra over a field can be *embedded* in an associative algebra ([**100**], [**19**, §3]).

We turn next to Jordan rings. Here we will get a similar result, which in this context is still more surprising. Recall that Jordan rings are modeled on the structure on the underlying additive group of an associative ring R given by the anti-commutator operation

$$(x, y) = xy + yx. \tag{25.34}$$

However, the identities used as the definition of Jordan rings are *not* a generating set for *all* identities satisfied in the additive group of an associative ring by the operation (25.34), but only for such identities of degree ≤ 4. Jordan rings which satisfy all the identities of the operation (25.34) form a proper subvariety, called the *semispecial* Jordan rings. The definition commonly used is that a semispecial Jordan ring is one which can be written as a *homomorphic image* of a Jordan ring *embeddable* in an associative ring via (25.34). Such embeddable Jordan rings are themselves called *special*. Jordan rings which themselves admit an associative multiplication such that (25.34) holds form a smaller class still. Our result will imply that for unital Jordan ring objects of $(\mathbf{Ring}_k^1)^{\mathrm{op}}$ the analogs of these four classes, i.e., the classes of arbitrary Jordan ring objects, semispecial Jordan ring objects, special Jordan ring objects, and Jordan ring objects admitting compatible associative ring structures, all coincide.

The details, now. A Jordan ring is defined [**101**] as a *commutative* but not necessarily associative ring whose multiplication $(\ ,\ )$ satisfies the identity

$$((x,x),(x,y)) = (x,((x,x),y)). \tag{25.35}$$

Let F be as in Theorem 25.1, m as in (25.5), and let us write (x,y) for $m(x,y)$. (Jordan theorists often simply write xy, for instance writing (25.35) as $x^2(xy) = x(x^2y)$.) By Lemma 25.27 and (25.5), if m is commutative we have

$$(as, bt) = a^*b(st) + b^*a(ts) \qquad (a, b \in A;\ s, t \in S). \tag{25.36}$$

In writing down the conditions on * corresponding to (25.35), it is not sufficient to consider the case $x = as$, $y = bt$, because (25.35) is not linear in x. It is not hard to see, however, that a substitution sufficiently general to take account of the degrees of x and y in (25.35) is $x = as + bt + cu$, $y = dv$. Moreover, when we make this substitution, we need only look at the coefficients of monomials *linear* in each of s, t, u, v. For these carry all the information carried by any monomials occurring; e.g., from the coefficient of $stuv$, we may obtain that of $sttv$ by substituting b for c throughout. Finally, by symmetry in s, t and u, we can restrict attention to those monomials in which these three letters occur in that order. There are four such monomials. These monomials, and the equations they yield, are

$$\begin{aligned} stuv&: & 2(a^*b)^*(c^*d) &= 2a^*((b^*c)^*d), \\ stvu&: & 2(a^*b)^*(d^*c) &= 2((a^*b)^*d)^*c, \\ svtu&: & 2(a^*d)^*(b^*c) &= 2a^*(d^*(b^*c)), \\ vstu&: & 2(d^*a)^*(b^*c) &= 2(d^*(a^*b))^*c. \end{aligned} \tag{25.37}$$

If we assume $\operatorname{char} k \neq 2$ and divide by 2, and relabel variables in each equation so that the left-hand-sides all become $(a^*b)^*(c^*d)$, we see that (25.37) says that the five ways of associating four factors a, b, c, d in the (possibly nonassociative) ring $(A, {}^*)$ all give the same value. This is slightly weaker than full associativity. For instance it is satisfied by any algebra in which all 4-fold products are zero, though 3-fold multiplication may not be associative. But now suppose F also has a neutral element operation, i.e., that A has an element ε satisfying

$$a = \varepsilon^*a + a^*\varepsilon \qquad (a \in A). \tag{25.38}$$

We claim that this forces * to be associative. For consider the *associator* of three elements of A

$$a^*(b^*c) - (a^*b)^*c. \tag{25.39}$$

By "4-fold associativity", if we put either $a^*\varepsilon$ or ε^*a for a in (25.39), the result vanishes, hence the same will happen if we apply the substitution (25.38), so

$*$ is indeed associative.

All that this argument used of (25.38) was the consequence, $A = A^*A$. Let us now show assuming the full strength of (25.38) that ε is central in the associative algebra $(A, {}^*)$. If we put $a = \varepsilon$ in (25.38), we get $\varepsilon = 2\varepsilon^*\varepsilon$. Hence

$$\begin{aligned} \varepsilon^* a - a^*\varepsilon &= 2(\varepsilon^*\varepsilon^* a - a^*\varepsilon^*\varepsilon) \\ &= 2(\varepsilon^*(\varepsilon^* a + a^*\varepsilon) - (\varepsilon^* a + a^*\varepsilon)^*\varepsilon) \ = \ 2(\varepsilon^* a - a^*\varepsilon) \end{aligned}$$

(last step by (25.38) again). Comparing the initial and final expressions, we conclude that the common value is zero, so ε is central as claimed. Given this centrality, (25.38) clearly says 2ε is a neutral element in $(A, {}^*)$.

Let us write

$$J\colon \mathbf{Ring}^1_{\mathbf{Z}[1/2]} \to \mathbf{Jordan}^1$$

for the functor associating to a unital associative ring S in which 2 is invertible (a unital associative $\mathbf{Z}[1/2]$-algebra) the Jordan ring with the same additive group, multiplication given by (25.34), and hence $1/2$ as neutral element (or as some authors prefer, multiplication given by (25.34) with a factor $1/2$ thrown in, and 1 as neutral element). Then the above discussion yields

THEOREM 25.40. *Suppose* $\operatorname{char} k \neq 2$. *Then any representable functor* $\mathbf{Ring}^1_k \to \mathbf{Jordan}^1$ *can be written as a composite*

$$\text{(25.41)} \qquad \mathbf{Ring}^1_k \xrightarrow{A\,\hat{\otimes}\,-} \mathbf{Ring}^1_{\mathbf{Z}[1/2]} \xrightarrow{J} \mathbf{Jordan}^1,$$

where $A \in \mathbf{LCpRing}^1_k$, *and* J *is defined as above. This construction induces an equivalence of categories*

$$\text{(25.42)} \qquad \mathbf{LCpRing}^1_k \xrightarrow{A \,\mapsto\, (25.41)} \mathbf{Rep}(\mathbf{Ring}^1_k, \mathbf{Jordan}^1). \ \square$$

Thus, as asserted earlier, any representable functor from $\mathbf{Ring}^1_k$ to $\mathbf{Jordan}^1$ actually lands in the subvariety of semispecial Jordan rings, and, indeed, in the quasivariety of special Jordan rings, and in the still smaller class of Jordan rings admitting associative structures.

(Note: In making the codomain of the functor $A \,\hat{\otimes}\, -$ in (25.41) $\mathbf{Ring}^1_{\mathbf{Z}[1/2]}$ rather than $\mathbf{Ring}^1$, we are simply "forgetting" slightly less of the k-vector-space structure than at the corresponding stage of previous constructions.)

25.43. Observe that in (25.41), and likewise earlier in (25.32), the functor indicated by the first arrow does not give the most general representable functor to the indicated category of associative rings. (Cf. Theorems 25.15 and 25.25.) If one replaces this functor with the most general associative ring-valued construction, i.e., one involving "opposite" as well as ordinary multiplication, the composite will clearly still be a functor to Jordan, respectively Lie rings. But the above Theorems tell us that, due to the way in which the functors L and J mix regular and opposite multiplications, this will give us nothing new – whatever composite functor one gets this way can *also* be obtained, and in a *unique* way, using an "^{op}-free" construction for the first factor.

For Jordan ring valued functors, as for associative ring valued functors, the nonunital case is not as clean as the unital. Let us write $\mathbf{Ring}^{(n)}$ for the subvariety of **NARing** defined by the weak associativity identities saying that all bracketings of a product of n elements in a fixed order yield the same value. Thus

$$\mathbf{Ring} = \mathbf{Ring}^{(3)} \subseteq \mathbf{Ring}^{(4)} \subseteq \ldots \subseteq \mathbf{Ring}^{(n)} \subseteq \ldots \subseteq \mathbf{NARing}.$$

In the next two Corollaries, the symbol J is used for two obvious modifications of the functor so denoted in (25.41), the common feature being that multiplication is defined by (25.34). The first of these records the results of the calculations we had made on Jordan-valued representable functors up to the point where we introduced the unitality assumption (25.38).

COROLLARY 25.44 (to the proof of Theorem 25.40). *Suppose* $\operatorname{char} k \neq 2$. *Then any representable functor* $\mathbf{Ring}^1_k \to \mathbf{Jordan}$ *is a composite*

$$\mathbf{Ring}^1_k \xrightarrow{A \,\hat{\otimes}\, -} \mathbf{Ring}^{(4)} \xrightarrow{J} \mathbf{Jordan}, \tag{25.45}$$

where $A \in \mathbf{LCpRing}^{(4)}_k$. *This construction induces an equivalence between* $\mathbf{LCpRing}^{(4)}_k$ *and the category of such representable functors. The same result holds with* $\mathbf{Ring}_k$ *in place of* $\mathbf{Ring}^1_k$. □

Finally, the degenerate cases we have till now excluded:

COROLLARY 25.46. *Suppose* $\operatorname{char} k = 2$. *Then there are no nontrivial representable functors* $\mathbf{Ring}^1_k \to \mathbf{Jordan}^1$. *Dropping the unitality condition on our Jordan rings, any representable functor* $\mathbf{Ring}^1_k \to \mathbf{Jordan}$ *can be written as a composite*

$$\mathbf{Ring}^1_k \xrightarrow{A \,\hat{\otimes}\, -} \mathbf{NARing}_{\mathbf{Z}/2\mathbf{Z}} \xrightarrow{J} \mathbf{Jordan}, \tag{25.47}$$

where $A \in \mathbf{LCpNARing}_k$; *and this construction induces an equivalence of categories* $\mathbf{LCpNARing}_k \simeq \mathbf{Rep}(\mathbf{Ring}^1_k, \mathbf{Jordan})$. *The same results hold with* $\mathbf{Ring}_k$ *in place of* $\mathbf{Ring}^1_k$.

PROOF. Putting $a = \varepsilon$ in (25.38), we get $\varepsilon = 0$, establishing the first assertion. In characteristic 2, all the relations (25.37) are vacuous, giving the remaining results. To see these directly, recall that any commutative operation one can put on a representable functor in characteristic 2 will satisfy $(x, x) = 0$ by the final sentence of Lemma 25.27, and hence in particular, will satisfy the Jordan identity (25.35). □

Jordan theorists are aware that the ordinary concept of Jordan algebra is too weak when 2 is not assumed invertible, and have devised some alternative versions of the concept, involving quadratic operations, that are equivalent to ordinary Jordan algebras when 2 is invertible but better behaved in the general case. We shall characterize quadratic operations on our representable **Ab**-valued functors in §54-§55, and sketch in §56 how these results might be applied to the study of representable functors to such generalized Jordan rings.

The occurrence of *associative* operations on our linearly compact vector space A in most of the preceding results is not as mysterious as it may seem. Note that in each term of (25.5), the order of a and b parallels that of s and t. It follows that when we expand some identity to be imposed on m using (25.5), and take the coefficient of a given multilinear monomial in s, t, u etc. (as in (25.{11, 12, 30, 37})) the variables a, b etc. will appear in the same order in each summand. Thus, assuming that we are dealing with one binary operation and no operations of other arities, such terms can differ only in bracketing and in the choices of * versus $_*$. If such a system of conditions is strong enough, it is likely to lead to the simplest law of this sort, namely the associative law, for each of the operations * and $_*$.

However, in the next section, we shall look at some weaker systems of identities, which lead to more complicated conditions.

In view of the importance of *linearly compact associative* k-algebras in these results, let us recall the structure theorem for such algebras:

(25.48) Any linearly compact *associative* k-algebra is the inverse limit of an inverse system of *finite-dimensional* associative k-algebras (namely, its quotients by open ideals), under the inverse limit of the discrete topologies.

Modulo the duality theory reviewed in §24 above, this is [**183**, Corollary 2.2.2].

To see that (25.48) is *not* a trivial corollary of the analogous statement for vector spaces, Proposition 24.4(d), let us now show that the corresponding statement is false for linearly compact Lie algebras, and hence for linearly compact not necessarily associative algebras in general:

25.49. *Example. A linearly compact Lie algebra* L *which is not an inverse limit of finite-dimensional Lie algebras.*

Let us take L to have underlying linearly compact vector space $D \oplus k^{\mathbf{N}}$, where D is a 1-dimensional (hence discrete) k-vector-space with a generator we shall denote d, and $k^{\mathbf{N}}$ is the linearly compact vector space of all sequences $(\xi_0, \xi_1, \ldots)$ $(\xi_i \in k)$. Let us make each of D and $k^{\mathbf{N}}$ a Lie algebra using the zero bracket operation (an "abelian Lie algebra"); it remains to describe the bracket of d with elements of $k^{\mathbf{N}}$. We define $[d, -]$ to carry $k^{\mathbf{N}}$ into itself by the *left shift* map:

$$[d, (\xi_0, \xi_1, \ldots)] = (\xi_1, \xi_2, \ldots). \tag{25.50}$$

From the fact that the shift is (trivially) a derivation on the abelian Lie algebra $k^{\mathbf{N}}$, it follows that (25.50) determines a Lie algebra structure on $D \oplus k^{\mathbf{N}}$; and from the fact that the shift operator is continuous, we see that the resulting Lie brackets are continuous.

Now suppose I is the kernel of a homomorphism of L to a finite-dimensional Lie algebra. Thus I is an open ideal in L, hence by the nature of the topology of $k^{\mathbf{N}}$, for some n, I must contain the subspace thereof consisting of all sequences with 0 for their first n terms. But being a Lie ideal, I must be closed under bracketing with d, and the closure under d of the space just described is all of $k^{\mathbf{N}}$. Hence the intersection of all such ideals I contains $k^{\mathbf{N}}$, showing that L is not an inverse limit of finite dimensional images.

(If $\operatorname{char} k = 0$, and we identify $k^{\mathbf{N}}$ with $k[[t]]$ by $(\xi_0, \xi_1, \ldots) \mapsto \Sigma\, \xi_i t^i/i!$, then d acts as differentiation of formal power series; thus we can "realize" our example as the Lie algebra of linear operators on $k[[t]]$ spanned by its multiplications, and the derivation d. For k of arbitrary characteristic we can get a similar "natural" realization of this Lie algebra by replacing $k[[t]]$ with a power series ring with "divided powers". Nevertheless, we can see from (25.48) that the above linearly compact Lie algebra cannot be embedded by a continuous homomorphism into a linearly compact *associative* algebra under Lie brackets.)

In [**27**], a very general sufficient condition will be developed for algebraic structure on an inverse limit A of "small" objects to arise from a representation of A as an inverse limit of "small" objects with algebraic structure. (Here "small" is not meant in the set-theoretic sense introduced in the last paragraph of §5, but as a generic term for conditions such as "finite" (on sets), "finite-dimensional" (on vector spaces), etc..) This result will embrace (25.48) and also such facts as that a *topological group* whose underlying topological space is a Stone space is an inverse limit of finite groups – though the corresponding statement for topological *lattices* is false. Roughly, [**27**] will put in more abstract and general form the argument found in [**104**, VI.2.6-9]. In fact, (25.48) can be obtained by a straightforward adaptation of the method of proof used in [**104**, *loc. cit.*], by replacing the open congruences of that argument by open ideals.

Combining (25.48) with our descriptions of representable functors to associative, Lie and Jordan algebras, we get inverse-limit descriptions of such functors. The criterion of [**27**] (or, appropriately modified, [**104**]) also shows that the analog of (25.48) holds for $\mathbf{LCpRing}_k^{(n)}$ for each n, and for linearly compact spaces with *two* associative operations * and $_*$ which "interact trivially" as described in Theorem 25.25. Thus we have:

COROLLARY 25.51. *The representable functors of Theorems 25.{15, 25, 31, 40}, and Corollary 25.44 (from* $\mathbf{Ring}_k^1$ *to* $\mathbf{Ring}^1$, $\mathbf{Ring}$, $\mathbf{Lie}$, $\mathbf{Jordan}^1$ *and* $\mathbf{Jordan}$) *can all be written as* inverse limits *of functors of the same sort in which the algebras* A *are finite-dimensional (so that linear compactness is vacuous and* $\hat{\otimes}$ *is just* $\otimes_k$).

Equivalently, if $\mathbf{V}$ *is any of the categories listed above (with* $\mathrm{char}\, k \neq 2$ *if* $\mathbf{V} = \mathbf{Jordan}$), *then any co-*$\mathbf{V}$ *object in* $\mathbf{Ring}_k^1$ *or* $\mathbf{Ring}_k$ *is a directed union of sub-co-*$\mathbf{V}$ *objects which are finitely generated as k-subalgebras.* □

25.52. *Example.* $k[[t]]$ is the inverse limit of the finite-dimensional algebras $k[t]/(t^n)$. Hence the operation $-\hat{\otimes}\, k[[t]]$ is the inverse limit of the constructions $-\otimes_k k[t]/(t^n)$, i.e., for all $S \in \mathbf{Ring}_k^1$, $S[[t]]$ is the inverse limit of the rings $S[t]/(t^n) = S \otimes_k k[t]/(t^n)$. This corresponds to the fact that in the representing coalgebra $k<x_1, x_2, \ldots>$, with the co-operations

$$\mathbf{a}(x_i) = x_i^\lambda + x_i^\rho, \qquad \mathbf{m}(x_i) = \Sigma_{0 \leq j \leq i}\, x_j^\lambda x_{i-j}^\rho,$$

the subalgebras $k<x_0, x_1, \ldots, x_n>$ $(n \geq 0)$ are all closed under the co-operations, and so constitute sub-co-rings.

In §32, we shall see that the above Corollary fails when the assumption that k

is a field is dropped. Note also that the list of results with which this Corollary begins does not include Corollary 25.46, the case of Jordan algebras in characteristic 2, because the analog of (25.48) is *not* valid for $\mathbf{LCpNARing}_k$, as shown by Example 25.49.

§26. Functors to other subvarieties of NARing.

In this section, we consider representable functors from $\mathbf{Ring}_k^1$ to still other subvarieties of **NARing**, and in the next section, some general questions this subject leads to. Subsequent sections will not depend on these two.

We begin by streamlining our method of computation. Consider any identity

$$(26.1) \qquad f(x_1, \ldots, x_r) \;=\; 0,$$

where f is a nonassociative polynomial, homogeneous of degree $d(i)$ in x_i for $i = 1, \ldots, r$, hence of total degree $d = d(1) + \ldots + d(r)$. (The assumption of homogeneity is for convenience; we shall eventually remove it.) We wish to determine the identities that must be satisfied by a system $(A, {}^*, {}_*)$ if (26.1) is to hold identically in all the rings $(A \hat{\otimes} S,\, m)$ for $S \in \mathbf{Ring}_k^1$, where m is constructed from * and ${}_*$ as in (25.5).

The basic procedure, which we have already seen, is to substitute for each x_i a sum of $d(i)$ decomposable elements

$$(26.2) \qquad x_i \;=\; a_{i1}\, s_{i1} + \ldots + a_{i\,d(i)}\, s_{i\,d(i)} \quad (1 \leq i \leq r,\ a_{ij} \in A,\ s_{ij} \in S\ (1 \leq j \leq d(i))),$$

then expand (26.1) using (26.2) and (25.5), and take the coefficient of each monomial *multilinear* in all the s_{ij}'s which occur in this expansion.

Now when we make the substitutions (26.2), the expression we get is symmetric, for each i, in the $d(i)$ terms $a_{i1}\, s_{i1}, \ldots, a_{i\,d(i)}\, s_{i\,d(i)}$. Hence when we take the coefficients of monomials

$$s_{i_1 j_1} \cdots s_{i_d j_d},$$

it actually suffices to consider *one* representative from each equivalence class of such monomials under permutations of the indeterminates that preserve each subfamily $\{s_{i1}, \ldots, s_{i\,d(i)}\}$ $(i = 1, \ldots, r)$. (Cf. discussion immediately preceding (25.37).) These equivalence classes are in one-one correspondence with formal associative monomials in the original variables $x_1, \ldots, x_r$, of degree $d(i)$ in each x_i. For example, in the context of (25.37), where the substitutions $x = as + bt + cu$, $y = dv$ lead to an alphabet of symbols $\{s, t, u, v\}$, divided into two subfamilies $\{s, t, u\}$ and $\{v\}$, the equivalence class of the associative monomial $svtu$, namely $\{svtu, svut, tvsu, tvus, uvst, uvts\}$ corresponds to the associative monomial $xyxx$ in the original variables x and y.

Let us also recall what we noted in 25.21: In every equation we obtain in this way, the variables appear in a fixed order in all terms; i.e., the terms differ only in bracketing and in the occurrences of * and ${}_*$. This means that if we leave blanks for the variables in these equations, no information is lost. Variables can be reinserted, following the rule that the same sequence of variables is used in every term of an equation; but we may find some different order of variables more convenient than the one given by our original description. (Cf. paragraph following

(25.37).)

This leads to the following form of our procedure. Given a nonassociative identity (26.1), homogeneous of degree $d(i)$ in each x_i, one first lists all possible *associative* monomials u homogeneous of the same degrees in these same variables. For each such u, one constructs an identity for objects $(A;\, ^*,\, _*)$ as follows. Look at each (nonassociative) monomial U occurring in f, and find all ways (if any) of preserving the orders of some of the multiplications it involves and reversing the orders of the others, so as to bring the indeterminates to precisely the order of the indeterminates in u. For each way that is found, write out the corresponding system of parentheses, with blanks in place of the variables, using * for nonreversed multiplications and $_*$ for reversed multiplications, and attach to it the coefficient of U in f. Add up all the terms so obtained, and finally, substitute into the blanks in each summand of this expression d distinct variable-symbols (say $a_1, \ldots, a_d$) in a uniform order.

This should be made clear by

26.3. *Example.* Recall the Jordan identity

$$((x, x), (x, y)) = (x, ((x, x), y)), \tag{26.4}$$

from which, in the presence of the commutative identity, we derived the identities (25.37). Let us now consider representable functors to *not necessarily commutative* rings satisfying (26.4). For illustrative purposes we shall go through only one-fourth of the computation required to characterize such functors, namely the derivation of the identity on $(A,\, ^*,\, _*)$ corresponding to the associative monomial

$$u = xyxx.$$

We first look at the left-hand term of (26.4). Abandoning the special notation "(,)", we write this $(xx)(xy)$. To get the factors into the order $xyxx$ we must get the "y" into the second place, hence the order of terms in the "outer" multiplication (i.e., xx times xy) must clearly be reversed, while the order within the factor that this reversal puts in the first place, xy, should be preserved. Either order is acceptable in the remaining factor, xx. Thus we get two terms

$$(\ ^*\ )_*(\ ^*\ ) + (\ ^*\ )_*(\ _*\ ).$$

In the other side of the Jordan identity (26.4), $x((xx)y)$, we see that if the outer multiplication (x times $(xx)y$) were reversed, the y could only appear in first or third place, so this multiplication must be left unreversed. Subject to this condition, we get y in the second place if and only if we use reversed order in the next-to-last multiplication (xx times y). Once again, the order of multiplication in the term xx makes no difference, so we again get two terms. Equating with those obtained before, we have

$$(\ ^*\ )_*(\ ^*\ ) + (\ ^*\ )_*(\ _*\ ) = (\ ^*(\ _*(\ ^*\ ))) + (\ ^*(\ _*(\ _*\ ))),$$

or, putting in variables:

$$(a^*b)_*(c^*d) + (a^*b)_*(c_*d) = (a^*(b_*(c^*d))) + (a^*(b_*(c_*d))).$$

(Under the commutativity condition ${}_* = {}^*$, assumed in our earlier consideration of (26.4), the above equation becomes, up to labeling of variables, the third equation of (25.37).)

We have not discussed how to handle identities (26.1) which may involve zeroary or unary operations in addition to the binary multiplication. When zeroary operations appear in f, they shift as do other terms under reversals of the binary multiplication. They are retained when our procedure deletes variable-symbols, and ignored when one classifies terms according to corresponding associative monomials in these variable-symbols. As a simple example, consider the left identity law, $\varepsilon x = x$, which is homogeneous of degree 1 in its unique variable x. The identity in $(A, \varepsilon, {}^*, {}_*)$ corresponding to the unique monomial $u = x$ of degree 1 is $(\varepsilon^*\ ) + (\ {}_*\varepsilon) = (\ )$, or, inserting a variable, $\varepsilon^* a + a_* \varepsilon = a$.

Unary linear operations appear as function-symbols φ that move around with their arguments when we reverse multiplications. We shall not consider any unary operations in our examples.

The assumption that (26.1) was homogeneous in each x_i was made only for convenience in explaining the above construction. If an identity is not homogeneous, one makes the calculations described above separately for *each* of its homogeneous components. To look at it another way, in the calculation we have described, we could as well have let u run over *all* monomials in the x's; but only those monomials having the same degree in each x_i as some term of f would have led to nonvacuous identities.

Below, we shall not discuss the details of such calculations, but simply record the results.

If $\mathbf{V}$ is a subvariety of **NARing**, we shall call a system $(A, {}^*, {}_*)$ a *proto-*$\mathbf{V}$*-algebra* if for every $S \in \mathbf{Ring}^1_k$, the ring $(A \otimes_k S, m)$, where m is defined by (25.5), lies in $\mathbf{V}$. (Mutatis mutandis for rings $(A \otimes_k S, m, z)$ if $\mathbf{V}$ also has a zeroary operation z, etc..) We shall likewise attach the prefix "proto-" to the names of identities, to get names for the corresponding families of identities in * and $_*$.

Note that in the above paragraph we wrote "$A \otimes_k S$" rather than "$A \hat{\otimes} S$". As observed at the end of §24, the identities satisfied by a completed tensor product of linearly topologized algebras are the same as those satisfied by the non-completed tensor product, so we can forget about topology and completions in the definition of "proto-$\mathbf{V}$-algebra"; though our motivation is of course the fact that representable functors from $\mathbf{Ring}^1_k$ to $\mathbf{V} \subseteq$ **NARing** correspond to constructions of *completed* tensor product with fixed *linearly compact* proto-$\mathbf{V}$ algebras.

Three of the simplest identities which arise in the theory of nonassociative rings are the *left alternative, right alternative*, and *flexible* laws. Each of these says that multiplication of three elements is associative if a certain two of them are equal. Here are these three identities and the corresponding proto-identities; the reader is encouraged to derive of some of the latter:

(26.5) *left alternative law*: $(xx)y = x(xy)$

(26.6) *proto-left alternative identities*:

$$(a^*b)^*c + (a_*b)^*c = a^*(b^*c) \qquad a_*(b_*c) + a_*(b^*c) = (a_*b)_*c$$
$$0 = a^*(b_*c) + (a^*b)_*c.$$

(26.7) *right alternative law*: $y(xx) = (yx)x$

(26.8) *proto-right alternative identities*:

$$a^*(b^*c) + a^*(b_*c) = (a^*b)^*c \qquad (a_*b)_*c + (a^*b)_*c = a_*(b_*c)$$
$$0 = a_*(b^*c) + (a_*b)^*c.$$

(26.9) *flexible law*: $x(yx) = (xy)x$

(26.10) *proto-flexible identities*:

$$a^*(b_*c) = a_*(b^*c) \qquad (a^*b)_*c = (a_*b)^*c$$
$$a^*(b^*c) + (a_*b)_*c = (a^*b)^*c + a_*(b_*c).$$

It is not hard to verify that the conjunction of any two of the identities (26.{5,7,9}) is equivalent to the condition that the 3-linear *associator* operation

$$(x, y, z) \mapsto x(yz) - (xy)z$$

is *alternating* in its three arguments (hence the term "alternative"); in particular, any two of these identities imply the third. A ring satisfying these three identities is called an *alternative ring*; thus, this can be characterized as a ring in which the associative law holds for any three elements which are linear combinations of two elements. This is in turn known to be equivalent to the formally stronger condition that the subring (or equivalently, the multiplicative subbinar) generated by any two elements is associative [**169**, Theorem 3.1]. (However, a binar in which associativity holds for all 3-tuples having two equal members need *not* have the property that all 2-generator subbinars are associative.) A well-known example of an alternative ring is the ring of Cayley numbers.

The *proto-alternative identities* are obtained by putting together the above three systems of identities (26.{6, 8, 10}); we may write the result as one extended equation:

(26.11) *proto-alternative identities*:

$$a^*(b_*c) = -(a^*b)_*c = (a_*b)_*c - a_*(b_*c)$$
$$= a_*(b^*c) = -(a_*b)^*c = (a^*b)^*c - a^*(b^*c).$$

Note that the proto-*associative* identities derived in the preceding section, (25.11) and (25.12), are equivalent to (26.11) with the one further equality "= 0" added at the end!

Let us write down the equation connecting those members of (26.11) that are single monomials, namely

$$a^*(b_*c) = -(a^*b)_*c = a_*(b^*c) = -(a_*b)^*c.$$

This equation says that a 3-fold product formed using one * and one $_*$ is invariant under transposing these operations, but changes sign on rearranging the parentheses. Let us test this condition for "coherence" when applied to the 4-fold product $a^*(b_*(c^*d))$. Shuffling parentheses whenever we can, and interchanging * and $_*$ whenever this is the only way to get something new, we get

$$\begin{aligned} a^*(b_*(c^*d)) &= -(a^*b)_*(c^*d) = ((a^*b)_*c)^*d = -(a^*(b_*c))^*d = -(a_*(b^*c))^*d \\ &= a_*((b^*c)^*d) = a^*((b^*c)_*d) = -a^*(b^*(c_*d)) = -a^*(b_*(c^*d)). \end{aligned}$$

Comparing the first and last terms we see that if $\operatorname{char} k \neq 2$,

$$a^*(b_*(c^*d)) = 0.$$

By the analogous calculation starting with $a_*(b^*(c_*d))$ we get

$$a_*(b^*(c_*d)) = 0.$$

From these equations and the preceding observation about rearranging *, $_*$ and parentheses, it is easy to deduce that *any* product of four or more terms which involves at least one * and at least one $_*$ is zero. Let us now write down the equality between the two unused terms of (26.11):

$$(a^*b)^*c - a^*(b^*c) = (a_*b)_*c - a_*(b_*c). \tag{26.12}$$

Observe that if we substitute for any of the variables in (26.12) a product e^*f, the right-hand side vanishes because it involves both * and $_*$, and the resulting identity says that two of the ways of bracketing a product of four elements under the operation * are equal. We get a similar identity on applying $-^*d$ or d^*- to both sides of (26.12); taken together these equations say that * satisfies the equations of associativity for all products of four terms. The corresponding arguments apply to $_*$.

We could summarize the above results as a rather complicated description of the general representable functor $\mathbf{Ring}^1 \to \mathbf{Alt}$, but let us content ourselves with noting a strong consequence, which in the unital case does lead (in conjunction with our earlier results) to a complete description of such functors:

PROPOSITION 26.13. *Let* **Alt** *denote the variety of alternative rings. Then if* $\operatorname{char} k \neq 2$, *any representable functor from* $\mathbf{Ring}^1_k$ *or* $\mathbf{Ring}_k$ *to* **Alt** *actually takes values in* $\mathbf{Alt} \cap \mathbf{Ring}^{(4)}$ *(defined in paragraph preceding Corollary 25.44).*

Hence, letting $\mathbf{Alt}^1$ *denote the variety of alternative unital rings, if* $\operatorname{char} k \neq 2$ *then any representable functor from* $\mathbf{Ring}^1_k$ to $\mathbf{Alt}^1$ *actually takes values in* $\mathbf{Ring}^1$. □

Perhaps more interesting than identities like these, that more or less reduce to familiar conditions, are those that define new sorts of objects $(A, {}^*, {}_*)$. Let us consider, for example, the proto-flexible identities (26.10). The first two of these say that in any "monomial" in a proto-flexible algebra, the *'s and $_*$'s can be permuted freely. The invariants under such permutations are the sequence of variables, their bracketing, and the total numbers of *'s and of $_*$'s. So let us adopt a notation based on these invariants, writing the common value of $a_*((b^*c)^*d) = a^*((b_*c)^*d) = a^*((b^*c)_*d)$, for instance, as $a((bc)d)^{*2}_{*1}$. To use the remaining equation of (26.10), which we previously rewrote as (26.12), let us rewrite it once more, this time as

(26.14) $$a^*(b^*c) - a_*(b_*c) = (a^*b)^*c - (a_*b)_*c.$$

This can be read as a weak kind of associativity. And in fact, it implies a general weak associative law:

LEMMA 26.15. *Suppose* m *and* n *are nonnegative integers, and* P *and* Q *are nonassociative monomials of degree* $m+n+3$ *that become the same associative monomial* $\bar{P} = \bar{Q}$ *when parentheses are dropped. Then*

(26.16) $$P^{*m+2}_{*n} - P^{*m}_{*n+2} = Q^{*m+2}_{*n} - Q^{*m}_{*n+2}$$

is an identity of proto-flexible algebras.

SKETCH OF PROOF. We can find a chain of nonassociative monomials, $P = P_1$, P_2, ..., $P_r = Q$, such that each P_{i+1} is obtained from P_i by application of associativity to a single pair of multiplications, which we will call the *distinguished pair* of multiplications of P_i. Now consider for each i the difference

(26.17) $$(P_i)^{*m+2}_{*n} - (P_i)^{*m}_{*n+2}.$$

Since by definition, a monomial P^{*u}_{*v} specifies only the number and bracketing of *'s and $_*$'s, and not which of the multiplications are which, we may insert these operations in the two terms of (26.17) in such a way that these two monomials differ only in that the distinguished pair of multiplications consists of two *'s in the first term, and of two $_*$'s in the second. Then an application of (26.14) to the distinguished pair shows that (26.17) equals $(P_{i+1})^{*m+2}_{*n} - (P_{i+1})^{*m}_{*n+2}$. Now (26.16) follows by induction on i. □

From this one can obtain the following construction for the *free* proto-flexible algebra F over k on a set X. Let A be the free algebra on X in a single nonassociative multiplication (denoted by juxtaposition) over the commutative associative polynomial algebra in two indeterminates, $k[s, t]$. Let I be the associator ideal of A – the kernel of the natural homomorphism into the free associative algebra $k[s, t]\langle X\rangle$, which is generated as an ideal by all elements $(ab)c - a(bc)$. Define on A the operations

(26.18) $$a^*b = s\,ab, \qquad a_*b = t\,ab.$$

These are easily seen to satisfy the first two proto-flexible identities. The induced operation on $A/(s^2-t^2)I$ is easily seen to satisfy (26.14), equivalently, the third proto-flexible identity, as well.

LEMMA 26.19. *For* X, A, I *as above, the* $(^*, {}_*)$*-*k*-subalgebra*

(26.20) $$F \subseteq A/(s^2-t^2)I$$

generated under the operations (26.18) *by the image of* X *is the free proto-flexible algebra on* X.

Further, if we let B_1 *denote the set of all expressions*

$$P^{*m}_{*n} \qquad \text{with } n = 0 \text{ or } 1 \tag{26.21}$$

where P *ranges over all* nonassociative *monomials in* X, *and* B_2 *denote the set of elements*

$$Q^{*m+2}_{*n} - Q^{*m}_{*n+2} \qquad \text{with } n \text{ arbitrary,} \tag{26.22}$$

where Q *ranges over any system of nonassociative* representatives *of all* associative *monomials in* X *(and where in both* (26.21) *and* (26.22), m *has the unique value, if any, consistent with the value of* n *and the length of the given monomial* P *or* Q), *then* $B_1 \cup B_2$ *is a* k*-basis for* F.

SKETCH OF PROOF. Given a system B_2 of representatives of the associative monomials in X, it follows from Lemma 26.15 that any formal $(^*, {}_*)$-polynomial in X can be reduced using the proto-flexible identities to a linear combination of elements of $B_1 \cup B_2$. The asserted conclusions will follow if we can show that this set, when evaluated in $A/(s^2 - t^2)I$, is linearly independent. To get this independence, one first notes that the canonical homomorphism from the free proto-flexible algebra to $(A/(s^2 - t^2)A, {}^*, {}_*)$ (note the larger denominator than in (26.20)) takes all elements of B_2 to 0, and B_1 to a linearly independent set, and then that the homomorphism from the same free algebra to $(A/I, {}^*, {}_*)$ (note that A/I is the free associative algebra $k[s, t]\langle X\rangle$) takes B_2 to a linearly independent set. □

The authors do not know whether any such nice results hold for the right or left proto-alternative identities. Perhaps on looking in higher degrees, they will, like the full system of proto-alternative identities (26.11), turn out to be "non-coherent" in a way that leads to high-degree proto-associativity. Or perhaps, like the proto-flexible identities, they will have a theory essentially of their own.

We have mentioned that a ring is alternative if and only if every subring (or multiplicative subbinar) generated by two elements is associative. The weaker condition that every subring (or multiplicative subbinar) generated by *one* element be associative is called *power-associativity*. It is known ([**2**], [**3**, Lemma 3]) that in the variety $\mathbf{NARing}_k$ for k a field of characteristic 0, the subvariety of power-associative algebras is defined by the two identities

$$x(xx) = (xx)x, \tag{26.23}$$

$$(xx)(xx) = x(x(xx)). \tag{26.24}$$

Since for any $(^*, {}_*)$-algebra $(A, {}^*, {}_*)$ and any $S \in \mathbf{Ring}^1_k$, the object $(A \otimes_k S, m) \in \mathbf{NARing}$ admits an $\mathbf{NARing}_k$ structure, it follows that for k of characteristic 0, the variety of proto-power-associative k-algebras may be defined by the proto-identities corresponding to (26.23) and (26.24). Because these identities each involve only one variable, each yields a single proto-identity. The identity (26.23) – which we note, incidentally, holds in any left alternative *or* right alternative *or* flexible ring – gives the proto-identity

$$a^*(b^*c) + a^*(b_*c) + (a^*b)_*c + (a_*b)_*c = \\ (a^*b)^*c + (a_*b)^*c + a_*(b^*c) + a_*(b_*c). \tag{26.25}$$

This equation can be written more compactly in terms of the pair of operations $a^\circ b = a^*b + a_*b$ and $a_\circ b = a^*b - a_*b$, which when $\operatorname{char} k \neq 2$ give the same information as * and $_*$. Let us do so, bringing all terms "of type $a(bc)$" to the left-hand side of (26.25) and all terms "of type $(ab)c$" to the right. Then (26.25) reduces to

$$a_\circ(b^\circ c) = (a^\circ b)_\circ c. \tag{26.26}$$

In this form the condition is quite convenient to work with. E.g., it is straightforward to obtain a normal form in a free algebra $(A, {}^\circ, {}_\circ)$ subject to this identity.

The proto-identity arising from the fourth degree identity (26.24) is unwieldy whether written in terms of * and $_*$ or in terms of $^\circ$ and $_\circ$, but in the presence of (26.26) it can be replaced by a less cumbersome formula in the latter operations. To see this, let us first note that in the presence of (26.23), the term $x(xx)$ in (26.24) can be rewritten $\frac{1}{2}(x(xx) + (xx)x)$. Multiplying by 2 (as we may, since we are assuming $\operatorname{char} k = 0$), and passing to the corresponding proto-identity, we have

$$2(a^\circ b)^\circ(c^\circ d) = a^*(b^\circ(c^\circ d) + (b^\circ c)^\circ d) + (a^\circ(b^\circ c) + (a^\circ b)^\circ c)_* d.$$

Now by the left-right symmetry of its definition, the condition of power-associativity must also be equivalent to the conjunction of (26.23) and the left-right dual of (26.24); hence in the presence of (26.26), the above equation must entail the formula one gets by interchanging * and $_*$ therein. Hence (again using the invertibility of 2 in k), it will be equivalent to the conjunction of its sum with that equation and its difference therefrom. The sum is

$$4(a^\circ b)^\circ(c^\circ d) = a^\circ(b^\circ(c^\circ d) + (b^\circ c)^\circ d) + (a^\circ(b^\circ c) + (a^\circ b)^\circ c)^\circ d, \tag{26.27}$$

while the difference,

$$0 = a_\circ(b^\circ(c^\circ d) + (b^\circ c)^\circ d) - (a^\circ(b^\circ c) + (a^\circ b)^\circ c)_\circ d,$$

is easily shown to be a consequence of (26.26). So when $\operatorname{char} k = 0$, the proto-power-associative identities reduce to (26.26) and (26.27).

The above is as far as the authors have investigated this area, but clearly it is just a "sampler" of the possibilities. Perhaps some readers will carry these investigations further. Some other interesting varieties of nonassociative rings are considered in [**57**]; two general references are [**169**] and [**140**]. A brief survey of the field is given in [**141**]; [**85**] is an exhaustive analysis of all inequivalent identities of degree 3; [**11**] is a survey of papers in the area reviewed in *Referativniy Zhurnal* between 1972 and 1978.

We remark that for **V** a variety of *unital* nonassociative rings, the free proto-**V** algebra on the empty set, i.e., the object generated under * and $_*$ by an element ε subject only to the left and right proto-unital laws

(26.28) $$\varepsilon^*a + a_*\varepsilon = a = a^*\varepsilon + \varepsilon_*a$$

may be infinite-dimensional – (26.28) alone is not strong enough to make ε generate a finite-dimensional algebra.

One can generalize these investigations and look at "proto-modules" and "proto-bimodules" over proto-algebras, at "proto-derivations" from a proto-algebra to a proto-bimodule, etc.. The definitions are straightforward – if in doubt, go back to Theorem 25.1, adjust it to operations among more than one object of $\mathbf{Rep}(\mathbf{Ring}_k^1, \mathbf{Ab})$, and see what identities characterize the desired objects.

§27. A Galois connection.

Let us formalize abstractly the idea with which we have been working in the last two sections.

DEFINITION 27.1. *We shall denote by* $\mathbf{P}$ *the variety of all objects* $(A, {}^*, {}_*)$, *where* A *is a k-vector-space and* *, ${}_*$ *are arbitrary k-bilinear operations on* A. *We shall denote by* $\mathbf{I}$ *the class of all formal identities*

(27.2) $$f = 0$$

for ordinary not necessarily associative rings (B, m), *i.e., abelian groups with a single bilinear operation* m *(which will often be abbreviated to juxtaposition).*

Given an object $(A, {}^*, {}_*) \in \mathbf{P}$ *and an identity* $(f = 0) \in \mathbf{I}$, *let us write* $(A, {}^*, {}_*)\ \tau\ (f = 0)$ *if for every* $S \in \mathbf{Ring}_k^1$, *the ring* $(A \otimes_k S, m)$, *with* m *defined by* (25.5), *satisfies the identity* $f = 0$.

Similarly, we will denote by $\mathbf{P}^\varepsilon$ *the variety of proto-unital objects* $(A, {}^*, {}_*, \varepsilon)$, *where* A, * *and* ${}_*$ *are as above and* ε *is a zeroary operation satisfying the proto-unital identities* (26.28), *we will denote by* $\mathbf{I}^1$ *the class of formal identities for not necessarily associative unital rings, and we will denote by* $\tau^1 \subseteq \mathbf{P}^\varepsilon \times \mathbf{I}^1$ *the relation analogous to* $\tau \subseteq \mathbf{P} \times \mathbf{I}$.

The relation τ yields what is known as a *Galois connection* between $\mathbf{P}$ and $\mathbf{I}$, and likewise τ^1 yields a Galois connection between $\mathbf{P}^\varepsilon$ and $\mathbf{I}^1$. Let us set up notation for these connections:

DEFINITION 27.3. *For every subclass* $X \subseteq \mathbf{P}$ *we define* $X^\# \subseteq \mathbf{I}$ *by*

$$X^\# = \{y \in \mathbf{I} \mid \forall x \in X,\ x\,\tau\,y\}.$$

Similarly, for $Y \subseteq \mathbf{I}$ *we define* $Y^\# \subseteq \mathbf{P}$ *by*

$$Y^\# = \{x \in \mathbf{P} \mid \forall y \in Y,\ x\,\tau\,y\}.$$

We will likewise denote by ${}^\#$ *the operators between subsets of* $\mathbf{P}^\varepsilon$ *and* $\mathbf{I}^1$ *defined analogously, using* τ^1 *in place of* τ.

Strictly speaking, this notation is ambiguous, since a given Y may be both a subset of $\mathbf{I}$ and a subset of $\mathbf{I}^1$, giving two possible interpretations of $Y^\#$. However, most of our discussion will be abstract, rather than referring to specific sets of identities, and where specific cases are mentioned, the context will make the meaning clear. We shall mostly refer to the nonunital case in our general statements, it being understood that the same observations apply to the other; we may or may not add "respectively $\mathbf{P}^{\varepsilon}$" or the like for emphasis. Note that the distinction between the unital and nonunital cases is one of whether the objects A are given structure making the rings $(A \otimes_k S, m)$ unital; the associative k-algebras S are *always* assumed unital in this section. Also observe that we are speaking of "subsets of $\mathbf{P}$" where we would, in precise usage, refer to subsets (or subclasses) of $\mathrm{Ob}(\mathbf{P})$. We leave it to the reader to choose exactly how to define "formal identities"; e.g., whether the f in (27.2) is to be understood as a string of symbols, or an element of a free object of **NARing**; this distinction will not affect our conclusions. For general results on Galois connections, cf. [**46**, pp.43-44], [**24**, §5.5]. The more usual notation for the operators constituting such a connection is $*$ or $^{\circ}$ rather than $^{\#}$, but those two symbols have already been used here.

As usual, composing our two operators, we get *closure operators* $^{\#\#}$ on $\mathbf{P}$ and $\mathbf{I}$ (respectively $\mathbf{P}^{\varepsilon}$ and $\mathbf{I}^1$). The closed subsets of $\mathbf{P}$ will be certain varieties of objects $(A, {}^*, {}_*)$; closed subsets of $\mathbf{I}$ will be certain equational theories of nonassociative rings. An obvious question is *which* subvarieties and *which* equational theories are closed.

We have seen some cases above where these operators acted nontrivially on varieties, respectively equational theories. Theorem 25.40 implies that the closure in $\mathbf{I}^1$ of the theory of *unital Jordan rings* contains the theory of *unital semispecial Jordan rings*. The latter theory is easily seen to be *closed* if $\mathrm{char}\, k = 0$: it is (essentially by the definition of semispecial Jordan ring) $X^{\#}$, where X is the singleton $\{(k, \cdot, \cdot, \frac{1}{2})\}$ ($\cdot$ denoting the multiplication of the field k). Hence our statement that the closure of the theory of unital Jordan rings contains this theory can be strengthened to say that it equals this theory. (If $\mathrm{char}\, k = p$ the closure under our Galois connection of the equational theory of Jordan rings is the equational theory of semispecial Jordan rings of characteristic p, which contains in particular the identity $px = 0$. A systematic development of this subject would describe our Galois connection in the characteristic p case as relating certain subvarieties of $\mathbf{P}$ and certain equational theories of not necessarily associative $\mathbf{Z}/p\mathbf{Z}$-algebras. For brevity, we will forego setting up the notation this requires, and will assume $\mathrm{char}\, k = 0$ in all examples where this point would otherwise come up.)

In the same way, we see from Corollary 25.23 that the closure of the theory of unital *commutative associative* rings is all of $\mathbf{I}^1$, that is, the theory of the trivial ring $\{0\}$, and from Proposition 26.13 that in characteristic 0 the closure of the theory of unital *alternative rings* is the theory of unital *associative* rings. Similarly, in characteristic 0 the theories of (unital or nonunital) associative rings and of Lie rings are closed, each being of the form $X^{\#}$ for an obvious singleton X.

This closure operator $^{\#\#}$ on $\mathbf{I}$ (or $\mathbf{I}^1$) is *algebraic* ([**46**, p.45], [**73**, p.24]), that is, the union of any chain of closed theories is closed. To see this, note that for every identity $(f = 0) \in \mathbf{I}$, the condition $(A, {}^*, {}_*)\ \tau\ (f = 0)$ on objects $(A, {}^*, {}_*)$ of $\mathbf{P}$ is equivalent to a finite conjunction of identities in * and $_*$, the "proto-identities" obtained from $(f = 0)$ by the construction of the preceding section. This conjunction is a first order sentence, hence the Compactness Theorem of Model

Theory [**46**, p.213], [**73**, p.242] tells us that an infinite family Y of such conditions on $(A, {}^*, {}_*)$ entails no such sentence not entailed by a finite subfamily of Y.

The class of *subvarieties* of **NARing** (or $\mathbf{NARing}^1$) whose *theories* are closed in $\mathbf{I}$ (or $\mathbf{I}^1$) under the above connection will form, not a closure system, but a structure of the dual sort. Let us call a variety of rings *stable* if it is defined by a theory closed under ${}^{\#\#}$; equivalently, if it is the variety generated by all rings $(A \otimes_k S, m)$ where $(A, {}^*, {}_*)$ ranges over some fixed subfamily $X \subseteq \mathbf{P}$, and S ranges over all of $\mathbf{Ring}_k^1$. Then we see that the class of stable subvarieties of **NARing** will be closed under taking joins of varieties, but not necessarily under taking intersections; and that for an arbitrary variety of rings $\mathbf{V}$, there must exist a largest stable subvariety contained in $\mathbf{V}$, but not necessarily a smallest stable subvariety containing it.

Let us clarify the above statement that what we have is a structure of the sort dual to that of a closure operator, since it is a bit of an oversimplification: A stable variety is, first, a *variety* of not necessarily associative rings, i.e., a class of not necessarily associative rings *closed* under the closure operator described in Birkhoff's Theorem. But we now also have an idempotent *decreasing* operator on the complete lattice of such varieties, carrying a variety $\mathbf{V}$ to the largest subvariety of $\mathbf{V}$ generated by classes of the form

$$\{(A \otimes_k S, m) \mid S \in \mathbf{Ring}_k^1\} \qquad (A \in \mathbf{P}); \tag{27.4}$$

and $\mathbf{V}$ is stable in our sense if and only if it is fixed by this operator. In other words, we have a *dual* closure operator *relative* to the lattice of varieties of rings, which is itself defined by an *ordinary* closure operator on classes of rings.

Let us note some examples of stable varieties of rings. From our above examples of closed theories, we see that if $\operatorname{char} k = 0$, the variety of all unital associative rings is stable, determined by $X = \{(k, \cdot, 0, 1)\} \subseteq \mathbf{P}^{\varepsilon}$ (where ${}^* = \cdot$, the ordinary multiplication of k, ${}_* = 0$, the zero multiplication, and $\varepsilon = 1$ the unit $1 \in k$). It is not hard to show similarly that the variety of all unital commutative not necessarily associative rings is determined by the class of objects $(A, {}^*, {}_*, \frac{1}{2})$ such that ${}^* = {}_*$, hence is stable.

However, the intersection of these two varieties, the variety of unital *commutative associative* rings, is not stable – we have noted that the largest stable variety it contains is the trivial variety, containing only the zero ring. A consequence of these observations is that there is no smallest stable variety *containing* the variety of unital commutative associative rings – it would have to be contained in both the variety of unital commutative rings and the variety of unital associative rings, hence would be precisely the variety of unital commutative associative rings, contradicting the observation that this is not stable.

Though as we have just seen, the class of stable varieties of rings is not closed under pairwise intersections, it follows easily from the algebraicity of our closure operator on $\mathbf{I}$ that it is closed under forming intersections of *chains*.

Turning to the objects $(A, {}^*, {}_*)$, we have seen that ${}^{\#\#}$ gives a *closure operator* on *subvarieties* of $\mathbf{P}$, so in this case we get on corresponding *theories* a dual closure operator relative to the complete lattice of all equational theories of such objects. Let us call the theory of a closed subvariety of $\mathbf{P}$ a *stable*

$(^*, {}_*)$-*theory*.

A $(^*, {}_*)$-theory is stable if and only if it is generated as an equational theory by the "proto-identities" corresponding to a set Y of ring identities. From our mode of construction of proto-identities, it is easy to see that the following two conditions will be satisfied by any stable $(^*, {}_*)$-theory T:

(27.5) T is generated by homogeneous multilinear identities in which the variables occur in the same order in every monomial. (Equivalently, if an identity $f = 0$ belongs to T, then so does the identity $f_{x_1, \dots, x_d} = 0$ for each string of distinct variables $x_1, \dots, x_d$, where $f_{x_1, \dots, x_d}$ denotes the sum of all terms of f which are multilinear in the variables $x_1, \dots, x_d$, and in which these occur in just that order, and no other variables occur.)

(27.6) T is closed under the operation on identities which interchanges * and ${}_*$, and simultaneously reverses the order of factors in each monomial (e.g., which takes $a^*(b^*(c_*d)) = 0$ to $((d^*c)_*b)_*a = 0$).

Here, however, is an example showing that these conditions are not sufficient to guarantee stability. Let T be the theory generated by the two associativity laws

(27.7) $$a^*(b^*c) = (a^*b)^*c, \qquad a_*(b_*c) = (a_*b)_*c.$$

Conditions (27.5) and (27.6) are satisfied, but let us note

LEMMA 27.8. *Any* $B \in \mathbf{NARing}_k$ *can be embedded in a k-algebra of the form* $(A \otimes_k S, m)$, *where* $(A, ^*, {}_*)$ *satisfies* (27.7).

PROOF. Given B, we define A to be $B \times B$ with operations given by

(27.9) $$(p, p')^*(q, q') = (0, pq), \qquad (p, p')_*(q, q') = (q'p', 0).$$

We see that in A any 3-fold product under * alone, or ${}_*$ alone, is zero, so in particular (27.7) holds. We now take $S = k$, and find that $(A \otimes_k k, m)$ is $B \times B$ with multiplication $(p, p')(q, q') = (p'q', pq)$. The diagonal map of B into $B \times B$ gives the desired embedding. □

From the above result we see that when $\operatorname{char} k = 0$, the theory T generated by the identities (27.7) cannot contain the proto-identities corresponding to any nontrivial ring identities. Hence

LEMMA 27.10. *The largest stable theory contained in the theory* T *generated by the identities* (27.7) *is the trivial theory; thus,* T *is not stable. Equivalently, the closure of the subvariety of* $\mathbf{P}$ *defined by* T *is all of* $\mathbf{P}$, *so that subvariety of* $\mathbf{P}$ *is not closed.*

Hence, conditions (27.5) *and* (27.6) *are not sufficient for the stability of a* $(^*, {}_*)$-*theory.* □

Let us sketch another way of getting this last statement. Note that the class of $(^*, {}_*)$-theories satisfying (27.5) and (27.6) is closed under intersection of theories.

(To see that (27.5) carries over to intersections, use the parenthesized formulation.) However, the following example shows that the class of stable $(^*, {}_*)$-theories is not closed under intersections. Consider the subvariety of **NARing** defined by the identity

$$x(yz) = 0. \tag{27.11}$$

The proto-identities corresponding to (27.11) are

$$a^*(b^*c) = 0, \quad a^*(b_*c) = 0, \quad (a^*b)_*c = 0, \quad (a_*b)_*c = 0. \tag{27.12}$$

We will not try to determine whether the variety generated by all rings $(A \otimes_k S, m)$, as A ranges over all $(^*, {}_*)$-algebras satisfying (27.12), is the variety of all rings satisfying (27.11), i.e., whether the theory generated by (27.11) is closed; what we will need to know is that its closure contains no new identities homogeneous of degree 3. This can be deduced using the object $(A, {}^*, {}_*)$ where A is 3-dimensional with basis $\{a, a^*a, (a^*a)^*a\}$ and all other products (including all products involving ${}_*$) are 0.

Now comparing (27.12) with (25.11) and (25.12), we see that the intersection of the $(^*, {}_*)$-theory determined by (27.11) with the theory of *proto-associative* algebras contains the degree-3 identities

$$a^*(b_*c) = 0 \qquad \text{and} \qquad (a^*b)_*c = 0. \tag{27.13}$$

The largest *stable* theory contained in this intersection will be the $(^*, {}_*)$-theory corresponding to the intersection of the theory of associative rings, which is closed, and the closed theory generated by (27.11). But we have observed that the latter theory contains essentially no degree-3 identities but (27.11) itself. Hence the theories of our two closed varieties of rings have no nontrivial degree-3 identities in common. Hence the largest *stable* $(^*, {}_*)$-theory in the intersection of our two stable $(^*, {}_*)$-theories has no nontrivial degree-3 identities; in particular, it does not contain (27.13), hence it is strictly smaller than the intersection of the two given $(^*, {}_*)$-theories.

We obtained (27.5) and (27.6) from properties of our procedure for constructing proto-identities from ring identities; the following alternative derivation is illuminating. Observe that we can factor the construction that takes $(A, {}^*, {}_*) \in \mathbf{P}$ and $S \in \mathbf{Ring}_k^1$ to $(A \otimes_k S, m) \in \mathbf{NARing}$ into two steps: First form a new object of $\mathbf{P}$, with underlying k-vector-space $A \otimes_k S$ and the two operations defined by

$$(as)^*(bt) = (a^*b)st, \qquad (as)_*(bt) = (a_*b)st.$$

Let us call this object $(A, {}^*, {}_*) \otimes_k S \in \mathbf{P}$. One then applies to this the "multiplication-mixing" construction, that takes an object $(B, {}^*, {}_*) \in \mathbf{P}$ to the object of **NARing** with additive group B and multiplication $x \cdot y = x^*y + y_*x$, which we will denote $(B, {}^*, {}_*)^{\mathrm{mix}} \in \mathbf{NARing}$. Thus, given $(A, {}^*, {}_*)$ and S, the object $(A \otimes S, m)$ with m as in (25.5) can be written

$$((A, {}^*, {}_*) \otimes_k S)^{\mathrm{mix}}.$$

Now the tensor product operation $\mathbf{P} \times \mathbf{Ring}_k^1 \to \mathbf{P}$ which forms the first of the above two steps is associative with ordinary tensor multiplication of objects of

$\mathbf{Ring}_k^1$:

$$((A,\ ^*,\ _*)\otimes_k S)\otimes_k T\ \cong\ (A,\ ^*,\ _*)\otimes_k (S\otimes_k T).$$

From this it is easily deduced that any closed subvariety of $\mathbf{P}$ must be closed under tensoring with all $S\in\mathbf{Ring}_k^1$. This condition on a subvariety of $\mathbf{P}$ is equivalent to condition (27.5) on its theory. If, further, we define the "reversal" construction on objects of $\mathbf{P}$ by $(A,\ ^*,\ _*)^{\mathrm{rev}}=(A,\ _*\mathrm{op},\ ^{*\mathrm{op}})$, then we have

$$((A,\ ^*,\ _*)\otimes_k S)^{\mathrm{mix}}\ \cong\ ((A,\ ^*,\ _*)^{\mathrm{rev}}\otimes_k (S^{\mathrm{op}}))^{\mathrm{mix}},$$

whence we can deduce that a closed subvariety of $\mathbf{P}$ must be closed under ^{rev}, and this is equivalent to condition (27.6) on the corresponding theory.

Let us pose formally the general problem we have been nibbling at:

PROBLEM 27.14. (i) *Characterize intrinsically those theories* $Y\subseteq\mathbf{I}$ *(respectively* $Y\subseteq\mathbf{I}^1$*) which are closed under our Galois connection; equivalently, the* stable *subvarieties* $\mathbf{V}\subseteq\mathbf{NARing}$ *(respectively* $\mathbf{NARing}^1$*).*
(ii) *Characterize intrinsically those subvarieties* $\mathbf{W}\subseteq\mathbf{P}$ *(respectively* $\mathbf{W}\subseteq\mathbf{P}^{\varepsilon}$*) which are closed under our Galois connection, equivalently, the* stable *equational* $(^*,\ _*)$*-theories.*

We remark that in addition to the "new" Galois connection we have introduced in this section, there are a couple of "standard" Galois connections that we have been using implicitly in our discussion; namely, the two relations "object satisfies identity", one between rings and members of $\mathbf{I}$, and one between objects of $\mathbf{P}$ and $(^*,\ _*)$-identities. The closed sets of each are called respectively "varieties" and "equational theories" of the kinds of objects in question. These standard Galois connections are represented by the vertical arrows in the diagram

(27.15)
$$\begin{array}{ccccc}
\mathbf{NARing}=\left\{\begin{array}{c}\text{rings}\\(R,\,m)\end{array}\right\} & \xleftarrow[(S\in\mathbf{Ring}_k^1)]{-\otimes S} & \left\{\begin{array}{c}\text{algebras}\\(A,\ ^*,\ _*)\end{array}\right\}=\mathbf{P}\\
\updownarrow & & \updownarrow\\
\mathbf{I}=\left\{\begin{array}{c}\text{identities}\\\text{for rings}\\(R,\,m)\end{array}\right\} & \xrightarrow[\text{construction}]{\text{proto-identity}} & \left\{\begin{array}{c}\text{identities}\\\text{for algebras}\\(A,\ ^*,\ _*)\end{array}\right\}
\end{array}$$

The dotted arrow on top represents the one-to-many map taking a $(^*,\ _*)$-algebra $(A,\ ^*,\ _*)$ to the class of constructed rings $(A\otimes S,\ m)$ $(S\in\mathbf{Ring}_k^1)$. This, combined with the relation "satisfies" on the left, induces our Galois connection between $\mathbf{P}$ and $\mathbf{I}$. It is easy to see using Birkhoff's Theorem (and the left exactness of $\otimes_k$ for k a field) that for any subvariety $\mathbf{V}\subseteq\mathbf{NARing}$, the set of algebras in $\mathbf{P}$

mapped into **V** by this composite map is also a subvariety, and so corresponds to an equational theory; this explains the existence of the lower arrow (also one-to-many) inducing the same Galois connection between **P** and **I** as the upper arrow.

How much of these ideas go over to the study of representable functors $\mathbf{C} \to \mathbf{D}$ for arbitrary varieties of algebras **C** and **D** in the sense of universal algebra?

What is special to the ring-theoretic context is the description of representable functors in terms of linearly compact algebras with two bilinear operations, and the subsequent observation that linear compactness was not essential to the study of identities satisfied. Given general varieties of algebras **C** and **D**, we can start drawing a diagram analogous to (27.15), putting **D** in the upper left-hand corner, the set of identities that can hold in objects of **D** at the lower left, and the class of representable functors $\mathbf{C} \to \mathbf{D}$ at the upper right; but since these functors cannot be identified with algebras, there is no obvious analog of the lower right-hand corner of (27.15). We can, however, still study the Galois connection between the upper right and lower left-hand corners of this diagram, getting concepts of "closed" subclasses of each; and we can call subvarieties of **D** with closed theories "stable"; these are the subvarieties generated by images of families of representable functors $\mathbf{C} \to \mathbf{D}$. Thus, in this general context one has an analog to the first half of Problem 27.14, but, so far as the authors can see, no very natural analog to the second half.

Incidentally, when we said at the beginning of the preceding section that we could drop the assumption that our objects $(A, {}^*, {}_*)$ or $(A, {}^*, {}_*, \varepsilon)$ had a linearly compact topology in our consideration of identities, we skirted over a subtle point. It is true that the same proto-identities which give the necessary and sufficient conditions on a $({}^*, {}_*)$-algebra A for all rings $A \otimes_k S$ to belong to a given variety **V** also give the necessary and sufficient conditions on a linearly compact $({}^*, {}_*)$-algebra A for all rings $A \hat{\otimes} S$ to belong to **V**; but it is not immediately evident whether there are enough linearly compact $({}^*, {}_*)$-algebras to distinguish all these varieties of such objects; in other words, it is not clear whether every stable variety of $({}^*, {}_*)$-algebras is generated by its members admitting linearly compact topologies.

This *is* in fact true in the nonunital case; more generally, every variety of $({}^*, {}_*)$-algebras determined by homogeneous identities (which, unless k is finite, means *every* variety of $({}^*, {}_*)$-algebras) is generated by objects admitting such a topology. This is because a free algebra on finitely many generators in such a variety has a natural grading, under which each homogeneous component is finite-dimensional, and the completion of this algebra with respect to that grading is linearly compact. However, for the unital case we are left with

PROBLEM 27.16. *Which subvarieties* $\mathbf{W} \subseteq \mathbf{P}^{\varepsilon}$ *are generated by objects that admit linearly compact structures?*

In particular, must every closed *subvariety of* $\mathbf{P}^{\varepsilon}$ *with respect to our Galois connection be the closure of a subvariety generated by a class of objects admitting such structure?*

If the answer to the latter question is negative, then along with the closure operators and dual closure operators introduced in this section, one would have to consider, in studying identities satisfied by the values of representable functors $\mathbf{Ring}_k^1 \to \mathbf{NARing}^1$, certain stronger operators, giving the closed subvarieties (in our sense) of $\mathbf{P}^\varepsilon$ generated by classes of linearly compact algebras.

We remark that the "universal algebra of k-vector spaces with one or more multilinear operations", which would embrace our objects $(A, {}^*, {}_*)$ as well as more classical single-multiplication algebras, has seen recent activity, in the introduction and study of the concept of an *operad*, essentially an abstract clone of k-multilinear operations. Cf. [**70**].

CHAPTER VI.

Representable functors from k-rings to rings.

In Chapter III, we characterized all representable functors from k-rings to abelian groups as being represented by tensor rings $k\langle M\rangle$ or $[k]\langle M\rangle$, with natural coaddition maps. In the last chapter, we characterized bilinear operations on such functors, and, in the special case where k was a field and our functors were represented by k-algebras, used this characterization to get precise descriptions of the representable functors to a large number of varieties of rings.

In theory, we might have studied such functors for general k and unrestricted representing object, by expressing the desired ring identities as properties of the bimodule map $\mathbf{m}: M \to M^{\lambda} \otimes_k M^{\rho} \oplus M^{\rho} \otimes_k M^{\lambda}$ determining the comultiplication on $k\langle M\rangle$ or $[k]\langle M\rangle$. Indeed, the calculations that we made instead, using linearly compact algebras, can be translated back into such terms. However, the authors find comultiplications and their coidentities difficult to grasp intuitively and calculate with, compared to operations and identities, and we doubt that working in the former context, we would have discovered results analogous to those obtained in §§25-27, e.g., that every representable functor to associative unital rings breaks up uniquely as the direct product of a functor based on the ordinary multiplication and a functor based on reversed multiplication, that every representable functor to Lie rings can be written as the composite of a functor to associative rings and the commutator-bracket construction, and that every functor to unital alternative rings is actually associative-ring valued.

What made the case of algebras over a field accessible was the duality between vector spaces and linearly compact vector spaces, which turned comultiplications on M back into multiplications on the dual space. In the next section, we shall develop an approach to the category k-$\mathbf{Bimod}^{\mathrm{op}}$, for k an arbitrary unital associative ring, which has formal properties like those of that vector-space duality, and will allow us to extend our earlier results to this general context.

From the point of view of calculations and proofs, it is unimportant that this way of viewing k-$\mathbf{Bimod}^{\mathrm{op}}$ is less naturally motivated than the identification of $\mathbf{Mod}_k^{\mathrm{op}}$ (k a field) with the category of linearly compact k-vector spaces. But in terms of the final statements of our results this is a drawback; hence in §29 we translate our descriptions of these representable functors back into the language of bimodule maps $\mathbf{m}: M \to M^{\lambda} \otimes_k M^{\rho} \oplus M^{\rho} \otimes_k M^{\lambda}$. This will bring us in contact with some concepts from the theory of Hopf algebras (though we shall not see Hopf algebras themselves until much later, in §43).

In these sections we shall avoid repeating the computations made in §§25-27, but will simply show how the context of those calculations can be generalized. We shall likewise not repeat all of our previous results, but only give the high points.

The last three sections of this chapter concern more diverse questions. In §30 we redevelop a result of Sweedler's which is equivalent to a description of all representable subfunctors of the forgetful functor k-$\mathbf{Ring}^1 \to \mathbf{Ring}^1$ for k a division ring. In §31 we consider the concept of the "image" of a morphism of representable functors, while §32 concerns what is called in Hopf algebra theory the "Fundamental Theorem on coalgebras".

§28. Element-chasing without elements.

Let k be a fixed ring with unit. To mimic the duality used in §§25-27, let us, for $M \in k$-**Bimod**, denote by $\hat{M}$ the corresponding object of k-$\mathbf{Bimod}^{\mathrm{op}}$ (usually denoted by the same symbol M), and let us write $\hat{\otimes}$: k-$\mathbf{Bimod}^{\mathrm{op}} \times k$-$\mathbf{Bimod}^{\mathrm{op}} \to k$-$\mathbf{Bimod}^{\mathrm{op}}$ for the functor corresponding to $\otimes_k$:

(28.1) $\hat{M} \hat{\otimes} \hat{N} =_{\mathrm{def}} \widehat{M \otimes_k N}$ ($M, N \in k$-**Bimod**; compare Corollary 24.25).

For full parallelism with the earlier situation, one might also write the hom-bifunctor k-$\mathbf{Bimod}^{\mathrm{op}} \times k$-$\mathbf{Bimod} \to \mathbf{Ab}$ as $\hat{\otimes}$ (cf. (24.21)). But since in the present context this does not admit a description closely resembling that of the bifunctor we just named $\hat{\otimes}$, we shall instead use "$\circ$":

(28.2) $$\hat{M} \circ N =_{\mathrm{def}} k\text{-}\mathbf{Bimod}(M, N) \in \mathbf{Ab}.$$

We now come to the crucial question of what to use in the role of "elements" of $\hat{M}$, corresponding to the $a, b, c, \ldots \in A = \hat{V}$ of §§25-27. Our earlier use of the duality of vector spaces given by the hom-functor $\mathbf{Mod}_k(-, k)$ suggests that we should associate to $\hat{M}$ the set k-$\mathbf{Bimod}(M, U)$ for an appropriate fixed k-bimodule U; say some injective. But in fact, the solution of greatest versatility is not to tie U down by a fixed choice, but to allow $\hat{M}$ to have various kinds of "elements". For *every* $U \in k$-**Bimod**, let us define a *U-element* of $\hat{M}$ to mean an element of k-$\mathbf{Bimod}(M, U)$. (This approach will be familiar to algebraic geometers.) Notationally

(28.3) $$a \in_U \hat{M} \text{ will mean } a \in k\text{-}\mathbf{Bimod}(M, U).$$

Having made this definition, it becomes natural to define "U-elements" of objects of k-**Bimod** as well:

(28.4) $$r \in_U M \text{ will mean } r \in k\text{-}\mathbf{Bimod}(U, M).$$

Note that U-elements of an object can be added and subtracted, though one cannot add a U-element to a V-element for $U \neq V$. The free k-bimodule on one generator is $k \otimes_{\mathbf{Z}} k$, hence $k \otimes_{\mathbf{Z}} k$-elements of M correspond to ordinary elements.

Clearly, these constructions are functorial:

(28.5) If $r \in_U M$ and $f \in k$-$\mathbf{Bimod}(M, M')$, we get an element $f(r) \in_U M'$.

(28.6) If $a \in_U \hat{M}$ and $g \in k$-$\mathbf{Bimod}^{\mathrm{op}}(\hat{M}, \hat{M}')$, we get an element $g(a) \in_U \hat{M}'$.

Furthermore, note that

(28.7) If $r \in_U M$, $s \in_V N$, we get an element $r \otimes s \in_{U \otimes_k V} M \otimes_k N$.

(28.8) If $a \in_U \hat{M}$, $b \in_V \hat{N}$, we get an element $a \otimes b \in_{U \otimes_k V} \hat{M} \hat{\otimes} \hat{N}$.

(28.9) If $a \in_U \hat{M}$, $s \in_U N$, we get an element $a \circ s \in \hat{M} \circ N$.

Here (28.7) and (28.8) use the functoriality of $\otimes$, while (28.9) uses composition of maps. (For consistency with the order in which we compose morphisms, it would have been preferable to have written $\hat{M}$ to the right of N in (28.2), and thus to have gotten $s \circ a$ rather than $a \circ s$ in (28.9). However, we are using the present order for notational parallelism with the order of factors we used in the completed tensor products $A \hat{\otimes} S$ of §§25-27.) The operations (28.7)-(28.9) are clearly themselves functorial, i.e., form commuting diagrams with the maps obtained as in (28.5)-(28.6).

Recall that for $L, M, N \in k$-**Bimod**, a k-bilinear map

$$*: L \times M \to N$$

is equivalent to a morphism of bimodules

$$\mathbf{m}: L \otimes_k M \to N.$$

We can use this observation to define the actions of such bilinear maps on our generalized elements. If $*$ is a bilinear map, and $\mathbf{m}$, as above, the corresponding bimodule map, then from elements $x \in_U L$, $y \in_V M$ we get, by (28.7) and (28.5), an element $x*y = \mathbf{m}(x \otimes y) \in_{U \otimes_k V} N$. Recalling in particular that the multiplication of a k-ring S, unital or nonunital, is a k-bilinear map, we see that from a U-element r and a V-element s of such an S, we get a "product", a $U \otimes_k V$-element of S, which we shall write rs.

Analogously, let us make

DEFINITION 28.10. *For* $\hat{L}$, $\hat{M}$, $\hat{N} \in k$-$\mathbf{Bimod}^{\mathrm{op}}$, *a "bilinear map"* $*: \hat{L} \times \hat{M} \to \hat{N}$ *will mean a morphism* $\mathbf{m}: \hat{L} \hat{\otimes} \hat{M} \to \hat{N}$ *(which, we recall, means a bimodule homomorphism* $N \to L \otimes_k M$).

Given such a $*$, *we define its action on generalized elements* $x \in_U \hat{L}$, $y \in_V \hat{M}$ *by setting* $x*y = \mathbf{m}(x \otimes y) \in_{U \otimes_k V} \hat{N}$ (*cf.* (28.8) *and* (28.6)).

In the above situation we shall sometimes call $\mathbf{m}$ "the morphism corresponding to the bilinear map $*$", though in fact under our Definition, $\mathbf{m}$ and $*$ are the same!

We might likewise define a bilinear map from $\hat{L} \times M$ to an abelian group G as a group homomorphism $\hat{L} \circ M \to G$, and make the corresponding convention on the action of such operations on generalized elements, but the concept will not be needed here.

In §§24-27, we used the fact that for a linearly compact vector space A and a general vector space V, sums of decomposable elements $a \otimes v$ were dense in $A \hat{\otimes} V$, so that a continuous map on this completed tensor product space could be described by giving its values on such sums, or, if linear, on decomposable elements alone. Here a simpler statement is true: *every* element of $\hat{M} \circ N$ is of the form $a \circ r$ $(a \in_U \hat{M},\ r \in_U N)$ for some U; for instance, for $U = M$ or N!

We can now repeat the proof of Theorem 25.1, and get

THEOREM 28.11. *Let* $k \in \mathbf{Ring}^1$, *and let* $F\colon k\text{-}\mathbf{Ring}^1 \to \mathbf{Ab}$ *be a representable functor. Then*

(i) *F can be written $S \mapsto A \circ S$ for a fixed object $A = \hat{M} \in k\text{-}\mathbf{Bimod}^{\mathrm{op}}$. Below, we will abbreviate $a \circ r$ to ar in writing elements of $F(S)$.*

(ii) *Zeroary operations $z\colon I \to F$ correspond to k-elements $\zeta \in_k A$, by the formula*

$$z_S = \zeta 1_S, \tag{28.12}$$

where $1_S \in_k S$ is the map $k \to S$ defining the k-ring structure of S.

(iii) *Linear operations $f\colon F \to F$ correspond to morphisms $\varphi\colon A \to A$, by the formula*

$$f(as) = \varphi(a)s. \tag{28.13}$$

(iv) *Bilinear operations $m\colon F \times F \to F$ correspond to pairs of bilinear maps* (*as defined in Definition 28.10*)

$$^{*}\colon A \times A \to A, \qquad {}_{*}\colon A \times A \to A, \tag{28.14}$$

via the formula

$$m(as, bt) = (a^{*}b)(st) + (b_{*}a)(ts). \tag{28.15}$$

*The same results hold for representable functors on the variety k-**Ring** of nonunital associative k-rings, except for* (ii): *There are no zeroary operations on such an F other than the zero operation, $z_S = 0$.* □

Let us display (28.15) diagrammatically. Let the bimodule maps giving * and ${}_{*}$ be $\mathbf{m}'$, $\mathbf{m}''\colon M \to M \otimes_k M$. Then if the generalized elements appearing in (28.15) are $a \in_U A$, $s \in_U S$, $b \in_V A$, $t \in_V S$, so that $as, bt \in A \circ S = k\text{-}\mathbf{Bimod}(M, S)$, then the element $m(as, bt) \in A \circ S$ described by the right-hand side thereof is the sum of the upper and lower branches of the diagram

$$\begin{array}{ccccccccc} & \overset{\mathbf{m}'}{\nearrow} & M^{\lambda} \otimes_k M^{\rho} & \xrightarrow{a \otimes b} & U \otimes_k V & \xrightarrow{s \otimes t} & S \otimes_k S & \searrow & \\ M & & & & & & & & S, \\ & \underset{\mathbf{m}''}{\searrow} & M^{\rho} \otimes_k M^{\lambda} & \xrightarrow{b \otimes a} & V \otimes_k U & \xrightarrow{t \otimes s} & S \otimes_k S & \nearrow & \end{array} \tag{28.16}$$

(where the rightmost arrows denote the map induced by the k-bilinear multiplication of S).

Since the formula (28.15) is formally identical with (25.5), in studying identities we may expand equations involving the multiplication m just as we did in §25-27. For instance, m will be associative for all S if and only if * and ${}_{*}$ satisfy (25.10) for all generalized elements a, b, c, s, t, u.

Can we justify in the present context the next step we took in §25, of isolating the coefficient of each monomial in s, t, u in (25.10), so as to get (25.11) and (25.12)? The justification there was that we could take for s, t, u the generators of a free associative algebra, $k<s, t, u>$. Here we can make a similar universal choice for s, t and u: Given any $a \in_{U_1} A$, $b \in_{U_2} A$, $c \in_{U_3} A$, take $S = k<U_1 \oplus U_2 \oplus U_3>$, and let $s \in_{U_1} S$, $t \in_{U_2} S$, $u \in_{U_3} S$ be the canonical inclusions of the bimodules U_1, U_2, U_3 in S. Then stu, sut, etc. become the inclusions of $U_1 \otimes_k U_2 \otimes_k U_3$, $U_1 \otimes_k U_3 \otimes_k U_2$, etc. in S. Since these tensor products are distinct summands in a direct sum decomposition of this tensor ring as a k-bimodule, a family of maps into these tensor product bimodules whose composites with those inclusions sum to zero must *each* be zero. The equations so obtained tell us that the necessary and sufficient conditions for the multiplication of our representable functor F to be associative are again that each of * and $_*$ be associative (this time on generalized elements), and each annihilate all values given by the other.

We can duplicate in this way all the arguments of §§25-27. Thus we have

THEOREM 28.17. *A representable functor* F: k-**Ring**$^1 \to$ **NARing** *or* **NARing**^{1}, *with representing object* $k<M>$ *and multiplication expressed as in Theorem 28.11*(iv), *will take values in a given subvariety* **V** *of* **NARing** *or* **NARing**^{1} *if and only if the* proto-identities *obtained by the method of §§25-27 from the identities of* **V** *are satisfied by the generalized elements of* $\hat{M}$ *under* * *and* $_*$. □

28.18. *Remarks on notation and language.* What we have called a U-element of $\hat{M}$ might have been called a $\hat{U}$-element, regarded as belonging to k-**Bimod**$^{\mathrm{op}}(\hat{U}, \hat{M})$.

In saying that a linear map $L \otimes_k M \to N$ is equivalent to a "bilinear map" $L \times M \to N$, we have used a terminology that is not universal: Many authors feel the term "bilinear" should logically entail only left linearity in the left variable and right linearity in the right variable, i.e., the first and third conditions in the display of Definition 10.4; they add the word "balanced" when they wish to assert that the middle condition of that display holds. However, we here understand this middle condition to be included in the definition of a bilinear map of bimodules. (Still another term sometimes used for this condition is "middle linearity".)

§29. $\otimes_k$-co-rings.

Theorem 28.17 is not as satisfying an answer to the problem of characterizing representable functors from varieties of k-rings ($k \in$ **Ring**^{1}) to varieties of rings as were the corresponding results of §§25-27 for functors on varieties of k-algebras (k a field), because an identity stated in terms of the U-elements of an object $\hat{M} \in k$-**Bimod**^{op}, as U ranges over all k-bimodules, is not as "concrete" a condition as an identity stated in terms of elements of a linearly compact vector space $\hat{V}$. So let us now translate our proto-identities back into statements about the "concrete" object M. Consider, for instance, one of the identities from (25.11), the associativity condition on *. In our present context this says

(29.1) $(a^*b)^*c = a^*(b^*c)$ $(a \in_{U_1} \hat{M},\ b \in_{U_2} \hat{M},\ c \in_{U_3} \hat{M};\ U_1, U_2, U_3 \in k\text{-}\mathbf{Bimod})$.

It is not hard to verify that this equation will hold for all $a,\ b,\ c$ as above if and only if it holds in the case where each of $U_1,\ U_2,\ U_3$ is M, and each of $a,\ b,\ c$ is the identity map of M. For notational convenience, let us label the domains of these three maps distinctly, writing

$$\begin{array}{lll} U_1 = M^{\lambda}, & & a = i^{\lambda}: M \cong M^{\lambda}, \\ U_2 = M^{\mu}, & & b = i^{\mu}: M \cong M^{\mu}, \\ U_3 = M^{\rho}, & & c = i^{\rho}: M \cong M^{\rho}. \end{array}$$

When we evaluate the two sides of (29.1) as $M^{\lambda} \otimes_k M^{\mu} \otimes_k M^{\rho}$-elements of $\hat{M}$, i.e., maps $M \to M^{\lambda} \otimes_k M^{\mu} \otimes_k M^{\rho}$, we find that they represent the two routes in the diagram

$$(29.2) \qquad \begin{array}{ccc} M & \xrightarrow{\quad \mathbf{m}' \quad} & M^{\lambda} \otimes_k M^{\rho} \\ {\scriptstyle \mathbf{m}'}\Big\downarrow & & \Big\downarrow{\scriptstyle \mathbf{m}' \otimes \mathrm{id}} \\ M^{\lambda} \otimes_k M^{\rho} & \xrightarrow{\quad \mathrm{id} \otimes \mathbf{m}' \quad} & M^{\lambda} \otimes_k M^{\mu} \otimes_k M^{\rho}. \end{array}$$

A k-bimodule M with a map $\mathbf{m}'$: $M \to M \otimes_k M$ such that the square (29.2) commutes is called, in the theory of Hopf algebras, a *coassociative co-ring*. (Hopf algebraists most often limit consideration to modules over a commutative ring k, frequently a field, where we are considering bimodules over general k; but the bimodule version of the concept is not unknown, cf. [**184**].) To avoid confusion with the sense of "co-ring" used in this work, let us add a distinguishing prefix:

DEFINITION 29.3. *By a* $\otimes_k$-co-ring *in* k-**Bimod** *we will mean a k-bimodule M given with a morphism* $\mathbf{m}$: $M \to M \otimes_k M$, *which we shall call its* comultiplication. *Such an M is said to be* coassociative *if the square* (29.2) (*with* $\mathbf{m}$ *for* $\mathbf{m}'$) *commutes.*

The unprefixed term "co-ring" (and generally "co-**V** object") will continue to have the meaning given it in §8. If we wished to put these two concepts on an equal footing, we could call the latter "$\amalg$-co-rings". Note, however, that the analogy between these two kinds of "co-ring" is not exact, because the kind defined in §8 have a coaddition map as well as a comultiplication, while in a $\otimes_k$-co-ring, the corresponding role is played by the additive structure of M rather than by another $\otimes_k$-co-operation. Cf. §9.1, second paragraph.

A *counit* on a $\otimes_k$-co-ring M will mean a map ε: $M \to k$ which makes the obvious triangles commute; in our dual language this is a k-element ε of $\hat{M}$ which satisfies $\varepsilon * a = a = a * \varepsilon$ for all $a \in_U \hat{M}$ (under the natural identifications $k \otimes_k U \cong U \cong U \otimes_k k$).

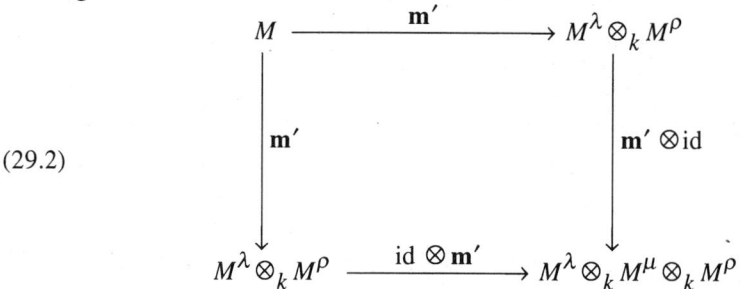

DEFINITION 29.4. *The category having for objects all coassociative (respectively coassociative counital) $\otimes_k$-co-rings, and for morphisms the bimodule maps making commuting diagrams with the comultiplication (and counit), will be denoted* $\otimes_k$**-co-Ring** *(respectively* $\otimes_k$**-co-Ring**^{1}).

Hopf algebraists have observed that if M is a coassociative counital $\otimes_k$-co-ring, and S an associative unital k-ring, then k-**Bimod**(M, S) becomes an *associative unital ring*, under operations which they call the *convolution* of the $\otimes_k$-co-ring structure of M with the k-ring structure of S [**183**, pp. 69 and 72]. In our terms, the convolution multiplication takes a pair as, bt of elements of that hom-set to the element $\mathbf{m}(a \otimes b)\,st$ (the upper path of (28.16)), and the convolution unit is defined by (28.12). These operations are the analogs, for the construction $\circ$, of the natural associative unital ring structure on a tensor product of two associative unital k-algebras. If the assumption of (co)associativity, respectively (co)unitality is dropped on either M or S, we still have a convolution multiplication, but it gives a not-necessarily-associative, respectively nonunital ring structure on $M \circ S$.

Let us now render some cases of Theorem 28.17 into the language of $\otimes_k$-co-rings. Throughout, k is a fixed associative unital ring.

THEOREM 29.5. *Every representable functor* k-**Ring**$^1 \to$ **Ring**^{1} *has the form*

$$S \mapsto k\text{-}\mathbf{Bimod}(M', S) \times k\text{-}\mathbf{Bimod}(M'', S)^{\mathrm{op}}, \tag{29.6}$$

where M' and M'' are coassociative counital $\otimes_k$-co-rings, i.e., k-bimodules given with coassociative $\otimes_k$-comultiplications $\mathbf{m}'$, $\mathbf{m}''$ and counit maps ε', ε'', and where the hom-groups on the right hand side of (29.6) *are made into unital rings by the* convolution *construction.*

If we write $M = M' \oplus M''$, then the representing k-ring for (29.6) *is the tensor ring $k\langle M\rangle$, and its coaddition is the morphism $\mathbf{a}_M$ defined in* (12.16-17). *The comultiplication is the map $k\langle M\rangle \to k\langle M^\lambda \oplus M^\rho\rangle$ defined to take $M = M' \oplus M''$ into*

$$(M'^{\lambda} \otimes_k M'^{\rho}) \oplus (M''^{\rho} \otimes_k M''^{\lambda}) \subseteq k\langle M^\lambda \oplus M^\rho\rangle$$

using the pair of maps $\mathbf{m}'$ and $\mathbf{m}''$, and the counit is the map $k\langle M\rangle \to k$ induced by the counit maps ε' and ε''.

This construction induces an equivalence of categories

$$(\otimes_k\text{-}\mathbf{co\text{-}Ring}^1 \times \otimes_k\text{-}\mathbf{co\text{-}Ring}^1)^{\mathrm{op}} \xrightarrow{(M',\ M'') \mapsto (29.6)} \mathbf{Rep}(k\text{-}\mathbf{Ring}^1,\ \mathbf{Ring}^1). \ \square \tag{29.7}$$

We have been detailed in the above statement; let us omit some analogous details in the next few results.

THEOREM 29.8. *Every representable functor* k-**Ring**$^1 \to$ **Lie** *can be written uniquely as a composite*

$$k\text{-}\mathbf{Ring}^1 \xrightarrow{S \mapsto k\text{-}\mathbf{Bimod}(M,\,S)} \mathbf{Ring} \xrightarrow{L} \mathbf{Lie}, \tag{29.9}$$

where M *is a coassociative* $\otimes_k$*-co-ring, and* L *is the functor appearing in* (25.33). □

THEOREM 29.10. *Suppose* 2 *is invertible in* k. *Then every representable functor* $k\text{-}\mathbf{Ring}^1 \to \mathbf{Jordan}^1$ *can be written uniquely as a composite*

$$k\text{-}\mathbf{Ring}^1 \xrightarrow{S \mapsto k\text{-}\mathbf{Bimod}(M,\,S)} \mathbf{Ring}^1_{\mathbf{Z}[1/2]} \xrightarrow{J} \mathbf{Jordan}^1, \tag{29.11}$$

where M *is a coassociative counital* $\otimes_k$*-co-ring, and* J *is the functor appearing in* (25.42). □

Let us also formulate in these terms our general description of representable functors to not necessarily associative rings (Theorem 23.9, Theorem 28.11 parts (i) and (iv)):

THEOREM 29.12. *Every representable functor* $k\text{-}\mathbf{Ring}^1 \to \mathbf{NARing}$ *has the form*

$$S \mapsto k\text{-}\mathbf{Bimod}(M,\,S), \tag{29.13}$$

where M *is a* k*-bimodule given with two (not necessarily coassociative)* $\otimes_k$*-comultiplications*

$$\mathbf{m}',\ \mathbf{m}'' \colon M \to M \otimes_k M,$$

and where the hom-set of (29.13) *is made a nonassociative ring using the* sum *of the convolution of the comultiplication* $\mathbf{m}'$ *with the multiplication of* S, *and the* opposite *of the convolution of the comultiplication* $\mathbf{m}''$ *with the multiplication of* S. □

The other results of §§25-27 can be translated similarly. Of course, when proto-identities other than associativity occur in these results, the corresponding coidentities on $\otimes_k$-co-rings must be described, whether diagrammatically or another way, for instance by a formula in universal elements in an appropriate ring $k\text{-}\mathbf{Bimod}(M,\, k\langle M \oplus M \ldots \oplus M\rangle)$.

When k is a commutative ring, corresponding results for functors on the subvarieties of *k-algebras*, $\mathbf{Ring}^1_k \subseteq k\text{-}\mathbf{Ring}^1$ and $\mathbf{Ring}_k \subseteq k\text{-}\mathbf{Ring}$ may, as in earlier sections, be deduced from these results. The $\otimes_k$-co-rings that correspond to such functors are those whose underlying k-bimodules belong to $\mathbf{Mod}_k$ (Definition 14.1). Let us give these objects a name for later reference:

DEFINITION 29.14. *For* k *a* commutative *ring, a* $\otimes_k$-co-ring *whose underlying* k*-bimodule is a* k*-module (i.e., has the same* k*-operations on both sides) will be called a* $\otimes_k$-coalgebra.

(Still more generally, one can, of course, deduce from our results on $\otimes_k$-co-rings the corresponding statements for k-centralizing $\otimes_K$-co-rings, where $K \in \mathbf{Ring}^1_k$.)

29.15. *Remarks on the technique of "generalized elements".* Though the results of this section are stated in terms of $\otimes_k$-co-rings, we are relying for the proofs on the method of the preceding section, using "U-elements" of $\hat{M}$ and S. Here are some observations on that method.

The bimodules U, V in the description of the multiplication m in (28.16) are logically unnecessary. If we had simply denoted the given elements of $A \bigcirc S$ as $x, y: M \to S$, then we could have replaced the two-step paths through $U \otimes_k V$ and $V \otimes_k U$ in that diagram by single arrows labeled $x \otimes y$ and $y \otimes x$. To put it another way, since U and V can always be universally taken to be copies of M, with a and b the identity map, we might as well have done this initially, then deleted these identity maps, and so simplified the diagram. So we seem to have unnecessarily complicated this work by a slavish imitation of the method of §§25-27.

But inserting the "unnecessary" bimodules U and V actually makes the situation easier to deal with computationally. It "separates" an element of $A \bigcirc S$ into two factors on which we let different kinds of multiplication act, that of A and that of S. So on second thought, by beginning with the special case of vector spaces over a field, and their duality with linearly compact vector spaces, we have been led to a "simplifying complication". This is the reverse of the common occurrence, where going to a more *general* case forces one to *eliminate* unnecessarily complications in one's approach!

Another point along the same lines. In §§25-27 when we encountered identities which were nonlinear in some variable (such as the Jordan identity (25.35)), we were forced to substitute for the variable in question not a single product, as, but a sum such as $as+bt+cu$. Since we have said that the present results are proved by the same computations as the earlier results, we have implicitly repeated this substitution in their proofs. But this is logically unnecessary, since we know that every element of $A \bigcirc S$ can be written $a \circ s$ ($a \in_U A$, $s \in_U S$, for some U). Would writing the general element in this way have simplified the development of our "proto-identities"? It appears not. This "simpler" substitution gives nonlinear identities, which in the present context are equivalent to their linearizations, and it is the linearized versions that are more convenient to work with. One gets these linearizations from the nonlinear identities by applying the latter to sums of elements. So here too, close imitation of §§25-27 works best.

Hopf algebraists might find the trick of computing with "generalized elements" useful. (That approach is also an alternative to the use of concretizations when proving in a general abelian category results originally obtained by element-chasing in categories of modules or abelian groups. For a discussion of how to get around non-exactness of the hom-functor in that context, see [**17**].)

As noted in the introductory remarks to this chapter, an alternative to the above development would have been to simply write down the diagrams or formulas in k, $k\langle M\rangle$, $k\langle M^\rho \oplus M^\lambda\rangle$, and $k\langle M^\rho \oplus M^\mu \oplus M^\lambda\rangle$ expressing the coassociativity and counitality (or other coidentities) of the co-operation on $k\langle M\rangle$ induced by $\mathbf{m}: M \to (M^\lambda \otimes_k M^\rho) \oplus (M^\rho \otimes_k M^\lambda)$, and to chase diagrams of tensor products of M till we came up with Theorems 29.{5, 8, 10}; in other words, to deal from the start only with the universal choices for U, V etc.. Though the authors felt such an approach unpromising, others may wish to see whether they *can* get an elegant development of these results in that way.

29.16. *"$\otimes_k$-co-rings" or "$\otimes_k$-cosemigroups"?* We noted earlier that the analogy between a $\otimes_k$-co-ring in k-**Bimod** and a co-ring (i.e., $\amalg$-co-ring) in a general category was imperfect, because the latter requires specification of a coaddition morphism **a**, while the additive structure of the former is simply the **Ab**-structure of the category k-**Bimod**. Might it not, then, be more appropriate to name what we have been calling $\otimes_k$-co-rings "$\otimes_k$-cosemigroups"?

As long as we are only dealing with the *coassociative* identity on the comultiplication, this is reasonable. However when we look at objects whose duals satisfy other identities than associativity, we find that some of these, such as the proto-alternative or proto-flexible identities, cannot be written down without referring to an additive structure, so that they admit no "semigroup-like" alternative to the "ring-like" formulation.

However, recall that an associative ring can itself be looked at in two ways: as a set R with an additive and a multiplicative structure related by several identities, or as an abelian group R with a morphism $R \otimes R \to R$ subject only to the associative identity, i.e., a $\otimes$-semigroup object in **Ab**. (The latter is the viewpoint taken in [**122**, p.4].) So the contrast between calling our objects "coassociative $\otimes_k$-co-rings" and "$\otimes_k$-cosemigroups" is not really fundamental.

We will encounter the idea that "a coassociative $\otimes_k$-co-ring structure is a kind of cosemigroup structure" again in §43.

§30. Subfunctors of forgetful functors, and a result of Sweedler.

In this section we shall examine representable subfunctors of the forgetful functor k-$\mathbf{Ring}^1 \to \mathbf{Ring}^1$, and prove a characterization of all such subfunctors in the case when k is a division ring, which we shall see is equivalent to a result of Sweedler [**184**] on $\otimes_k$-co-rings.

We begin with some background on the concept of *subfunctor*.

If $G\colon \mathbf{C} \to \mathbf{D}$ is a functor, in whose codomain category $\mathbf{D}$ one has a concept of "subobject" of an object (in particular, if $\mathbf{D}$ is a variety of algebras, in which case we shall understand "subobject" to mean "subalgebra"), then one may define a *subfunctor* of G to mean a functor $F\colon \mathbf{C} \to \mathbf{D}$ given with a morphism $f\colon F \to G$ such that for every object C of $\mathbf{C}$, $f(C)\colon F(C) \to G(C)$ is the inclusion of a subobject, $F(C) \subseteq G(C)$. Thus if, as is the case in a variety of algebras, the *embeddings* of $\mathbf{D}$ (the maps in $\mathbf{D}$ that up to isomorphism are the inclusions of subobjects) are precisely the monomorphisms, then the subfunctors of a functor G are, *up to isomorphism*, the functors F given with morphisms to G whose values at all objects are monomorphisms.

The following facts about this concept are easily verified. (The last statement below uses the observation that an *epimorphism* in a category $\mathbf{C}$ is a morphism $S \to R$ such that the induced maps $\mathbf{C}(R, X) \to \mathbf{C}(S, X)$ are *one-to-one* for all X in $\mathbf{C}$.)

LEMMA 30.1. *Let* **D** *be a category which admits pullbacks (limits over the diagram* $\substack{\cdot\to\\ \cdot\to}\cdot$*).*

Then a morphism $a\colon A \to B$ *of* **D** *is a* monomorphism *if and only if the pullback of the diagram* $\substack{A \xrightarrow{a} \\ A \xrightarrow[a]{}} B$ *is given by* A, *with the identity morphism into the two copies of itself in this diagram. If* **C** *is any other category, then the functor category* $\mathbf{D}^{\mathbf{C}}$ *also has pullbacks, which can be constructed "objectwise". Hence the monomorphisms in this functor category are those morphisms which give monomorphisms on all objects.*

Now suppose **D** *is a variety of algebras and* **C** *has finite colimits. Then* $\mathbf{Rep}(\mathbf{C}, \mathbf{D})$ *is closed in* $\mathbf{D}^{\mathbf{C}}$ *under taking pullbacks. Hence the monomorphisms of* $\mathbf{Rep}(\mathbf{C}, \mathbf{D})$ *are those morphisms that give monomorphisms (i.e., one-to-one maps) on all objects. Hence, finally, if* $G\colon \mathbf{C} \to \mathbf{D}$ *is a functor represented by a co-***D** *object* R, *then the* representable subfunctors *of* G *may be characterized (up to natural isomorphism) as represented by* **D***-coalgebras* S *in* **C** *given with coalgebra morphisms* $R \to S$ *which are* epimorphisms *of underlying* **C***-objects.* □

We remark that a general subfunctor F of a representable functor $G\colon \mathbf{C} \to \mathbf{D}$ need not be representable. For example, among the four subfunctors of the identity functor of **Ab** given by $F_1(A) = nA \subseteq A$ ($n>1$ a fixed integer), $F_2(A) = \{a \in A \mid na = 0\}$, $F_3(A)$ = the torsion subgroup of A, and $F_4(A)$ = the intersection of all kernels of homomorphisms $A \to \mathbf{Z}$, only F_2 is representable.

Now let k be any associative unital ring, and consider the forgetful functor $k\text{-}\mathbf{Ring}^1 \to \mathbf{Ring}^1$. Like any underlying-set-preserving functor between varieties, this is representable, with representing coalgebra having for underlying object the free algebra on one generator in the domain variety. In this case, that is the free k-ring R on one generator, which we shall denote g. We see that the ring co-operations are given by

$$\begin{aligned} \mathbf{a}(g) &= g^{\lambda} + g^{\rho}, \\ \mathbf{m}(g) &= g^{\lambda} g^{\rho}, \\ \mathbf{i}(g) &= -g, \\ \mathbf{z}(g) &= 0, \\ \varepsilon(g) &= 1. \end{aligned}$$

This free k-ring $k{<}g{>}$ can be described as the tensor ring $k{<}M{>}$ on the free k-bimodule M on one generator g, and this bimodule, we recall, has the form $M = k \otimes_{\mathbf{Z}} k$:

$$R = k{<}g{>} = k{<}M{>} = k{<}k \otimes_{\mathbf{Z}} k{>} \qquad \text{where } g = 1 \otimes 1.$$

When we express the co-$\mathbf{Ring}^1$ structure of R in terms of a pair of $\otimes_k$-co-rings M' and M'' as in Theorem 29.5, we find that $M' = M$ and $M'' = \{0\}$, since this structure does not involve the opposite multiplication. The $\otimes_k$-co-operations on M are seen to be given by

(30.2) $$\mathbf{m}(g) \;=\; g \otimes g,$$

(30.3) $$\varepsilon(g) \;=\; 1.$$

Let us make the convention that in writing elements of this ring $k\langle M\rangle$, and hence in particular, in writing elements of its subbimodules

$$M \;=\; k \otimes_{\mathbf{Z}} k \quad \text{and} \quad M \otimes_k M \;=\; (k \otimes_{\mathbf{Z}} k) \otimes_k (k \otimes_{\mathbf{Z}} k) \;\cong\; k \otimes_{\mathbf{Z}} k \otimes_{\mathbf{Z}} k,$$

we shall express these in terms of the k-ring generator g. E.g., we shall write $\Sigma\, \alpha_i\, g\, \beta_i$ rather than $\Sigma\, \alpha_i \otimes \beta_i$, and $\Sigma\, \alpha_i\, g\, \beta_i\, g\, \gamma_i$ rather than $\Sigma\, \alpha_i \otimes \beta_i \otimes \gamma_i$ (α_i, β_i, $\gamma_i \in k$).

By the preceding observations, representable subfunctors of the forgetful functor k-$\mathbf{Ring}^1 \to \mathbf{Ring}^1$ will correspond to epimorphs of this representing object $k\langle g\rangle$. Now ignoring the co-ring structure on this object, its epimorphs can be quite diverse, and in particular need not be surjective. For instance, the extension of $k\langle g\rangle$ gotten by universally adjoining multiplicative inverses to any family of elements $a(g)$, $b(g)$, ... is an epimorph, but generally not a surjective image of $k\langle g\rangle$. (This represents the set-valued functor which carries every k-ring S to the set of elements $s \in S$ such that $a(s)$, $b(s)$, ... are all invertible in S.)

However, when we restrict attention to representable subfunctors which give additive subgroups, i.e., to representable subfunctors of the forgetful functor to $\mathbf{Ab}$, then by Theorem 13.15, the representing object must have the form $k\langle N\rangle$, and the morphism of representing objects will be induced by a bimodule map

(30.4) $$M \;\to\; N.$$

It is easy to see that (30.4) induces an epimorphism of rings if and only if it is an epimorphism of bimodules, which means a surjective homomorphism, so in this case we can take $N = M/J$ for some subbimodule J of M.

For our subfunctor to be *unital subring* valued, the subbimodule J must have the properties that the comultiplication $\mathbf{m}$ of M induces a comultiplication $M/J \to (M/J) \otimes_k (M/J)$, and the counit ε induces a counit $M/J \to k$. This happens if and only if

(30.5) $$\mathbf{m}(J) \;\subseteq\; J \otimes M + M \otimes J,$$

(30.6) $$\varepsilon(J) \;=\; \{0\}.$$

For any $\otimes_k$-co-ring M, Hopf-theorists call a subbimodule $J \subseteq M$ satisfying (30.5) and (if M is counital) (30.6) a *coideal* of M. Summarizing our observations about subfunctors of the forgetful functor, we have

(30.7) Representable subfunctors F of the underlying unital ring functor on k-$\mathbf{Ring}^1$ are precisely the functors induced by $\otimes_k$-co-rings M/J, where M is $k \otimes_{\mathbf{Z}} k$, made a $\otimes_k$-co-ring as in (30.2-3), and J is a coideal thereof. In this situation, the functor F is described by

$$F(S) \;=\; \{s \in S \mid (\forall\; \Sigma\, \alpha_i g \beta_i \in J)\;\; \Sigma\, \alpha_i s \beta_i = 0\}.$$

What are examples of such quotient $\otimes_k$-co-rings of $M = k \otimes_{\mathbf{Z}} k$? An important class of these can be discovered formally, by noting that the bimodule-theoretic computation showing that $k \otimes_{\mathbf{Z}} k$ is a $\otimes_k$-co-ring does not depend on the fact that the tensor product is taken over $\mathbf{Z}$. Given any subring $L \subseteq k$, the bimodule

$$k \otimes_L k, \tag{30.8}$$

which is the quotient of $k \otimes_{\mathbf{Z}} k$ obtained by imposing the bimodule relations

$$\alpha g \ = \ g \alpha \qquad (\alpha \in L), \tag{30.9}$$

is also a $\otimes_k$-co-ring with co-operations (30.2-3). This $\otimes_k$-co-ring can be seen to yield the functor which associates to every $S \in k\text{-}\mathbf{Ring}^1$ the *centralizer* in S of $L \subseteq k$. That the subbimodule $J \subseteq M$ generated by $\{\alpha g - g\alpha \mid \alpha \in L\}$ is a coideal is also straightforward to verify directly:

$$\begin{aligned} \mathbf{m}(\alpha g - g\alpha) \ &= \ \alpha g g - g g \alpha \\ &= \ (\alpha g - g\alpha)g + g(\alpha g - g\alpha) \in (J \otimes M) + (M \otimes J), \\ \varepsilon(\alpha g - g\alpha) \ &= \ \alpha 1 - 1\alpha \ = \ 0. \end{aligned}$$

Do distinct subrings $L \subseteq k$ give distinct quotient $\otimes_k$-co-rings $k \otimes_L k$, and do all quotient $\otimes_k$-co-rings of $k \otimes_{\mathbf{Z}} k$ have this form?

Since $k \otimes_L k$ is the k-bimodule generated by a universal L-centralizing element, the question of whether this object determines L (which is the content of the first of the above questions) can be reformulated, "For every subring $L \subseteq k$, is the ring of elements of k centralizing $1 \otimes 1 \in k \otimes_L k$ precisely L?" The answer is, in general, no. For instance, suppose $\alpha \in L$ is invertible in k but its inverse does not lie in L. Then in $k \otimes_L k$, if we take the equation $\alpha g = g\alpha$ and multiply on both sides by α^{-1}, we get $g\alpha^{-1} = \alpha^{-1} g$; so the element $\alpha^{-1} \notin L$ also centralizes g.

This looks as though it has something to do with epimorphisms. Actually, what is involved is a more general concept, which we recall.

DEFINITION 30.10 (Isbell [**92**]). *Let* $\mathbf{C}$ *be a variety of algebras (or more generally, a full subcategory of such a variety which is closed under limits). If* $A, B \in \mathbf{C}$ *and* A *is a subalgebra of* B, *then the* dominion *of* A *in* B *means the subalgebra* D *consisting of those elements* $d \in B$ *with the property that every pair of morphisms of* B *into a common object* E *of* $\mathbf{C}$ *which agree on* A *also agree at* d; *equivalently, the intersection* D *of all subalgebras of* B *which are difference kernels of pairs of maps in* $\mathbf{C}$, *and which contain* A; *equivalently, the* least *subalgebra* $D \subseteq B$ *which is a difference kernel of a pair of maps in* $\mathbf{C}$, *and which contains* A.

In particular, a morphism in $\mathbf{C}$ is an *epimorphism* if and only if the dominion of its image is its whole codomain; but the concept of dominion in $\mathbf{C}$ does not conversely reduce to that of epimorphism in $\mathbf{C}$. For a fixed object B, the operation "dominion of – in B" is a closure operator on subobjects of B; we shall call its fixed subobjects "dominion-closed subalgebras of B". We see that these are precisely the subalgebras that are difference kernels of pairs of homomorphisms on B. (For more on the dominion construction, see [**92**]-[**96**].)

In our ring-theoretic context, if $\alpha \in L \subseteq k$ is invertible in k, its inverse will belong to the dominion of L in k, by uniqueness of inverses. Our observation that α^{-1} centralizes $1 \otimes 1 \in k \otimes_L k$ if $\alpha \in L$ now follows from a standard characterization of dominions in $\mathbf{Ring}^1$, namely the first assertion of:

LEMMA 30.11. *Let k be an associative unital ring, and L a subring. Then the dominion of L in k is equal to the centralizer in k of the element $1 \otimes 1$ in the k-bimodule $k \otimes_L k$.*

Thus, if we denote this dominion $L' \supseteq L$, the natural surjective map $k \otimes_L k \to k \otimes_{L'} k$ is an isomorphism.

PROOF. (After [**175**, proof of Proposition 1.1].) First, assuming $\alpha \in k$ centralizes $1 \otimes 1 \in k \otimes_L k$, we want to show that α belongs to the dominion of L in k. By the universal property of $k \otimes_L k$, we know that α will centralize any L-centralizing element of any k-bimodule. Now given ring homomorphisms f, g of k into a ring R, we may make R a k-bimodule by letting each $\xi \in k$ act on R on the left by left multiplication by $f(\xi)$, and on the right by right multiplication by $g(\xi)$. If f and g agree on L, left and right multiplication of $1 \in R$ by elements of L give the same values, so 1 is L-centralizing, hence by our preceding observation, it also centralizes α, i.e., $f(\alpha) = g(\alpha)$. As f and g were arbitrary homomorphisms agreeing on L, this shows that α lies in the dominion of L in k.

To show that elements *not* in the centralizer of $1 \otimes 1$ do *not* belong to the dominion of L, we shall construct a pair of ring homomorphisms whose difference kernel is precisely this centralizer. To do this, let us make $k \oplus (k \otimes_L k)$ into a ring by letting the second summand have square zero, and defining the remaining cases of multiplication by the ring-structure of k and the k-bimodule structure of $k \otimes_L k$. For our two homomorphisms of k into this ring, we use $c \mapsto c + 0$ and the conjugate of this map by the invertible element $1 + (1 \otimes 1)$. It is immediate that the difference kernel of these two maps is the centralizer in k of $1 \otimes 1$, as claimed.

The final assertion follows because the relations in the description of $k \otimes_{L'} k$ that are not in the description of $k \otimes_L k$ are the equations $\alpha \otimes 1 = 1 \otimes \alpha$ $(\alpha \in L' - L)$, but we have just seen that these are indeed satisfied in the latter bimodule. □

Since a dominion-closed subring L of a ring k is closed under taking inverses, a dominion-closed subring of a division ring is a division ring. On the other hand, a division ring L is dominion-closed in *any* overring k. For given $\alpha \in k - L$, we may take a right L-basis B of k containing $\{1, \alpha\}$. Then $B \otimes B$ is a right L-basis of $k \otimes_L k$, with distinct members $\alpha \otimes 1$ and $1 \otimes \alpha$, hence by the above Lemma, α is not in the dominion of L. These two observations give

LEMMA 30.12. *If k is a division ring, its dominion-closed subrings are precisely its division subrings. Hence in view of Lemma 30.11, the* distinct *quotient $\otimes_k$-co-rings of $k \otimes_{\mathbf{Z}} k$ of the form $k \otimes_L k$ (L a subring of k) correspond to the distinct* division subrings $L \subseteq k$. □

Note the qualification "of the form $k \otimes_L k$" – we have not yet addressed the question of whether every quotient $\otimes_k$-co-ring of $k \otimes_{\mathbf{Z}} k$ has this form. Sweedler [**184**, Lemma 2.2] proves that this is indeed true for k a division ring. We shall now give his proof, and then some examples showing that for general k, there can exist other sorts of quotient $\otimes_k$-co-rings.

Let k be a division ring, and N any quotient $\otimes_k$-co-ring of $k \otimes_{\mathbf{Z}} k$, i.e., any $\otimes_k$-co-ring generated as a k-bimodule by a single element g satisfying (30.2) and (30.3). (In the language of Hopf theorists, an element g of a $\otimes_k$-co-ring satisfying these relations is called *grouplike*.) Let $L \subseteq k$ denote the centralizer in k of $g \in N$, a division subring of k. Thus N is a homomorphic image of the k-bimodule $k \otimes_L k$. If it is not an isomorphic image, let

(30.13) $$\alpha_1 g \beta_1 + \ldots + \alpha_t g \beta_t = 0 \qquad (t \geq 1,\ \alpha_1, \ldots, \alpha_t, \beta_1, \ldots, \beta_t \in k)$$

be a relation holding in N which does not hold in $k \otimes_L k$, i.e., which is not a consequence of the k-bimodule relations (30.9), and let this relation be chosen to minimize t. Thus, all α_i and β_i are nonzero. Since k is a division ring, we can, by appropriate left and right multiplication of (30.13), assume that $\alpha_1 = \beta_1 = 1$.

Condition (30.3) implies that $g \neq 0$ in N, hence $t > 1$. We claim that (30.13) is the "unique" right k-linear relation satisfied by $\alpha_1 g, \ldots, \alpha_t g \in N$, i.e., that the right k-vector space of right k-linear relations satisfied by these elements is 1-dimensional, spanned by this relation. For if we had a relation linearly independent of this one, then by taking linear combinations of that relation and (30.13), we could get two relations each involving fewer than t terms, and summing to (30.13). But by our minimality assumption, any relation involving fewer than t terms is a consequence of the relations (30.9), hence (30.13) would also be a consequence of these relations, contradicting our assumption.

In particular, the elements $\alpha_2 g, \ldots, \alpha_t g \in N$ are right k-linearly independent. Let us now look at $N \otimes_k N$, and write down two consequences of (30.13) in that bimodule. On the one hand, applying $- \otimes g$ we get $\alpha_1 g \beta_1 g + \ldots + \alpha_t g \beta_t g = 0$. On the other hand, applying the map $\mathbf{m}$, we get $\alpha_1 g g \beta_1 + \ldots + \alpha_t g g \beta_t = 0$. Subtracting, we get $\alpha_1 g(\beta_1 g - g\beta_1) + \ldots + \alpha_t g(\beta_t g - g\beta_t) = 0$. Since $\beta_1 = 1$, the first term of this sum is zero. The right k-linear independence of $\alpha_2 g, \ldots, \alpha_t g$ now implies that each of the coefficients $\beta_2 g - g\beta_2, \ldots, \beta_t g - g\beta_t$ is also 0. (This is because the right k-linearly independent elements $\alpha_i g$ $(i = 2, \ldots, n)$ can be incorporated into a right k-basis of N, so the corresponding right k-submodules $\alpha_i g \otimes N$ give independent summands in $N \otimes_k N$, each isomorphic to N. This basis-argument is one of several places where we have used the fact that k is a division ring.) This says that each of $\beta_2, \ldots, \beta_t$ belongs to L, and we know that $\beta_1 = 1$ does. Hence modulo consequences of (30.9), the relation (30.13) is equivalent to

(30.14) $$(\alpha_1 \beta_1 + \ldots + \alpha_t \beta_t) g = 0.$$

But we have noted that $g \neq 0$, hence (30.14) implies that $\alpha_1 \beta_1 + \ldots + \alpha_t \beta_t = 0$, i.e., (30.14) is trivial. Hence (30.13) is a consequence of (30.9), a contradiction, showing that $N \cong k \otimes_L k$.

Translating this result on $\otimes_k$-co-rings back to a statement about representable functors, we have

THEOREM 30.15 (after Sweedler [**184**, Lemma 2.2]). *Let* k *be a division ring. Then every representable subfunctor of the forgetful functor* k-**Ring**$^1 \to$ **Ring**^{1} *has the form*

$$S \mapsto \text{centralizer of } L \text{ in } S,$$

for a unique division subring $L \subseteq k$. □

In the above proof, the counit was used only to argue that in the representing $\otimes_k$-co-ring N we had $g \neq 0$. Hence dropping the assumption of a counit, we have

COROLLARY 30.16. *For* k *again a division ring, the representable subfunctors of the forgetful functors* k-**Ring**$^1 \to$ **Ring** *and* k-**Ring** $\to$ **Ring** *are the centralizer-subring functors described above, and also the trivial subfunctor* $S \mapsto \{0\} \subseteq S$. □

When k is not a division ring, other quotient $\otimes_k$-co-rings of $k \otimes_{\mathbf{Z}} k$ may exist. Suppose, for instance, that k contains a nontrivial idempotent element e. Then any k-ring S can be decomposed as a 2×2 "matrix" of summands

$$\begin{pmatrix} eSe & eS(1-e) \\ (1-e)Se & (1-e)S(1-e) \end{pmatrix}.$$

Note that for $s \in S$, the condition

$$(1-e)se = 0 \qquad (30.17)$$

says that the above "matrix" representation of s has the form $\begin{pmatrix} * & * \\ 0 & * \end{pmatrix}$. The elements s with this property form a unital subring of S, giving a representable subfunctor of the identity, corresponding to the coideal J of $k \otimes_{\mathbf{Z}} k$ generated as a subbimodule by the element

$$(1-e)ge.$$

This is not in general a centralizer subring. For example, if E is a field, and we take for k *either* the full 2×2 matrix ring $M_2(E)$, *or* the ring $E \times E$ regarded as the ring of diagonal matrices in $M_2(E)$, and let $S = M_2(E)$ in either case, then the subring of S determined by (30.17) (with $e = e_{11}$) is the ring of upper triangular 2×2 matrices over E, which is not a centralizer subring of $M_2(E)$. (In this situation, there are also various non-counital quotients of $k \otimes_{\mathbf{Z}} k$, corresponding to constructions such as $\begin{pmatrix} * & 0 \\ 0 & 0 \end{pmatrix}$, $\begin{pmatrix} 0 & * \\ 0 & 0 \end{pmatrix}$ and $\begin{pmatrix} * & * \\ 0 & 0 \end{pmatrix}$.) This failure of the analog of Theorem 30.15 for $k = E \times E$ and $M_2(E)$ is noteworthy, since these two examples can be looked at as being "as close to division rings as one can get". This leads us to ask

QUESTION 30.18. (i) *For what rings* k *is it true that the quotient* $\otimes_k$*-co-rings of* $k \otimes_{\mathbf{Z}} k$ *all have the form* $k \otimes_L k$ *for subrings (necessarily dominion subrings)* $L \subseteq k$*?*

(ii) *For what rings* k *can one obtain a reasonable characterization of all quotient* $\otimes_k$*-co-rings of* $k \otimes_{\mathbf{Z}} k$*? (Can one do so for simple Artinian rings? For von Neumann regular rings? For commutative integral domains?)*

(iii) *What can be said if one generalizes these problems, and considers representable subfunctors of the forgetful functor* K-$\mathbf{Ring}_k^1 \to K_0$-$\mathbf{Ring}^1$, *where* k *is a commutative ring,* K *a k-algebra, and* K_0 *a subring of* K*; i.e., if one seeks to describe all quotient* $\otimes_{K_0}$*-co-rings of* $K \otimes_k K$*?*

(Concerning part (i), it is easy to deduce from Theorem 30.15 a slight generalization of that Theorem, saying that the desired conclusion holds if k is a product of finitely many division rings of distinct characteristics. In the context of (iii), one gets an analogous statement with the kernel in k of the k-algebra structure-map replacing the characteristic.)

§31. Images of morphisms.

We begin this section, like the preceding, with some considerations on general varieties of algebras.

Suppose $\mathbf{V}$ and $\mathbf{W}$ are varieties, $F,\ G\colon \mathbf{V} \to \mathbf{W}$ two functors, and $f\colon F \to G$ a morphism between them. Then the *image* of f, i.e., the functor taking each $V \in \mathbf{V}$ to $f(F(V)) \subseteq G(V)$, will be a subfunctor of G. However, even if F and G are representable, this image functor may not be representable. Indeed, one of the examples of nonrepresentable subfunctors noted immediately after Lemma 30.1 was the image of the endomorphism "multiplication by n" of the identity functor of the category of abelian groups.

One may attempt to get an "approximation" of the image of f by a representable functor, by using for representing object the image of the morphism $\mathbf{f}$ representing f. This object, being a homomorphic image, and hence an epimorph, of the object representing G will represent a *set*-valued subfunctor of G, but it may not yield a $\mathbf{W}$-valued functor. The condition for it to do so is noted below.

(When we speak of a "set-valued subfunctor of the algebra-valued functor G", we of course mean, more precisely, a subfunctor of the composite of G with the underlying set functor on its codomain. We will be imprecise in this way about the distinction between an algebra-valued functor and its composite with the underlying-set functor in this section when there is no danger of confusion, to avoid cumbersome statements.)

LEMMA 31.1. *Let* $F,\ G\colon \mathbf{V} \to \mathbf{W}$ *be representable functors between varieties of algebras, represented by coalgebras whose underlying* $\mathbf{V}$*-objects we shall denote* R *and* S *respectively, and let* $f\colon F \to G$ *be a morphism of functors, corresponding to a morphism of coalgebras* $\mathbf{f}\colon S \to R$. *Let* H *be the functor* $\mathbf{V} \to \mathbf{Set}$ *represented by* $\mathbf{f}(S) \subseteq R$, *identified with a set-valued subfunctor of* G; *thus, for every object* A *of* $\mathbf{V}$, $H(A) \subseteq G(A)$ *can be described as the set of those elements of* $G(A)$ *whose coordinates satisfy all* $\mathbf{V}$*-relations satisfied, for all* $B \in \mathbf{V}$, *by the corresponding coordinates of* $f(F(B))$. *Then the following conditions are equivalent:*

(i) *For every object* A *of* $\mathbf{V}$, *the set* $H(A) \subseteq G(A)$ *is closed under the* $\mathbf{W}$*-operations of* $G(A)$ *(i.e.,* H *can be regarded as a* $\mathbf{W}$*-valued subfunctor of* G).

(ii) *The co-*$\mathbf{W}$ *structure of* S *induces a co-*$\mathbf{W}$ *structure on* $\mathbf{f}(S) \subseteq R$.

Moreover, these conditions are always satisfied for zeroary *and* unary *(co)operations.* □

We now consider varieties of rings:

LEMMA 31.2. *Suppose, in the situation of the preceding Lemma, that* $\mathbf{V}$ *is* k-$\mathbf{Ring}^1$ *for* k *an associative unital ring, or* $\mathbf{Ring}^1_k$ *for* k *an associative commutative unital ring, and that* $\mathbf{W}$ *is* $\mathbf{Ab}$, $\mathbf{NARing}$, $\mathbf{NARing}^1$ *or some subvariety of one of these. Let us write the representing objects for* F *and* G *as* $R = k<M>$, $S = k<N>$, *with the coaddition of Theorem 13.15 and, in the co-ring cases, comultiplication of the form described in Theorem 29.12.*

Then the equivalent conditions of the preceding Lemma will hold for the abelian-group *structure of* H *if and only if, writing* L *for the image bimodule* $\mathbf{f}(N) \subseteq M$, *the map* $k<L> \to k<M>$ *induced by* $\mathbf{f}$ *is an embedding.*

In this case, the conditions of that Lemma will also hold for the ring *structures of our functors.*

PROOF. By Theorem 13.15, the image of $k<N>$ in $k<M>$ admits a co-$\mathbf{Ab}$ structure compatible with that of $k<N>$ if and only if it has the form $k<L>$ for some k-bimodule L and the map of $k<N>$ into it is induced by a k-bimodule homomorphism $N \to L$. Clearly, L must then be the bimodule named in the Lemma, and the first assertion follows.

To verify the second assertion, let us denote by

$$\mathbf{m}'_M : M \to M^\lambda \otimes_k M^\rho, \qquad \mathbf{m}''_M : M \to M^\rho \otimes_k M^\lambda$$

$$\mathbf{m}'_N : N \to N^\lambda \otimes_k N^\rho, \qquad \mathbf{m}''_N : N \to N^\rho \otimes_k N^\lambda$$

the $\otimes_k$-co-operations determining the comultiplications of $R = k<M>$ and $S = k<N>$. Then it will suffice to show, assuming the map $k<L> \to k<M>$ is an embedding, that $\mathbf{m}'_N$ and $\mathbf{m}''_N$ induce $\otimes_k$-co-operations $\mathbf{m}'_L$ and $\mathbf{m}''_L$ on L. Letting $\mathbf{m}_N$ denote either of the former $\otimes_k$-co-operations and $\mathbf{m}_M$ the corresponding co-operation on M, the fact that $\mathbf{f}$ is a morphism of co-rings gives us commutativity of the diagram

$$\begin{array}{ccccc} N & \longrightarrow & L & \longrightarrow & M \\ \Big\downarrow \mathbf{m}_N & & & & \Big\downarrow \mathbf{m}_M \\ N \otimes_k N & \longrightarrow & L \otimes_k L & \longrightarrow & M \otimes_k M. \end{array}$$

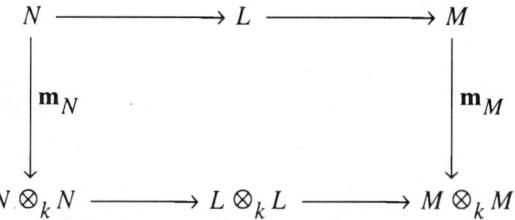

Here by definition the horizontal arrows on top are respectively onto and one-to-one. The first horizontal arrow on the bottom is thus also onto. Moreover, the second horizontal arrow on the bottom, being a homogeneous component of the map $k\langle L\rangle \to k\langle M\rangle$, which we have assumed is an embedding, must be one-to-one. Hence the two vertical arrows induce a third arrow $\mathbf{m}_L$ between the middle terms of our diagram, as desired. □

Note that in the situation of the above Lemma, even if the first condition is *not* satisfied, so that the least **Set**-valued subfunctor of G containing $f(F)$ is not **Ab**-valued, the tensor ring $k\langle L\rangle$ (where $L = \mathbf{f}(N) \subseteq M$) will still represent the least **Ab**-valued representable subfunctor of G containing $f(F)$. If our functors are ring-valued, we may then ask whether *this* subfunctor of G is subring-valued. The condition for this to be true is noted in

LEMMA 31.3. *Let F, G: k-*$\mathbf{Ring}^1 \to \mathbf{NARing}$ *be representable functors, let $f\colon F \to G$ be a morphism, let $\mathbf{f}\colon k\langle N\rangle \to k\langle M\rangle$ be the corresponding morphism of representing objects, and let $L = \mathbf{f}(N) \subseteq M$. Then the comultiplication of $k\langle N\rangle$ induces a comultiplication on $k\langle L\rangle$ (equivalently, the additive-subgroup-valued subfunctor of G represented by $k\langle L\rangle$ is subring-valued) if and only if* $\ker(\mathbf{f}\mid N)$ *is a coideal in N with respect to each of the two $\otimes_k$-co-operations $\mathbf{m}'_N$ and $\mathbf{m}''_N$ which together determine the co-ring structure of $k\langle N\rangle$.* □

Let us now give an example showing how all of the above conditions can fail, after which we shall determine those rings k such that these conditions hold for all F, G and f.

EXAMPLE 31.4. Let k be a commutative principal ideal domain and p an irreducible element of k. Let F be the functor associating to each associative k-algebra S the set of elements of S annihilated by p^2, made a ring using the zero multiplication:

$$\zeta\cdot\zeta' = 0.$$

Let G, on the other hand, be the functor associating to each S the set of ordered pairs of elements annihilated by p, with the multiplication

$$(\xi, \eta)(\xi', \eta') = (0, \xi\xi'). \tag{31.5}$$

Each of these operations is associative "by default", all threefold products being 0. If we map F to G by defining $f(\zeta) = (p\zeta, 0)$, we find this to be a homomorphism:

$$f(\zeta)f(\zeta') = (p\zeta, 0)(p\zeta', 0) = (0, p^2\zeta\zeta') = (0, 0) = f(0) = f(\zeta\cdot\zeta').$$

The image of f is the functor associating to S the set of all pairs of the form $(\xi, 0)$ such that ξ is both a multiple of p in S, and satisfies $p\xi = 0$, and this image-set, of course, forms a subring of $G(S)$. But the condition of having first coordinate a multiple of p is not equivalent to a **V**-relation in the coordinates of our element. Rather, the least set-valued subfunctor H of G containing $f(F)$ and expressible by relations in the coordinates is the one that takes each S to the set of pairs $(\xi, 0)$ with $p\xi = \xi^2 = 0$. This set is in general closed neither under addition nor under multiplication. The least **Ab**-valued subfunctor containing these values takes S to the set of pairs $(\xi, 0)$ with $p\xi = 0$, but this is still not in general closed under the multiplication (31.5); for example, it is not when $S = k/pk$ with the ordinary multiplication.

From the $\otimes_k$-coalgebra viewpoint, F corresponds to the $\otimes_k$-coalgebra M presented as a k-module by one generator z and one relation $p^2 z = 0$, with comultiplication $\mathbf{m}_M(z) = 0$, while the $\otimes_k$-coalgebra N corresponding to G is presented by two generators x, y and relations $px = py = 0$, with comultiplication $\mathbf{m}_N(x) = 0$, $\mathbf{m}_N(y) = x \otimes x$. The module homomorphism $\mathbf{f}\colon N \to M$ defined by

$$\mathbf{f}(x) = pz, \qquad \mathbf{f}(y) = 0$$

is a $\otimes_k$-coalgebra homomorphism, the nontrivial verification being:

$$\begin{aligned}(\mathbf{f}\otimes\mathbf{f})(\mathbf{m}_N(y)) &= (\mathbf{f}\otimes\mathbf{f})(x\otimes x) = pz\otimes pz \\ &= p^2 z\otimes z = 0 = \mathbf{m}_M(0) = \mathbf{m}_M(\mathbf{f}(y)).\end{aligned}$$

Now if we let $L = \mathbf{f}(N) \subseteq M$, we see that L may be presented as a k-module by one generator $w = \mathbf{f}(x)$ and one relation $pw = 0$. It follows that $w \otimes w$ is nonzero in $L \otimes_k L$, but the image of this element in $L \otimes L \subseteq k\langle M\rangle$ is $pz \otimes pz$, which as noted above is 0, so the embedding of L in M does not induce an embedding of $k\langle L\rangle$ in $k\langle M\rangle$. Moreover, a comultiplication $\mathbf{m}$ on L compatible with that of N would have to carry $\mathbf{f}(y) = 0$ to $\mathbf{f}(x) \otimes \mathbf{f}(x) \neq 0$, a contradiction. In particular, (30.5) fails for the element $y \in \ker(\mathbf{f} \mid N)$, so this kernel is not a coideal.

In the opposite direction, it is known that when k is a field, the kernel of a homomorphism of $\otimes_k$-coalgebras is a coideal [**183**, Theorem 1.4.7(b)]. We can also see that an embedding of k-vector-spaces induces an embedding of tensor algebras. Thus, the conditions of Lemmas 31.2 and 31.3 are both satisfied for functors on $\mathbf{Ring}^1_k$ when k is a field. Let us now determine the most general rings K, k such that the corresponding results hold for functors on K-$\mathbf{Ring}^1_k$.

Recall that a ring K is called *von Neumann regular* if for every $a \in K$ there exists $b \in K$ such that $aba = a$. This is equivalent to the existence of such an element which also satisfies $bab = b$ (as one sees by replacing the original b by bab). When both of these equations hold, we shall call a and b *pseudoinverses* of one another. Let us note that a pseudoinverse b of a central element a of a ring K is also central in K. For if $x \in K$, we have

$$\begin{aligned}bx - xb &= babx - xbab = bbxa - axbb = bbxaba - abaxbb \\ &= a(babxb - bxbab) = a(bxb - bxb) = 0.\end{aligned}$$

(Caveat: it is not true that all elements centralizing an arbitrary element a will centralize a pseudoinverse b of a; this is true if and only if a and b commute.)

We will need the following extension of a standard result:

LEMMA 31.6. *Let K be an associative unital ring, and C its center. Then the following conditions are equivalent.*

(i) *K is von Neumann regular.*

(ii) *Tensor multiplication over K of right K-modules with left K-modules is left exact in the first variable. (I.e., tensoring a one-to-one map of right K-modules with a fixed left K-module gives a one-to-one map of abelian groups.)*

(iii) *Tensor multiplication over K of K-bimodules with K-bimodules (to get K-bimodules) is left exact in the first variable.*

(iv) *Tensor multiplication over K of C-centralizing K-bimodules with C-centralizing K-bimodules is left exact in the first variable.*

(ii′)-(iv′) *The tensor multiplication bifunctors referred to in* (ii)-(iv) *respectively are left exact in the* second *variable.*

PROOF. The equivalence of (i) and (ii) is well known in the form, "K is von Neumann regular if and only if every left K-module is flat" [**49**, Theorem 6.6.5, p.242]. (Note that "left exactness" has nothing to do with left modules, but refers to the left end of a short exact sequence.) Clearly (ii)⇒(iii)⇒(iv). We shall prove (iv)⇒(ii) by showing that given a counterexample to (ii), i.e., a one-to-one map of right K-modules $A_0 \to A$, a left K-module B, and a nonzero element $x \in A_0 \otimes_K B$ which goes to 0 in $A \otimes_K B$, we can manufacture a similar example for C-centralizing K-bimodules.

This is easy if C is itself von Neumann regular. In that case, we give each of our left or right modules the C-module structure on the other side that makes it C-centralizing, and tensor over C with K, getting a map of K-bimodules $K \otimes_C A_0 \to K \otimes_C A$ and a K-bimodule $B \otimes_C K$. The left exactness of tensor multiplication of C-modules implies on the one hand that the above map is an embedding, and on the other hand that our original one-sided modules embed in these bimodules and their various tensor products embed in the corresponding tensor products of these bimodules. We conclude that the image of x in $(K \otimes_C A_0) \otimes_K (B \otimes_C K)$ is nonzero, but goes to zero in $(K \otimes_C A) \otimes_K (B \otimes_C K)$, which is the required example.

On the other hand, if C is not von Neumann regular, there is an element $a \in C$ having no pseudoinverse in C. Since pseudoinverses in K of central elements are central, a has no pseudoinverse in K either. This means that $Ka^2 \neq Ka$. If we now consider the tensor product of the inclusion map of K-bimodules $Ka \subseteq K$ with the K-bimodule K/Ka, we see that the element $a \otimes \bar{1} \in Ka \otimes_K (K/Ka)$ is nonzero, but goes to zero in $K \otimes_K (K/Ka)$, again giving a counterexample to (iv).

The equivalence of these conditions with (ii′)-(iv′) follows from the left-right symmetry of (i). □

We shall now show that von Neumann regularity of K is the necessary and sufficient condition for the results we want concerning images of representable functors to hold. We begin with

LEMMA 31.7. *Let K be an associative unital algebra over a commutative ring k. Then the following conditions are equivalent.*

(i) *K is von Neumann regular.*

(ii) *Every one-to-one map $L \to M$ of k-centralizing K-bimodules induces a one-to-one homomorphism $K<L> \to K<M>$.*

(iii) *The kernel of every homomorphism of not necessarily coassociative k-centralizing $\otimes_K$-co-rings is a coideal.*

(iii′) *The kernel of every homomorphism of* coassociative *k-centralizing $\otimes_K$-co-rings is a coideal.*

PROOF. Assuming (i), an inductive application of Lemma 31.6(iii) shows that a one-to-one bimodule map as in (ii) induces a one-to-one map $L^{\otimes n} \to M^{\otimes n}$ for each n. Taking the direct sum, we get (ii). The last statement of Lemma 31.2 gives (ii)⇒(iii), and (iii)⇒(iii′) is clear.

To prove the final implication (iii′)⇒(i), it will suffice to show that when K is *not* von Neumann regular, there exists a morphism $\mathbf{f}\colon N \to M$ of *coassociative k-centralizing* $\otimes_K$-co-rings whose kernel is not a coideal. Note that if we write C for the center of K as in the preceding Lemma, a C-centralizing $\otimes_K$-co-ring will in particular be k-centralizing; it will thus suffice to construct a C-centralizing example.

Assuming K not von Neumann regular, the preceding Lemma tells us that there exists a one-to-one morphism $A_0 \to A$ of C-centralizing K-bimodules, a C-centralizing K-bimodule B, and a nonzero element $x \in A_0 \otimes_K B$, which goes to zero under the induced map $A_0 \otimes_K B \to A \otimes_K B$. Let us define the C-centralizing K-bimodules

$$M = A \oplus B, \qquad N = (K \otimes_C K) \oplus A_0 \oplus B.$$

We define the comultiplication $\mathbf{m}_M$ of M to be zero, while we let the comultiplication $\mathbf{m}_N$ of N be zero on $A_0 \oplus B$, but carry the generator $1 \otimes 1$ of $K \otimes_C K$ to $x \in A_0 \otimes_K B \subseteq N \otimes_K N$. Both of the resulting $\otimes_K$-co-rings are coassociative by default. If we now let $\mathbf{f}\colon N \to M$ take A_0 into A by the given embedding of bimodules, act as the identity on B, and annihilate $K \otimes_C K$, we find that $\mathbf{f}$ is a $\otimes_K$-co-ring homomorphism with $\mathbf{f}(N) \cong A_0 \oplus B$. Essentially as in Example 31.4, the element $1 \otimes 1 \in \ker(\mathbf{f})$ shows that (30.5) does not hold, so $\ker(\mathbf{f})$ is not a coideal. □

(Note that k appears in conditions (ii)-(iii′), but not in condition (i). Thus, the former conditions actually depend on K alone, and not on k.)

Since every *field* is von Neumann regular, the above results include, in particular, the result quoted earlier, that for k a field the kernel of a homomorphism of $\otimes_k$-coalgebras is a coideal.

The result (i)⇒(iii′) above applies, in particular, to homomorphisms of *counital* $\otimes_K$-co-rings; the coideal one gets in such a situation will respect the counit (i.e., satisfy (30.6)), since, as we have mentioned, zeroary and unary operations go over to images without difficulty. To get the converse implication for the counital case, we would like a way to turn the counterexample-construction used in proving (iii′)⇒(i) into a counital counterexample. Since we got associativity of the represented functor by using a multiplication that was zero on threefold products, that example is certainly not counital as it stands. But we shall now apply to it a construction of "adjoining a counit".

If $(M, \mathbf{m})$ is any noncounital $\otimes_K$-co-ring, let us form the K-bimodule $M^1 \cong K \oplus M$, write g for the image of $1 \in K$, and define a comultiplication $\mathbf{m}^1$ on M^1 by

$$\mathbf{m}^1(g) = g \otimes g, \qquad \mathbf{m}^1(x) = g \otimes x + x \otimes g + \mathbf{m}(x) \quad (x \in M). \tag{31.8}$$

We find that $(M^1, \mathbf{m}^1)$ is coassociative if and only if $(M, \mathbf{m})$ is, and that the map $M^1 \to K$ taking g to 1 and M to 0 is a counit. Moreover, given a $\otimes_K$-co-ring morphism $\mathbf{f}\colon N \to M$ whose kernel is not a coideal, the induced morphism $\mathbf{f}^1\colon N^1 \to M^1$ will have the same property; namely, taking $x \in \ker(\mathbf{f})$ such that $\mathbf{m}_N(x) \notin \ker(\mathbf{f}) \otimes N + N \otimes \ker(\mathbf{f})$, it is easy to see that x has the same property with respect to $\mathbf{f}^1$. We deduce

COROLLARY 31.9. *The equivalent conditions of Lemma 31.7 are also equivalent to the variants of* (iii), (iii′) *stated for* counital *$\otimes_K$-co-rings.* □

The above results were stated for a single $\otimes_K$-comultiplication on a K-bimodule M. Recalling that two such $\otimes_K$-comultiplication structures satisfying appropriate coidentities determine a (coassociative or general) co-ring structure on the tensor ring $K\langle M\rangle$, we deduce

THEOREM 31.10. *Let* $\mathbf{V}$ *be one of* **NARing**, **NARing**^{1}, **Ring**, *or* **Ring**^{1}. *Then for any associative unital algebra* K *over a commutative ring* k, *the following conditions are equivalent.*

(i) *K is von Neumann regular.*

(ii) *Every morphism $\mathbf{f}\colon S \to R$ of co-$\mathbf{V}$ objects in K-$\mathbf{Ring}^1_k$ induces a co-$\mathbf{V}$ structure on its image, $\mathbf{f}(S) \subseteq R$.*

(ii′) *For every morphism $f\colon F \to G$ of representable $\mathbf{V}$-valued functors on K-$\mathbf{Ring}^1_k$, the least representable set-valued subfunctor $H \subseteq G$ containing the image of f is in fact subring-valued.* □

31.11. *Further observations on adjunction of units.* If M is a non-counital $\otimes_K$-co-ring, and F the representable ring-valued functor it determines, it is not hard to see that the functor determined by the $\otimes_K$-co-ring M^1 defined above takes a K-ring S to the unital ring $C_K(S) \oplus F(S)$, where $C_K(S)$ denotes the centralizer of K in S. (This is more often written $C_S(K)$; we are using the reversed notation $C_K(S)$ to emphasize the functoriality in S, the ring K here being fixed.) The following exercise sheds some light on this construction.

EXERCISE 31.12. *Let $K \in \mathbf{Ring}^1$, and let $C_K(-)$ denote the functor taking every K-ring S to the centralizer of K in S, as above.*

(i) *Show that any representable functor $F\colon K$-$\mathbf{Ring}^1 \to \mathbf{Ab}$ admits a natural* *structure of* C_K*-bimodule.* (*This bimodule structure is not given by a set of unary operations on* F, *because* C_K *is not a fixed ring but a ring-valued functor; rather, it is given by two bilinear maps of representable functors; cf.* §*23.*)

(ii) *Let* I *denote the forgetful functor* K-**Ring**$^1 \to$ **Ring**^{1}. *Why cannot one obtain in the same way a structure of* I-bimodule *on* F?

(iii) *Suppose* $m: F \times G \to H$ *is a bilinear map of functors* K-**Ring**$^1 \to$ **Ab**, *corresponding to a map* $\mathbf{m}: N \to L \otimes_K M \oplus M \otimes_K L$ *of K-bimodules. Show that* m *is bilinear* (*in the sense discussed in Remark 28.18*) *with respect to the* C_K*-bimodule structures on these functors if and only if the* $M \otimes_K L$*-component of* $\mathbf{m}$ *is zero* (*i.e.,* m *does not involve "reversed multiplication"*).

(iv) *State the analogous bilinearity condition which holds if and only if the* $L \otimes_K M$*-component of* $\mathbf{m}$ *is zero.*

(v) *Show that any* **Ring**^{1}*-valued functor on* K-**Ring**^{1} *extends to a* $C_K \times (C_K)^{\mathrm{op}}$-**Ring**^{1}*-valued functor.*

(vi) *Show that* $C_K \times (C_K)^{\mathrm{op}}$ *is the initial object of* **Rep**(K-**Ring**^{1}, **Ring**^{1}), *thus generalizing Corollary 25.22.*

If we consider similar questions for **NARing**-valued or even **Ring**-valued (i.e., nonunital associative-valued) functors F, we find that we cannot in general put a $C_K \times (C_K)^{\mathrm{op}}$-ring structure on F, because F does not split into parts that use only "regular", and only "reversed" multiplications. However, we can still define an "adjunction of unit" construction carrying **Rep**(K-**Ring**^{1}, **NARing**) to **Rep**(K-**Ring**^{1}, **NARing**^{1}), using the direct product with C_K, or with $(C_K)^{\mathrm{op}}$, or various other functors, because the identities of **NARing**^{1} do not include the associativity conditions required by the definition of a bilinear map of bimodules, and hence by that of a C_K-ring. Neither of the choices C_K or $(C_K)^{\mathrm{op}}$ is particularly canonical; and indeed, the initial object of **Rep**(K-**Ring**^{1}, **NARing**^{1}) appears to be quite complicated; we hope to examine it in [**26**].

No "adjunction of unit" construction can carry **Rep**(K-**Ring**^{1}, **Ring**) into **Rep**(K-**Ring**^{1}, **Ring**^{1}), even in the case where K is a field and our representing objects are K-algebras; for an object of the former category can have a multiplication whose two components * and $_*$ have nondisjoint images, while an object of the latter category cannot. Fortunately, in getting our unital associative example above, we did not need a construction that carried *every* $F \in$ **Rep**(K-**Ring**^{1}, **Ring**) to a member of **Rep**(K-**Ring**^{1}, **Ring**^{1}). Since we applied that construction to a co-ring with no "opposite" component, i.e., with $_* = 0$, it was sufficient to have a construction which worked in this case.

31.13. *Further observations on the approximation of the image of a morphism of representable functors by a representable functor.* At the beginning of this section, we noted that given a morphism $f: F \to G$ of representable functors between arbitrary varieties, corresponding to a morphism $\mathbf{f}: S \to R$ of representing objects, one could consider the set-valued functor represented by $\mathbf{f}(S) \subseteq R$ as a representable "approximation" of the image of F in G. Let us note that there exists a still better representable approximation, gotten by using, not the universal *surjective image* of S through which $\mathbf{f}$ factors, but the universal *epimorph* with this property, i.e., the largest subalgebra of R into which $\mathbf{f}$ maps S epimorphically. This is what Isbell calls the *stable dominion* of $\mathbf{f}(S)$ in R. It can be constructed as the intersection of a decreasing chain of subalgebras $R_{(\alpha)} \supseteq \mathbf{f}(S)$, where α ranges over the ordinals of cardinality $\leq$ card(R):

$$\begin{aligned} R_{(0)} &= R, \\ R_{(\alpha+1)} &= \text{dominion of } \mathbf{f}(S) \text{ in } R_{(\alpha)}, \\ R_{(\beta)} &= \textstyle\bigcap_{\alpha\in\beta} R_{(\alpha)} \quad \text{when } \beta \text{ is a limit ordinal;} \end{aligned}$$

or, alternatively, as the subalgebra of R generated by all the subalgebras which contain $\mathbf{f}(S)$ and are epimorphs thereof. This subalgebra will represent the smallest *representable set-valued* subfunctor of G containing the image of F.

What can one say about dominions and stable dominions of homomorphisms of tensor rings $K<N> \to K<M>$ induced by bimodule maps $N \to M$? It turns out that the dominion of such a homomorphism may be larger than its image, but the stable dominion is always precisely the image.

For a case of the former sort, let K be a commutative integral domain and c a nonzero noninvertible element of K, let $M = N =$ the free K-module on one generator x, and let us map N to M by sending x to cx. Then the degree-2 component of the image of $K<N>$ in $K<M>$ is generated by c^2x^2, but it is easy to verify that cx^2 is in the dominion of this image.

Our claim that the *stable* dominion of such a map is, nevertheless, just its image is a case of a more general result:

LEMMA 31.14. *If $R = \bigoplus R_i$ is a unital ring graded by the nonnegative integers, and $S = \bigoplus S_i$ a graded subring, then the stable dominion of S in R (both as rings and as graded rings) is the subring generated by S and the stable dominion of S_0 in R_0.*

In particular, the stable dominion of a homomorphism of tensor rings $K<N> \to K<M>$ induced by a homomorphism of K-bimodules $N \to M$ is equal to the image of that map.

PROOF. We shall prove the first assertion, from which the second follows.

Replacing S by the subring of R generated by S and the stable dominion of S_0 in R_0, we may assume that S_0 is its own stable dominion in R_0. It is then not hard to see that the stable dominion of S in R will be contained in $S_0 \oplus \bigoplus_{i>0} R_i$, so we can without loss of generality replace R by this graded subring. (Note that we cannot, a priori, simply replace R by the stable dominion of S, because we do not know that this stable dominion is a graded subring.) Thus we are reduced to the case where $S_0 = R_0$, and must prove that S is then its own stable dominion in R.

Suppose inductively that for some $j > 0$ we know that the stable dominion of S in R is contained in

$$(\textstyle\bigoplus_{i<j} S_i) \oplus (\bigoplus_{i\geq j} R_i).$$

Consider two maps of this ring into the factor-ring

$$(\bigoplus_{i<j} S_i) \oplus R_j/S_j,$$

namely the factor-map, and the same map modified to be 0 on the degree-j component. These agree on the image of S, but disagree on every element whose degree-j component is not in S_j, hence the stable dominion of S can contain no element of the latter sort. This establishes our inductive condition for j, completing the proof. □

It sometimes happens that the image of a morphism of representable functors is itself representable, so that the various "approximations" we have been studying coincide with the image itself. The following result characterizes this rather special situation.

LEMMA 31.15. *Suppose* $\mathbf{f}\colon S \to R$ *is a morphism in a category* $\mathbf{C}$. *Then the corresponding morphism* $f\colon h_R \to h_S$ *of representable functors* $\mathbf{C} \to \mathbf{Set}$ *has the property that its* image *is representable by an object* T *if and only if* $\mathbf{f}$ *can be written as the composite of an epimorphism* $S \to T$ *and a left invertible map* $T \to R$. *If* $\mathbf{C}$ *is a variety of algebras, this is equivalent to saying that the stable dominion* T *of* $\mathbf{f}(S)$ *is a retract of* R.

SKETCH OF PROOF. This is essentially the conjunction to two results: (i) a morphism $h_R \to h_T$ of representable functors gives surjective maps on all objects if and only if the corresponding morphism $T \to R$ is left invertible, and (ii) a morphism $h_T \to h_S$ of representable functors gives one-to-one maps on all objects if and only if the corresponding morphism $S \to T$ is an epimorphism. The "$\Rightarrow$" direction of (i) is proved by applying the hypothesis to $1_T \in h_T(T)$; the other direction holds because left invertibility of the morphism $T \to R$ implies right invertibility of all induced morphisms $h_R(X) \to h_T(X)$. As we observed earlier, (ii) is a restatement of the definition of epimorphism. □

Let us now consider the case where $\mathbf{f}$ is a morphism of coalgebras in a variety. Then f will be a morphism of algebra-valued functors, hence its image will be subalgebra-valued. Applying the criterion of the above Lemma, we get

COROLLARY 31.16. *Suppose* $\mathbf{V}$ *and* $\mathbf{W}$ *are varieties of algebras, and* $\mathbf{f}\colon S \to R$ *a morphism of* $\mathbf{V}$*-coalgebra objects of* $\mathbf{W}$. *If the stable dominion* T *of* $\mathbf{f}(S)$ *in* R *is a retract of* R *as an object of* $\mathbf{W}$, *then the* $\mathbf{V}$*-coalgebra structures of* R *and* S *induce a* $\mathbf{V}$*-coalgebra structure on* T. □

Combining this Lemma and Corollary with our result $\mathbf{Rep}(K\text{-}\mathbf{Ring}_k^1, \mathbf{Ab}) \cong K\text{-}\mathbf{Bimod}_k{}^{\mathrm{op}}$, and recalling that retracts of bimodules are direct summands, we get the equivalence of conditions (a) and (b) of the next Corollary. We then characterize this situation ring-theoretically in (c) and (d).

COROLLARY 31.17. *Let* k *be a commutative ring and* K *a unital* k*-algebra. Then the following conditions are equivalent:*

(a) *For every morphism* $f\colon F \to G$ *of representable functors* $K\text{-}\mathbf{Ring}_k^1 \to \mathbf{Ab}$, *the image functor* $f(F) \subseteq G$ *is also representable.*

(b) *In the abelian category* $K\text{-}\mathbf{Bimod}_k$, *short exact sequences split; i.e., subbimodules are direct summands.*

(c) *The k-algebra* $K \otimes_k K^{\mathrm{op}}$ *is semisimple Artinian.*

(d) K *is semisimple Artinian, each of its simple Artinian factors* K_i *is finite-dimensional over its center, and each of these centers is a finite separable algebraic extension of its subfield generated by the image of* k.

PROOF. We have noted the equivalence of (a) and (b). Regarding objects of K-$\mathbf{Bimod}_k$ as left $(K \otimes_k K^{\mathrm{op}})$-modules, and recalling that the rings over which all short exact sequences of left modules split are the semisimple Artinian rings (cf. [**48**, Theorem 5.3.7, p.176]) we get the equivalence of (b) and (c).

To prove (c)$\Rightarrow$(d), let us first note that for any k-algebra K, if to each left ideal $I \subseteq K$ we associate the left ideal $I \otimes K^{\mathrm{op}} \subseteq K \otimes_k K^{\mathrm{op}}$, this mapping is one-to-one. For under the natural left $K \otimes_k K^{\mathrm{op}}$-module structure on K, we clearly have $(I \otimes K^{\mathrm{op}})K = I$. Hence if $K \otimes_k K^{\mathrm{op}}$ is left Artinian, K must also be. Likewise, if $K \otimes_k K^{\mathrm{op}}$ is semisimple, K must also be, for a nonzero nilpotent two-sided ideal $I \subseteq K$ would induce a nilpotent two-sided ideal $I \otimes K^{\mathrm{op}} \subseteq K \otimes_k K^{\mathrm{op}}$, which by the above observation would be nonzero. So assuming (c), we see that K is semisimple Artinian.

Now if K_i is one of the simple factors of K, then (c) implies in particular that $K_i \otimes_k K_i^{\mathrm{op}}$ is semisimple Artinian; from this we see that it will suffice to prove our implication under the assumption that K is simple Artinian. Next, writing $K = M_n(K_0)$ for K_0 a division ring, we see that (c) implies that $K_0 \otimes_k K_0{}^{\mathrm{op}}$ is semisimple Artinian, so we are reduced to proving (c)$\Rightarrow$(d) when K is a division ring. Finally, letting k' denote the subfield of the center of the division ring K generated by the image of k, we see that $K \otimes_k K^{\mathrm{op}}$ can be identified with $K \otimes_{k'} K^{\mathrm{op}}$, so we are reduced to the case where the base-ring k is a field.

So assume K a division ring and k a field. If K were infinite-dimensional over k, it would contain a subfield L infinite-dimensional over k (cf. [**114**, Theorem 15.8(4)$\Rightarrow$(5), p.255]). But for an infinite extension-field L/k, the tensor product $L \otimes_k L$ is non-Artinian, and since $K \otimes_k K^{\mathrm{op}}$ is free as a left module over $L \otimes_k L$, it would also be left non-Artinian. So K is a finite-dimensional division algebra over k. If its center C were inseparable over k, then the central subring $C \otimes_k C$ of $K \otimes_k K$ would have nilpotent elements, contradicting semisimplicity of the latter ring. This completes the proof of (d).

Finally, let us show (d)$\Rightarrow$(c). For this it will clearly suffice to show that given two *simple* Artinian k-algebras K_1 and K_2 whose centers are finite separable extensions of their subfields generated by the images of k, the algebra

$$K_1 \otimes_k K_2$$

is semisimple Artinian (possibly zero). Now if the kernels of the maps of k into K_1 and K_2 are different, this tensor product is zero, since the image of some element of k will be zero in one factor and invertible in the other. In the contrary case, let k' denote the field of fractions of the common image of k in K_1 and K_2, and let C_1, C_2 denote the centers of these algebras. Then

$$K_1 \otimes_k K_2 \cong K_1 \otimes_{k'} K_2 \cong (K_1 \otimes_{k'} C_2) \otimes_{(C_1 \otimes_{k'} C_2)} (C_1 \otimes_{k'} K_2). \tag{31.18}$$

We see that each of $K_1 \otimes_{k'} C_2$, $C_1 \otimes_{k'} K_2$ is a central semisimple algebra over

$C_1 \otimes_{k'} C_2$ finitely generated as a module. Since by assumption C_1 and C_2 are finite separable extensions of k', their tensor product $C_1 \otimes_{k'} C_2$ is a finite direct product of fields, say $F_1 \times \ldots \times F_m$ (cf. [**203**, Theorem III.39, p.195] or [**48**, Theorem 5.7.4, p.194]). For each F_i, the corresponding factor in the right-hand side of (31.18) is a tensor product of finite-dimensional central simple F_i-algebras, hence is itself one, hence (31.18) is semisimple, as required. □

On the other hand, for *any* nonzero K and k, there exist morphisms of representable *set-valued* functors on K-$\mathbf{Ring}_k^1$ whose images are not representable; for instance, the morphism represented by the inclusion $K[x^2] \to K[x]$ of polynomial algebras.

We have not found the above result (c)⇔(d) in the literature, though there are some overlapping results, especially in the case where k is a field. Cf. [**147**, §§10.1-10.7], [**162**, §7c], and [**33**, §7.3, Theorem 1; §7.5, Cor. 2 to Prop. 6, and §7.6, Cor. 4 to Theorem 3]. For k a field, the k-algebras characterized in these conditions are what are called the *separable* k-algebras.

§32. A non-locally-finite $\otimes_k$-coalgebra.

We cited in (25.48) the result that a linearly compact associative algebra over a field k is a topological inverse limit of finite-dimensional associative algebras. We can now state this in the form usually given, again without proof:

(32.1) ''*Fundamental Theorem on coalgebras*'' ([**183**, Theorem 2.2.1 and Corollary 2.2.2], cf. (25.48) above). For k a field, any coassociative $\otimes_k$-coalgebra is locally finite-dimensional; i.e., is a directed union of its finite-dimensional $\otimes_k$-subcoalgebras.

Example 25.49 showed that this statement becomes false if ''coassociative'' is replaced by ''co-Lie''. Let us now show that it also becomes false for coassociative coalgebras if k is not assumed a field, but allowed to be a more general commutative ring; i.e., that in this situation, a coassociative $\otimes_k$-coalgebra M need *not* be a directed union of $\otimes_k$-subcoalgebras finitely generated as k-modules. In fact, we shall show that it need not be a *direct limit* of any system of $\otimes_k$-coalgebras finitely generated as k-modules. This statement is stronger than the former, since the argument that makes two such statements equivalent in many other mathematical contexts is based on looking at the *images* of the given system of objects in the limit object, but as we saw in the preceding section, images of homomorphisms of $\otimes_k$-coalgebras need not admit $\otimes_k$-coalgebra structures.

To construct our example, let k again be a commutative principal ideal domain and $p \in k$ an irreducible element. Let $M \in \mathbf{Mod}_k$ be the k-module presented by generators x_i $(i = 1, 2, \ldots)$ and relations $p^2 x_i = 0$. Define a comultiplication $\mathbf{m}$: $M \to M \otimes_k M$ by $\mathbf{m}(x_i) = p\,x_{i+1} \otimes x_{i+1}$. Thus the corresponding representable functor associates to every k-ring S the additive group of all sequences

(32.2) $$(\xi_1, \xi_2, \ldots) \qquad (\xi_i \in S)$$

satisfying

(32.3) $$p^2 \xi_i = 0 \qquad (i = 1, 2, \ldots),$$

with the multiplication

(32.4) $$(\xi_1, \xi_2, \ldots)(\eta_1, \eta_2, \ldots) = (p\xi_2\eta_2, \ p\xi_3\eta_3, \ \ldots).$$

The two ways of multiplying out a 3-fold product both give 0 because of (32.3), so the multiplication is associative, equivalently, $\mathbf{m}$ is coassociative.

To show that $(M, \mathbf{m})$ is not a direct limit of $\otimes_k$-coalgebras finitely generated as k-modules, we shall show that if $(M', \mathbf{m}')$ is any $\otimes_k$-coalgebra with M' finitely generated as a k-module, then every homomorphism $\mathbf{f}$ from $(M', \mathbf{m}')$ to $(M, \mathbf{m})$ satisfies $\mathbf{f}(M') \subseteq pM$. Indeed, since M' is finitely generated, there exists i such that

(32.5) $$\mathbf{f}(M') \subseteq kx_1 + \ldots + kx_i + pM.$$

Let i be the least such value. Assuming $\mathbf{f}(M') \not\subseteq pM$, we have $i > 0$. If we apply $\mathbf{m}$ to an element of $\mathbf{f}(M')$ in which the coefficient of x_i is not divisible by p, we get an element of $\mathbf{f}(M') \otimes \mathbf{f}(M') \subseteq M \otimes_k M$ in which the coefficient of $x_{i+1} \otimes x_{i+1}$, though divisible by p, is not divisible by p^2. But from (32.5) we see that the coefficient of $x_{i+1} \otimes x_{i+1}$ in any element of $\mathbf{f}(M') \otimes \mathbf{f}(M')$ is divisible by p^2, a contradiction.

The recipe for the above example might be summarized as follows: Take a noncoassociative counterexample to the Fundamental Theorem on coalgebras over the field k/pk (in the above case, the coalgebra with comultiplication $\mathbf{m}(x_i) = x_{i+1} \otimes x_{i+1}$). Lift this to a coalgebra over k/p^2k, then multiply the co-operation by p, to make it coassociative by default.

We can get a counital version of the above example by adjoining a counit, as in the examples of the preceding section. Writing the resulting $\otimes_k$-coalgebra $M^1 = kx_0 \oplus M$, one proves that every morphism $\mathbf{f}$ of a counital $\otimes_k$-coalgebra M' into M^1 satisfies $\mathbf{f}(M') \subseteq kx_0 + pM$, by essentially the same argument used above.

In these examples, as in the initial example of the preceding section, the assumption that k is a principal ideal domain has been made only to provide familiarity. The above examples go over if k is any non-von Neumann regular commutative ring, and p any non-pseudoinvertible element. We shall not try to determine necessary and sufficient conditions on a not necessarily commutative ring k for the analog of (32.1) to hold. In **[27]** it will be proved that if k is a semisimple Artinian ring, then any coassociative $\otimes_k$-co-ring is the directed union of a system of $\otimes_k$-sub-co-rings finitely generated as k-bimodules.

CHAPTER VII.

Representable functors from rings to general groups and semigroups.

So far, the representable functors on varieties of associative rings that we have studied have been to structures having an *abelian* group or semigroup operation. In this chapter we shall describe what we know about functors to general groups and semigroups.

In §33, we motivate, and state without proof, a characterization of all representable functors $\mathbf{Ring}_k^1 \to \mathbf{Group}$ for k a field, to be proved in [**82**]. In §34, we prove characterizations due to Berstein [**31**] (though the formulations here are somewhat more general than his) of all representable functors from the category of *connected graded k-rings* (definition recalled below) to $\mathbf{Binar}^e$, to $\mathbf{Semigp}^e$, and to $\mathbf{Group}$.

We then return to ungraded rings, and devote four sections to trying to understand representable functors $\mathbf{Ring}_k^1 \to \mathbf{Semigp}^e$ for k a field. These are quite diverse, but we develop a class of constructions which may conceivably include them all.

In §39 we show that for every nonlocal commutative integral domain k, there exist functors F: $\mathbf{Ring}_k^1 \to \mathbf{Semigp}$ such that $F(k)$ is nonempty, but contains no idempotent element, and we also raise a question about subfunctors of forgetful functors.

§33. Functors to Group.

Let k be a field. We have seen that the only representable functors from $\mathbf{Ring}_k^1$ to $\mathbf{Ab}$ are the "obvious" ones, which we can describe as the constructions of completed tensor product with fixed linearly compact vector spaces; and that this result leads, in turn, to characterizations of representable functors to categories such as $\mathbf{Ring}^1$, $\mathbf{Lie}$ and $\mathbf{Jordan}^1$, in each case as a construction of completed tensor product with a linearly compact algebra, followed by an appropriate "obvious" construction. We would like to similarly characterize the representable functors from $\mathbf{Ring}_k^1$ to $\mathbf{Group}$, but it is not clear what the "obvious" class of constructions is. Hence let us begin by gathering examples.

The simplest representable functor from $\mathbf{Ring}_k^1$ to $\mathbf{Group}$ that is not $\mathbf{Ab}$-valued is the *group of invertible elements* construction. More generally, for every $n > 0$, GL_n: $\mathbf{Ring}_k^1 \to \mathbf{Group}$ is representable (an old acquaintance from §1!) Some related constructions are the group of invertible upper triangular $n \times n$ matrices, and the subgroup of such matrices with all diagonal entries 1 ("strictly upper triangular" matrices). There are also constructions intermediate between these last two, e.g., the group of 2×2 matrices of the form $\begin{pmatrix}1 & a\\ 0 & b\end{pmatrix}$ with b invertible.

Another construction is the functor associating to a k-algebra S the multiplicative group of formal power series $f\in S[[t]]$ with invertible constant term. By adding the restriction that the constant term be 1, and/or that the linear term be 0, we get representable subfunctors of this functor.

We see that a number of these examples arise by applying the group-of-invertible-elements construction $\mathbf{Ring}^1 \to \mathbf{Group}$ to various representable functors $\mathbf{Ring}^1_k \to \mathbf{Ring}^1$, such as the forgetful functor, the $n\times n$ matrix ring functor, the upper triangular matrix ring functor, and the formal power series functor. Can this description be generalized so as to cover our other examples – including the underlying additive group functor?

It will be easier to see the answer after we have looked at another question. Since the category **Group** has a unique zeroary operation, we know from the results of §11 that $\mathbf{Rep}(\mathbf{Ring}^1, \mathbf{Group}) \simeq \mathbf{Rep}(\mathbf{Ring}, \mathbf{Group})$. What is the functor on nonunital rings corresponding under this equivalence to the group-of-invertible-elements functor on unital rings?

The group-of-invertible-elements functor may be factored

(33.1) $$\mathbf{Ring}^1 \xrightarrow{F_{\mathrm{mult}}} \mathbf{Semigp}^e \xrightarrow{U} \mathbf{Group},$$

where F_{mult} is the underlying multiplicative semigroup functor, and U the construction giving the group of invertible elements (units) of a semigroup. Since $\mathbf{Semigp}^e$ already has a unique zeroary operation, we can focus our question on the first step in this diagram, and ask for the functor $\mathbf{Ring} \to \mathbf{Semigp}^e$ corresponding to $F_{\mathrm{mult}}\colon \mathbf{Ring}^1 \to \mathbf{Semigp}^e$.

The functor F_{mult} is represented by the free unital ring on one generator, $\mathbf{Z}[x]$, with comultiplication and counit

(33.2) $$\begin{aligned}\mathbf{m}(x) &= x^\lambda x^\rho,\\ \varepsilon(x) &= 1.\end{aligned}$$

The corresponding co-$\mathbf{Semigp}^e$ object of **Ring** will be the kernel of the augmentation ε, which is the nonunital ring $[\mathbf{Z}][w]$ freely generated by $w = x-1$. If we put $x = 1+w$ in (33.2), we get

(33.3) $$\begin{aligned}\mathbf{m}(w) &= w^\lambda + w^\rho + w^\lambda w^\rho,\\ \varepsilon(w) &= 0.\end{aligned}$$

So the functor from nonunital rings to semigroups with neutral element which we are seeking can be described as associating to a ring S its underlying set, with the operation

(33.4) $$s\circ t = s + t + st,$$

and neutral element 0. This operation is known as *quasimultiplication*, and elements of S invertible with respect to it are called *quasiinvertible* elements. Some elementary properties of this operation are recalled in

LEMMA 33.5. *Let* $S \in \mathbf{Ring}$. *Then* (33.4) *defines an associative operation on* S, *with neutral element* 0. *If* S *has a unit element* 1, *then the semigroup with neutral element* $(S, \circ, 0)$ *is isomorphic to the multiplicative semigroup* $(S, \cdot, 1)$ *via the correspondence*

$$s \mapsto 1+s. \tag{33.6}$$

For general S *if, as in §10, we let* S^1 *denote the ring obtained by adjoining a unit to* S, *then* $(S, \circ, 0)$ *is isomorphic to the multiplicative subsemigroup of* S^1 *consisting of elements of the form* $1+s$ $(s \in S)$, *again via* (33.6). □

(In the definition of quasimultiplication, the term st of (33.4) is often given the opposite sign [**99**, p.8]. The semigroup so described is isomorphic to the one described here, by the correspondence $s \mapsto -s$.)

We now see that the previously mentioned class of representable functors $\mathbf{Ring}^1_k \to \mathbf{Group}$ which act by first applying to $\mathbf{Ring}^1_k$ some representable functor to *unital* rings, and then taking this to **Group** by the group-of-invertible-elements construction, can be generalized by starting with a *not necessarily* unital ring-valued representable functor $A \hat{\otimes} -$, and following it by the group-of-*quasiinvertible*-elements construction.

This generalization covers all the examples noted above! For instance, the group of strictly upper triangular invertible $n \times n$ matrices over a unital ring S can be described as the group of quasiinvertible elements of the nonunital ring $A \hat{\otimes} S$, where A is the nonunital algebra of strictly upper $n \times n$ triangular matrices over k (in this context meaning upper triangular matrices with *zeroes* along the diagonal). The group of invertible matrices $\begin{pmatrix} 1 & a \\ 0 & b \end{pmatrix}$ over S is likewise obtained by taking for A the algebra $\begin{pmatrix} 0 & k \\ 0 & k \end{pmatrix}$. The multiplicative group of formal power series f which agree with the constant series 1 up to the t^n term is obtained by taking for A the ideal $t^n k[[t]] \subseteq k[[t]]$. Finally, the representable functors to *abelian* groups, characterized in Theorem 23.9(i), are the cases of the above construction in which the linearly compact k-vector space A is given the zero multiplication.

In fact, the second author has proved (though the proof will not be given here)

THEOREM 33.7 ([**82**, in preparation]). *Let* k *be a field. Then any representable functor* $\mathbf{Ring}^1_k \to \mathbf{Group}$ *can be written as a composite*

$$\mathbf{Ring}^1_k \xrightarrow{A \hat{\otimes} -} \mathbf{Ring} \xrightarrow{Q} \mathbf{Semigp}^e \xrightarrow{U} \mathbf{Group}, \tag{33.8}$$

where $A \in \mathbf{LCpRing}_k$, Q *is the quasimultiplicative semigroup functor* (33.4), *and* U *the group-of-invertible-elements functor. This yields an equivalence of categories:*

$$\mathbf{LCpRing}_k \xrightarrow{A \mapsto (33.8)} \mathbf{Rep}(\mathbf{Ring}^1_k, \mathbf{Group}). \tag{33.9}$$

The same result holds with $\mathbf{Ring}_k$ *in place of* $\mathbf{Ring}^1_k$. □

It seems likely that the analog of the above Theorem holds with k-$\mathbf{Ring}^1$ in place of $\mathbf{Ring}^1_k$ for arbitrary $k \in \mathbf{Ring}^1$, where, of course, A must be replaced by a $\otimes_k$-co-ring, and $A \hat{\otimes} -$ by the construction of §29. If technical difficulties can be overcome, this generalized statement will be proved in [**82**].

33.10. Note that the first arrow in (33.8) does not give the most general representable functor from $\mathbf{Ring}_k^1$ to **Ring** (cf. Theorem 25.25), and that if we replace it by a more general such functor, the composite will still be a representable functor $\mathbf{Ring}_k^1 \to \mathbf{Group}$. By the above Theorem, however, it must always be possible to express this composite, after some "change of coordinates", in the form (33.8). We made similar observations when considering representable functors from $\mathbf{Ring}_k^1$ to **Lie** and $\mathbf{Jordan}^1$ (remark 25.43). Let us here record briefly the form that this change of coordinates takes for the construction of the above Theorem.

Let $F\colon \mathbf{Ring}_k^1 \to \mathbf{Ring}$ be induced by an object $(A, {}^*, {}_*)$ as described in Theorem 25.25. For $S \in \mathbf{Ring}_k^1$, let $^{\bullet}$ and $_{\bullet}$ denote the bilinear operations on $F(S)$ determined by the conditions

$$(33.11) \qquad as^{\bullet}bt \;=\; (a^*b)\,st, \qquad as_{\bullet}bt \;=\; (a_*b)\,st.$$

Thus, each of $^{\bullet}$ and $_{\bullet}$ is associative, each annihilates the range of the other on both sides, and the multiplication of $F(S)$ is

$$(33.12) \qquad xy \;=\; x^{\bullet}y + y_{\bullet}x.$$

The group of elements of $F(S)$ quasiinvertible with respect to this operation can be described as having for elements all pairs $(x, \bar{x})$ such that $x, \bar{x} \in F(S)$ are mutually quasiinverse, i.e., satisfy

$$(33.13) \qquad x + \bar{x} + x^{\bullet}\bar{x} + \bar{x}_{\bullet}x \;=\; 0 \;=\; \bar{x} + x + \bar{x}^{\bullet}x + x_{\bullet}\bar{x}.$$

Consider now a new multiplication on $F(S)$, given by

$$(33.14) \qquad x \# y \;=\; x^{\bullet}y - x_{\bullet}y.$$

The associativity and mutual annihilation conditions on $^{\bullet}$ and $_{\bullet}$ that make (33.12) associative also make (33.14) associative. A calculation shows that the group of quasiinvertible elements of our original ring structure (33.12) is isomorphic to the group of quasiinvertible elements of this new structure via the map

$$(33.15) \qquad \varphi(x, \bar{x}) \;=\; (x + \bar{x}_{\bullet}x,\;\; \bar{x} + x_{\bullet}\bar{x}),$$

which has inverse

$$(33.16) \qquad \psi(u, \tilde{u}) \;=\; (u - u_{\bullet}\tilde{u},\;\; \tilde{u} - \tilde{u}_{\bullet}u).$$

Now (33.14) involves no reversed multiplication. Indeed, it is the operation on $F(S)$ gotten by applying the construction of Theorem 25.25 to the system

$$(33.17) \qquad (A,\;\; ({}^*) - ({}_*),\;\; 0).$$

Thus we have the desired isomorphism of our group-valued functor with one of the form (33.8).

(Hints for the reader wishing to check the assertions that φ and ψ are mutually inverse, and yield homomorphisms from one group to another: First, by (33.13), the right hand side of (33.15) can also be written $(-\bar{x} - x^{\bullet}\bar{x},\; -x - \bar{x}^{\bullet}x)$, and one similarly gets an alternative expression for (33.16). If one calls the original expressions for φ and ψ the "$_\bullet$-forms" and the above variants the "$^\bullet$-forms" of φ and ψ, one sees that when substituting a component of φ or ψ into an expression in terms of $^\bullet$ and $_\bullet$, it is most convenient to use the $_\bullet$-form when substituting into a $^\bullet$-product, and the $^\bullet$-form when substituting into a $_\bullet$-product, so as to take advantage of the mutual annihilation of $_\bullet$ and $^\bullet$. Secondly, by applying $x_\bullet -$ to the left hand equality of (33.13) and $-_\bullet x$ to the right hand side, and subtracting, one sees that $x_\bullet \bar{x} = \bar{x}_\bullet x$. Similarly, $x^\bullet \bar{x} = \bar{x}^\bullet x$, $u_\bullet \tilde{u} = \tilde{u}_\bullet u$, and $u^\bullet \tilde{u} = \tilde{u}^\bullet u$.)

33.18. For comparison, let us consider the corresponding question for functors to *Lie* rings. Given any representable functor $\mathbf{Ring}_k^1 \to \mathbf{Ring}$, say induced by an object $(A, {}^*, {}_*)$, let us compose it with the commutator brackets functor $\mathbf{Ring} \to \mathbf{Lie}$, and ask for the expression for the resulting functor $\mathbf{Ring}_k^1 \to \mathbf{Lie}$ in the form described in Theorem 25.31, i.e., the form that starts with a functor $\mathbf{Ring}_k^1 \to \mathbf{Ring}$ involving no reversed multiplication. In this case, no change of coordinates will be called for, because the underlying additive group functor $\mathbf{Ring}_k^1 \to \mathbf{Ab}$ has an essentially unique description $S \mapsto A \mathbin{\hat{\otimes}} S$; we merely have to find an associative operation to put on these tensor products which leads to the same Lie structure, but involves no reversed multiplication. This turns out again to be the operation (33.14). This suggests that the connection between the right and left ends of the chain of equivalences

$$(33.19) \qquad \mathbf{Rep}(\mathbf{Ring}_k^1, \mathbf{Group}) \xleftrightarrow{(33.10)} \mathbf{LCpRing}_k \xleftrightarrow{(25.32)} \mathbf{Rep}(\mathbf{Ring}_k^1, \mathbf{Lie})$$

is a particularly natural one. It can, indeed, be considered analogous to the relation between a Lie group and its associated Lie algebra.

§34. Functors on connected graded rings.

Although the proof of Theorem 33.7 (which we did not give) is of the same order of difficulty as that of Theorem 13.15, which occupied §§12-13 of this work, and although, as we shall see in subsequent sections, all that we have at present for representable functors from rings to general semigroups are some bizarre examples and conjectures, there is in the literature a group of results answering some similar questions, obtained comparatively easily by Berstein [**31**]. These concern representable functors on categories of *connected graded* k-algebras. In this section we shall develop them (in the more general context of k-rings) and note some consequences.

We should first say a word about what we understand by a graded ring. Under the most commonly used definition, this means a ring R given with a decomposition as a direct sum of additive subgroups, $R = \bigoplus_i R_i$ (where i most often ranges over the nonnegative integers, though other grading groups and semigroups are used), in such a way that, writing the grading semigroup additively, one has

$$(34.1) \qquad R_i R_j \subseteq R_{i+j} \text{ for all } i,\ j,$$

and if we are dealing with unital rings, also

(34.2) $$1 \in R_0.$$

We followed this standard definition when a grading made a brief appearance in §13. However when working with graded rings per se, we prefer to formally define a graded ring R to be a family of additive groups (R_i) (corresponding to the summands in the conventional definition), together with a family of bilinear multiplication maps

$$R_i \times R_j \to R_{i+j},$$

which satisfy an associative law on each 3-tuple of groups, R_h, R_i, R_j, and where, if R is to be unital, an element $1 \in R_0$ is also given satisfying the neutral element laws. (Cf. [**30**, §3].) Thus, the "elements" of our graded rings correspond to the "homogeneous elements" of the standard definition. The "nonhomogeneous elements" of the standard definition become the elements of an associated *nongraded* ring, which we may denote

(34.3) $$\mathrm{Sum}(R) = \bigoplus_i R_i.$$

Thus, the concept of "putting a grading" on an ordinary ring S becomes, in our language, that of giving an isomorphism with a ring of the form $\mathrm{Sum}(R)$ (R a graded ring in our sense).

An admitted pedagogic disadvantage of our approach is that graded rings become "many-sorted algebras" [**32**] [**87**], i.e., objects consisting of a family of sets $(R_i)_{i \in I}$ indexed by a specified index-set I, and operations $\alpha: R_{i_1(\alpha)} \times \ldots \times R_{i_{n(\alpha)}(\alpha)} \to R_{j(\alpha)}$ among them; and though many-sorted algebras are in most respects as well-behaved as one-sorted algebras, they tend, understandably, to be omitted for the sake of simplicity from elementary presentations of universal algebra. However, once many-sorted algebras are admitted, we see that our approach makes the category of rings graded by a given semigroup a *variety* of many-sorted algebras, while graded rings as conventionally defined cannot be made a variety unless the grading semigroup is finite (cf. remark 58.2 below). We shall follow the many-sorted approach to graded rings in this section. A morphism $f: R \to S$ in any variety of many-sorted algebras means, of course, a system of maps $(f_i: R_i \to S_i)_{i \in I}$ respecting the operations. An "element" of a many-sorted algebra will be understood as shorthand for an element of one of its components.

If k is an ungraded unital associative ring, a *graded k-ring* will mean a graded ring in which each component is a k-bimodule, and the multiplication maps are k-bilinear.

A unital graded k-ring R graded by the semigroup $\mathbf{N}$ of nonnegative integers is called *connected* if its degree-0 component R_0 is (up to isomorphism of k-rings) precisely k. This terminology comes from algebraic topology, where the cohomology ring of a space, $H^*(X, k)$, has this property if and only if X is a connected topological space. The condition of connectedness does not define a subvariety of the variety of unital graded k-rings, but it is easy to see that the category of connected graded k-rings is equivalent to the variety of *nonunital* k-rings graded by the *positive* integers, via the correspondence $(R_i)_{i \geq 0} \mapsto (R_i)_{i > 0}$. We shall therefore work mainly with the latter variety.

DEFINITION 34.4. *For the remainder of this section,* k *will be a fixed unital associative ungraded ring. We shall denote by* k-$\mathbf{GrRing}^{(>0)}$ *the variety of nonunital* k*-rings with grading by the positive integers, and call these objects* connected *graded nonunital* k*-rings.*

Symbols formed by attaching $\mathbf{Gr}...^{(>0)}$ *to the names of other categories, e.g.,* k-$\mathbf{GrBimod}^{(>0)}$, $\mathbf{GrMod}_k^{(>0)}$, *will similarly denote the varieties of bimodules, modules, etc. graded by the positive integers, regarded as many-sorted algebras.*

When there is no danger of confusion, the word "graded" may occasionally be omitted in referring to a graded ring, bimodule, etc..

Connected graded k-rings are, as noted, natural objects to the topologist. But why, from the algebraic point of view, will it be easier to prove results about representable functors on categories of such objects than about functors on categories of ungraded rings, or rings with arbitrary gradings by the nonnegative integers? Rings with arbitrary gradings by the nonnegative integers are at least as difficult to study as ungraded rings, because they can have arbitrary ungraded rings as their degree-0 components. On the other hand, connected graded rings carry no information in that component, and we shall see that the structure in the higher components can be handled conveniently by induction, because whenever a product lies in R_i, the factors come from lower-degree components.

In preparation for the proof of Berstein's result, let us recall the graded analogs of some standard constructions on bimodules and rings. These are valid for any grading semigroup, abelian or nonabelian, though we shall for convenience use additive notation.

Products and coproducts (direct sums) of graded k-bimodules are constructed componentwise. The tensor product of two graded k-bimodules M and N is the graded k-bimodule $M \otimes_k N$ defined by:

$$(M \otimes_k N)_i \;=\; \textstyle\bigoplus_{i'+i''=i}\; M_{i'} \otimes_k N_{i''}. \tag{34.5}$$

The nonunital tensor ring on a graded k-bimodule M (called the nonunital tensor *algebra* when k is commutative and M has the same k-module structure on each side) is given by a construction formally identical to (12.14):

$$[k]\langle M\rangle \;=\; M \oplus (M \otimes_k M) \oplus \ldots, \tag{34.6}$$

but with $\otimes$ and $\oplus$ interpreted as above; and this construction is left adjoint to the forgetful functor from graded k-rings to graded k-bimodules.

Coproducts of nonunital graded k-rings similarly have the form given in (12.2) for ungraded k-rings, namely

$$\textstyle\coprod_{\alpha\in I} R^{\alpha} \;=\; \bigoplus_{(\alpha_1,\ldots,\alpha_h)} R^{\alpha_1} \otimes_k \ldots \otimes_k R^{\alpha_h} \quad (h \geq 1,\ \alpha_i \in I,\ \alpha_i \neq \alpha_{i+1}). \tag{34.7}$$

The category of nonunital k-rings graded by a given semigroup is, like k-$\mathbf{Ring}$, pointed, the initial-final object being the zero ring; hence every object R of this category has a unique zeroary co-operation. The condition for a binary co-operation $\mathbf{m}\colon R \to R^{\lambda} \amalg R^{\lambda}$ to have this zeroary co-operation as a coneutral element is as in (12.9):

(34.8) $\mathbf{m}(x) = x^\rho + x^\lambda + \mathbf{b}(x)$ $(x \in R)$, where $\mathbf{b}(x)$ is a sum of products each involving factors from both R^λ and R^ρ.

We now restrict attention to the case where the grading semigroup is the positive integers. Interestingly, a major step in Berstein's work uses the coneutral law (34.8) alone; it effectively determines the structure of the general co-$\mathbf{Binar}^e$ object R of k-$\mathbf{GrRing}^{(>0)}$. Recall that if R is an object of a variety, representing a functor F, then a generating set for R can be thought of as a system of "coordinates" for elements of the objects $F(S)$. One generally would like to find coordinate functions which are reasonably "independent". We claim that this can be done in the present situation by selecting the system of all coordinate functions which have the property that the coordinate of the product of two elements $\xi, \eta \in F(S)$ under the binary operation of $F(S)$, when written as a ring-theoretic expression in the coordinates of ξ and η in the same system, involves no monomial of degree > 1 of the particular form

(a coordinate of η) (a coordinate of ξ) (a coordinate of η) (a coordinate of ξ) ... .

But we have stated this characterization of the functions which are to comprise our system of coordinates in terms of that same system of coordinates! It is here that induction with respect to the grading will first come in. Let us now give the formal statement.

Given $(R, \mathbf{m})$ satisfying (34.8), we define a graded k-subbimodule $M = (M_i)$ of R by induction on i: Assuming $M_1 \subseteq R_1, \ldots, M_{i-1} \subseteq R_{i-1}$ have been defined, let M_i consist of all $x \in R_i$ such that, writing $\mathbf{m}(x) = x^\rho + x^\lambda + \mathbf{b}(x)$ as in (34.8), $\mathbf{b}(x)$ lies in the sum of all products $M_{i_1}^{\alpha_1} \ldots M_{i_h}^{\alpha_h}$, with $i_1 + \ldots + i_h = i$, and $(\alpha_1, \ldots, \alpha_h)$ ranging over sequences of λ's and ρ's *not* of the form $(\rho, \lambda, \rho, \lambda, \ldots)$ (i.e., not an alternating sequence beginning with ρ. Note that this excludes more than one sequence, since the length of the sequence is not necessarily i. On the other hand, it does not exclude all terms lying in products $R^\rho R^\lambda R^\rho \ldots$, since, for example, a factor R^ρ in such a product will contribute terms in $M^\rho M^\rho$, $M^\rho M^\rho M^\rho$ etc. as well as in M^ρ.) Sequences involving two or more successive λ's or ρ's *are* allowed here, since M is not in general closed under multiplication. However, sequences of only λ's or only ρ's will not occur, in view of (34.8). Thus, M is the largest k-subbimodule of R such that

(34.9) $\mathbf{b}(M) \subseteq \Sigma_{(\alpha_1, \ldots, \alpha_h)} M^{\alpha_1} \ldots M^{\alpha_h}$ where $(\alpha_1, \ldots, \alpha_h)$ ranges over all strings of λ's and ρ's involving both symbols, and not of the form $(\rho, \lambda, \rho, \lambda, \ldots)$.

We can now prove

THEOREM 34.10 (after [**31**, Theorem 1.1]). *Let* $(R, \mathbf{m})$ *be a co-*$\mathbf{Binar}^e$ *object of* k-$\mathbf{GrRing}^{(>0)}$ *(so that* $\mathbf{m}\colon R \to R^\lambda \amalg R^\rho$ *is a map satisfying* (34.8)). *Let* M *be the graded k-subbimodule of* R *defined above. Then* R *is the tensor algebra* $[k]\langle M\rangle$. *(Precisely, the natural map* $[k]\langle M\rangle \to R$ *is an isomorphism.)*

Conversely, given any object M *of* k-$\mathbf{GrBimod}^{(>0)}$, *we get a co-*$\mathbf{Binar}^e$ *structure on* $[k]\langle M\rangle$ *by taking any bimodule-map* $\mathbf{b}\colon M \to [k]\langle M^\lambda \oplus M^\rho\rangle$

satisfying (34.9), *defining* $\mathbf{m}$ *on* M *by* (34.8), *and extending* $\mathbf{m}$ *to a k-ring homomorphism* $[k]\langle M\rangle \to [k]\langle M^\lambda \oplus M^\rho\rangle$.

PROOF. The second paragraph is immediate. Indeed, we see that the indicated construction gives a co-$\mathbf{Binar}^e$ structure even without the final restriction in (34.9) on strings $(\rho,\lambda,\rho,\lambda,\dots)$; but when we have proved the main assertion of the Theorem, we will be able to say that if we have such a coalgebra which does not satisfy that restriction, then a change of coordinates will convert it to one that does.

To prove the main assertion, assume $(R, \mathbf{m})$ given, and $M \subseteq R$ defined as above. We shall first show that M generates R. For this purpose we need the observation

(34.11) Given $x_1,\dots,x_h \in M$ $(h>1)$, the element $\mathbf{b}(x_1 \dots x_h) \in R^\lambda \amalg R^\rho$ can be written as a sum of terms $y_1^{\alpha_1}\dots y_j^{\alpha_j}$ satisfying $j \geq h$ and $y_1,\dots,y_j \in M$, so that the only term with $j=h$ and $(\alpha_1,\dots,\alpha_j) = (\rho,\lambda,\rho,\lambda,\dots)$ is $x_1^\rho x_2^\lambda x_3^\rho \dots$.

We get this by applying to $\mathbf{m}(x_1 \dots x_h)$ the fact that $\mathbf{m}$ is a ring homomorphism, and to its factors $\mathbf{m}(x_i)$ the formulas (34.8) and (34.9). (The last condition of (34.9) is not used here.)

Now assume inductively that the subring of R generated by M contains $R_1,\ \dots,\ R_{i-1}$, and let $x \in R_i$. Then $\mathbf{b}(x) \in (R^\lambda \amalg R^\rho)_i$ is a sum of products of elements of degrees $<i$ in R^λ and R^ρ, to which our inductive assumption applies. Hence applying that assumption to these lower-degree terms, we can write $\mathbf{b}(x)$ as an element of a sum of the form shown in (34.9), except that some terms indexed by strings of the form $(\rho,\lambda,\rho,\lambda,\dots)$ may occur.

If one or more such terms do occur, let h be the least length of any string of superscripts $(\rho,\lambda,\rho,\lambda,\dots)$ occurring in such a term. Then we see from (34.11) that by subtracting from x an appropriate sum of products of elements of M, we may cancel all terms with this string of superscripts in the modified $\mathbf{b}(x)$. Thus we have increased h by at least one. Repeating this process, we can drive h upward until there are no such terms with strings of superscripts of length $\leq i$, the degree of x. But $(R^\lambda \amalg \mathrm{R}^\rho)_i$ can contain no products of $>i$ elements, so the modified $\mathbf{b}(x)$ has *no* summands indexed by strings of the form $(\rho,\lambda,\rho,\lambda,\dots)$; that is, our modified x belongs to M. Since our successive modifications were by sums of products of members of M, we conclude that the original value of x belonged to the subring generated by M. Thus, R_i is contained in the subring so generated, completing our inductive proof that that subring is R.

Let us pause a moment and compare this argument with the proof of (13.11). Both were based on the idea that the components of an element $\mathbf{b}(x)$ consisting of factors with superscripts alternating between ρ and λ could be eliminated by subtracting from x the element of R obtained from this term by "dropping superscripts". However, in the present proof, we worked *upward* from the *shortest* such terms – we could not control longer terms until the shorter ones were eliminated, because in (34.8) only the lowest degree terms are known exactly. This upward induction terminated because of our grading. In the proof in §13, we used downward induction, which was possible because the definition of $M = R_1$

excluded higher degree terms. There, we could not work from below, for we only had a "filtration from above" (defined by $\deg(x) = \mathrm{ht}(\mathbf{a}(x))$), not a grading.

Returning to our proof, we now have a canonical surjective homomorphism

$$q\colon\ [k]\langle M\rangle \ \to\ R,$$

which we wish to prove one-to-one. Suppose inductively that q is one-to-one on all homogeneous components of degree $<i$, and consider an element $x \in ([k]\langle M\rangle)_i$. Let $x = x_{(1)} + \ldots + x_{(i)}$, where each $x_{(h)}$ lies in the h-fold tensor product in the decomposition (34.6):

(34.12) $$x_{(h)} \in (M \otimes_k \ldots \otimes_k M)_i \qquad (h \text{ factors}).$$

(In [**31**], this decomposition of $[k]\langle M\rangle$ by number of tensor factors "M" is treated as a second grading, making $[k]\langle M\rangle$ a bigraded ring, but we shall not adopt that formalism.)

If x lies in the kernel of q, we have

(34.13) $$q(x_{(1)}) + q(x_{(2)} + \ldots + x_{(i)}) \ =\ 0,$$

hence

(34.14) $$\mathbf{b}(q(x_{(1)})) + \mathbf{b}(q(x_{(2)} + \ldots + x_{(i)})) \ =\ 0.$$

Note that (34.14) is an equation in $(R^{\rho} \amalg R^{\lambda})_i$, which by (34.7) is the direct sum of $(R^{\lambda})_i$, $(R^{\rho})_i$, and summands which are *tensor products* of lower-degree components of R^{ρ} and R^{λ}. By our inductive hypothesis, lower-degree components of R can be identified via q with the corresponding components of $[k]\langle M\rangle$, hence the corresponding summands of $(R^{\rho} \amalg R^{\lambda})_i$ can be identified with the bimodules

(34.15) $$(M^{\alpha_1} \otimes_k \ldots \otimes_k M^{\alpha_h})_i \qquad (\alpha_j \in \{\rho,\, \lambda\}).$$

Let us now look at the components of (34.14) in summands (34.15) indexed by strings $(\alpha_1, \ldots, \alpha_h)$ $(h \geq 1)$ of the form $(\rho, \lambda, \rho, \lambda, \ldots)$. Since $x_{(1)} \in M$, the first term of (34.14) contributes no term to any of these components, by (34.9). On the other hand, if any of $x_{(2)}, \ldots, x_{(i)}$ are nonzero, let $x_{(h)}$ be the first such nonzero term. We see from (34.11) that $\mathbf{b}(q(x_{(h)}))$ will be the unique term contributing a nonzero component in the summand indexed by the string $(\rho, \lambda, \rho, \lambda, \ldots)$ of length h, and that this component will be precisely the image of $x_{(h)}$ under the natural isomorphism $M \otimes_k M \otimes_k \ldots \cong M^{\rho} \otimes_k M^{\lambda} \otimes_k \ldots$. This gives a contradiction to (34.14). Hence $x_{(2)}, \ldots, x_{(i)}$ are all zero; thus (34.13) says $q(x_{(1)}) = 0$. But on M, q is an inclusion map, so $x_{(1)} = 0$. Hence $x = 0$, showing that q is one-to-one, and completing the proof of the Theorem. □

We remark that in defining our generating subbimodule $M \subseteq R$, the decision (in which we follow Berstein) to make the strings "excluded" in (34.9) those of the form $(\rho, \lambda, \rho, \lambda, \ldots)$ is fairly arbitrary – any choice of one string of each length $h \geq 2$, other than $(\lambda, \lambda, \ldots, \lambda)$ or $(\rho, \rho, \ldots, \rho)$, would have worked. (E.g., we might have excluded the string of length h consisting of a single ρ followed by $h-1$ λ's.)

For an example of what the conditions of the above Theorem mean, suppose k

is a field, M the graded k-vector-space with a basis of two elements $x_1 \in M_1$ and $x_3 \in M_3$, $R = [k]<M>$, i.e., the free graded k-algebra on x_1 and x_3, and F the (temporarily set-valued) functor represented by R. Thus, for any connected graded k-algebra S, the set $F(S)$ can be described as $S_1 \times S_3$. Suppose now that $\mathbf{m}$ is a co-operation on R satisfying (34.8). Then we can see from degree considerations that $\mathbf{b}(x_1)$ must be 0, while $\mathbf{b}(x_3)$ will be an expression $f(x_1^\lambda, x_1^\rho)$ containing only monomials that involve both x_1^λ and x_1^ρ. Assuming that $\mathbf{b}$ satisfies (34.9), f will also not contain the monomial $x_1^\rho x_1^\lambda x_1^\rho$. Thus, the binary operation on sets $F(S)$ induced by such a co-operation $\mathbf{m}$ on R has the form

$$(34.16) \qquad (\xi_1, \xi_3)(\eta_1, \eta_3) = (\xi_1 + \eta_1,\ \xi_3 + \eta_3 + f(\xi_1, \eta_1)).$$

where f can be any homogeneous polynomial of degree 3 not involving ξ_1^3, η_1^3 or $\eta_1 \xi_1 \eta_1$; the coefficients of the other five monomials being arbitrary. Of the above three exclusions, the first two correspond to the left and right neutral laws with respect to the element (0, 0), while the third normalizes our choice of coordinates.

We saw in §15 that for k a field, the question, ''Is the underlying k-algebra of every co-**AbBinar**^{e} object of **Ring**$_k$ free?'' was equivalent to an open problem on retracts of free algebras. In that context, such a result could not be expected to hold for noncoabelian co-**Binar**^{e} objects: it is not even true for noncoabelian cogroups, since the ''group of quasiinvertible elements'' functor is represented by a nonfree algebra. But we have just seen that in the category of connected graded k-rings, the corresponding result holds for arbitrary co-**Binar**^{e} objects; so we have more than we need to deduce

COROLLARY 34.17. *Any retract in* k-**GrRing**$^{(>0)}$ *of a tensor ring* $[k]<M>$ *is isomorphic to a tensor ring* $[k]<N>$, *where* N *is a retract of* M *in* k-**GrBimod**$^{(>0)}$.

PROOF. By the argument used in §15, a retract R of $[k]<M>$ admits a binary co-operation with coneutral element (in fact, a cocommutative one, but we don't need that fact). Hence by Theorem 34.10, R is isomorphic to $[k]<N>$ for some graded bimodule N. Now note that the functor k-**GrRing**$^{(>0)} \to k$-**GrBimod**$^{(>0)}$ given by $R \mapsto R/RR$, i.e., taking a graded k-ring to its universal image with zero multiplication, carries a tensor k-ring $[k]<M>$ to M, and being a functor, it carries retractions to retractions. Hence, as $[k]<N>$ is a retract of $[k]<M>$, the graded bimodule N must be a retract of M. □

The above result leaves open the question of whether every retraction of $[k]<M>$ to $[k]<N>$ is actually a composite of the morphism induced by a retraction of M to N, and automorphisms of these two graded rings.

Returning to our example with operation (34.16), let us note that if we write

$$(34.18) \qquad (\zeta_1, \zeta_3) = (\xi_1, \xi_3)(\eta_1, \eta_3) = (\xi_1 + \eta_1,\ \xi_3 + \eta_3 + f(\xi_1, \eta_1)),$$

then it is possible to solve for (η_1, η_3) in terms of (ξ_1, ξ_3) and (ζ_1, ζ_3). For we can solve for η_1 in the equation $\zeta_1 = \xi_1 + \eta_1$, and then, knowing η_1, we can solve for η_3 in $\zeta_3 = \xi_3 + \eta_3 + f(\xi_1, \eta_1)$. This approach works for the general functor of the sort we have characterized, and gives the first assertion of the next Corollary, from which the second follows immediately. (The definition of a *loop* was recalled preceding Problem 15.3.)

COROLLARY 34.19. *Let* $F \in \mathbf{Rep}(k\text{-}\mathbf{GrRing}^{(>0)}, \mathbf{Binar}^e)$. *Then*

(i) *For every object* S *of* k-**GrRing**$^{(>0)}$ *and element* $\xi \in F(S)$, *the left and right "translation" maps* $\xi \cdot -,\ - \cdot \xi \colon F(S) \to F(S)$ *are both bijections. Hence*

(ii) F *extends uniquely to a representable functor* k-**GrRing**$^{(>0)} \to$ **Loop**. □

Having described the co-**Binar**^{e} objects of our category, we are ready to consider the condition of coassociativity on such objects.

THEOREM 34.20 (after [**31**, Theorem 1.2]). *Let* $(R, \mathbf{m})$ *be a co-***Binar**^{e} *object of* k-**GrRing**$^{(>0)}$, *and let us, using Theorem 34.10, identify* R *with a tensor k-ring* $[k]\langle M\rangle$ *in such a way that the map* $\mathbf{b} = \mathbf{m} - i^\lambda - i^\rho$ *satisfies* (34.9). *Then* $\mathbf{m}$ *is coassociative if and only if it satisfies the conditions*

(a) $\mathbf{b}(M) \subseteq M^\lambda M^\rho$,

and, using (a) *to regard the pair* $(M, \mathbf{b})$ *as a graded* $\otimes_k$*-co-ring* (*cf. Definition 29.3*)

(b) *the graded* $\otimes_k$*-co-ring* $(M, \mathbf{b})$ *is coassociative.*

PROOF. It is straightforward to verify that if (a) holds, then coassociativity of the comultiplication $\mathbf{m}$ on R in k-**GrRing**$^{(>0)}$ is equivalent to coassociativity of the $\otimes_k$-co-operation $\mathbf{b}$ on M in k-**GrBimod**$^{(>0)}$, i.e., to (b). (This is a coalgebra version of the observation that a bilinear operation on an abelian group is associative if and only if the corresponding "quasimultiplication" is so.) Hence to prove the Theorem, it suffices to show that if $\mathbf{m}$ is coassociative, (a) holds.

So suppose that $\mathbf{m}$ is coassociative, i.e., that in notation analogous to that of (12.20),

$$\mathbf{m}_\lambda \mathbf{m} = \mathbf{m}_\rho \mathbf{m}. \tag{34.21}$$

The common value of the two sides of the above equality is a map $R \to R^\lambda \amalg R^\mu \amalg R^\rho$; let us denote it $\mathbf{mm}$. Suppose inductively that for some h the image of M under $\mathbf{b}$ has zero component in all products $M^{\alpha_1} \dots M^{\alpha_j}$ $(j < h)$ indexed by strings $(\alpha_1, \dots, \alpha_j)$ of λ's and ρ's other than (λ, ρ). Let us prove that it also has zero component in all such products of h factors.

The result is immediate for $h = 1, 2$ by the definition of $\mathbf{b}$ and the choice of M, so assume $h \geq 3$. Note that *if* the statement we are trying to prove is true, then on applying either side of (34.21) to elements $x \in M$, we may get nonzero terms in $M^\lambda M^\mu M^\rho$, but the components of $\mathbf{mm}(x)$ in all other products (of arbitrary lengths) involving all three factors M^λ, M^μ, M^ρ will be zero. Hence let J_h denote the set of all strings $\beta = (\beta_1, \dots, \beta_h)$ with entries in $\{\lambda, \mu, \rho\}$, in which all three indices occur, and such that $\beta \neq (\lambda, \mu, \rho)$ if $h = 3$. Let us similarly write I_h for the set of strings consisting of λ's and ρ's only, which have length h and involve both these indices. For $\beta \in J_h$, let $\beta^{\mu \mapsto \lambda} \in I_h$ denote the string obtained from β by replacing all occurrences of μ by λ, and $\beta^{\mu\mapsto\rho}\in I_h$ the string obtained similarly by replacing μ's by ρ's. We now claim that the component of $\mathbf{m}_\lambda\mathbf{m}(x)$ in the tensor product of M's indexed by $\beta\in J_h$ arises from the action of the linear part of $\mathbf{m}_\lambda$ (which we note satisfies $\mathbf{m}_\lambda(y^\lambda)=y^\lambda+y^\mu+$ "higher" terms for $y\in M$) on the M^λ-factors of the component of $\mathbf{m}(x)$ indexed by $\beta^{\mu\mapsto\lambda}$. Indeed, by our inductive assumption, the only components of $\mathbf{m}(M)$ of degree $<h$ are those of degree 1 and the (λ,ρ) component in degree 2. Terms of degree h arising from the action of the degree-h component of $\mathbf{m}_\lambda$ on the degree-1 component of $\mathbf{m}(x)$ will not involve the index ρ, while the only terms arising by the action of the (λ,μ) component of $\mathbf{m}_\lambda$ on the λ-factor in the (λ,ρ) component of $\mathbf{m}(x)$ lie in $M^\lambda M^\mu M^\rho$. Since $\beta\in J_h$, neither of these sorts of terms can be in the summand indexed by β, so only the terms arising from the action of the linear part of $\mathbf{m}_\lambda$ on the λ-factors in the degree-h component of $\mathbf{m}(x)$ can lie in that summand; and we see that, more particularly, only the $\beta^{\mu\mapsto\lambda}$-component of $\mathbf{m}(x)$ can contribute to that summand, proving our claim.

In notation analogous to that introduced in Definition 12.21 and used in (12.24), what this shows is that for $x\in M$ and $\beta\in J_h$, $f(\mathbf{mm})^\beta(x)=f\,\mathbf{m}^{\beta^{\mu\mapsto\lambda}}(x)$. (As in that section, f is the "superscript-dropping" map.) The same argument applied to the right-hand side of (34.21) gives $f(\mathbf{mm})^\beta(x)=f\,\mathbf{m}^{\beta^{\mu\mapsto\rho}}(x)$. Hence for every $\beta\in J_h$,

(34.22) $$f\,\mathbf{m}^{\beta^{\mu\mapsto\rho}}\,|\,M \;=\; f\,\mathbf{m}^{\beta^{\mu\mapsto\lambda}}\,|\,M.$$

Now if we order the set $\{\lambda,\,\rho\}$ by making $\lambda<\rho$, and partially order I_h by componentwise comparison, we see that

(34.23) two strings $\alpha,\ \alpha'\in I_h$ can be written $\beta^{\mu\mapsto\lambda}$ and $\beta^{\mu\mapsto\rho}$ respectively for some $\beta\in J_h$ if and only if $\alpha<\alpha'$,

– except in the case $h=3$. In that case, our exclusion of (λ,μ,ρ) from J_3 has the effect that the pair of strings $\alpha=(\lambda,\lambda,\rho)$, $\alpha'=(\lambda,\rho,\rho)$ cannot be so written. However, we may make (34.23) valid for $h=3$ by modifying the partial ordering of I_3 so as to delete from the diagram of this ordering the edge representing the relation $(\lambda,\lambda,\rho)<(\lambda,\rho,\rho)$. Now (34.22) and (34.23) tell us that $f\,\mathbf{m}^\alpha|M$ is unchanged as α ranges over any connected component of the partially ordered set I_h. But I_h is easily seen to be connected – its picture has the form of an h-cube with top and bottom vertices deleted, and, if $h=3$, one of the remaining edges deleted as well, which still does not disconnect it. Hence the maps $f\,\mathbf{m}^\alpha|M$ are equal for all $\alpha\in I_h$.

Now the definition of M tells us that one of these maps, the one with α of the form $(\rho,\lambda,\rho,\lambda,...)$, is zero. Hence they are all zero, establishing condition (a), and completing the proof of the Theorem. □

The above characterization of cosemigroup objects of k-**GrRing**$^{(>0)}$ is translated below into a description of the corresponding representable functors. There $\otimes_k$-**GrCoRing**$^{(>0)}$ denotes the category of graded bimodules $M\in k$-**GrBimod**$^{(>0)}$ given with coassociative morphisms $M\to M\otimes_k M$. The first and third sentences (containing the two displays) are straightforward to verify; the second sentence (the radicality assertion) follows from Corollary 34.19, and the final sentence follows from the equivalence between the categories of unital and nonunital connected graded k-rings. Note that the construction of the first sentence may be thought of as splitting the co-operation $\mathbf{m}$ into a coaddition $i^\lambda + i^\rho$ and a comultiplication $\mathbf{b}$, then putting these back together via the quasimultiplication construction.

THEOREM 34.24. *In the situation of Theorem 34.20, the functor represented by* $(R, \mathbf{m})$ *can be described as the composite*

$$(34.25)\qquad k\text{-}\mathbf{GrRing}^{(>0)} \xrightarrow{k\text{-}\mathbf{GrBimod}^{(>0)}(M,\,-)} \mathbf{Ring} \xrightarrow{Q} \mathbf{Semigp}^e,$$

where the hom-sets $k\text{-}\mathbf{GrBimod}^{(>0)}(M, S)$ *are made into rings using their natural additive group structure and the* convolution *(cf. §29, paragraph following Definition 29.4) of the* $\otimes_k$*-comultiplication* $\mathbf{b} = \mathbf{m} - i^\lambda - i^\rho$ *with the multiplication of* S, *and where* Q *is the quasimultiplicative semigroup construction* (33.4). *The rings arising under the first functor are all radical; equivalently, all the semigroups given by the above composite functor "are groups". The above construction induces equivalences of categories:*

$$(34.26)\qquad \begin{array}{ccc} (\otimes_k\text{-}\mathbf{GrCoRing}^{(>0)})^{\mathrm{op}} & & \mathbf{Rep}\,(k\text{-}\mathbf{GrRing}^{(>0)},\ \mathbf{Group}) \\ \big\downarrow\!\!\xrightarrow[M\mapsto(34.25)]{} & \mathbf{Rep}\,(k\text{-}\mathbf{GrRing}^{(>0)},\ \mathbf{Semigp}^e) & \xleftarrow[\text{forget}]{}\!\!\big\downarrow \end{array}$$

The corresponding statements hold with the variety $k\text{-}\mathbf{GrRing}^{(>0)}$ *replaced by the equivalent category of connected* unital *graded k-rings.* □

Suppose we now let k be a field, and restrict attention to functors on graded k-algebras. a field. Can we get descriptions of these functors in terms of completed tensor products, like the results of Chapter V?

At first sight, things seem to go wrong. If M and N are graded k-vector-spaces, then $\mathbf{GrMod}_k(M, N) \cong \prod_i \mathbf{Mod}_k(M_i, N_i) \cong \prod_i (\hat{M}_i \hat{\otimes} N_i)$, which is not the completed tensor product of N with some fixed graded vector space.

Note, however, that the category of graded k-vector spaces can be enriched by allowing, in addition to morphisms that preserve degrees of elements, also morphisms that add a positive or negative constant to all degrees. These generalized morphisms form a graded abelian group, graded by the full additive group of integers, in which our "old" morphisms constitute the component of degree 0; and this full graded abelian group can indeed be described as the completed graded tensor product $\hat{M} \hat{\otimes} N$. Here we are denoting by $\hat{M}$ the graded vector space of generalized morphisms (in the above sense) $M \to k$, where k is the graded vector space having k in degree 0 and zero in all other degrees, and we are understanding a *completed* tensor product to have for ith component the direct *product* of the completed tensor products of appropriate components of the given objects (in place of the direct *sum* in an ordinary tensor product (34.5)).

If M is graded by the positive integers, then the above dual object $\hat{M}$ is graded by the *negative* integers. Let us use the notation $\mathbf{Gr}\ldots^{(<0)}$ for categories of objects so graded. Then we can deduce

COROLLARY 34.27. *Let k be a field. Then any representable functor* **GrRing**$_k^{(>0)} \to$ **Semigp**^{e} *can be described as the composite*

(34.28) $\quad$ **GrRing**$_k^{(>0)} \xrightarrow{A \hat{\otimes} -}$ **GrRing** $\xrightarrow{R \mapsto R_0}$ **Ring** $\xrightarrow{Q}$ **Semigp**^{e},

where A is a fixed negative-integer-graded associative nonunital linearly compact k-algebra. As noted in Theorem 34.24, the rings $(A \hat{\otimes} S)_0$ obtained via the first two steps will always be radical, equivalently, the semigroups given by the above composite functor "are groups".

The above construction induces equivalences of categories:

(34.29) $\quad$ **LCpGrRing**$_k^{(<0)} \qquad\qquad$ **Rep**(**GrRing**$_k^{(>0)}$, **Group**)
$\qquad\qquad \hookrightarrow$ **Rep**(**GrRing**$_k^{(>0)}$, **Semigp**^{e}) $\hookleftarrow \qquad \square$

It would be of interest to investigate whether results as good as those we have proved above hold for functors on categories of rings graded by other semigroups than the positive integers. Things probably go over easily to gradings by any semigroup admitting a semigroup homomorphism into the positive integers. At the opposite extreme, a grading by the 1-element semigroup is vacuous, so we cannot get anything better in that case than in the case of ungraded rings. Restricting attention to gradings by "the elements strictly greater than the neutral element in an ordered semigroup with neutral element" will not help, because a ring graded by the 1-element semigroup can also be described as graded by the positive elements of the additive semigroup $\{0, +\infty\}$. However, if we take the semigroup of elements greater than the neutral element in an ordered semigroup with *cancellation*, there may be hope. Let us pose, as a test case,

QUESTION 34.30. *Can Theorems 34.10 and 34.20 be generalized to the case of representable functors on the category of rings graded by the additive semigroup of positive rational (or even real) numbers?*

To state our next question, let k-**GrRing**$^{(\geq 0)}$ denote the category of (not necessarily connected) k-rings graded by the nonnegative integers. As noted earlier, we cannot expect general results for coalgebras in this category better than those holding for coalgebras in k-**Ring**, since any coalgebra in k-**Ring** yields a coalgebra in this category living entirely in degree 0. But we can ask for "relative" results:

PROBLEM 34.31. *Suppose R is an object of k-**GrRing**$^{(\geq 0)}$ given with a structure of cogroup (or cosemigroup, or cobinar with coneutral element) in this category. Suppose the cogroup (cosemigroup, etc.) structure on the component R_0 is "known". Can we obtain results characterizing the possible structures of the whole coalgebra R relative to this smaller coalgebra, which, in the case where $R_0 = \{0\}$, reduce to some of the results proved in this section?*

Schlomiuk [171] points out another relationship between the ungraded and the graded situations. Note that if $(R, \mathbf{m})$ is a cogroup in k-**Ring**, we have a (decreasing) *filtration* of R:

$$R \supseteq R^2 \supseteq R^3 \supseteq \dots . \tag{34.32}$$

(If one thinks in terms of the corresponding cogroup objects in the variety of *unital* k-rings, this is the filtration by powers of the kernel of the co-neutral-element.) We can thus form the associated connected graded k-ring $\mathrm{gr}(R)$, where $\mathrm{gr}(R)_i = R^i/R^{i+1}$. Now if k is a field, then the functor gr carries coproducts of filtered k-algebras to coproducts of graded k-algebras. It is easily deduced that a cogroup object of $\mathbf{Ring}_k$, or equivalently a cogroup object of $\mathbf{Ring}_k^1$, yields a cogroup object in $\mathbf{GrRing}_k^{(>0)}$. Since the graded ring associated with the filtration (34.32) is generated by its degree-1 component, the graded vector space M determining it lives wholly in degree 1, from which we can see that the map **b** of Theorem 34.20 is zero; so the resulting cogroup will be coabelian. If we apply the functor Sum (34.3) to this cogroup, we get a coabelian cogroup in $\mathbf{Ring}_k^1$, which can be thought of as the "tangent space" to the original cogroup.

We remark that the observation used above, that gr respects coproducts, fails if k is not a field. For instance, let $k = \mathbf{Z}$, and let $R = [\mathbf{Z}][x]$, under any filtration such that x has degree 1, x^2 has degree > 2, px has degree 2, and p^2x has degree > 3. We find that $\mathrm{gr}(R)_1$ and $\mathrm{gr}(R)_2$ are each cyclic $\mathbf{Z}$-modules of order p, with generators $\bar{x}$ and $\overline{px}$ respectively. One can deduce that in $\mathrm{gr}(R)^\lambda \amalg \mathrm{gr}(R)^\rho$, the product $\overline{px}^\lambda\, \overline{px}^\rho$ is a nonzero element of the homogeneous component of degree 4. But in $R^\lambda \amalg R^\rho$, the product $(px^\lambda)(px^\rho) = (p^2x^\lambda)x^\rho$ must have degree $\geq 4+1 = 5$, so the image of $\overline{px}^\lambda\, \overline{px}^\rho$ in $\mathrm{gr}(R^\lambda \amalg R^\rho)_4$ is 0. One can modify this example to get one in which the filtration has the form (34.32) for some k-algebra R, but we do not know whether this can happen for an R which is the underlying k-algebra of a cogroup in $\mathbf{Ring}_k$.

§35. Functors from k-algebras to semigroups: some examples.

The remaining sections of this chapter will consider $\mathbf{Semigp}^e$-valued functors on categories of ungraded rings. The complications are surprisingly great even under the best of assumptions; hence, except in the last section, we shall restrict attention to functors on k-algebras, where k is a field.

Clearly, one rich class of representable functors $\mathbf{Ring}_k^1 \to \mathbf{Semigp}^e$ is given by the group-valued functors characterized in §33. When we regard these as semigroup-valued, the result cited in that section say that they have the form

$$\mathbf{Ring}_k^1 \xrightarrow{A \hat{\otimes} -} \mathbf{Ring} \xrightarrow{Q} \mathbf{Semigp}^e \xrightarrow{U'} \mathbf{Semigp}^e, \tag{35.1}$$

where U' denotes the *semigroup*-of-invertible-elements functor. By dropping that last step, we can get another large class, the functors of the form

$$\mathbf{Ring}_k^1 \xrightarrow{A \hat{\otimes} -} \mathbf{Ring} \xrightarrow{Q} \mathbf{Semigp}^e.$$

However, for this class, we cannot say that our use of functors not involving reverse multiplication in the first step is as good as the most general functor $\mathbf{Ring}_k^1 \to \mathbf{Ring}$ would be. For instance, given a k-algebra S, the multiplicative semigroup of S and the multiplicative semigroup of S^{op} are in general nonisomorphic. So we may generalize the above class of constructions to

(35.2) $$\mathbf{Ring}_k^1 \xrightarrow{\text{as in Theorem 25.25}} \mathbf{Ring} \xrightarrow{Q} \mathbf{Semigp}^e.$$

Unfortunately, while the distinction between "regular" and "reversed" multiplication is needed in some cases, it is excessive in others. Though we have just observed that the functors associating to a k-algebra S the quasimultiplicative semigroups of elements of S, of elements of S^{op}, and of quasiinvertible elements of S (which up to isomorphism can be described as the *multiplicative* semigroups of elements of S, of elements of S^{op}, and of invertible elements of S) are nonisomorphic, if we let $F(S)$ denote the nonunital ring of strictly upper triangular $n\times n$ matrices over S for some fixed $n>2$, we see that $F(S)$ is nilpotent, hence radical, hence that the three functors taking S to the quasimultiplicative semigroups of elements of $F(S)$, of elements of $F(S^{\mathrm{op}})$, and of quasiinvertible elements of $F(S)$ are all isomorphic. Thus, distinct constructions of the forms (35.1) and (35.2) can give isomorphic functors. We will return to this problem later.

Our next example will be put together from two constructions of the form (35.2). Let us, for familiarity's sake, write these using ordinary multiplication of matrices with some entries "1", rather than quasimultiplication in nonunital rings. The first construction will be the functor associating to S the multiplicative semigroup of matrices

(35.3) $$\begin{pmatrix} \alpha & \xi \\ 0 & 1 \end{pmatrix} \qquad (\alpha,\ \xi \in S).$$

The second will take S to the *opposite* of the semigroup of matrices

(35.4) $$\begin{pmatrix} 1 & \xi \\ 0 & \beta \end{pmatrix} \qquad (\beta,\ \xi \in S).$$

Let us write these matrix-symbols as 3-tuples, suppressing the common "0" but showing the "1"s, so that the multiplicative laws of the two semigroups are:

$$\begin{aligned}(\alpha_1,\ 1,\ \xi_1)(\alpha_2,\ 1,\ \xi_2) &= (\alpha_1\alpha_2,\ 1,\ \xi_1+\alpha_1\xi_2),\\ (1,\ \beta_1,\ \xi_1)(1,\ \beta_2,\ \xi_2) &= (1,\ \beta_2\beta_1,\ \xi_1+\xi_2\beta_1).\end{aligned}$$

It turns out that these operations can be pasted together to give an operation on $S\times S\times S$, which is easily shown to be associative, and to have neutral element $(1,\ 1,\ 0)$:

(35.5) $$(\alpha_1,\ \beta_1,\ \xi_1)(\alpha_2,\ \beta_2,\ \xi_2) = (\alpha_1\alpha_2,\ \beta_2\beta_1,\ \xi_1+\alpha_1\xi_2\beta_1).$$

This differs in an important way from the operations we have seen so far: its expression (at least in terms of the indicated coordinates) involves a term of degree >2. To express this fact in coalgebra language, let us write the free algebra representing this functor as $R = k\langle a,\ b,\ x\rangle$, with comultiplication

$$\mathbf{m}(a) = a^\lambda a^\rho, \qquad \mathbf{m}(b) = b^\rho b^\lambda, \qquad \mathbf{m}(x) = x^\lambda + a^\lambda x^\rho b^\lambda.$$

In the language introduced in §12, $\mathbf{m}(x)$ has *height* 3; equivalently, x has *degree* 3. We claim in fact that R is not generated by elements of degrees < 3. For if we compose $\mathbf{m}$ with the endomorphism of $R^\lambda \amalg R^\rho$ which fixes a^λ, x^ρ and b^λ, while sending a^ρ, x^λ and b^ρ to 0, we see that the image of $\mathbf{m}(R)$

is the subalgebra of $R^\lambda \amalg R^\rho$ generated by $a^\lambda x^\rho b^\lambda$. This subalgebra certainly cannot be generated by elements of height <3, and the endomorphism we used is nonincreasing on height, since it acts on $R = k<a^\lambda, b^\lambda, x^\lambda, a^\rho, b^\rho, x^\rho>$ by fixing some monomials and annihilating the rest. Our claim is easily deduced from these observations.

Since this example is based on the fact that one can multiply an element of a ring by other elements on two different sides, one might expect that it is as far as one can go – that any cosemigroup object should be generated by elements of degree ≤ 3. When one of the present authors so conjectured, however, the other found a counterexample. Namely, on 6-tuples of elements of a unital k-algebra S, the operation

$$(35.6)\quad \begin{aligned}&(\alpha_1, \beta_1, \gamma_1, \xi_1, \eta_1, \zeta_1)(\alpha_2, \beta_2, \gamma_2, \xi_2, \eta_2, \zeta_2)\\ &= (\alpha_1\alpha_2,\ \beta_2\beta_1,\ \gamma_1\gamma_2,\ \xi_1 + \alpha_1\xi_2\beta_1,\ \beta_2\eta_1\gamma_2 + \eta_2,\ \zeta_1\gamma_2 + \alpha_1\zeta_2 - \alpha_1\xi_2\eta_1\gamma_2)\end{aligned}$$

is associative with neutral element $(1, 1, 1, 0, 0, 0)$, and the corresponding cosemigroup, $R = k<a, b, c, x, y, z>$, cannot be generated by elements of degree < 4, as one may see with the help of the endomorphism of $R^\lambda \amalg R^\rho$ which sends to 0 all the named generators other than a^λ, x^ρ, y^λ, c^ρ.

What is going on here?

To help understand these constructions, let us compose them with the group-of-invertible-elements functor $\mathbf{Semigp}^e \to \mathbf{Group}$, since we know that the resulting functors will have to be of the sort characterized in Theorem 33.7.

The invertible elements of the two semigroups described above are those tuples (α, β, ξ), respectively $(\alpha, \beta, \gamma, \xi, \eta, \zeta)$, such that α, β and γ are invertible. On such elements, one finds that one can make a change of coordinates so that the multiplication takes the form of multiplication of invertible upper triangular 2×2, respectively 3×3 matrices over S. Specifically, an invertible element (α, β, ξ) of the first semigroup corresponds to the matrix

$$(35.7)\qquad \begin{pmatrix} \alpha & \xi\beta^{-1} \\ 0 & \beta^{-1} \end{pmatrix},$$

(compare this with (35.3) on the one hand, and the *inverse* of (35.4), $\begin{pmatrix} 1 & \xi\beta^{-1} \\ 0 & \beta^{-1} \end{pmatrix}$ on the other – inverse because of the opposite multiplication used on matrices (35.4)), while the element $(\alpha, \beta, \gamma, \xi, \eta, \zeta)$ of the second semigroup corresponds to the matrix

$$(35.8)\qquad \begin{pmatrix} \alpha & \xi\beta^{-1} & \zeta + \xi\beta^{-1}\eta \\ 0 & \beta^{-1} & \beta^{-1}\eta \\ 0 & 0 & \gamma \end{pmatrix}.$$

The situation can be thought of as follows. Suppose we take a pair of upper triangular 2×2 (respectively 3×3) matrices over a ring S whose diagonal entries are invertible, write each in the form (35.7) (respectively (35.8)), multiply them together (the result, of course, again having invertible diagonal entries), and write this *product* in the form (35.7) (respectively (35.8)), with appropriate ring elements in place of α, β, ξ (respectively $\alpha, \dots, \zeta$). Then it turns out that these coordinates of the product are *polynomial* expressions in the corresponding coordinates $\alpha_1, \dots, \xi_2$ (respectively $\alpha_1, \dots, \zeta_2$) of the factors, i.e., can be expressed in terms of them without using inverses. (This is a delicate property of the forms of the expressions (35.7) and (35.8). The reader can check, for instance, that it is lost if the upper right hand entry of (35.7) is replaced by ξ, or that of (35.8) by ζ.) If we write out the formulas by which these coordinates of the product are expressed in terms of those of the factors, these determine an operation which is associative and unital even when the diagonal coordinates of the factors are not assumed invertible. (One can deduce this using the fact that invertibility of those coordinates forces no additional polynomial relations; that is, that the free algebras $k<a, b, x>$ and $k<a, b, c, x, y, z>$ embed in their localizations $k<a, a^{-1}, b, b^{-1}, x>$, respectively $k<a, a^{-1}, b, b^{-1}, c, c^{-1}, x, y, z>$, and hence that the same is true for the coproduct of 3 copies of each of these free algebras, which is involved in the verification of associativity.) The semigroups defined by (35.5) and (35.6) may thus be regarded as "renormalizations" of ordinary upper triangular matrix semigroups.

One can in fact find triangular matrices analogous to (35.7) and (35.8) of arbitrary size, with alternating variables and inverse variables along the diagonal, which yield, in the analogous manner, cosemigroup structures on free algebras whose expression requires *arbitrarily high degree* polynomials.

Can we boil this infinite progression of constructions down to something in "closed form"?

In fact, we can. The first step is to rearrange the rows and columns of our matrices so as to collect at one end of the diagonal the "uninverted" diagonal terms, and at the other, the "inverted" diagonal entries. When this is done, each of the matrices in the series (35.7), (35.8), ... (of which we have only shown the first two) is found to have the form

$$\begin{pmatrix} W + XZ^{-1}Y & XZ^{-1} \\ Z^{-1}Y & Z^{-1} \end{pmatrix}, \tag{35.9}$$

where W, X, Y and Z are certain blocks of indeterminates and zeroes. In particular, Z is a square *upper triangular* matrix. The arbitrarily high degree terms in the examples subsumed by (35.9) turn out to arise in the computation of Z^{-1}.

Let us abbreviate (35.9) as

$$\begin{bmatrix} W & X \\ Y & Z \end{bmatrix}.$$

If we multiply two matrices of this form together, and write the product in the same form, we discover that

$$\begin{bmatrix} W_1 & X_1 \\ Y_1 & Z_1 \end{bmatrix} \begin{bmatrix} W_2 & X_2 \\ Y_2 & Z_2 \end{bmatrix} = \begin{bmatrix} W_1(I+X_2Y_1)^{-1}W_2 & X_1 + W_1(I+X_2Y_1)^{-1}X_2Z_1 \\ Z_2(I+Y_1X_2)^{-1}Y_1W_2 + Y_2 & Z_2(I+Y_1X_2)^{-1}Z_1 \end{bmatrix} \tag{35.10}$$

where the I's denote identity matrices of appropriate sizes. Here the entries of $(I+X_2Y_1)^{-1}$ and $(I+Y_1X_2)^{-1}$ are polynomials in the entries of X_2 and Y_1, because the forms of our matrices of indeterminates and zeroes are such that $X_2 Y_1$ and $Y_1 X_2$ are *strictly* upper triangular matrices! So, for instance, the upper left hand entry of the right hand side of (35.10) can be written $W_1 W_2 - W_1 X_2 Y_1 W_2 + W_1 X_2 Y_1 X_2 Y_1 W_2 - \ldots$, and this sum terminates in a number of steps depending on the size of our matrices, because $X_2 Y_1$ is nilpotent. Thus, as claimed, the coordinates of the right hand side of (35.10) are polynomials in the coordinates of the factors of the left-hand side.

We remark that the off-diagonal terms of the right hand side of (35.10) are more symmetric than they appear, because of the identities

$$(I+X_2 Y_1)^{-1} X_2 = X_2 (I+Y_1 X_2)^{-1}, \qquad (I+Y_1 X_2)^{-1} Y_1 = Y_1 (I+X_2 Y_1)^{-1}.$$

Though in each of the constructions we have discussed, there has been some finite bound on the degrees of expressions needed to describe the semigroup operation, we can go to a limit, a semigroup of "infinite upper triangular matrices". For this construction there will be no such bound, though of course each coordinate of a product will still be describable by an expression of finite degree in the coordinates of the factors.

We have not described the explicit forms of the matrix blocks to be used for W, X, Y, Z in (35.9), because these block matrix constructions will be subsumed in a general construction to be described in the next section. But the reader can, with a little thought, fill in such details now.

§36. Jacobson radicals, and a general construction.

We now want to describe a general construction which uses the formula (35.10) to define representable functors to semigroups.

Let us first note that an associative unital or nonunital ring A with a formal matrix decomposition $A = \begin{pmatrix} A_{11} & A_{12} \\ A_{21} & A_{22} \end{pmatrix}$, i.e., with a decomposition $A = A_{11} \oplus A_{12} \oplus A_{21} \oplus A_{22}$ such that $A_{hi} A_{i'j} \subseteq \delta_{ii'} A_{hj}$, is essentially the same as an object of $(\mathbf{Z}\times\mathbf{Z})$-$\mathbf{Ring}^1$, respectively $(\mathbf{Z}\times\mathbf{Z})$-$\mathbf{Ring}$. For, letting e_1, e_2 denote the elements $(1, 0)$, $(0, 1)$ of $\mathbf{Z}\times\mathbf{Z}$, we may regard each $(\mathbf{Z}\times\mathbf{Z})$-ring A as decomposed into the summands $A_{ij} = e_i A e_j$, and conversely regard any such decomposition of a ring as a $(\mathbf{Z}\times\mathbf{Z})$-ring structure.

Let us recall some characterizations of the Jacobson radical of a ring A (which we shall often shorten to "the radical of A" in this and the next section) [**99**, p.9]:

(36.1)
$$\begin{aligned} J(A) &= \{c \in A \mid AcA \text{ consists of quasiinvertible elements}\} \\ &= \{c \in A \mid Ac \text{ consists of quasiinvertible elements}\} \\ &= \{c \in A \mid cA \text{ consists of quasiinvertible elements}\} \\ &= \text{the largest two-sided (left, right) ideal of } A \\ &\qquad \text{consisting of quasiinvertible elements.} \end{aligned}$$

Now suppose A is a unital $(\mathbf{Z}\times\mathbf{Z})$-ring such that $A_{12} \subseteq J(A)$. Then for any $X \in A_{12}$ and $Y \in A_{21}$, we see from (36.1) that $1+XY$ and $1+YX$ are both invertible in A, hence that e_1+XY and e_2+YX are invertible in A_{11} and A_{22} respectively, considered as rings with units e_1 and e_2. Thus, we can now use

(35.10) to define a binary operation on A.

The results of §33 suggest, however, that for greatest generality we should not assume A unital, but instead note that without this assumption, the formula (35.10) still defines an operation on elements of the form $1+x$ of the extended ring A^1 such that $x \in A$, which *in the unital case* gives a structure isomorphic to the one just described. Here, when A is a $(\mathbf{Z}\times\mathbf{Z})$-ring, we understand A^1 to mean $(\mathbf{Z}\times\mathbf{Z}) \oplus A$, made a unital $(\mathbf{Z}\times\mathbf{Z})$-ring in the natural way. We shall write "$1+A$" for the set of elements $1+x$ $(x \in A)$; we may also at times write this set $\begin{pmatrix} 1+A_{11} & A_{12} \\ A_{21} & 1+A_{22} \end{pmatrix}$ (understanding a "1" in the (i, i) position to mean e_i). We see that $\begin{bmatrix} 1 & 0 \\ 0 & 1 \end{bmatrix} \in 1+A$ is a neutral element for the operation (35.10). What is not evident is whether that operation is always associative. Before we prove that it is, let us clarify a few points we have brushed over.

First, in our application of (35.10), we are no longer interpreting expressions of the form $\begin{bmatrix} W & X \\ Y & Z \end{bmatrix}$ as abbreviations for (35.9). This would not make sense, for although we have assumed A_{12} to lie in the radical of A, we have not made this assumption about A_{22}, hence elements $Z \in 1+A_{22}$ need not be invertible. Rather, we are now using $\begin{bmatrix} W & X \\ Y & Z \end{bmatrix}$ as a symbol for the element $\begin{pmatrix} W & X \\ Y & Z \end{pmatrix} \in 1+A$ when we consider this set as given with the operation (35.10), rather than the ring multiplication of A^1. This operation makes sense as long as $A_{12} \subseteq J(A)$.

Second, an esthetic point. The condition $A_{12} \subseteq J(A)$, which we have assumed, is in fact equivalent to its mirror-image condition $A_{21} \subseteq J(A)$. For let us write the two-sided ideal generated by A_{12} as Ae_1Ae_2A. (Since A is nonunital, e_1, $e_2 \in \mathbf{Z}\times\mathbf{Z}$ are here operators; cf. Definition 10.4.) We see that the *square* of this ideal is contained in the two-sided ideal Ae_2Ae_1A generated by A_{21}, hence if the latter ideal lies in $J(A)$ so does the former. (For I an ideal, $I^2 \subseteq J(A) \Rightarrow I \subseteq J(A)$.) The converse holds by the same argument. (Thus the condition might be written symmetrically as $A_{12}+A_{21} \subseteq J(A)$. One can also show it equivalent to either of the conditions $A_{12}A_{21} \subseteq J(A_{11})$, $A_{21}A_{12} \subseteq J(A_{22})$; cf. [**133**, Lemma 4.1].)

Finally, let us note that the category of associative rings A with a formal 2×2 matrix decomposition in which the off-diagonal components are in the Jacobson radical can be considered a variety; for instance by taking as operations the operations of $(\mathbf{Z}\times\mathbf{Z})$-**Ring**, together with a binary operation carrying a pair of elements x, y to the quasiinverse of xe_1ye_2 (cf. (36.1)). Let us denote this variety $(\mathbf{Z}\times\mathbf{Z})$-**Ring**$^{\mathbf{d}}$. (The $^{\mathbf{d}}$ stands for "diagonal, modulo the radical" – not great notation, but at least mnemonic.) Now observe that the structure of binary operation and neutral element that we have put on the set of elements $1+A$ for any $A \in (\mathbf{Z}\times\mathbf{Z})$-**Ring**$^{\mathbf{d}}$ using (35.10) is functorial. Since this set is, up to a natural bijection, the underlying set of A, we have described a representable functor from $(\mathbf{Z}\times\mathbf{Z})$-**Ring**$^{\mathbf{d}}$ to sets with a binary operation with neutral element.

We now wish to prove this operation associative. This can be done directly, by a several-page calculation, but let us look for a conceptual proof instead.

We saw in the preceding section that the associativity of the operations (35.5) and (35.6) could be considered a consequence of the fact that when certain entries of the given 3-tuples or 6-tuples were invertible, these tuples could be mapped to matrices in such a way that our operations corresponded to ordinary matrix multiplication, and that, moreover, these invertibility conditions forced no polynomial relations among the entries. The latter statement followed from the fact that a free associative algebra could be embedded in a larger associative algebra in which certain of the free generators became invertible.

Now similarly, for $A \in (\mathbf{Z}\times\mathbf{Z})\text{-}\mathbf{Ring}^{\mathbf{d}}$, the operation (35.10) on the set of those elements $\begin{bmatrix} W & X \\ Y & Z \end{bmatrix} = \begin{bmatrix} 1+w & x \\ y & 1+z \end{bmatrix} \in 1+A$ such that Z is invertible, i.e., z is quasiinvertible, corresponds to ring multiplication of the corresponding elements (35.9), and so is associative. We want to deduce from this that associativity holds for arbitrary elements of $1+A$.

To bring us into more familiar territory, let us note that any object $A \in (\mathbf{Z}\times\mathbf{Z})\text{-}\mathbf{Ring}^{\mathbf{d}}$ can be embedded in another object of this category of the form $\begin{pmatrix} B & J(B) \\ B & B \end{pmatrix}$, where B is an ordinary associative ring (for instance, by taking for B the underlying ring of A); so it suffices to prove associativity of (35.10) for objects of this form. Now there exists an associative ring B with a family of 12 elements, $w_1, x_1, y_1, z_1;\ w_2, x_2, y_2, z_2;\ w_3, x_3, y_3, z_3,$ universal for the property that x_1, x_2, x_3 are in the radical of B – this can be constructed by starting with the free ring $[\mathbf{Z}]\langle w_1, \ldots, z_3\rangle$, and recursively adjoining quasiinverses to elements of the ideal generated by $\{x_1, x_2, x_3\}$. Clearly our desired associativity result will follow if we can prove the associativity of (35.10) on the single 3-tuple $\begin{bmatrix} 1+w_1 & x_1 \\ y_1 & 1+z_1 \end{bmatrix}$, $\begin{bmatrix} 1+w_2 & x_2 \\ y_2 & 1+z_2 \end{bmatrix}$, $\begin{bmatrix} 1+w_3 & x_3 \\ y_3 & 1+z_3 \end{bmatrix}$ of matrices over this ring B. Moreover, from what we have said, this will follow if B can be embedded in a ring in which z_1, z_2, z_3 become quasiinvertible, and x_1, x_2, x_3 remain in the radical.

We can obtain such an embedding of B with the help of a generalization of the description of *free radical rings* given in [**45**]. A full proof of this generalization would constitute too much of a digression, but we will sketch here, for the reader familiar with [**45**] (and some of the methods of [**47**]) how to adapt the arguments of that paper. We are indebted to P. M. Cohn for pointing out this application of the technique of [**45**].

PROPOSITION 36.2 (after [**45**]). *Let $P \supseteq Q$ be a set and a subset, let K be a commutative ring, and let $[K]\langle P \mid Q \subseteq J\rangle$ denote the associative K-algebra with a map of the set P into it universal for the property that the image of the subset Q lies in the radical.*

Then if K is a principal ideal domain, *$[K]\langle P \mid Q \subseteq J\rangle$ can be identified with the subalgebra of the formal power series algebra $[K]\langle\langle P\rangle\rangle$ generated by the image of P under the K-algebra operations, and the operation of taking quasiinverses of elements of the ideal generated by Q.*

SKETCH OF PROOF. Here are the modifications to make in [**45**]:

In Proposition 1, replace the last paragraph by: "Moreover, given an ideal $H \subseteq R$, the set $\bar{R}$ of all u_1 satisfying an equation (2) with $A \in H_n$ is a

K-subalgebra containing R. If every such matrix has a quasiinverse over S, then H is contained in the radical of $\bar{R}$, and $\bar{R}$ is the least subalgebra of S with this property''. The adaptation of the proof is straightforward.

In the statement of Proposition 2, delete the word ''radical'' from the first line, and replace the condition ''$\varepsilon(A)$ is invertible'' by ''A is invertible''.

Throughout §4, which deals with the concept of the ''free radical K-algebra S on a K-algebra R'' (which we would prefer to call the ''universal radical K-algebra S on R''), use the more general concept of the K-algebra S with a homomorphism $R \to S$ universal for the property that a specified ideal $H \subseteq R$ has image in $J(S)$. (Although this construction is no longer the left adjoint of a forgetful functor among varieties, it is easy to verify that it exists; cf. our description of the universal B in the discussion preceding this Proposition.) Modify the statement of Theorem 4 by assuming we are given an ideal H of the K-algebra R, replacing the hypothesis that S is radical by the condition $H \subseteq J(S)$, keeping unchanged the hypothesis that R is totally inert in S, and letting the conclusion be that in this situation the least subalgebra $\bar{R} \subseteq S$ such that $H \subseteq J(\bar{R})$ is the universal K-algebra with a homomorphism from R which carries H into the radical. Again, the proof goes over without difficulty.

Now in the context of the present Proposition, our definition of $[K]{<}P \mid Q \subseteq J{>}$ makes it the universal K-algebra with a map from the free algebra $[K]{<}P{>}$ into it under which the ideal H generated by Q is carried into the radical. If K is a field then, as noted in [**45**], $K{<}P{>}$ is totally inert in $K{\ll}P{\gg}$, so the modification of [**45**, Theorem 4] we have sketched is applicable and gives the desired conclusion.

Unfortunately, the proof of [**45**, Theorem 5], which generalizes this result to the case of K a general integral domain, appears to contain an error: It is asserted at [**45**, p.144, line 3] that the equation at the top of that page means that a is right linearly dependent on the last $n-1$ columns of $A-I$ over R', but referring to [**45**, Proposition 2] we see that this should be ''over F''', and the remainder of the proof then fails.

However, for K a principal ideal domain, this result can be obtained in another way, though we must make one more modification in the earlier sections of [**45**]. Where [**45**, §3] defines the concept of a subset X being *n-inert* in a ring R, in terms of properties of arbitrary families (a_λ) and (b_μ) of elements of ${}^nR$ and R^n respectively, let us restrict attention to the situation where the families (a_λ) and (b_μ) are finite, and call the corresponding property *n-semiinertia*. A subring of a ring which is n-semiinert for all n will be called totally semiinert. It is easy to see that the one use of total inertia made in the proof of [**45**, Theorem 4] only requires total semiinertia.

We now claim that, writing k for the field of fractions of K, we can complete the proof of the present Proposition if we know that $[K]{<}P{>}$ is totally semiinert in $[k]{<}P{>}$. Indeed, this, together with the total (semi)inertia of $[k]{<}P{>}$ in $[k]{\ll}P{\gg}$ noted above, implies that $[K]{<}P{>}$ is totally semiinert in $[k]{\ll}P{\gg}$, hence by the generalization of [**45**, Theorem 4] just sketched, the natural map of $[K]{<}P \mid Q \subseteq J{>}$ into $[k]{\ll}P{\gg}$ is an embedding. Furthermore, P lies in $[K]{\ll}P{\gg}$, which is a K-subalgebra of $[k]{\ll}P{\gg}$ closed under quasiinverses; hence the image of $[K]{<}P \mid Q \subseteq J{>}$ lies in this subalgebra, and so has the form asserted.

The total semiinertia of $[K]<P>$ in $[k]<P>$, essentially a version of Gauss's Lemma, is established in the proof of [**47**, Theorem 5.5.12, p. 260] (the actual work is done in the proof of the auxiliary result [**47**, Lemma 4.6.3, p. 220]); except that the definitions used there concern a still weaker concept of inertia, like the definition of n-semiinertia given above but with the added restriction that the families (a_λ) and (b_μ) each have exactly n elements. However, all that the arguments given in [**47**] use is that the families are finite, so the proof actually gives total semiinertia, which, as noted, completes the proof of this Proposition.

(Some further corrections and additions to [**45**]: P.138, line 4, S^2 should be S^1. P.140, line 5, "if is" should be "if a is". P.131, lines 3 and 4 from bottom: A better reference can now be given: [**47**, Theorem 2.9.15, p.133], [**47**] being the second edition of the work cited in [**45**] as [**6**]. P.142, first diagram: the diagonal and vertical arrows should be labeled f and f' respectively. Same page, last line: $-a$ should be a.) □

We do not know whether the above Proposition remains true if the assumption that K is a principal ideal domain is weakened to "integral domain", or dropped.

In any case, the Proposition tells us that the ring B we are interested in embeds in $[\mathbf{Z}]\ll w_1, \ldots, z_3\gg$, which is a radical ring; hence in this overring, the x_i continue to lie in the radical and the z_i become quasiinvertible, as desired. (A. I. Valitskas has informed us that the embeddability of a ring of the form $[K]<P \mid Q \subseteq J>$ in $[K]<P \mid P \subseteq J>$, which is what we are using, can also be proved by the methods of his paper [**192**].)

We can now conclude

COROLLARY 36.3. *For every* $A \in (\mathbf{Z}\times\mathbf{Z})\text{-}\mathbf{Ring}^{\mathbf{d}}$ *the formula* (35.10) *defines an associative operation on* $1+A \subseteq A^1$, *with* 1 *as neutral element. Since the set* $1+A$ *is in natural bijective correspondence with* A, *this defines a representable functor*

$$(\mathbf{Z}\times\mathbf{Z})\text{-}\mathbf{Ring}^{\mathbf{d}} \to \mathbf{Semigp}^{e}. \quad \square \tag{36.4}$$

The reader can easily write down the formula for the binary operation on A corresponding to the operation (35.10) on $1+A$, using ring operations and quasiinverses; we will not do so here because there seems to be no gain in transparency.

The constructions of (35.5), (35.6), and the further cases referred to in the preceding section were clearly based on composing various representable functors

$$\mathbf{Ring}^1_k \to (\mathbf{Z}\times\mathbf{Z})\text{-}\mathbf{Ring}^{\mathbf{d}} \tag{36.5}$$

with the functor (36.4). To characterize a large class of functors (36.5), we need some further observations.

It is easy to see from (36.1) that any *surjective* ring homomorphism $f\colon A \to A'$ carries the radical $J(A)$ into $J(A')$. Hence given an inverse system of surjective homomorphisms $A_i \to A_j$ $(i, j \in I,\ i \leq j)$, the radicals $J(A_i)$ also form an inverse system. The inverse limit of this system will be an ideal of the inverse limit of the

A_i's, and by the uniqueness of quasiinverses, will consist of quasiinvertible elements. Thus,

(36.6) $\varprojlim_I J(A_i) \subseteq J(\varprojlim_I A_i)$ for an inverse system of surjective homomorphisms of rings A_i.

Now suppose A is a linearly compact k-algebra. Writing it as the inverse limit of its finite-dimensional continuous homomorphic images A_i (see (25.48), equivalently (32.1)), we can apply (36.6). Moreover, not only the natural maps among the A_i, but also the natural map from A to each of these images is surjective. Hence the image of $J(A)$ is contained in each $J(A_i)$, so in this case we have equality in (36.6). Now because each algebra A_i is finite-dimensional, its Jacobson radical is nilpotent, hence for every associative k-algebra S, the ideal $J(A_i) \otimes_k S$ is nilpotent, so $J(A_i) \otimes_k S \subseteq J(A_i \otimes_k S)$. We deduce

$$\begin{aligned} J(A) \,\hat{\otimes}\, S &= (\varprojlim J(A_i)) \,\hat{\otimes}\, S = \varprojlim \,(J(A_i) \otimes_k S) \\ &\subseteq \varprojlim J(A_i \otimes_k S) = J(\varprojlim \,(A_i \otimes_k S)) = J(A \,\hat{\otimes}\, S). \end{aligned} \tag{36.7}$$

(Here the second and last steps use the fact that *completed* tensor product commutes with inverse limits; cf. Lemma 24.23. At the next-to-last step, (36.6) gives us "$\subseteq$" which is all we need, but equality actually holds, for the same reason as in the paragraph following (36.6).) In other words, the ideal of $A \,\hat{\otimes}\, S$ induced by the closed ideal $J(A) \subseteq A$ lies in the radical of $A \,\hat{\otimes}\, S$.

Now consider linearly compact k-algebras A with 2×2 matrix decompositions, such that the component A_{12} (equivalently A_{21}) lies in the radical. Ordinary k-algebras with 2×2 decomposition form the variety $(k\times k)$-$\mathbf{Ring}_k$ (see Definition 14.3), and we will again indicate the property that the off-diagonal components are in the radical with a superscript $^{\mathbf{d}}$; hence, denoting linear compactness in our usual way, the category of such topological algebras may be written $(k\times k)$-$\mathbf{LCpRing}_k^{\mathbf{d}}$. The operation of completed tensor product with a member of this category carries k-algebras to rings with formal 2×2 matrix decompositions, and (36.7) tells us that these will again have off-diagonal summands in the radical. Thus, a class of representable functors $\mathbf{Ring}_k^1 \to \mathbf{Semigp}^e$ containing those discussed at the end of the preceding section consists of the composites

$$\mathbf{Ring}_k^1 \xrightarrow{\;A\,\hat{\otimes}\,-\;} (\mathbf{Z}\times\mathbf{Z})\text{-}\mathbf{Ring}^{\mathbf{d}} \xrightarrow[\text{to elements } 1+x]{\text{(35.10) applied}} \mathbf{Semigp}^e, \tag{36.8}$$

for $A \in (k\times k)$-$\mathbf{LCpRing}_k^{\mathbf{d}}$.

We end this section by recording two miscellaneous observations that may give some reader an insight into the to us still mysterious operation (35.10). First, the expression (35.9) which led us to that operation can be written

$$\begin{pmatrix}1\\0\end{pmatrix} W\,(1\;\;0) + \begin{pmatrix}X\\1\end{pmatrix} Z^{-1}\,(Y\;\;1). \tag{36.9}$$

Second, the upper right hand corner of the right-hand side of (35.10) involves all the entries of the first factor, but only the upper right hand corner entry X_2 of the second factor. (The lower left-hand corner entry has the opposite properties.)

Hence we have an action of the semigroup defined by (35.10) on the set A_{12}, by a sort of "noncommutative fractional linear transformations".

§37. Representable functors to semigroups: toward some conjectures.

The class of constructions (36.8) still does not give all representable functors $\mathbf{Ring}^1_k \to \mathbf{Semigp}^e$! Though it includes the functors taking the unital ring S to the multiplicative semigroup of S on the one hand, and to that of S^{op} on the other, it does not give the functor taking S to its semigroup of invertible elements. We can get this by composing the multiplicative semigroup functor with the semigroup-of-invertible-elements functor; indeed, we can compose the former functor with arbitrary representable functors $\mathbf{Semigp}^e \to \mathbf{Semigp}^e$ (characterized in §20), getting such constructions as "pairs consisting of an element of S and a multiplicative right inverse", etc..

We can mix this "one-sided inverse" idea with our construction (36.8). For instance, a representable subfunctor of the functor with operation (35.5) (or (35.6)) is defined by the condition that β be a right (or left) inverse of α.

The natural generalization of these examples seems to be the following construction: Let A be an object of $(k\times k)$-$\mathbf{LCpRing}^{\mathbf{d}}_k$, let B, C be objects of $\mathbf{LCpRing}^1_k$ having trivial radical, and let us be given continuous surjective k-algebra homomorphisms

$$(37.1)\qquad f_1\colon A_{11} \to B, \qquad f_2\colon A_{22} \to B, \qquad g_1\colon A_{11} \to C, \qquad g_2\colon A_{22} \to C.$$

Note that the statement that B and C have trivial radicals is equivalent to the condition that the above surjective maps annihilate the radicals of their domain rings. (The implication "$\Leftarrow$" uses the facts that (36.6) is an equality for the inverse system of finite dimensional continuous homomorphic images of a linearly compact algebra, and that if a finite-dimensional algebra has trivial radical, so does any homomorphic image.) Now for any $S \in \mathbf{Ring}^1_k$, f_1 induces a map $A_{11} \hat{\otimes} S \to B \hat{\otimes} S$ of completed tensor product rings, which in turn induces a unital ring homomorphism $(A_{11} \hat{\otimes} S)^1 \to B \hat{\otimes} S$. Let us call this $\bar{f}_1$, and define $\bar{f}_2$, $\bar{g}_1$, $\bar{g}_2$ analogously. Then, when we make $1 + A \hat{\otimes} S$ a semigroup via (35.10), the subset

$$(37.2)\qquad \left\{ \begin{bmatrix} W & X \\ Y & Z \end{bmatrix} \in 1 + A \hat{\otimes} S \;\middle|\; \bar{f}_1(W)\bar{f}_2(Z) = 1 \ (\text{in } B \hat{\otimes} S),\ \bar{g}_2(Z)\bar{g}_1(W) = 1 \ (\text{in } C \hat{\otimes} S) \right\}$$

is a subsemigroup. This follows from the fact that modulo $J(A) \hat{\otimes} S$, the operation (35.10) reduces to

$$\begin{bmatrix} W_1 & 0 \\ 0 & Z_1 \end{bmatrix} \begin{bmatrix} W_2 & 0 \\ 0 & Z_2 \end{bmatrix} = \begin{bmatrix} W_1 W_2 & 0 \\ 0 & Z_2 Z_1 \end{bmatrix}.$$

QUESTION 37.3. *Do the subfunctors* (37.2) *of functors* (36.8) *give, up to isomorphism, all representable functors from* $\mathbf{Ring}^1_k$ *to* $\mathbf{Semigp}^e$?

A natural complement to this question is whether instances of this construction based on nonisomorphic data give nonisomorphic functors. The answer, unfortunately, is no. For instance, if we write k° for the 1-dimensional k-vector space made a k-algebra using zero multiplication, while continuing to write k for the same k-vector space with its normal k-algebra structure, or with no k-algebra structure, we find that the underlying additive group functor $\mathbf{Ring}_k^1 \to \mathbf{Semigp}^e$ arises in exactly *four* ways as an instance of (36.8), namely, through each of the following choices of $A \in (k\times k)$-$\mathbf{LCpRing}_k^{\mathbf{d}}$.

$$(37.4) \qquad \begin{pmatrix} k^\circ & 0 \\ 0 & 0 \end{pmatrix}, \quad \begin{pmatrix} 0 & k \\ 0 & 0 \end{pmatrix}, \quad \begin{pmatrix} 0 & 0 \\ k & 0 \end{pmatrix}, \quad \begin{pmatrix} 0 & 0 \\ 0 & k^\circ \end{pmatrix}.$$

(These rings, having zero multiplication, are radical, so the further refinement of (37.1-37.2) is vacuous – the only possibilities for B and C are $\{0\}$).

We can exclude this particular case of ambiguity by imposing on the class of constructions we are considering the restriction that A_{22} have a unit, which acts as the identity on the appropriate sides of A_{12} and A_{21}; this condition excludes all but the first of the above descriptions of the underlying additive group functor. (The k-algebra-with-unit A_{22} may be the trivial algebra, with $1 = 0$, as indeed it is in the first of the above examples; but if it is trivial, all modules over it are zero, eliminating the middle two cases.)

It appears likely that every functor of the form (37.2) is isomorphic to one which satisfies this condition; the idea is as follows: The linearly compact algebra A_{22} of (37.2) will have a maximal idempotent e. This induces a 2×2 matrix decomposition of A_{22}, with the unital ring $eA_{22}e$ in the lower right hand corner. The upper left hand corner, which within $(A_{22})^1$ can be written $(1-e)A_{22}(1-e)$, will be a linearly compact algebra without idempotent elements, hence radical. Thus the quasimultiplicative semigroup of the latter ring is a group, and essentially unaffected by the "opposite" construction, so $(1-e)A_{22}(1-e)$ can (we suspect) be removed from A_{22} and grafted onto A_{11} in such a way as to leave unchanged the semigroups (37.2).

However, the condition that A_{22} be unital is unpleasantly asymmetric; moreover, the above setup still leaves infinitely many ways of describing some functors. As an example of an unnatural way to describe the semigroup-of-invertible elements functor, for instance, we may let $A \in (k\times k)$-$\mathbf{LCpRing}_k^{\mathbf{d}}$ be given by $A_{11} = k\times k\times k$, $A_{22} = k$, $A_{12} = A_{21} = 0$, and then choose B, C, f_i and g_i so that our functor selects from $A \mathbin{\hat{\otimes}} S = S\times S\times S\times S^{\mathrm{op}}$ the semigroup of all (x_1, x_2, x_3, y) such that y is a right inverse to x_1 and to x_2, but a left inverse to x_3. These conditions are easily seen to imply that $x_1 = x_2 = x_3$ is an invertible element, with inverse y.

A better approach may be to go a bit further, and replace our $(k\times k)$-rings by $(k\times k\times k)$-rings such that A_{11} and A_{33} are *both* unital, and are each involved in at most "one-sided" inverse conditions, while A_{22}, which is not required to be unital, is obligatorily subject to the group-of-invertible-elements construction. (The "one-sided only" restriction referred to above could be formalized as the statement that the kernels of f_1 and g_1 should sum to A_{11}, and the analogous condition on the kernels of the maps defined on A_{33}.)

Assuming that such machinations can make our construction bijective between isomorphism classes of defining data and of functors, there remains the question of characterizing morphisms among such functors. Observe that the second of the objects in (37.4) has a natural inclusion into the object $\begin{pmatrix} k & k \\ 0 & k \end{pmatrix} \in$ $(k\times k)$-$\mathbf{LCpRing}_k^{\mathbf{d}}$, which leads to a morphism from the underlying-additive-group functor on $\mathbf{Ring}_k^1$ to the functor determined by the latter object. But this morphism is not induced by any morphism of $(k\times k)$-rings from our representative $\begin{pmatrix} k^\circ & 0 \\ 0 & 0 \end{pmatrix}$ of the family (37.4) to this ring; and this problem persists when we translate our description of these functors into the $(k\times k\times k)$-ring formalism sketched above. So it appears that to get a description of the full category of functors $\mathbf{Ring}_k^1 \to \mathbf{Semigp}^e$ of the sort we have constructed, one must devise a non-obvious concept of morphism on our defining objects. This suggests that our description of these functors is itself somehow not the natural one, and indeed that this class most likely does not give the most general representable functor $\mathbf{Ring}_k^1 \to \mathbf{Semigp}^e$.

§38. Density of invertible elements.

Having said enough about the difficulties of characterizing the general representable functor $\mathbf{Ring}_k^1 \to \mathbf{Semigp}^e$, let us note some restricted cases that might be easier to study. Clearly, one way of getting a handle on functors $G\colon \mathbf{Ring}_k^1 \to \mathbf{Semigp}^e$ is to compose them with the group-of-invertible-elements functor $U\colon \mathbf{Semigp}^e \to \mathbf{Group}$, and make use of the description of representable functors $\mathbf{Ring}_k^1 \to \mathbf{Group}$ given by Theorem 33.7. For this to be effective, we would like the resulting group-valued "subfunctor" of G to in some sense be "dense" in G. The next definition makes this concept precise.

Definition 38.1. *Let* $\mathbf{C}$ *be a category and* $f\colon F \to G$ *a morphism in* $\mathbf{Rep}(\mathbf{C}, \mathbf{Set})$. *Then* f *will be said to "have dense image" if it is an epimorphism in* $\mathbf{Rep}(\mathbf{C}, \mathbf{Set})$, *equivalently, if the corresponding morphism of representing objects,* $\mathbf{f}\colon R_G \to R_F$, *is a monomorphism in* $\mathbf{C}$.

If $\mathbf{C}$ *has finite coproducts and* $\mathbf{V}$ *is a variety of algebras, a morphism* $f\colon F \to G$ *in* $\mathbf{Rep}(\mathbf{C}, \mathbf{V})$ *will be said to have dense image if its composite with the forgetful functor* $\mathbf{V} \to \mathbf{Set}$ *has dense image. In the special case where* $\mathbf{V}$ *is* $\mathbf{Semigp}^e$, *and* F *is the composite of* G *with the "subsemigroup of invertible elements" functor, and* $f\colon F \to G$ *the natural inclusion, we shall say that "invertible elements are dense in* G*" if this inclusion morphism has dense image.*

Note that if $\mathbf{C}$ is also a variety of algebras, then to say that $\mathbf{f}\colon R_G \to R_F$ is a monomorphism is to say that it is one-to-one. Hence in this situation, the condition that f have dense image can be interpreted as saying that if we choose some system of coordinates for elements of objects $G(S)$, i.e., a system of generators for R_G, then the coordinates of elements of the image-algebras $f(F(S)) \subseteq G(S)$ do not satisfy identically for all S any relations other than those satisfied by the coordinates of elements of the whole algebras $G(S)$.

An illustration of the effect of the condition that invertible elements be dense is

Lemma 38.2. *Let* $\mathbf{C}$ *be a category with finite coproducts, let* $A, B\colon \mathbf{C} \to \mathbf{Semigp}^e$ *be representable functors, and let* $f\colon A \to B$, $g\colon A \to \mathrm{op} \circ B$ *be morphisms of functors, where* $\mathrm{op}\colon \mathbf{Semigp}^e \to \mathbf{Semigp}^e$ *is the opposite-*

semigroup functor. Suppose that g *is an elementwise* right *inverse to* f, *i.e., that for all* $S \in \mathbf{C}$, $x \in A(S)$ *one has*

$$f(x)\,g(x) \;=\; e \tag{38.3}$$

in $B(S)$. *Then if invertible elements are dense in* A, g *is an elementwise* two-sided *inverse to* f, *i.e., for all* $x \in A(S)$ *one also has*

$$g(x)\,f(x) \;=\; e. \tag{38.4}$$

Proof. Let U denote the underlying set functor of $\mathbf{Semigp}^e$, and consider the morphism $h\colon UA \to UB$ of representable functors $\mathbf{C} \to \mathbf{Set}$ given by

$$h(x) \;=\; g(x)f(x) \qquad (x \in A(S),\;\; S \in \mathbf{C}).$$

When x is an invertible element of $A(S)$, then $f(x)$ is an invertible element of $B(S)$. Hence from (38.3) we deduce that (38.4) holds for all invertible x, which says that the morphism $h\colon UA \to UB$ agrees with the constant morphism $x \mapsto e$ on such x. This constitutes a family of equations satisfied by the coordinates of all invertible elements of semigroups $A(S)$, so if invertible elements are dense in A, the same equations hold for *all* elements of $A(S)$. □

Thus, in studying semigroup-valued functors on k-algebras, we may hope to avoid complications arising from one-sided inverse conditions such as we encountered in the preceding section by requiring that invertible elements be dense. Before we make a formal conjecture of this sort, let us note another possible way to exclude that phenomenon. One-sided inverses that are not two-sided inverses are often regarded as pathological in ring theory, and there is a name for rings that avoid this behavior in a robust fashion:

Lemma 38.5. *For* $R \in \mathbf{Ring}^1_k$, *the following conditions are equivalent:*

(a) *For every integer* n, *the multiplicative semigroup of* $n \times n$ *matrices over* R *satisfies the condition*

$$xy = I \;\;\Rightarrow\;\; yx = I. \tag{38.6}$$

(b) *For every finite-dimensional associative unital k-algebra* A, *the multiplicative semigroup of* $A \otimes_k R$ *satisfies* (38.6).

(c) *For every linearly compact associative unital k-algebra* A, *the multiplicative semigroup of* $A \hat{\otimes} R$ *satisfies* (38.6).

(b′), (c′) *Like* (b), (c) *but for* nonunital A, *and quasimultiplicative semigroups.*

The k-algebras R *satisfying these equivalent conditions, called* weakly finite k-algebras, *form a subquasivariety* (*cf.* §9.2) *of* $\mathbf{Ring}^1_k$. □

(This concept is generally defined for rings, and not just algebras over a field k, using condition (a) above. We are restricting attention to k-algebras here to make the other conditions meaningful. It is easy to show that for an arbitrary ring R, condition (a) above is equivalent to the condition that no free R-module of finite rank be isomorphic to a proper direct summand of itself; equivalently, that no finitely generated projective R-module be isomorphic to a proper direct summand of itself. The term "weakly finite" is based on this property, in view of the characterization of *finite sets* as sets that cannot be put in bijective correspondence with proper subsets of themselves.)

Note that the object representing the construction which takes a k-algebra S to the semigroup $\{(\alpha, \beta) \in S\times S \mid \alpha\beta = 1\}$, namely the k-algebra $k\langle a, b \mid ab = 1\rangle$, is *not* weakly finite, by condition (a) above with $n = 1$, $x = a$, $y = b$. We now ask

QUESTION 38.7. *Are some or all of the following conditions on a representable functor* $G\colon \mathbf{Ring}_k^1 \to \mathbf{Semigp}^e$ *equivalent?*

(a) *Invertible elements are dense in* G.

(b) *The representing k-algebra* R *for* G *is weakly finite.*

(c) G *can be expressed in the form* (37.2), *where moreover* $B = C$, $f_1 = g_1$, *and* $f_2 = g_2$ *(so that every right inverse condition is matched by the corresponding left inverse condition).*

We remark that using the results of [**16**, §2], one can show that the class of weakly finite k-algebras is closed under forming coproducts in $\mathbf{Ring}_k^1$, hence by Lemma 14.2 above, the class of all representable functors $\mathbf{Ring}_k^1 \to \mathbf{Semigp}^e$ with weakly finite representing k-algebras can be identified with the class of all representable functors from the quasivariety of weakly finite k-algebras to $\mathbf{Semigp}^e$.

We suggested earlier that some of the complications of the construction of the preceding sections might be avoided by requiring that one diagonal component of our defining object $A \in (k\times k)\text{-}\mathbf{LCpRing}_k^{\mathbf{d}}$ be unital. Does the case where they are both unital, i.e., where A is unital, have particularly nice properties? In this case, the semigroup $\begin{bmatrix} 1+A_{11}\hat{\otimes} S & A_{12}\hat{\otimes} S \\ A_{21}\hat{\otimes} S & 1+A_{22}\hat{\otimes} S \end{bmatrix}$ is isomorphic to $\begin{bmatrix} A_{11}\hat{\otimes} S & A_{12}\hat{\otimes} S \\ A_{21}\hat{\otimes} S & A_{22}\hat{\otimes} S \end{bmatrix}$, made a semigroup using the same formal operation (35.10). Clearly this semigroup has a *zero* element, $\begin{bmatrix} 0 & 0 \\ 0 & 0 \end{bmatrix}$ (an element 0 satisfying the identities $0x = 0 = x0$). On the other hand, any nontrivial condition of the sort imposed on the right hand side of (37.2) excludes this zero element. Writing $(k\times k)\text{-}\mathbf{LCpRing}_k^{\mathbf{d},1}$ for the variety of unital objects of $(k\times k)\text{-}\mathbf{LCpRing}_k^{\mathbf{d}}$, and $\mathbf{Semigp}^{e,0}$ for the variety of semigroups with neutral element and zero element, we may ask

QUESTION 38.8. *Can every representable functor* $\mathbf{Ring}_k^1 \to \mathbf{Semigp}^{e,0}$ *be written in the form* (36.8), *where* $A \in (k\times k)\text{-}\mathbf{LCpRing}_k^{\mathbf{d},1}$? *(In this case, as noted above, one can formulate this construction in terms of* (35.10) *applied directly to* $A\hat{\otimes} S$, *rather than to* $1+A\hat{\otimes} S$.)

If this and the implication (c)⇒(a) of Question 38.7 both hold, then one has the curious result that the presence of a functorial *zero* element implies density of *invertible* elements. (Note that (36.8) corresponds to the case where condition (c) of Question 38.7 is trivially satisfied because "B" and "C" are zero.)

We should make it clear that we have no *good* reason to believe that the mysterious formula (35.10) will in some form underlie all representable functors $\mathbf{Ring}_k^1 \to \mathbf{Semigp}^e$. It simply describes a class of "exotic" constructions that the authors accidentally discovered. It could be one of a much larger family of such constructions.

38.9. *Remarks on the concept of a morphism of representable functors having dense image.* Let us note the relation between that property (defined at the beginning of this section) and some similar conditions.

Suppose $\mathbf{C}$ is a category with finite coproducts, $\mathbf{V}$ a variety, and $f\colon F \to G$ a morphism of representable functors $\mathbf{C} \to \mathbf{V}$. Let $U_{\mathbf{V}}\colon \mathbf{V} \to \mathbf{Set}$ denote the underlying-set functor on $\mathbf{V}$. Then we have the following implications, where, we recall, the condition "f has dense image" is equivalent to the statement in the upper right-hand corner, while as noted in the proof of Lemma 31.15, the condition in the upper left-hand corner is equivalent to "f is right invertible":

$$\begin{array}{ccccc}
\begin{matrix}(\forall S\in \mathbf{C})\ U_{\mathbf{V}}f(S)\ \text{is}\\ \text{a surjection of sets}\end{matrix} & \Longleftrightarrow & \begin{matrix}U_{\mathbf{V}}f\ \text{is an}\\ \text{epimorphism in } \mathbf{Set}^{\mathbf{C}}\end{matrix} & \Longrightarrow & \begin{matrix}U_{\mathbf{V}}f\ \text{is an epimorphism}\\ \text{in } \mathbf{Rep}(\mathbf{C},\mathbf{Set})\end{matrix}\\
\Downarrow & & & & \Downarrow\\
\begin{matrix}(\forall S\in \mathbf{C})\ f(S)\ \text{is an}\\ \text{epimorphism in } \mathbf{V}\end{matrix} & \Longleftrightarrow & \begin{matrix}f\ \text{is an}\\ \text{epimorphism in } \mathbf{V}^{\mathbf{C}}\end{matrix} & \Longrightarrow & \begin{matrix}f\ \text{is an epimorphism}\\ \text{in } \mathbf{Rep}(\mathbf{C},\mathbf{V}).\end{matrix}
\end{array} \tag{38.10}$$

Here the rightward and downward implications are all straightforward. The reversibility of the left hand rightward implication on each line is proved by the dual of the argument noted for monomorphisms at the beginning of §30; but since representable functors do not respect pushouts, as they do pullbacks, this reversibility does not carry over to the right hand rightward implications. Indeed, for counterexamples, we can take $\mathbf{V} = \mathbf{Set}$, so that the upper and lower levels fall together, and let f be the functor represented by any monomorphism $\mathbf{f}\colon R_G \to R_F$ of $\mathbf{C}$ that is not left invertible.

For an example showing that the downward implications are not reversible, let us recall that for any $S \in \mathbf{Ring}^1$, the inclusion of 2×2 matrix rings, $\begin{pmatrix} S & S\\ 0 & S\end{pmatrix} \subseteq \begin{pmatrix} S & S\\ S & S\end{pmatrix}$ is an epimorphism in $\mathbf{Ring}^1$. (This can be deduced from the observation that if a moduie-homomorphism $e_{12}\colon Re_{11} \to Re_{22}$ of left ideals of a ring R has a 2-sided inverse e_{21}, this inverse is unique.) Hence letting F be the functor $S \mapsto \begin{pmatrix} S & S\\ 0 & S\end{pmatrix}$ and G the functor $S \mapsto \begin{pmatrix} S & S\\ S & S\end{pmatrix}$, the inclusion $F \subseteq G$ satisfies the left hand condition, and hence all the conditions, on the bottom line. But the coordinates of all elements of the image of this inclusion satisfy the equation $s_{21} = 0$, so this image is not dense in the sense of Definition 38.1, and the right hand condition, hence all the conditions, of the upper line fail.

The upper left, upper right, and lower right corners of the diagram (38.10) can be translated into conditions on the morphism $\mathbf{f}\colon R_G \to R_F$ representing f; namely "$\mathbf{f}$ is left invertible in $\mathbf{C}$", "$\mathbf{f}$ is a monomorphism in $\mathbf{C}$" and "$\mathbf{f}$ is a monomorphism in the category of $\mathbf{V}$-coalgebra objects of $\mathbf{C}$". We do not know any convenient translation of the lower-left-hand condition.

Note that our example of the nonreversibility of the downward implication on the right side of (38.10) shows that a monomorphism **f** in the category of **V**-coalgebras in a variety **C** need not be one-to-one on underlying objects.

The problem we encountered in §36, of showing that an algebra universal for the property that certain elements belonged to the radical could be *embedded* in an algebra in which certain other elements also belonged to the radical, was essentially a question of density of the image of a morphism of representable functors, and we used the result in the same way as in Lemma 38.2: to establish certain relations for general elements of a representable functor by verifying them for well behaved elements. We shall encounter further questions involving density in the next chapter.

§39. An idempotentless example, and a question on subfunctors.

We shall note in this section an interesting phenomenon that can occur when a couple of the restrictions we have been assuming for the last four sections are dropped. We end with a question, unrelated except that it also discards some of these hypotheses.

Let us first drop the assumption that k be a field, and allow it to be an arbitrary commutative ring. Then for any $c \in k$, we have a representable functor F_c: $\mathbf{Ring}^1_k \to \mathbf{Ring}$ that takes a k-algebra S to the ring with the same underlying additive group, and the multiplication $x \cdot y = cxy$. (Up to isomorphism, this multiplication depends on c only up to associates. When k is a field there are thus essentially two cases, $c = 0$ and $c = 1$, which correspond to ring structures discussed in earlier sections.) If we follow this by the quasimultiplicative semigroup functor Q, we get a $\mathbf{Semigp}^e$-valued functor. If S is a k-algebra on which c is not a zero-divisor, we see that the ring $F_c(S)$ is isomorphic to the nonunital subring $cS \subseteq S$, and so $QF_c(S)$ is isomorphic to the multiplicative semigroup $\{1+cx \mid x \in S\}$.

Let us now, in addition, drop the restriction that the semigroups we construct have neutral elements. Then we can, alternatively, compose F_c with the multiplicative semigroup functor, getting a construction which, when c is a non-zero-divisor on S, gives a semigroup isomorphic to $\{cx \mid x \in S\}$.

Can we get representable functors similarly describable in terms of the multiplication of elements $e+cx$ for other $e \in k$? Yes. If e is any element of k which is idempotent modulo c, say satisfying $e^2 = e+cq$, then we see that

$$(e+cx)(e+cy) \;=\; e^2+c(ex+ey+cxy) \;=\; e+c(q+ex+ey+cxy),$$

and it follows that if c is not a zero-divisor on S, the operation

$$x \cdot y \;=\; q + ex + ey + cxy \tag{39.1}$$

gives a semigroup structure on the underlying set of S, isomorphic to the multiplicative semigroup of elements $e+cx$ of S. It is an easy computation that even without the restriction to k-algebras S on which c acts as a non-zero-divisor, this operation is associative; hence it gives a representable semigroup-valued functor. Moreover, the map $x \mapsto e+cx$ is still a *homomorphism* from this functor to the underlying-multiplicative-semigroup functor, whose image is the semigroup of elements congruent to e modulo c.

It follows that if our idempotent residue class $[e] \in k/ck$ does not contain an idempotent element of k, then the semigroup we have constructed can contain no idempotent. This contrasts with what we proved about functors k-**Ring**$^1 \to$ **AbSemigp** in Proposition 18.9. Let us now show that one can get elements c and e giving such an example whenever k is a commutative integral domain that is *not* local:

LEMMA 39.2. *If $k \neq \{0\}$ is any commutative integral domain which is not local, then there exists a representable functor F: $\mathbf{Ring}_k^1 \to$ **Semigp** such that $F(k)$ is nonempty but has no idempotent element, equivalently, such that F admits zeroary operations, but no such operation which is idempotent-valued.*

PROOF. Since k is not local, we can find two distinct maximal ideals P_0, $P_1 \subseteq k$. Choose $e \in k$ having image 0 in k/P_0 and 1 in k/P_1, and let $c = e^2 - e$. Then we see that $c \in P_0 \cap P_1$, and by choice of c, the image of e is idempotent in k/ck.

Now since the images of e in k/P_0 and k/P_1 are not both 0, nor both 1, the image of e in k/ck is neither 0 nor 1. This means that the multiplicative semigroup $\{e + cx \mid x \in k\} \subseteq k$ contains neither 0 nor 1. Since 0 and 1 are the only idempotent elements of the integral domain k, this semigroup contains no idempotents. But as shown above, this nonempty semigroup is isomorphic to $F(k)$ for a certain representable functor F: $\mathbf{Ring}_k^1 \to$ **Semigp**. □

On the other hand, we shall see in §42 that all fields, and *some,* but not all, commutative local rings k, do have the property that whenever F: $\mathbf{Ring}_k^1 \to$ **Semigp** is a representable functor such that $F(k)$ is nonempty, then $F(k)$ contains an idempotent.

Finally, a question which involves weakening the hypotheses of the preceding sections in a different way, namely, considering k-rings rather than k-algebras. Recall that in §30 we showed that for k a division ring, all subfunctors of the forgetful functor F: k-**Ring**$^1 \to$ **Ring**^{1} were centralizer-subring functors C_L, for L a division subring of k.

QUESTION 39.3. *Let k be a division ring, and let us write the multiplicative-semigroup functor k-**Ring**$^1 \to$ **Semigp**^{e} as a composite*

$$k\text{-}\mathbf{Ring}^1 \xrightarrow{F} \mathbf{Ring}^1 \xrightarrow{G} \mathbf{Semigp}^e,$$

where F forgets the action of k, and G forgets the additive structure. Then does every nontrivial representable subfunctor of this composite GF have one of the forms

$$GC_L \quad or \quad UGC_L,$$

*where $C_L \subseteq F$ is a centralizer functor as in §30, and U: **Semigp**$^e \to$ **Semigp**^{e}*

is the semigroup-of-invertible-elements functor? (I.e., can every subfunctor of GF *be obtained by taking a subfunctor of* F, *following it by* G, *and then applying a subfunctor of the identity functor of* **Semigp**^{e}*?)*

CHAPTER VIII.

Representable functors on categories of commutative associative algebras.

The subject of this chapter has been much more studied than those of the preceding (and the remaining) chapters, and there are many experts in the field. The present authors are not among these. Nevertheless, it seemed desirable to include this material, for the sake of perspective. The choice of subjects in some of the sections of this chapter may be a bit arbitrary and idiosyncratic, but we hope the reader will find them of interest. Some of the results proved below were discovered in the course of preparing this write-up, and we do not know which of these are new.

Let $k \in \mathbf{Comm}^1$, and let $\mathbf{Comm}^1_k \subseteq \mathbf{Ring}^1_k$ denote the variety of commutative associative unital k-algebras. The category $(\mathbf{Comm}^1_k)^{\mathrm{op}}$ is, up to equivalence, what algebraic geometers call the category of *affine schemes* over k. Cogroups in $\mathbf{Comm}^1_k$ are thus equivalent to *group* objects in this category, called *affine algebraic groups*, or *affine group schemes*, and these constitute a major area of study in algebraic geometry [**196**], [**51**, Chapter III], [**161**]. Affine algebraic *rings*, *semigroups* etc. are also studied, [**136**, §26], [**154-156**], though less intensively. We shall for the most part not assume any knowledge of algebraic geometry, though a few remarks in this chapter will be aimed at the reader familiar with these concepts. (Incidentally, algebraic geometers differ in the degree of generality they give the term "affine scheme". Some require the corresponding commutative k-algebras to be finitely generated, and when this condition is removed, speak of "pro-affine" schemes; e.g., [**158**]. However, it is natural for us to work with the full variety $\mathbf{Comm}^1_k$, so we shall not assume such a restriction when we do make algebraic-geometric remarks.) As we shall see in §43, the objects that represent functors from $\mathbf{Comm}^1_k$ to $\mathbf{Semigp}^e$ or $\mathbf{Group}$ are also instances of what are called *bialgebras* and *Hopf algebras*.

We will begin this chapter with a survey, in §40, of some phenomena that occur among functors on $\mathbf{Comm}^1_k$ but not among functors on $\mathbf{Ring}^1_k$, including counterexamples to the analogs of several results of earlier chapters. In the opposite direction, we prove in §41 a result on identities satisfied by subfunctors, which holds in the commutative but not in the noncommutative context. §42 again considers idempotents in semigroup-valued functors, to which we return briefly again in §48. §43 concerns bialgebras and Hopf algebras. In the remaining sections, §§44-47, we sketch in more detail some classes of examples, including co-ring structures on rings of integral polynomials and rings of linearly recursive sequences.

§40. Some easy examples.

In this section we present a smorgasbord of easily described examples, in order to give the reader some familiarity with the rich landscape of representable functors on $\mathbf{Comm}_k^1$, illustrate differences in behavior between these and functors on the categories of noncommutative algebras we have been considering, and point out the failure of some conjectures one might naively make about these functors. We start with some trivial observations.

40.1. The multiplicative structure of a commutative ring is a *commutative* semigroup. This immediately leads to two kinds of counterexamples to the analog of Theorem 13.15. First, the underlying multiplicative semigroup functor $\mathbf{Comm}_k^1 \to \mathbf{AbSemigp}^e$ is representable, but *not* group valued. Secondly, the subfunctor thereof giving the group of units is abelian group-valued, but not of the sort described in that Theorem. Note that if k is a field of characteristic $p \neq 0$, Theorem 13.15 implies that representable functors $\mathbf{Ring}_k^1 \to \mathbf{Ab}$ always yield abelian groups of exponent p, but the group-of-units functor on $\mathbf{Comm}_k^1$ does not have this property. If we compose this functor with the representable functor $\mathbf{Ab}(\mathbf{Q}, -)$: $\mathbf{Ab} \to \mathbf{Ab}$, we get a $\mathbf{Q}$-vector-space-valued functor on $\mathbf{Comm}_k^1$ which is nontrivial (i.e., does not take all k-algebras to the zero vector space) regardless of char(k).

The above constructions can, of course, be preceded by that of completed tensor product with any $A \in \mathbf{LCpComm}_k^1$ if k is a field, or in the case of a general commutative ring k, by the analogous constructions in the spirit of Theorem 29.5, yielding larger classes of examples. And as in §33, one can generalize this by allowing nonunital A if one replaces ''group of invertible elements'' by ''group of quasiunits''.

40.2. Since the above characteristic-changing functors arise from multiplicative structures, it is an additional surprise that such constructions can occur as the *additive* structures of *ring*-valued representable functors. For instance, for arbitrary $k \in \mathbf{Comm}^1$, the functor W associating to each $S \in \mathbf{Comm}_k^1$ the multiplicative group of formal power series over S with constant term 1,

$$1 + \xi_1 t + \xi_2 t^2 + \ldots$$

can be given a second ''multiplication'' operation, so that the objects $W(S)$ become commutative unital rings with 1, with the original multiplication of formal power series as their *addition*. We will sketch this construction in §44 below.

40.3. The existence of certain morphisms among representable functors leads to important new subfunctors of known functors:

The *determinant* map det: $\mathrm{GL}_n \to \mathrm{GL}_1$ allows one to define the representable subfunctor $\mathrm{SL}_n \subseteq \mathrm{GL}_n \in \mathbf{Rep}(\mathbf{Comm}^1, \mathbf{Group})$.

The *transpose* map $(\ )^{\mathrm{tr}}$ from the $n \times n$ matrix ring functor $M_n \in \mathbf{Rep}(\mathbf{Comm}^1, \mathbf{Ring}^1)$ to $\mathrm{op} \circ M_n$ (where op: $\mathbf{Ring}^1 \to \mathbf{Ring}^1$ denotes the opposite ring functor) leads to representable functors such as the *orthogonal* group functor

$$R \mapsto \{A \in M_n(R) \mid A A^{\mathrm{tr}} = I = A^{\mathrm{tr}} A\}$$

and the *symplectic* group functor

$$R \mapsto \{A \in M_{2n}(R) \mid A J A^{\mathrm{tr}} = I = A^{\mathrm{tr}} J A\},$$

where $J = \begin{pmatrix} 0 & I \\ -I & 0 \end{pmatrix}$. (However, as we shall note in §57, the transpose morphism does have an analog for matrices over *rings with involution*, hence these two group constructions also have analogs on the category of such rings.)

If p is a prime, then on any $S \in \mathbf{Comm}^1_{\mathbf{Z}/p\mathbf{Z}}$, the *Frobenius* map $\varphi\colon x \mapsto x^p$ is a ring endomorphism. For k a field of characteristic p and n a positive integer, this leads to such constructions as the functor associating to every $S \in \mathbf{Comm}^1_k$ the subgroup $\ker(\varphi^n)$ of the additive group of S. Unlike all nontrivial representable functors we have seen before, this is represented by a finite-dimensional algebra, $k[x]/(x^{p^n})$; such constructions are called by algebraic geometers *finite* group schemes. An example of a noncommutative group-valued functor with the same property is the functor taking S to the set of $r \times r$ matrices over S whose entries all lie in $\ker(\varphi^n)$, regarded as a group under quasimultiplication.

40.4. If $S \in \mathbf{Comm}^1$, the set of *idempotent* elements of S forms a Boolean ring $B(S)$, under the operations 0, 1, ordinary multiplication, and the Boolean addition

(40.5) $$x \dot{+} y \;=\; x - 2xy + y.$$

This functor,

(40.6) $$B\colon\ \mathbf{Comm}^1 \;\to\; \mathbf{Bool}^1$$

is represented by $\mathbf{Z} \times \mathbf{Z}$, the commutative ring with a universal idempotent. We can now construct representable functors from $\mathbf{Comm}^1$ into other varieties $\mathbf{V}$ by composing (40.6) with representable functors

(40.7) $$\mathbf{Bool}^1 \;\to\; \mathbf{V}.$$

And such representable functors are quite abundant. This follows from the well-known equivalence (Stone duality) between $(\mathbf{Bool}^1)^{\mathrm{op}}$ and the category of totally disconnected compact Hausdorff spaces (also called *Stone spaces* or *Boolean spaces*) [**104**, Introduction and §II.4]. Under this duality, $\mathbf{V}$-objects in the category of Stone spaces, i.e., topological $\mathbf{V}$-algebras X with totally disconnected compact Hausdorff underlying spaces, must clearly correspond to *co-$\mathbf{V}$-objects* of $\mathbf{Bool}^1$. In particular, a *finite* $\mathbf{V}$-algebra X taken with the discrete topology is compact, hence each such algebra yields a co-$\mathbf{V}$-algebra in $\mathbf{Bool}^1$. One finds that the functor (40.7) determined by any $\mathbf{V}$-algebra X with Stone topology takes the Boolean ring $\mathbf{Z}/2\mathbf{Z}$ to the algebra X.

This construction allows one to create tailor-made counterexamples. For instance, if one should conjecture that there were no nontrivial representable functors $\mathbf{Comm}^1_k \to \mathbf{Comm}^1_{\mathbf{Z}/p\mathbf{Z}}$ for k of characteristic 0, or of prime characteristic distinct from p, one has only to compose the forgetful functor

$\mathbf{Comm}_k^1 \to \mathbf{Comm}^1$ with (40.6), and then with the functor $\mathbf{Bool}^1 \to \mathbf{Comm}_{\mathbf{Z}/p\mathbf{Z}}^1$ determined by any finite nontrivial ring in the latter variety, to see that this is not so. Similarly, if on looking at (40.6) and noting that Boolean rings have a natural distributive lattice structure, one were to conjecture that all representable functors from rings to lattices were distributive lattice valued, or at least modular lattice valued, one could get a counterexample by taking a finite nonmodular lattice L, and composing (40.6) with the functor $\mathbf{Bool}^1 \to \mathbf{Lattice}$ that it determines.

(For S a noncommutative ring, the *central* idempotents of S form a Boolean ring, which one may regard as a generalization of $B(S)$. But this construction, though useful, is not a functor, let alone a representable one, hence does not yield such representable functors on $\mathbf{Ring}_k^1$.)

We remark, for those familiar with the viewpoint of algebraic geometry, that for X a topological $\mathbf{V}$-algebra whose underlying topological space is totally disconnected compact Hausdorff, the representable functor $\mathbf{Comm}^1 \to \mathbf{V}$ characterized above (i.e., the construction of first applying (40.6), and then composing with the functor $\mathbf{Bool}^1 \to \mathbf{V}$ determined by X) can be described as taking each $S \in \mathbf{Comm}^1$ to the $\mathbf{V}$-algebra of all continuous X-valued functions on $\mathrm{Spec}(S)$. In particular, if X is a finite $\mathbf{V}$-algebra, this is the algebra of *locally constant* X-valued functions on $\mathrm{Spec}(S)$.

PROBLEM 40.8. *Let k be an algebraically closed field. Characterize those varieties $\mathbf{V}$ for which every representable functor $\mathbf{Comm}_k^1 \to \mathbf{V}$ factors as*

$$(40.9) \qquad \mathbf{Comm}_k^1 \xrightarrow{\text{forget}} \mathbf{Comm}^1 \xrightarrow{B} \mathbf{Bool}^1 \longrightarrow \mathbf{V}.$$

We have assumed k algebraically closed in this question because for more general fields k, there exist representable functors from $\mathbf{Comm}_k^1$ to $\mathbf{Bool}^1$ itself which do not factor as above. For example, if L is a finite nontrivial separable field extension of k, the construction

$$(40.10) \qquad \mathbf{Comm}_k^1 \xrightarrow{-\otimes_k L} \mathbf{Comm}^1 \xrightarrow{\quad B \quad} \mathbf{Bool}^1$$

cannot be factored as in (40.9), since any functor which *can* be so factored must take the inclusion $k \to L$ to an isomorphism, but (40.10) does not, since $L \otimes_k L$ has nontrivial idempotents.

If k is not a field, there exist still other variants of (40.9). For example, for every $a \in k$, the idempotent elements of $S \in \mathbf{Comm}_k^1$ annihilated by a form a *nonunital* Boolean ring (with operations 0, $\dot{+}$ as in (40.5), and multiplication). The opposite of the category **Bool** of such rings can be identified with the category of Stone spaces with a distinguished basepoint, and the latter category also admits nontrivial $\mathbf{V}$-objects for a great many $\mathbf{V}$ (e.g., **Group**), yielding further examples of representable functors from $\mathbf{Comm}_k^1$ to such varieties. Still more generally, given elements $a, b \in k$, the representable functor associating to $S \in \mathbf{Comm}_k^1$ the set $\{x \in S \mid x^2 = x,\ \ ax = 0,\ \ b(1-x) = 0\}$ is closed under the operations of multiplication, and of 3-term Boolean addition, $x \dot{+} y \dot{+} z$, constituting a structure of what might be called "Boolean ring without 0 or 1". The opposite of the category of such objects is equivalent to the category of Stone spaces with *two* distinguished points, distinct if the space has more than one point. (A generalization of Stone duality which includes, in particular, the two equivalences just mentioned is given in [**9**].) A nontrivial **V**-object of this category means a **V**-algebra with a Stone topology, together with two distinct one-element subalgebras. Easy examples of such algebras with finite discrete underlying set exist if **V** is, for instance, the variety of lattices, of semigroups, or of heaps (§22).

40.11. It is not hard to verify that the functor (40.9) induced as described above by an n-element **V**-algebra X is represented by a **V**-coalgebra in $\mathbf{Comm}^1_k$ having for underlying k-algebra the n-fold direct product $k\times \ldots \times k$. Thus, we have further examples of *finite* affine algebra schemes, as defined in the last paragraph of 40.3, supplementing the examples mentioned there, which were based on the Frobenius map. These new examples suggest that representable functors whose underlying k-algebras are free as k-modules of rank n resemble n-element objects, leading to the following question of Frans Oort:

QUESTION 40.12 ([**189**, p.5, *Remark*]). *Suppose* $F\colon \mathbf{Comm}^1_k \to \mathbf{Group}$ *is a representable functor, whose representing algebra* R *is free of rank* $n<\infty$ *as a* k*-module. Must* F *take values in the variety of groups satisfying the identity* $x^n = e$*?*

The general case of the above question can be reduced to the case where k is local, artinian, and of positive characteristic; an affirmative answer is known if F is **Ab**-valued [**189**, p.4, Theorem], or if k is an integral domain, or if n is invertible in k. On the other hand, examples are known where such functors do *not* satisfy *all* the identities of groups of order n. For example, any group of order p^2 is abelian, but if k is of characteristic p, a functor with p^2-dimensional representing object that is not **Ab**-valued [**189**, p.6, *Remark*] is the construction taking S to the group of matrices $\begin{pmatrix} 1 & \xi \\ 0 & 1+\beta \end{pmatrix}$, where $\beta,\ \xi\in S$ satisfy $\beta^p = \xi^p = 0$.

40.13. For the reader familiar with some algebraic geometry, we mention also (though it falls outside the stated topic of this chapter) the phenomenon of *nonaffine* algebraic groups – additional group objects that appear when one replaces the category $(\mathbf{Comm}^1_k)^{\mathrm{op}}$ of affine schemes by the larger category of general schemes over k; notably *abelian varieties* [**51**, Chapters IV-V, VII-VIII], [**137**]. We remark that though there have been attempts to set up "noncommutative algebraic geometries" [**142**], [**12**], we know of none which have developed far enough to show analogs of this phenomenon.

Though a subscheme of an affine scheme need not be affine, it can be deduced from [**166**, Theorem 12, p. 429] that if the underlying scheme of an algebraic group is embeddable in an affine scheme, then it is affine. (Actually, that is proved in the context of classical algebraic geometry; we leave it to the experts to say whether the proof is applicable to the modern scheme-theoretic context.)

We do not know the answer to

QUESTION 40.14. *Do there exist connected nonaffine* $\mathbf{Ring}^1$*-schemes?*

§41. Identities and equational subfunctors.

We noted in §30 that a representable subfunctor F of a representable set-valued functor G on a category $\mathbf{V}$ will be represented by an *epimorph* R of the object S representing G, and that if $\mathbf{V}$ is a variety of algebras, a particularly convenient class of epimorphs of an object are its *homomorphic images*. The functor represented by a homomorphic image R of the algebra S representing G is determined by imposing additional equations on the coordinates of the tuples describing the values of G; hence let us call a subfunctor F of G *equational* if it has this form. (Examples: Let G: $\mathbf{Semigp}^e \to \mathbf{Set}$ be the underlying-set functor, which is represented by the free semigroup $\langle x \rangle$. Then G has, inter alia, the *equational* subfunctor giving the set of idempotent elements, and the *non*equational representable subfunctor giving the set of invertible elements. These are represented respectively by the epimorphs $\langle x \mid x^2 = x \rangle$ and $\langle x, y \mid xy = yx = 1 \rangle$ of $\langle x \rangle$.)

Now suppose G is a representable functor from a variety $\mathbf{V}$ to $\mathbf{Set}$, X is an object of $\mathbf{V}$, and A is any subset of $G(X)$. Then there will exist both a *least representable* subfunctor $E \subseteq G$ such that $A \subseteq E(X)$, and a *least equational* subfunctor $F \subseteq G$ such that $A \subseteq F(X)$. (Thus, $E \subseteq F \subseteq G$.) Indeed, if we denote by S the object representing G, the latter functor, F, is represented by the image of S in $\prod_A X$ under the homomorphism given by the family A of maps $S \to X$; equivalently, for $T \in \mathbf{V}$, $F(T)$ is the subset of $G(T)$ defined by those equations in the coordinates that hold for the coordinates of all elements of $A \subseteq G(X)$. The object representing E (which we will only occasionally be concerned with here) can likewise be described as the *stable dominion* of the image of S in $\prod_A X$.

Interesting properties of the subset $A \subseteq G(X)$ may or may not be inherited by these induced subfunctors E and F of G. In particular, if the values of G admit an algebra structure, and A is a subalgebra of $G(X)$, these subfunctors may or may not be subalgebra-valued, and if they are, they may or may not satisfy the identities satisfied by A. For a negative example, suppose k is a field and G any nontrivial representable functor $\mathbf{Ring}^1_k \to \mathbf{Ring}^1$. Then we can find commutative subrings $A \subseteq G(X)$, but there is no commutative-ring valued representable subfunctor of G by Corollary 25.23, so, in particular, the least equational and the least representable set-valued subfunctors of G satisfying $A \subseteq F(X)$ are not commutative-ring valued.

However, when we consider functors on the variety $\mathbf{Comm}^1_k$, algebraic geometry looks at the least equational subfunctor containing a set of values as something like the "closure" of that set in a topological space. (The algebraic-geometric entities corresponding to the homomorphic images S/I of an object $S \in \mathbf{Comm}^1_k$, and hence to the equational subfunctors of the functor represented by S, are known as the *closed subschemes* Spec(S/I) of Spec(S).) In such a situation, one would indeed expect the properties of being a subalgebra and of satisfying given identities to carry over to this closure. We shall prove such a result below under appropriate hypotheses on k and X, without assuming any knowledge of algebraic geometry.

(The reader will note a similarity between this question and that of §31, where we looked at the least representable or the least equational set-valued subfunctor H of a representable algebra-valued functor G which contained the image of a given morphism of functors $F \to G$, and asked when this would inherit the property of being subalgebra-valued. The subset A considered here is not of that form, but the questions are parallel.)

Since we will be proving a positive result rather than exhibiting examples, we will need to work explicitly with the coalgebras representing our functors; hence we recall that for any commutative ring k, the coproduct construction on $\mathbf{Comm}^1_k$ is given by the *tensor product* of k-algebras.

We begin with a Lemma and Corollary on tensor products of k-modules, which, when applied to k-algebras, and translated into algebraic geometric language, will say that given a suitable family of affine schemes, and in each of these a dense set of k-valued points, the product of these sets is dense in the product of the schemes. (This concept of "density" is related to the one discussed in §38, except that again, we were concerned there with density of a subfunctor, and here with density of a subset.)

LEMMA 41.1. *Let k be a field, and S, T be k-vector-spaces. Suppose we are given subsets $A \subseteq \mathbf{Mod}_k(S, k)$ and $B \subseteq \mathbf{Mod}_k(T, k)$ such that*

$$\bigcap_{a \in A} \ker(a\colon S \to k) = \{0\} \quad \textit{and} \quad \bigcap_{b \in B} \ker(b\colon T \to k) = \{0\}.$$

Then on $S \otimes_k T$ one has

$$\bigcap_{a \in A,\ b \in B} \ker(a \otimes b\colon S \otimes_k T \to k) = \{0\}.$$

PROOF. We need to show that given a nonzero element $u \in S \otimes_k T$, there exist $a \in A$, $b \in B$ such that $u \notin \ker(a \otimes b)$. Let us write

$$u = s_1 \otimes t_1 + \ldots + s_r \otimes t_r. \tag{41.2}$$

Assume this expression for u chosen to minimize r. Then we see that since k is a field, $s_1, \ldots, s_r \in S$ will be k-linearly independent, as will $t_1, \ldots, t_r \in T$.

Now if we restrict the linear functionals $a \in A$ to the subspace of S spanned by $s_1, \ldots, s_r$, their kernels still have trivial intersection, hence as this space is r-dimensional, we can choose $a_1, \ldots, a_r \in A$ whose kernels on this space have trivial intersection; i.e., such that the map $s_1 k + \ldots + s_r k \to k^r$ induced by $(a_1, \ldots, a_r)$ is an isomorphism. Similarly, we can find $b_1, \ldots, b_r \in B$ inducing an isomorphism $t_1 k + \ldots + t_r k \to k^r$. One sees by passing to the bases of $s_1 k + \ldots + s_r k$ and $t_1 k + \ldots + t_r k$ dual to $\{a_1, \ldots, a_r\}$ and $\{b_1, \ldots, b_r\}$ that the r^2-tuple of maps $a_i \otimes b_j$ induces an isomorphism

$$(s_1 k + \ldots + s_r k) \otimes (t_1 k + \ldots + t_r k) \to k^{r^2}.$$

In particular, at least one member of that r^2-tuple of maps must be nonzero on the nonzero element (41.2), as claimed. □

By induction we can go from the case of two tensor factors S and T

considered above to that of n factors $S_1, \dots, S_n$. It is also not hard to obtain from the above result, in which k is a field, a similar result for k an arbitrary commutative integral domain, if we assume our given k-modules *flat*, since a tensor product of flat modules is flat, hence torsion-free, and if k is an integral domain with field of fractions K, and S a torsion-free k-module, then the canonical map $S \to S \otimes_k K$ is an embedding. The resulting statement is given in the next Corollary. The case that will be of most importance to us will be that in which the morphisms in the given families A_i take values in k, but we lose nothing by proving this and the next few results for general K-valued maps. (The $n = 0$ case of this result takes a little thought, but is easy to check: There, $\prod A_i$ is a singleton, the hypothesis is vacuous, and the conclusion just says that the canonical map from k to its field of fractions K is an embedding.)

COROLLARY 41.3. *Let k be a commutative integral domain and K its field of fractions. Suppose $S_1, \dots, S_n$ $(n \geq 0)$ are flat k-modules, and suppose that for each $i \leq n$ we are given a subset $A_i \subseteq \mathbf{Mod}_k(S_i, K)$ such that*

$$\bigcap_{a \in A_i} \ker(a\colon S_i \to K) = \{0\}.$$

Then on $S_1 \otimes_k \dots \otimes_k S_n$ we have

$$\bigcap_{(a_1, \dots, a_n) \in \prod A_i} \ker(a_1 \otimes \dots \otimes a_n\colon S_1 \otimes_k \dots \otimes_k S_n \to K) = \{0\}. \ \square$$

We can now deduce our first result on algebra-valued representable functors.

COROLLARY 41.4. *Let k be a commutative integral domain, with field of fractions K, and suppose F is a representable functor from $\mathbf{Comm}^1_k$ to any variety $\mathbf{V}$, whose representing object $R \in \mathbf{Comm}^1_k$ is flat as a k-module. Let A be a sub-$\mathbf{V}$-algebra of $F(K)$ such that $\bigcap_{a \in A} \ker(a\colon R \to K) = \{0\}$. Then the objects $F(S) \in \mathbf{V}$ satisfy all identities that A does; in other words, if we denote by $\mathbf{U}$ the subvariety of $\mathbf{V}$ generated by A, then the original functor F is $\mathbf{U}$-valued.*

PROOF. Suppose f, f' are derived n-ary operations of $\mathbf{V}$ such that $f = f'$ is an identity of A; we claim that it is an identity of F as well. Indeed, if the corresponding co-operations $\mathbf{f}, \mathbf{f}'\colon R \to R \otimes_k \dots \otimes_k R$ are distinct, let us take $r \in R$ such that $\mathbf{f}(r) - \mathbf{f}'(r) \neq 0$. Then by the preceding Corollary, there exist $a_1, \dots, a_n \in A$ such that $a_1 \otimes \dots \otimes a_n$ assumes a nonzero value at $\mathbf{f}(r) - \mathbf{f}'(r)$. The distinct values $(a_1 \otimes \dots \otimes a_n)\mathbf{f}(r)$ and $(a_1 \otimes \dots \otimes a_n)\mathbf{f}'(r)$ can be written $f(a_1, \dots, a_n)(r)$ and $f'(a_1, \dots, a_n)(r)$, so $f \neq f'$ as operations on A, a contradiction. $\square$

We are now ready to prove our result about the least equational subfunctor $F \subseteq G$ containing a subalgebra $A \subseteq G(k)$, or more generally, a subalgebra $A \subseteq G(K)$. In the course of the proof we will want to apply Corollary 41.4 to this subfunctor F; but we need a way of guaranteeing that the ring R representing F is flat as a k-module. Since the elements of R are separated by the set A of homomorphisms into K, R will be torsion-free as a k-module. Hence it will suffice to assume that k belongs to the class of integral domains all of whose torsion-free modules are flat. We claim that these are the *Prüfer domains*. In [**35**,

Ch.VII, §2, Exercise 12], fourteen equivalent characterizations of the latter class of rings are given; of these, condition (δ) states that finitely generated torsion-free k-modules are projective. This immediately implies that all torsion-free k-modules are flat, which is what we need here. Toward the end of this section, we shall establish the converse, and some further equivalent conditions. Since the class of Prüfer domains is unfamiliar even to many commutative ring theorists, we begin the statement of the next result by referring to some more familiar classes of commutative rings.

As in §31, when we refer to a ''set-valued subfunctor'' of an algebra-valued functor G, what we mean is a subfunctor of the composite of G with the underlying-set functor of its codomain.

PROPOSITION 41.5. *Let* $k \in \mathbf{Comm}^1$ *be a field, more generally a principal ideal domain, still more generally a Dedekind domain, or, most generally, a Prüfer domain, and* K *its field of fractions. Suppose* G *is a representable functor from* $\mathbf{Comm}^1_k$ *to any variety* $\mathbf{V}$, *let* $S \in \mathbf{Comm}^1_k$ *denote the underlying object of its representing coalgebra, and suppose* A *is any* $\mathbf{V}$*-subalgebra of* $G(K)$.

Denote by F *the least equational set-valued subfunctor of* G *such that* $A \subseteq F(K)$, *namely the subfunctor represented by* $R = S/(\bigcap_A \ker a)$; *and let* $\mathbf{U}$ *be the subvariety of* $\mathbf{V}$ *generated by* $A \in \mathbf{V}$.

Then the $\mathbf{V}$*-operations on* G *induce operations on* F, *and the resulting* $\mathbf{V}$*-structure on* F *is in fact* $\mathbf{U}$*-valued.*

PROOF. Let q denote the quotient homomorphism $S \to R$, and for any nonnegative integer n, let $q^{\otimes n}$ denote the induced homomorphism of n-fold tensor powers, $S \otimes_k \dots \otimes_k S \to R \otimes_k \dots \otimes_k R$. Let $\bar{A} = \{\bar{a} \mid a \in A\}$ denote the family of homomorphisms $R \to K$ induced by the elements of A.

Suppose $\mathbf{f}_S\colon S \to S \otimes_k \dots \otimes_k S$ is an n-ary co-operation of S. To get our first assertion, we must show that there exists a morphism $\mathbf{f}_R$ making a commuting diagram

$$\begin{array}{ccc} S & \xrightarrow{\ \mathbf{f}_S\ } & S \otimes_k \dots \otimes_k S \\ \downarrow{\scriptstyle q} & & \downarrow{\scriptstyle q^{\otimes n}} \\ R & \xrightarrow{\ \mathbf{f}_R\ } & R \otimes_k \dots \otimes_k R. \end{array}$$

To do this, it suffices to show that for any $s \in S$, if $q^{\otimes n}\mathbf{f}_S(s) \neq 0$, then $q(s) \neq 0$ in R. Now by Corollary 41.3, applied to n copies of R and the system of maps $\bar{A}$, the assumption that $q^{\otimes n}\mathbf{f}_S(s) \neq 0$ implies that for some $\bar{a}_1, \dots, \bar{a}_n \in \bar{A}$,

$$(41.6) \quad 0 \neq (\bar{a}_1 \otimes \dots \otimes \bar{a}_n) q^{\otimes n}\mathbf{f}_S(s) = (a_1 \otimes \dots \otimes a_n)\mathbf{f}_S(s) = f(a_1, \dots, a_n)(s).$$

By hypothesis, A is a $\mathbf{V}$-subalgebra of $G(k)$, so $f(a_1, \dots, a_n) \in A$. Thus, s is not in the kernel of all elements of A, so $q(s) \neq 0$, completing the proof that a co-$\mathbf{V}$-structure is induced on R.

The coalgebra R and the family $\bar{A}$ of homomorphisms $R \to K$ satisfy the hypotheses of Corollary 41.4, so that Corollary yields our final assertion. □

The above result concerns an equational subfunctor F minimal for the condition $A \subseteq F(K) \subseteq G(K)$. The corresponding result for an F minimal for a condition of the form $A \subseteq F(k) \subseteq G(k)$ can be deduced therefrom, using the observation that if F is any equational subfunctor of any representable set-valued functor G on a variety $\mathbf{W}$, and $k \subseteq K$ are any objects of $\mathbf{W}$, we have

(41.7) $$F(k) = G(k) \cap F(K) \quad \text{as subsets of } G(K).$$

Thus, given $A \subseteq G(k)$, the least equational subfunctor F of G such that $A \subseteq F(k)$ is also the least equational subfunctor such that $A \subseteq F(K)$, so Corollary 41.4 is applicable to this functor. (The equality (41.7) is not in general true for nonequational representable subfunctors, as can be seen by applying the set-of-invertible-elements subfunctor F of the underlying-set functor G: $\mathbf{Comm}^1 \to \mathbf{Set}$ to an integral domain k and its field of fractions K.)

As a typical application of the above Proposition, suppose $G \in \mathbf{Rep}(\mathbf{Comm}^1_k, \mathbf{Ring}^1)$, let a be any element of $G(k)$, and let $A \subseteq G(k)$ be the subring generated by a. Then A is commutative, hence the least equational set-valued subfunctor $F \subseteq G$ containing A is *commutative* ring valued.

In subsequent sections, we shall make some further applications of the above result; at this point let us give examples showing the need for various conditions in the statement. We have already noted that the above consequence, "ring-valued representable functors have commutative ring valued representable subfunctors", is not true for functors on $\mathbf{Ring}^1_k$; on the contrary, a nontrivial $\mathbf{Ring}^1$-valued representable functor on $\mathbf{Ring}^1_k$ *cannot* have a representable $\mathbf{Comm}^1$-valued subfunctor. Hence the analog of Proposition 41.5 for functors on $\mathbf{Ring}^1_k$ is false for all nontrivial k. The next example will show that even the first half of the conclusion of that Proposition, saying that the least equational set-valued subfunctor $F \subseteq G$ containing a subalgebra $A \subseteq G(k)$ is closed under the algebra operations of G, does not hold for functors on $\mathbf{Ring}^1_k$.

Let k be an infinite field, let $G \in \mathbf{Rep}(\mathbf{Ring}^1_k, \mathbf{Ring}^1)$ be the direct product of the underlying ring functor and its opposite, represented by the free algebra $k\langle x, y\rangle$ with appropriate co-operations, and let $A \subseteq G(k)$ be the diagonal, $\{(c, c) \mid c \in k\} \subseteq k \times k$. This is a subring because k is commutative. In the homomorphic image of $k\langle x, y\rangle$ determined by this set A, we will clearly have $x = y$, and, in fact, this image will be precisely $k\langle x, y \mid x = y\rangle \cong k[x]$, since specializations of x to elements of the infinite field k separate points of $k[x]$. Hence for any $T \in \mathbf{Ring}^1_k$, $F(T) = \{(s, s) \mid s \in T\}$. But this subfunctor is not closed under the indicated ring multiplication, since a product $(s, s)(t, t) = (st, ts)$ does not lie in $F(T)$ if $s, t \in T$ do not commute.

Returning to functors on $\mathbf{Comm}^1_k$, the next Lemma shows that Proposition 41.5 can fail if the hypotheses on k are weakened. Note that in case (i) of this set of examples, k is a principal ideal *ring*, but not a domain, while case (ii) includes the most familiar examples of integral domains that are not Prüfer domains, including the polynomial ring $\mathbf{Z}[x]$.

LEMMA 41.8. *Suppose that either*

(i) $k = L[\varepsilon]/(\varepsilon^2)$, *where* L *is a field, or*

(ii) k *is a unique factorization domain that is* not *a principal ideal domain.*

Then there exist

(a) *a variety* $\mathbf{V}$, *a representable* $\mathbf{V}$*-valued functor* G *on* $\mathbf{Comm}_k^1$, *and a sub-*$\mathbf{V}$*-algebra* $A \subseteq G(k)$, *such that the least equational set-valued subfunctor* $F \subseteq G$ *satisfying* $A \subseteq F(k)$ *is not closed under the operations of* $\mathbf{V}$, *and*

(b) *a variety* $\mathbf{V}$, *a representable* $\mathbf{V}$*-valued functor* G *on* $\mathbf{Comm}_k^1$, *and a sub-*$\mathbf{V}$*-algebra* $A \subseteq G(k)$ *such that the least equational set-valued subfunctor* $F \subseteq G$ *satisfying* $A \subseteq F(k)$ *is closed under the operations of* $\mathbf{V}$, *but the resulting algebra-valued subfunctor does not satisfy all the identities of* A.

Moreover, in all combinations of cases except possibly (i)(b), *we can take* $\mathbf{V} = \mathbf{Group}$.

Sketch of Proof. (i)(a) Let $G\colon \mathbf{Comm}_k^1 \to \mathbf{Group}$ be the functor taking a k-algebra T to the underlying additive group of $T\times T$, and let $A \subseteq G(k)$ be the subgroup $\varepsilon k\times\varepsilon k$. One finds that the least equational set-valued subfunctor F of G such that $F(k)$ contains A is given by $F(T) = \{(s,t) \mid 0 = \varepsilon s = \varepsilon t = s^2 = st = t^2\}$. If we now let T be the k-algebra presented by two generators s and t, and all the above relations except $st = 0$, we find that $(s, 0)$ and $(0, t)$ belong to $F(T)$, but their sum does not.

(i)(b) Let $\mathbf{V} = \mathbf{Semigp}$, let G be the functor taking a k-algebra to its underlying multiplicative semigroup, and let $A = \varepsilon k \subseteq G(k)$. Then one finds that F is given by $F(T) = \{s\in T \mid \varepsilon s = s^2 = 0\}$. This is closed under the semigroup operation of G, but though A satisfies the identity $wx = yz$, this does not hold in all objects $G(T)$. Indeed, for T as in our proof of (i)(a), we have $st \neq 0\cdot 0$.

(ii)(a) This time let $G(T) = T\times T\times T$ with the group multiplication

$$(s,t,u)(s',t',u') = (s+s',\ t+t',\ u+u'+(st'-s't)). \tag{41.9}$$

Since k is a unique factorization domain but not a principal ideal domain, we can find two relatively prime elements a, b of that ring which generate a proper ideal. Let $A = \{(ca, cb, 0) \mid c\in k\}$. It is easy to verify that this is a subgroup of $G(k)$ isomorphic to the additive group of k, and one deduces using the relative primality of a and b that the least equational set-valued subfunctor $F \subseteq G$ containing A is given by

$$F(T) = \{(s,t,0) \mid bs = at\}.$$

But if we now let $T = k/(a,b)$, we see that $(1_T, 0, 0)$ and $(0, 1_T, 0)$ belong to $F(T)$, while their product under (41.9) has nonzero third coordinate, and so does not.

(ii)(b) Again let $G(T) = T\times T\times T$, but now let us define

$$(s,t,u)(s',t',u') = (s+s',\ t+t',\ u+u'+st'). \tag{41.10}$$

Taking a, b as in (ii)(a), this time let

$$A = \{(ca, cb, d) \mid c, d\in k\},$$

which we note is a *commutative* subgroup of $G(k)$. We find that

$$F(T) = \{(s,t,u) \mid bs = at\},$$

which certainly defines a subgroup of $G(T)$, but taking the same T as in our proof of (ii)(a), we see that our pair of elements $(1_T, 0, 0)$, $(0, 1_T, 0)$ fails to commute. □

From these examples one can also deduce that Corollaries 41.3 and 41.4 fail if the flatness hypothesis is deleted, since the Prüfer condition in Proposition 41.5 is only used to get the flatness needed to apply these Corollaries. (However, one can prove the analogs of Corollary 41.4 and Proposition 41.5 with k an arbitrary integral domain, and F and G representable functors on the *quasivariety* of *torsion-free* commutative k-algebras.)

It would be of interest to know whether there exists an example of (i)(b) above with $\mathbf{V} = \mathbf{Group}$.

Finally, let us show that even for k a field, the conclusion of Proposition 41.5 fails if we let A be a subalgebra of $G(L)$ for some extension field L of k, rather than a subalgebra of $G(k)$ itself.

LEMMA 41.11. *Let k be a field and L a nontrivial finite Galois extension field of k. Then there exists a representable functor $G\colon \mathbf{Comm}^1_k \to \mathbf{Ab}$ and a subgroup $A \subseteq G(L)$ such that, letting F denote the least equational set-valued subfunctor of G such that $A \subseteq F(L)$, the subset $F(L) \subseteq G(L)$ is not a subgroup.*

PROOF. Let $G(T)$ be the underlying additive group of $T\times T$. Take any $c \in L-k$, and let $A = \{(cs, s) \mid s \in L\} \subseteq G(L)$. Let us form the polynomial over L in two indeterminates

$$g(x, y) \;=\; \textstyle\prod_{\alpha\in \mathrm{Aut}_k(L)} (x - \alpha(c)y),$$

and note that this is $\mathrm{Aut}_k(L)$-invariant, hence lies in $k[x, y]$. Clearly every element $(s, t) \in A$, and hence also every $(s, t) \in F(T)$, satisfies $g(s, t) = 0$. Now $(c,1) \in A \subseteq F(L)$, and $F(L)$ is necessarily closed under the action of $\mathrm{Aut}_k(L)$, hence $(\alpha(c),1) \in F(L)$ for all $\alpha \in \mathrm{Aut}_k(L)$. But subtracting two such elements, we get a nonzero element of the form $(d,0)$, which cannot lie in $F(L)$, since $g(d,0) \neq 0$. □

41.12. *Remark.* To see why subalgebras of $G(k)$ and $G(K)$ are "good" in a way that subalgebras of $G(L)$ are not, note that k is initial in $\mathbf{Comm}^1_k$. Hence a subalgebra $A \subseteq G(k)$ has a canonical homomorphic image in each algebra $G(T)$, giving a (generally nonrepresentable, but nevertheless) *algebra*-valued subfunctor $F_0 \subseteq G$. Hence it is plausible that under reasonable conditions, the least equational subfunctor containing this F_0 should also be algebra-valued and satisfy the identities of A; and one finds that this subfunctor is also the least equational subfunctor containing A.

In the more general case $A \subseteq G(K)$ where K is the field of fractions of k, note that K, being an epimorph of the initial object k, is quasi-initial in $\mathbf{Comm}^1_k$, i.e., has *at most* one morphism to each object. From this it is not hard to see that we get a subfunctor $F_0 \subseteq G$ by taking every object T to the image of A in $G(T)$ under the map induced by this unique morphism when the latter exists, and to the least subalgebra of $G(T)$ (which is the image of the least subalgebra of $G(k)$) in the contrary case. Since k embeds in K, the least subalgebra of $G(k)$ embeds in the least subalgebra of $G(K)$, which is a subalgebra of A, hence every value of this functor F_0 satisfies the identities of A; so again it is reasonable that the same should be true of the least equational set-valued subfunctor $F \subseteq G$ containing it; and again one can prove that this coincides with the least equational subfunctor containing A.

But if we start with a subalgebra A of a general object $G(L)$, the least (not necessarily representable) set-valued subfunctor of G containing this is the functor associating to each T the union of the images $G(f)(A)$ as f ranges over all morphisms $L \to T$, and even for $T = L$, this union is typically not a subalgebra; so we have no reason to expect the least equational set-valued subfunctor containing it to be subalgebra-valued.

If we compare the "good" cases noted above with the topic of §31, namely the least equational set-valued subfunctor containing the image F_0 of a morphism of algebra-valued representable functors, we see that both can now be stated in the following form: Given a general algebra-valued subfunctor F_0 of a representable algebra-valued functor G, let F be the least equational set-valued subfunctor of G which contains F_0. Under what conditions will this functor also be subalgebra-valued, and satisfy the identities satisfied by the algebras $F_0(T)$? The question might be worth studying in this general form.

Let us now go back and show, as promised, that the condition "every torsion-free k-module is flat" is a characterization of Prüfer domains among commutative integral domains k.

PROPOSITION 41.13. *The 14 equivalent conditions* (α)-(ν) *and* (σ) *on a commutative integral domain* k *by which Prüfer domains are characterized in* [**35**, *Ch. VII, §2, Exercise 12*], *of which we recall in particular*

(β) *For every maximal ideal* $m \subseteq k$, *the localization* k_m *is a valuation ring,*

and

(δ) *Every finitely generated torsion-free k-module is projective,*

are also equivalent to each of

(τ) *Every torsion-free k-module is flat,*

(υ) *Every tensor product of torsion-free k-modules is torsion-free,*

(φ) *For every pair of ideals* $I,\ J \subseteq k$, *the natural surjection of k-modules* $I \otimes_k J \to IJ$ *is an isomorphism.*

PROOF. We shall prove (δ) $\Rightarrow$ (τ) $\Rightarrow$ (υ) $\Rightarrow$ (φ) $\Rightarrow$ (β).

(δ) $\Rightarrow$ (τ) because every module is a direct limit of finitely generated modules, and a direct limit of projective modules is flat.

(τ) $\Rightarrow$ (υ) because a tensor product of flat modules is flat, and a flat module over an integral domain is torsion-free.

(υ) $\Rightarrow$ (φ): Let I and J be ideals of k. If one of these is zero, the conclusion of (φ) is clear, so assume both nonzero. Then I, J and IJ are rank-1 torsion-free k-modules (where "rank" means maximal number of k-linearly independent elements), and assuming (υ), we see that $I \otimes_k J$ is, also. But any *surjective* homomorphism of torsion-free modules of *equal finite ranks* over a commutative integral domain is one-one, hence an isomorphism.

(φ) $\Rightarrow$ (β): We shall prove this in contrapositive form, showing that if (β) fails then so will (φ); specifically, that there will exist an ideal $I \subseteq k$ such that the natural map $I \otimes_k I \to I^2$ is not one-to-one. Assuming (β) false, let m be a maximal ideal of k such that the local ring k_m is not a valuation ring. Then we can find elements $x, y \in k_m$ neither of which divides the other in this ring. By clearing denominators, we can assume that $x, y \in k$. Let $I = xk + yk$. From the fact that neither x nor y divides the other in k_m, we can deduce that in the (k/m)-vector-space I/mI, their images $\bar{x}$ and $\bar{y}$ are linearly independent. Hence in $(I/mI) \otimes_k (I/mI)$, the element $\bar{x} \otimes \bar{y} - \bar{y} \otimes \bar{x}$ is nonzero, hence $x \otimes y - y \otimes x$ is nonzero in $I \otimes_k I$. But the image of this element in I^2 is $xy - yx = 0$, so $x \otimes y - y \otimes x$ is a nonzero element of the kernel of the indicated natural map. □

The technique by which the last implication was proved can be used to generalize case (ii) of Lemma 41.8 to any integral domain that is not a Prüfer domain. One can also extend case (i) of that Lemma to any commutative ring k with nilpotent elements. (Key observation: If I is a nonzero ideal of k which satisfies $I^2 = 0$, then in $T = (k/\mathrm{Ann}\, I)[s, t \mid s^2 = t^2 = 0]$, s and t each satisfy all relations satisfied by members of I, but their product st is nonzero.)

Proposition 41.5 as stated is only meaningful for k an integral domain, since it refers its field of fractions K. However, the case of greatest interest is that in which $A \subseteq G(k)$, and in this case the conclusion of the Proposition is meaningful for arbitrary k. The authors have not tried to determine the precise class of commutative rings k with zero-divisors for which the conclusion of the Proposition then holds. Reasonable candidates are the classes determined by conditions (β), (δ), and (φ) above. Condition (δ) defines the class of semihereditary commutative rings; the characterization of this class in [**14**, Theorem 4.1] (see also [**15**, §6]) shows that (δ) is strictly stronger than (β).

The three exercises below give some easy results about the general situation where

(41.14) $G\colon \mathbf{W} \to \mathbf{V}$ is a representable functor among varieties of algebras, I is the initial object of $\mathbf{W}$, A is a subalgebra of $G(I)$, and F is the least equational set-valued subfunctor of G such that $A \subseteq F(I)$.

The first will show that there is a close connection between the existence of the two types of counterexample called (a) and (b) in Lemma 41.8.

EXERCISE 41.15. *Let* G, A *and* F *be as in* (41.14), *and let* n *be a nonnegative integer. Let* $G^n\colon \mathbf{W} \to \mathbf{V}$ *be the direct product of* n *copies of the functor* G, *and let* $F' \subseteq G^n$ *be the least equational set-valued subfunctor such that* $A^n \subseteq F'(I)$. (*Thus,* $F' \subseteq F^n$.)

(i) *Suppose* F *is closed under the operations of* $\mathbf{V}$, *but does not satisfy some identity in* n *variables holding in* A. *Show that* F' *will be properly smaller than* F^n.

(ii) *Deduce, under the hypothesis of* (i), *that if* $\mathbf{V}$ *is the variety of groups, or more generally, has some binary operation* $*$ *with a neutral element* e *(given by a zeroary operation), then regarding* G^n *as a* $\mathbf{V}$*-valued functor,* F' *is* not *subalgebra-valued.* (*Hint: show that the subalgebra generated under* $*$ *by* $F'(T)$ *contains* $F^n(T)$).

Next, a positive result.

EXERCISE 41.16. *Show that in the situation of* (41.14), F *will always be closed under all derived* unary *operations of* $\mathbf{V}$, *and more generally, under all maps of the form* $x \mapsto f(x, a_2, \ldots, a_n)$, *where* f *is a derived n-ary operation of* $\mathbf{V}$ *and* $a_2, \ldots, a_n \in A$; *and that* F *will satisfy all identities in such unary operations satisfied by* A.

Finally, even when the set-valued subfunctor F of G is not closed under the operations of $\mathbf{V}$, one has

EXERCISE 41.17. *Show that in the situation of* (41.14), *there always exists a least equational* subalgebra-valued *subfunctor* $H \subseteq G$ *such that* $A \subseteq H(I)$.

Proposition 41.5 arose naturally in the context of functors on varieties $\mathbf{Comm}^1_k$, but we may ask

QUESTION 41.18. *For what sorts of varieties* $\mathbf{W}$ *other than varieties* $\mathbf{Comm}^1_k$ *(*k *a Prüfer domain) is it true that in every instance of* (41.14), *the functor* F *is necessarily closed under the operations of* $\mathbf{V}$ *and satisfies all identities of* A *?*

We have been studying *equational* subfunctors of representable functors, because homomorphic images of an algebra are easier to describe than general epimorphs. As noted at the beginning of this section, the same kinds of questions might be studied for the least (*not*-necessarily-equational) representable subfunctor $E \subseteq G$ containing a given subalgebra of $G(k)$. However, it appears that there will be fewer positive results for this class of questions. E.g., if k is a field, $G\colon \mathbf{Comm}^1_k \to \mathbf{Ab}$ the underlying additive group functor, and A the whole group $G(k)$, one can obtain from the results of [**40**] a description of E, and verify that it is not closed under the operations of G. As with equational subfunctors, one can look at the corresponding questions in the case where A is replaced by a nonrepresentable algebra-valued subfunctor of G; perhaps additional hypotheses on such a subfunctor could be found that would lead to positive results in this case.

Again, whether or not the subfunctor E constructed as above is algebra-valued, we have the analog of Exercise 41.17: there will also be a least *representable algebra-valued* subfunctor $E' \subseteq G$ containing A. Note that if in some situation we know (e.g., by Proposition 41.5) that the least representable *equational* subfunctor F containing A is subalgebra-valued, and satisfies certain identities, then E' will also satisfy those identities, since it will be a subfunctor of F.

§42. Idempotents again.

In earlier sections, we obtained two contrasting results on existence of idempotent constants in semigroup-valued functors on varieties $\mathbf{Ring}_k^1$, depending on whether the semigroups were required to be commutative: Proposition 18.9 showed for any $k \in \mathbf{Comm}^1$ that if $F: \mathbf{Ring}_k^1 \to \mathbf{AbSemigp}$ is representable and $F(k) \neq \varnothing$, then $F(k)$ contains an idempotent, while Lemma 39.2 showed that for every nonlocal commutative integral domain k, there exist representable functors $F: \mathbf{Ring}_k^1 \to \mathbf{Semigp}$ such that $F(k) \neq \varnothing$ but $F(k)$ contains no idempotent.

We claim, however, that for functors on varieties $\mathbf{Comm}_k^1$ where k is a Prüfer domain, there can be no such difference between the **Semigp**- and the **AbSemigp**-valued cases. Indeed, since every one-generator semigroup is commutative, every nonempty semigroup contains a nonempty commutative subsemigroup. Hence by an application of Proposition 41.5, a positive result on the existence of idempotents for **AbSemigp**-valued functors implies the corresponding result for **Semigp**-valued functors. (The reverse implication is trivial.)

In addition to the above equivalence, we have a relation between results for functors on $\mathbf{Comm}_k^1$ and on $\mathbf{Ring}_k^1$, since every functor on the latter category restricts to a functor on the former. In particular, from the functor constructed in the proof of Lemma 39.2, we obtain by restriction a representable functor $F: \mathbf{Comm}_k^1 \to \mathbf{Semigp}$ such that $F(k)$ is nonempty but idempotentless. If k is a Prüfer domain, we can also conclude as above that this restriction contains an **AbSemigp**-valued functor with the same properties.

Actually, it is easy to see that the restriction to $\mathbf{Comm}_k^1$ of the functor of Lemma 39.2 is itself abelian-semigroup-valued, so we do not need the assumption that k is a Prüfer domain for the last deduction. Let us also note that the proof of Lemma 39.2 used the assumption that k was an integral domain only to conclude that it had no nontrivial multiplicative idempotents, i.e., was *directly indecomposable* (not a direct product of two nontrivial rings). Hence we can assert

COROLLARY 42.1 (to proof of Lemma 39.2). *If k is a commutative integral domain (or more generally, a directly indecomposable commutative ring) which is not local, then there exists a representable functor $F: \mathbf{Comm}_k^1 \to \mathbf{AbSemigp}$ such that $F(k)$ is nonempty, but contains no idempotent.* □

What if k *is* local? A special case is answered by the following result, due essentially to Radford. (We shall see a more self-contained proof of the same result in §48.)

LEMMA 42.2 (after [**158**, Corollary 4]). *Let k be a* field, *and F a representable functor from either $\mathbf{Ring}_k^1$ or $\mathbf{Comm}_k^1$ to either* **Semigp** *or* **AbSemigp**. *Then if $F(k)$ is nonempty, it contains an idempotent.*

SKETCH OF PROOF. Let us first consider the case of $F: \mathbf{Comm}_k^1 \to \mathbf{AbSemigp}$. Following the trick of [**158**, Corollary 4], we "adjoin a neutral element to F"; that is, we take the ring R representing F, form its direct product with k, and make the result a cosemigroup in such a way that the projection map $R \times k \to k$ acts as a co-neutral-element. Let us call the resulting functor $G: \mathbf{Comm}_k^1 \to \mathbf{AbSemigp}^e$.

(Note that if a ring $T \in \mathbf{Comm}_k^1$ has no nontrivial idempotents, then every homomorphism $R \times k \to T$ factors either through the projection to R or through the projection to k; it follows that for such T, $G(T)$ is precisely the semigroup obtained by adjoining a neutral element to $F(T)$.)

Now [**158**, Theorem 1] shows that if k is a field and G is any representable functor $\mathbf{Comm}_k^1 \to \mathbf{AbSemigp}^e$ such that $G(k)$ contains no idempotent other than the neutral element e, then G admits an inverse operation making it group-valued. But in a semigroup with an adjoined neutral element, the only element that is invertible is that adjoined element, hence the $G(k)$ constructed in the preceding paragraph can be a group only if it is the trivial group; in other words $F(k)$ has no idempotents only if it is empty.

Since k is a field, hence in particular a Prüfer domain, this result for functors $\mathbf{Comm}_k^1 \to \mathbf{AbSemigp}$ implies the same result for functors $\mathbf{Comm}_k^1 \to \mathbf{Semigp}$ (by the "commutative subsemigroup" trick noted above). From these two cases, we get the corresponding results for functors with domain $\mathbf{Ring}_k^1$ by restricting such functors to the subvariety $\mathbf{Comm}_k^1$. □

We have been considering conditions on representable functors among various pairs of categories, and implications among these conditions are turning out to be important, so before going further, let us set up a notation that will allow us to name these conditions concisely.

Definition 42.3. *Suppose* $\mathbf{V}$ *is a variety of algebras, with initial object* I, *and* $\mathbf{S}$ *is a variety of semigroups. Then we shall say that*

$$\text{Idp.-in-n.e.}(\mathbf{V}, \mathbf{S})$$

(standing for "idempotent in nonempty") holds if, for every representable functor $F\colon \mathbf{V} \to \mathbf{S}$ *such that* $F(I)$ *is nonempty,* $F(I)$ *contains an idempotent element.*

Since composites of representable functors are representable, we clearly have

Lemma 42.4. *Let* $\mathbf{V}$, $\mathbf{V}'$ *denote arbitrary varieties, not necessarily of the same type, and* $\mathbf{S}$, $\mathbf{S}'$ *varieties of semigroups. Then*

(i) *If there exists a representable functor* $G\colon \mathbf{V} \to \mathbf{V}'$ *which carries the initial object of* $\mathbf{V}$ *to the initial object of* $\mathbf{V}'$, *then*

$$\text{Idp.-in-n.e.}(\mathbf{V}, \mathbf{S}) \Rightarrow \text{Idp.-in-n.e.}(\mathbf{V}', \mathbf{S}).$$

(ii) *If* $\mathbf{S}' \subseteq \mathbf{S}$, *then*

$$\text{Idp.-in-n.e.}(\mathbf{V}, \mathbf{S}) \Rightarrow \text{Idp.-in-n.e.}(\mathbf{V}, \mathbf{S}'). \ \square$$

Above, we in effect applied Lemma 42.4(i) to the forgetful functor $\mathbf{Comm}_k^1 \to \mathbf{Ring}_k^1$ and Lemma 42.4(ii) to the inclusion $\mathbf{AbSemigp} \subseteq \mathbf{Semigp}$. We can also apply Lemma 42.4(i) to functors among various categories $\mathbf{Comm}_k^1$ (respectively, $\mathbf{Ring}_k^1$), getting

COROLLARY 42.5. *Let* k *be a commutative ring and* $\mathbf{S}$ *a variety of semigroups. Then*

(i) *If* $E\colon \mathbf{Comm}_k^1 \to \mathbf{Comm}^1$ *is a representable functor, and we let* $K = E(k)$, *then*

$$\text{Idp.-in-n.e.}(\mathbf{Comm}_k^1, \mathbf{S}) \Rightarrow \text{Idp.-in-n.e.}(\mathbf{Comm}_K^1, \mathbf{S}).$$

(ii) *If* M *is a cocommutative coassociative counital* $\otimes_k$*-coalgebra* (*in notation analogous to that of §29, an object of* $\otimes_k$-**co-Comm**^{1}), *and* K *is the commutative ring* $\mathbf{Mod}_k(M, k)$ (*in the notation of §28,* $K = \hat{M} \circ k$; *in more conventional notation,* $K = \hat{M}$), *then*

$$\text{Idp.-in-n.e.}(\mathbf{Ring}_k^1, \mathbf{S}) \Rightarrow \text{Idp.-in-n.e.}(\mathbf{Ring}_K^1, \mathbf{S}).$$

(iii) *For given* k, *the class of rings* K *for which the hypothesis of* (i) *holds includes the class for which the hypothesis of* (ii) *holds, which in turn includes the class of commutative* k*-algebras* K *which are finitely generated and projective as* k*-modules.*

PROOF. Observe that if $\mathbf{V}$ is any category with initial object I, and $E\colon \mathbf{V} \to \mathbf{W}$ any functor, then E induces a functor E' from $\mathbf{V}$ to the category $\mathbf{W}'$ whose objects are objects of $\mathbf{W}$ given with morphisms of $E(I)$ into them, and whose morphisms are the obvious commuting triangles. If $\mathbf{W}$ was a variety, then this $\mathbf{W}'$ can also be regarded as a variety; if E was representable then so is E', since representability of a functor depends only on its composite with the underlying-set functor; and E' by construction carries initial object to initial object. In particular, letting $\mathbf{W} = \mathbf{Comm}_k^1$, so that $I = k$, and writing $K = E(k)$, we see that $\mathbf{W}' = \mathbf{Comm}_K^1$. Assertion (i) follows immediately by Lemma 42.4(i).

In the situation of (ii), if we denote by $E\colon \mathbf{Ring}_k^1 \to \mathbf{Ring}^1$ the functor $R \mapsto \hat{M} \circ R$, we see that the values of this functor similarly become K-algebras, so that Lemma 42.4(i) is again applicable.

Noting that for M as in (ii) the functor $\hat{M} \circ -$ restricted to $\mathbf{Comm}_k^1$ takes values in $\mathbf{Comm}^1$, we get the first inclusion of (iii). Finally, if K is a k-algebra which is finitely generated and projective as a k-module, then the dual k-module $\mathbf{Mod}_k(K, k)$ becomes a $\otimes_k$-coalgebra M, such that $K \cong \hat{M} \circ k$, giving the second inclusion. (The functor E in this case is isomorphic to $- \otimes_k K$.) □

We can use this Corollary to extend both the positive and the negative results noted earlier. On the one hand, from Radford's positive result for the case where k is a field, we get the corresponding statement for any commutative ring which is the image of a field k under a representable functor $\mathbf{Comm}_k^1 \to \mathbf{Comm}^1$. Such rings include formal power series rings in any number of indeterminates over fields (these are of the form described in case (ii) above), and also examples such as the ring of p-adic integers, which, as will be noted in §44 below, arises by applying a representable functor to the field $\mathbf{Z}/p\mathbf{Z}$. Since the latter functor changes the characteristic of this ring, it cannot arise under case (ii) above, showing that the class of K given by case (i) is strictly larger than that given by (ii).

On the other hand, we can use the above Corollary to extend the negative results of Corollary 42.1 and Lemma 39.2. In fact, point (iii) above, and ultimately the

Lemma and Corollary, were inspired by the observation of M. Takeuchi (personal communication) that if a commutative local ring k has a commutative extension ring K which is free of finite rank as a k-module, and contains no nontrivial idempotents, but is *not* local, then one can use the construction of Lemma 39.2 to get (in our present notation) counterexamples to Idp.-in-n.e.($\mathbf{Ring}^1_k$, $\mathbf{Semigp}$) and Idp.-in-n.e.($\mathbf{Comm}^1_k$, $\mathbf{AbSemigp}$).

We claim that the commutative local rings k admitting such an overring K are precisely the *non-Henselian* local rings. Indeed, by [**160**, Proposition I.5(2)], a local ring k is Henselian if and only every commutative k-algebra L which is free of finite rank as a k-module is a finite direct product of local rings. (The reader not previously familiar with Henselian local rings may take this as a definition.) From this it is immediate that if k is Henselian it has no overring K as in the preceding paragraph. On the other hand, if k is non-Henselian, let L be a counterexample to the criterion from [**160**] just cited. Being of finite rank over k, this L is a direct product of finitely many directly indecomposable k-algebras. Each of these, being a direct summand of a free k-module, will be a projective, hence a free k-module. Now by choice of L at least one of these must be non-local, giving the desired K.

We summarize the consequences of Corollary 42.5 sketched above in

PROPOSITION 42.6. *Let k be a directly indecomposable commutative ring. Then the following implications hold:*

$$\begin{array}{ccccc} & & \text{Idp.-in-n.e.}(\mathbf{Ring}^1_k, \mathbf{Semigp}) & & \\ & \nearrow & & \searrow & \\ (k = E(\mathit{field})) \ \Rightarrow \ \text{Idp.-in-n.e.}(\mathbf{Comm}^1_k, \mathbf{Semigp}) & & & & (k \ \text{is Henselian local}), \\ & \searrow & & \nearrow & \\ & & \text{Idp.-in-n.e.}(\mathbf{Comm}^1_k, \mathbf{AbSemigp}) & & \end{array}$$

where the leftmost condition means "there exists a field L and a representable functor E: $\mathbf{Comm}^1_L \to \mathbf{Comm}^1$ such that $k = E(L)$". (The hypothesis that k is directly indecomposable is needed only for the two implications into the rightmost condition.) □

We could, of course, have added the condition Idp.-in-n.e.($\mathbf{Ring}^1_k$, $\mathbf{AbSemigp}$) to the far right in the above diagram; but we already know that this holds for all k (Proposition 18.9).

We do not know examples distinguishing most of the conditions of the above Proposition, so we ask

QUESTION 42.7. *Are some or all of the four conditions forming a parallelogram in the above diagram equivalent?*

Are any of them equivalent to a property of approximability, in some sense, by algebras satisfying the leftmost condition?

Of course, we know that the implication into the bottom condition of the parallelogram is reversible if k is a Prüfer domain. (In the present context, with k local, that means a valuation ring.) Note that if we could find a Henselian local ring k and a representable functor E: $\mathbf{Comm}^1_k \to \mathbf{Comm}^1$ such that $E(k)$ was an integral domain but not Henselian local, then by Corollary 42.5(i), k would be a counterexample to the reverse of the implication out of the bottom condition of the parallelogram.

Turning back to Corollary 42.5, we may ask whether converses to any of the parts of that result hold. The converse to the first displayed implication is false. To get a counterexample, let p be a prime, and let E: $\mathbf{Comm}^1_{\mathbf{Z}/p\mathbf{Z}} \to \mathbf{Comm}^1$ be the functor taking a commutative $\mathbf{Z}/p\mathbf{Z}$-algebra S to the inverse limit of the system $\ldots \xrightarrow{\varphi} S \xrightarrow{\varphi} S \xrightarrow{\varphi} S$, where φ is the Frobenius map $x \mapsto x^p$. This functor is represented by the $\mathbf{Z}/p\mathbf{Z}$-algebra $\mathbf{Z}/p\mathbf{Z}[x, x^{1/p}, x^{1/p^2}, \ldots]$. If we take $k = L[t]$ where L is a field of characteristic p, we see that k is a nonlocal integral domain, hence Idp.-in-n.e.($\mathbf{Comm}^1_k$, $\mathbf{Semigp}$) does not hold. But the composite of the forgetful functor $\mathbf{Comm}^1_k \to \mathbf{Comm}^1_{\mathbf{Z}/p\mathbf{Z}}$ with E carries k to the largest perfect subfield K of L; and, being a field, this satisfies Idp.-in-n.e.($\mathbf{Comm}^1_K$, $\mathbf{Semigp}$).

There is a counterexample to the converse of the second displayed implication of Corollary 42.5 of a trivial sort: if $k = K_1 \times K_2$, where K_1 and K_2 are such that Idp.-in-n.e.($\mathbf{Ring}^1_{K_1}$, $\mathbf{Semigp}$) holds but Idp.-in-n.e.($\mathbf{Ring}^1_{K_2}$, $\mathbf{Semigp}$) does not, then taking $M = K_1$, with its natural $\otimes_k$-algebra structure, we get $K \cong K_1$. We then find that the left-hand side of the display in (ii) fails because K_2 is "bad", while the right-hand side holds since K_1 is "good". However, we do not know whether the converse to that display holds if we restrict attention to the case where k is an integral domain and K nontrivial, or, more generally, to the case where M is faithful as a k-module.

There is a true converse to an observation related to the last assertion of part (iii) of that Corollary and to Question 42.7. The observation, first, is that if $k \subseteq K$ are commutative local rings, with K free of finite rank as a k-module, then

$$k \text{ is Henselian} \Rightarrow K \text{ is Henselian.}$$

This follows immediately from the characterization of Henselian local rings quoted earlier. The proof of the converse, i.e., the reverse implication, was shown to us by H. W. Lenstra, Jr.; we sketch it for the reader familiar with the Henselian condition.

To show k Henselian, it suffices to prove that given a monic polynomial $f(x)$ over the local ring k, and a factorization of its image over the residue field of k into two relatively prime factors, this lifts to a factorization of f in $k[x]$. Now assuming K is Henselian, we *can* lift the factorization over the residue field to a factorization in $K[x]$; say $f = gh$; so it suffices to show that whenever we have a factorization over K of a polynomial f over k into monic factors g and h which, modulo the maximal ideal of K, are relatively prime and have coefficients in k, then the coefficients of these factors actually lie in k. Now given such f, g and h, it is not hard to show that we may replace k and K by subrings which are "finitely generated as local rings", hence are Noetherian, while preserving our hypotheses. A Hensel's Lemma type argument also shows that the coefficients of g and h can be approximated by members of k modulo every power of the maximal ideal of K. Thus their images in the finitely generated k-module K/k are "divisible" by every power of the maximal ideal of k. Hence by the Krull

Intersection Theorem, these images are zero, proving the desired result.

Turning back further, to Lemma 42.2, we remark that the method of proof of that Lemma can in fact be used to show that if k is a field, and F: $\mathbf{Comm}_k^1 \to$ **AbSemigp** is *any* representable functor other than the one represented by the zero k-algebra, then $F(k)$ has an idempotent. In other words, the condition "$F(k)$ is nonempty" is in this situation equivalent to the condition "$F(T)$ is nonempty for some nonzero k-algebra T". The point is that in the proof of the Lemma, we threw away information when we argued that $G(k)$ must be trivial, rather than noting this for all $G(T)$. However, the above result does not carry over to functors F: $\mathbf{Comm}_k^1 \to \mathbf{Semigp}$; this can be seen by taking any set-valued representable functor F with $F(k)$, but not all $F(T)$, empty, and making the sets $F(T)$ into semigroups using the "left zero" multiplication $a*b = a$. We also note that the above generalized statement for functors F: $\mathbf{Comm}_k^1 \to \mathbf{AbSemigp}$ does not hold for any integral domains k other than fields; for if k is an integral domain distinct from its field of fractions K, then K can be made a quasi-trivial co-**AbSemigp** object of $\mathbf{Comm}_k^1$ (Definition 16.8), which gives the required counterexample.

It would clearly be of interest to study the conditions Idp.-in-n.e.($\mathbf{V}$, $\mathbf{S}$) for more general $\mathbf{V}$ and $\mathbf{S}$ than those considered above. (One class of cases is trivial: if $\mathbf{V}$ is a pointed category, then Idp.-in-n.e.($\mathbf{V}$, $\mathbf{S}$) holds for all $\mathbf{S}$.)

We remark that these conditions Idp.-in-n.e.($\mathbf{V}$, $\mathbf{S}$) may be thought of as analogs of the observation that every finite nonempty semigroup has an idempotent element, and of the stronger result that every nonempty compact Hausdorff topological semigroup has an idempotent ([**88**, Proposition A-1.16]). This suggests

QUESTION 42.8. *Can one establish any general principle (or at least, any other classes of results) to the effect that if certain systems of equations have solutions in every nonempty finite algebra (or in every nonempty compact Hausdorff topological algebra) of a variety* $\mathbf{V}$, *then they also have solutions in* $F(k)$ *for every representable functor* F: $\mathbf{Comm}_k^1 \to \mathbf{V}$ *with* $F(k) \neq \varnothing$ (k *a field); or analogous statements for functors on some other "good" varieties?*

A related sort of question is suggested by the result of Radford used in the proof of Lemma 42.2: For what varieties $\mathbf{V}$ (with initial object I) and varieties $\mathbf{S}$ of semigroups with neutral element is it true that if $F(I)$ has no idempotent other than e, then F is group-valued? (This, too, is an analog of a result true for finite semigroups and compact Hausdorff semigroups.)

For simplicity, we have throughout this section compared results on varieties $\mathbf{Comm}_k^1$ with results on varieties $\mathbf{Ring}_k^1$; but let us end with a couple of easy observations about functors on the more general sort of variety K-$\mathbf{Ring}_k^1$ considered in earlier chapters.

COROLLARY 42.9. *Let* $k \in \mathbf{Comm}^1$ *and* $K \in \mathbf{Ring}_k^1$ *be such that* Idp.-in-n.e.(K-$\mathbf{Ring}_k^1$, **Semigp**) *holds. Then for any maximal commutative k-subalgebra* C *of* K, *the condition* Idp.-in-n.e.($\mathbf{Ring}_C^1$, **Semigp**) *holds. In particular, if such a* C *has a directly indecomposable direct factor* L, *then* L *is* *a Henselian local ring. More generally, the same conclusions hold if* C *is the intersection of any nonempty family of maximal commutative subalgebras of* k.

PROOF. The intersection C of a nonempty family of maximal commutative subalgebras of K will clearly be the centralizer of a subset $X \subseteq K$ containing C. (Namely, $X =$ the union of those subalgebras. Indeed, intersections of nonempty families of maximal commutative subalgebras of K are the same as subrings of K which are centralizers of subsets containing themselves, and which are thus commutative; we are using the former description as a convenient way of naming the latter class of subrings.) The construction taking each object S of $K\text{-}\mathbf{Ring}_k^1$ to the centralizer in S of this subset $X \subseteq K$ will be a representable functor $K\text{-}\mathbf{Ring}_k^1 \to \mathbf{Ring}_C^1$, carrying the initial object K of $K\text{-}\mathbf{Ring}_k^1$ to the initial object C of $\mathbf{Ring}_C^1$. Lemma 42.4(i) now gives the conclusion Idp.-in-n.e.($\mathbf{Ring}_C^1$, $\mathbf{Semigp}$). If C has a direct factor L, then $L = eC$ where e is an idempotent of C. Now the construction taking an object T of $\mathbf{Ring}_C^1$ to the annihilator in T of $1-e$ is a representable functor to $\mathbf{Ring}_L^1$, carrying initial object to initial object, and the same Lemma shows that Idp.-in-n.e.($\mathbf{Ring}_L^1$, $\mathbf{Semigp}$) holds. Hence if L is directly indecomposable, it is Henselian local by Proposition 42.6. □

In the role of $\mathbf{S}$ we have considered only the varieties $\mathbf{Semigp}$ and $\mathbf{AbSemigp}$; but we know no counterexamples to a strong generalization of Proposition 18.9:

QUESTION 42.10. *For* $K \in \mathbf{Ring}^1$, *does* Idp.-in-n.e.($K\text{-}\mathbf{Ring}^1$, $\mathbf{S}$) *hold for* every *proper subvariety* $\mathbf{S} \subset \mathbf{Semigp}$ *?*

§43. Bialgebras and Hopf algebras.

As we noted earlier, the coproduct in $\mathbf{Comm}_k^1$ of two algebras A and B is their tensor product algebra $A \otimes_k B$, the coprojection maps being given by $a \mapsto a \otimes 1$ and $b \mapsto 1 \otimes b$. This fact leads to a surprisingly symmetric description of co-$\mathbf{AbSemigp}^e$ objects R of $\mathbf{Comm}_k^1$. Given the underlying *k-module* structure of R, we see that its *unital ring* structure and *co-semigroup-with-neutral-element* structure are together given by four maps:

(43.1)

multiplication	*comultiplication*
$\mu\colon R \otimes_k R \to R$	$\mathbf{m}\colon R \to R \otimes_k R$
(satisfying associative and commutative laws)	(satisfying coassociative and cocommutative laws)
unit	*counit*
$\eta\colon k \to R$	$\varepsilon\colon R \to k$
(satisfying neutral law with respect to μ)	(satisfying coneutral law with respect to $\mathbf{m}$).

The conditions saying that the co-operations $\mathbf{m}$ and ε are homomorphisms of unital k-algebras, and not just of k-modules, also show this symmetry. Briefly, they say that

(43.2) Each of the maps on one side of (43.1) "respects" each of the maps on the other side.

(The reader who has not seen these observations before should think this through.)

If R is in fact a co-**Ab** object of $\mathbf{Comm}^1_k$, there is one more co-operation, a coinverse

$$\mathbf{i}\colon R \to R. \tag{43.3}$$

The coinverse laws for $\mathbf{i}$ with respect to $\mathbf{m}$ and ε show the same symmetry; they assert the commutativity of the diagram

$$\begin{array}{ccccc} & R\otimes_k R & \xrightarrow{\mathrm{id}_R\otimes \mathbf{i}} & R\otimes_k R & \\ \mathbf{m}\nearrow & & & & \searrow\mu \\ R & \xrightarrow{\quad\varepsilon\quad} & k & \xrightarrow{\quad\eta\quad} & R. \\ \mathbf{m}\searrow & & & & \nearrow\mu \\ & R\otimes_k R & \xrightarrow[\mathbf{i}\otimes\mathrm{id}_R]{} & R\otimes_k R & \end{array} \tag{43.4}$$

One also wants $\mathbf{i}$ to respect μ and η, but this turns out to follow from (43.4) [**183**, Prop. 4.0.1(1)-(2)].

A k-module R given with structure described by (43.1)-(43.4), i.e., a co-**Ab** object of $\mathbf{Comm}^1_k$, is called a *commutative and cocommutative Hopf algebra* over k [**183**]. If (43.3) and thus (43.4) are omitted, R is called a *commutative and cocommutative bialgebra* (because of the combined *algebra* and $\otimes_k$-*coalgebra* structures. What we are calling $\mathbf{m}$ is frequently written Δ and called the *diagonal map*, while $\mathbf{i}$, if present, is usually written S and called the *antipode*; but it would be inconvenient for us to switch to these notations here.)

A fundamental example is the functor $\mathbf{Comm}^1_k \to \mathbf{AbSemigp}^e$ giving the underlying multiplicative semigroup of a commutative k-algebra. This is represented by the polynomial algebra $k[t]$, with co-operations determined by the conditions

$$\mathbf{m}(t) = t^{\lambda}t^{\rho} = t\otimes t, \qquad \varepsilon(t) = 1. \tag{43.5}$$

Given arbitrary $F\in \mathbf{Rep}(\mathbf{Comm}^1_k, \mathbf{AbSemigp}^e)$, with representing object R, a morphism of functors from F to the above multiplicative semigroup functor corresponds to a morphism from $k[t]$ with cosemigroup structure (43.5) to $(R, \mathbf{m}, \varepsilon)$. Such a morphism will be determined by the image of t in R, which can be any element $x\in R$ satisfying $\mathbf{m}(x) = x\otimes x$, $\varepsilon(x) = 1$. As mentioned in §30, an element x of an arbitrary $\otimes_k$-co-ring with these properties is called a *grouplike* element.

To characterize general co-$\mathbf{Semigp}^e$ and co-**Group** objects of $\mathbf{Comm}^1_k$, one of course deletes the cocommutativity condition on $\mathbf{m}$ in (43.1), losing some symmetry, and getting the concept of a *commutative* but *not necessarily cocommutative* bialgebra or Hopf algebra.

If one further deletes the condition of *commutativity* on μ, one gets the general definitions of "bialgebra" and "Hopf algebra". Note that though in this case R is in general a noncommutative k-algebra, the comultiplication is still taken to be a map $R \to R \otimes_k R$, not a map into the coproduct of two copies of R in $\mathbf{Ring}_k^1$. Hence, despite the restored symmetry of these concepts, noncommutative Hopf algebras and bialgebras do not have obvious interpretations in terms of representable functors.

The symmetry considered so far has been formal. Suppose now, however, that k is a field and R a Hopf algebra or bialgebra over k, and we form the dual vector space $A = \hat{R}$. The multiplication of R induces a map $\hat{R} \to \hat{R} \hat{\otimes} \hat{R}$, which we may call a $\hat{\otimes}$*-comultiplication*. Similarly, the comultiplication of R induces a $\hat{\otimes}$*-multiplication*, and the unit and counit induce a counit and unit for A. By the symmetry of our definitions, the resulting structure on A is precisely one of *linearly compact Hopf algebra or bialgebra*, which will be *commutative* if and only if R is *cocommutative*, and vice versa. (Caveat: though the $\hat{\otimes}$-multiplication of $\hat{R}$ may be regarded as a $\otimes_k$-multiplication which is continuous with respect to the linearly compact topology, the $\hat{\otimes}$-comultiplication is not in general a $\otimes_k$-comultiplication, since the image of $\hat{R}$ may not lie in $\hat{R} \otimes_k \hat{R} \subseteq \hat{R} \hat{\otimes} \hat{R}$. So a linearly compact bialgebra is not a bialgebra, but an analog of that concept. We shall say more about this point in §47. Of course, if the original bialgebra was finite-dimensional, the linearly compact topology on its dual is discrete, so the dual will indeed again be a bialgebra.)

If R is a commutative (not necessarily cocommutative) bialgebra, representing a functor

$$F\colon \mathbf{Comm}_k^1 \to \mathbf{Semigp}^e,$$

so that $\hat{R}$ is a cocommutative (not necessarily commutative) linearly compact bialgebra, which we shall denote A, we would like to use the identification $\mathbf{Mod}_k(R, S) \cong A \hat{\otimes} S$ noted in §24 to help us study the functor F. However, $F(S)$ is smaller than $\mathbf{Mod}_k(R, S)$:

$$F(S) = \mathbf{Comm}_k^1(R, S) \subseteq \mathbf{Mod}_k(R, S) \cong A \hat{\otimes} S. \tag{43.6}$$

Note that the rightmost term of this inclusion has a structure of (not necessarily commutative) k-algebra. It is not hard to verify that the inclusion of the left-hand set in the right-hand set is an embedding of the former, as a semigroup, in the multiplicative semigroup of the latter ring.

Can we characterize this subsemigroup $F(S)$ within the ring $A \hat{\otimes} S$?

The easiest case is that in which $S = k$. An element $a \in \mathbf{Mod}_k(R, k) = A$ is a k-algebra homomorphism if and only if it respects the multiplication and neutral element; these conditions translate to say that under the $\hat{\otimes}$-coalgebra structure of A, $\mathbf{m}(a) = a \otimes a$ and $\varepsilon(a) = 1$; in other words, that a is a *grouplike* element of this $\hat{\otimes}$-coalgebra.

The next easiest case is that in which S is a finite-dimensional field extension of k. Here finite dimensionality implies that $A \hat{\otimes} S$ can be written $A \otimes_k S$. The structure of linearly compact k-bialgebra on A immediately induces a structure of linearly compact S-bialgebra on $A \otimes_k S$. It then turns out that $F(S) \subseteq A \otimes_k S$ can be characterized as the semigroup of grouplike elements of $A \otimes_k S$ with respect to this linearly compact S-bialgebra structure. It seems likely that for general S, $A \hat{\otimes} S$ can be given some sort of topological bialgebra structure over S with respect to which the subsemigroup $F(S)$ can be characterized in an analogous way.

Here is a result we can prove directly from (43.6), which, like the results of §41, constitutes a property of representable functors on $\mathbf{Comm}_k^1$ not satisfied by representable functors on $\mathbf{Ring}_k^1$.

LEMMA 43.7. *Let k be a field, and $F \in \mathbf{Rep}(\mathbf{Comm}_k^1, \mathbf{Semigp}^e)$. Then for all $S \in \mathbf{Comm}_k^1$, and $x, y \in F(S)$, one has*

(43.8) $$xy = e \ \Rightarrow\ yx = e.$$

PROOF. By (43.6) and the two sentences that follow, it suffices to prove (43.8) in the more general case where x, y are arbitrary members of $A \hat{\otimes} S$, and A is any object of $\mathbf{LCpRing}_k^1$ (not necessarily having a $\hat{\otimes}$-coalgebra structure). We know that such an A is an inverse limit of finite dimensional associative algebras ((25.48), cf. (32.1)), so we are reduced to the case where A is finite-dimensional over k. But a finite-dimensional algebra embeds in a matrix algebra $M_n(k)$, so it suffices to recall that (43.8) holds in every matrix ring $M_n(S)$ over a commutative ring S ([**44**, Theorem 2.6], [**48**, Proposition 4.4.6 (ii) and Theorem 4.4.7 (ii) pp. 143-144]; cf. [**115**, Proposition XIII.4.16, p. 518]). □

It seems likely that the above Lemma is valid for any commutative ring k, but the proof would certainly have to be modified. An example showing that the analog of the above Lemma fails for functors on $\mathbf{Ring}_k^1$, is given by the underlying multiplicative semigroup functor on that category, when applied to $S = k\langle x, y \mid xy = 1\rangle$.

Let us motivate a conjecture which we suspect might also be provable by Hopf algebra methods. Let k be a field and R a cogroup in $\mathbf{Comm}_k^1$, and again let $A = \hat{R} = \mathbf{Mod}_k(R, k)$. By (25.48), A, regarded as a linearly compact k-algebra, is an inverse limit of finite-dimensional k-algebras A', and for each such A' one knows that $A'/J(A')$ is a finite product of finite-dimensional simple k-algebras. One can deduce that $A/J(A)$ is a (possibly infinite) direct product of finite-dimensional simple algebras. If R is cocommutative, so that A is commutative, and if k is algebraically closed, $A/J(A)$ will thus be a direct product of copies of k. Now the group of units of $A \hat{\otimes} S$ ($S \in \mathbf{Comm}_k^1$) will be an extension of the quasimultiplicative group of $J(A) \hat{\otimes} S$ by the group of units of $(A/J(A)) \hat{\otimes} S$. Here $J(A)$ is an inverse limit of nilpotent k-algebras, and for such algebras, the structure of the quasimultiplicative group is strongly influenced by the characteristic of the base field k. From such considerations we suspect that one should be able to prove

CONJECTURE 43.9. *Let* k *be an algebraically closed field. If* $\operatorname{char} k = p > 0$, *let* $\mathbf{Ab}'$ *denote* $\mathbf{Mod}_{\mathbf{Z}[p^{-1}]}$, *while if* $\operatorname{char} k = 0$, *let us take* $\mathbf{Ab}' = \mathbf{Mod}_{\mathbf{Z}/n\mathbf{Z}}$ *for an arbitrary positive integer* n. *In either case, we conjecture that every representable functor* $\mathbf{Comm}_k^1 \to \mathbf{Ab}'$ *has the form*

$$\mathbf{Comm}_k^1 \xrightarrow{\text{group of units}} \mathbf{Ab} \xrightarrow{\mathbf{Ab}(G,\,-)} \mathbf{Ab}', \tag{43.10}$$

for some $G \in \mathbf{Ab}'$, *and that this gives an equivalence of categories* $\mathbf{Rep}(\mathbf{Comm}_k^1, \mathbf{Ab}') \cong \mathbf{Ab}'^{\mathrm{op}}$.

Note that the composite functor (43.10) is represented by the group algebra kG. If $G \in \mathbf{Ab}'$ is a torsion group and k algebraically closed, then kG is generated by idempotents; one can deduce that if $\operatorname{char} k = 0$, a functor (43.10) is of the form (40.9); thus Conjecture 43.9 is related to Problem 40.8.

If the above conjecture is true, then by studying bilinear maps among such functors, one may be able to get characterizations of representable functors from $\mathbf{Comm}_k^1$ to appropriate varieties of rings.

In previous sections we have at times considered sub*groups* of semigroups. One can also look at sub*semigroups* of groups. W. D. Nichols proves in [**139**] by Hopf algebra methods that (inter alia) any nonempty algebraic subsemigroup of an affine algebraic group over a field k is again an algebraic *group*. (This is used in getting the result of [**158**] cited in the proof of Lemma 42.2.)

We remark that the strange construction of representable semigroup-valued functors on associative rings that we discovered in §§35-37 leads to exotic functors on commutative algebras as well. For example, we may start with the functor $\mathbf{Comm}_k^1 \to (k\times k)\text{-}\mathbf{Ring}_k^{\mathbf{d},1}$ given by

$$S \mapsto \begin{bmatrix} S[[t]] & tS[[t]] \\ S[[t]] & S[[t]] \end{bmatrix},$$

and give the resulting sets the semigroup structure described by (35.10); or to keep things simpler, we might replace $S[[t]]$ above by $S[t]/(t^n)$ for some n. So far as we have been able to learn, these constructions and the corresponding bialgebras have not been noted before.

43.11. Let us end this section with a brief sketch of another important role that $\otimes_k$-coalgebras and bialgebras play in ring theory. Let k be a commutative ring, R and S objects of $\mathbf{Ring}_k$, and $(C, \mathbf{m})$ a $\otimes_k$-coalgebra. Let us define a $(C, \mathbf{m})$-*parametrized* (or when there is no danger of ambiguity, C-*parametrized*) map $R \to S$ to mean a k-bilinear map $C \times R \to S$ (which we shall write as taking $c \in C$, $x \in R$ to $c(x) \in S$) with the property that if $a \in C$ with $\mathbf{m}(a) = b_1 \otimes c_1 + \dots + b_n \otimes c_n \in C \otimes_k C$, then for all $x, y \in R$,

$$a(xy) \;=\; b_1(x)c_1(y) + \dots + b_n(x)c_n(y). \tag{43.12}$$

Easy examples: If C is the $\otimes_k$-coalgebra free as a k-module on one generator f, with $\mathbf{m}(f) = f \otimes f$, then (43.12) says that $f(xy) = f(x)f(y)$, and we see that a C-parametrized map $R \to S$ is equivalent to a *k-algebra homomorphism* $R \to S$, namely $x \mapsto f(x)$. If C is the $\otimes_k$-coalgebra free as a k-module on two generators

f and d, with comultiplication again taking f to $f \otimes f$, and taking d to $d \otimes f + f \otimes d$, then a C-parametrized map $R \to S$ is equivalent to a k-algebra homomorphism $f\colon R \to S$ together with a derivation $d\colon R \to S$ with respect to that homomorphism. (For simplicity, we are abusing notation and using the same symbol for the elements f, d of these coalgebras, and the homomorphism and derivation they induce.)

We see from these examples that a $\otimes_k$-coalgebra can be thought of as encoding a "product law" (43.12) for a family of k-module maps from one k-algebra to another. If R and S are in fact unital k-algebras, and C has a counit ε, one generally adds to (43.12) a "unitality" condition, which can be thought of as describing the action of members of C on the "empty product":

(43.13) $$a(1_R) = \varepsilon(a)1_S \qquad \text{for all } a \in C.$$

(In the preceding two examples, C does admit a counit ε, namely the map which takes f to 1, and, in the second case, d to 0. We see that in these cases, (43.13) states the usual unitality conditions for homomorphisms and derivations.)

Let us now consider the special situation $R = S$. Here a C-parametrized family as discussed above consists of maps of R into itself, so it is natural to supplement our "product law" (43.12) with a "composition law", specifying the composite of the actions of each pair of elements of C, (and also, perhaps, with an "identity law", specifying an element of C which should act as the identity map). This corresponds to a *multiplication* μ on C (and possibly a unit η), supplementing the comultiplication $\mathbf{m}$. Assuming appropriate compatibility conditions, $(C, \mu, \mathbf{m})$ will then be a bialgebra in the sense defined at the beginning of this section (and η and ε, if present, a unit, respectively a counit). A map $C \times R \to R$ satisfying the conditions sketched above with respect to both $\mathbf{m}$ and μ is called an *action* of this bialgebra on the k-algebra R. For example, a group algebra kG has a natural bialgebra structure such that an action of this bialgebra on a k-algebra R is equivalent to the standard concept of an action of the group G on R by k-algebra automorphisms. (Incidentally, in this case if k is an integral domain, the original group G can be characterized within kG as the set of elements satisfying $\mathbf{m}(x) = x \otimes x$, $\varepsilon(x) = 1$. This is the origin of the term "grouplike" for such elements in a general $\otimes_k$-coalgebra.) If G is finite, we recall that the linear dual of kG will again be a bialgebra; an action of this bialgebra corresponds to a *G-grading* of R. There is also a concept of a *coaction* of a bialgebra C on an algebra R, a map $R \to R \otimes_k C$ satisfying appropriate conditions; in terms of this concept, one finds that a grading by an *arbitrary* group G can be described as a coaction of the group algebra kG, while for *finite* G, an action of G by automorphisms can be described as a coaction of the dual bialgebra. (As usual, these finiteness restrictions can be gotten around if we are willing to introduce topological structure.)

The universal enveloping algebra $k[L]$ of a Lie algebra L similarly has a bialgebra structure, an action of which on a k-algebra corresponds to the classical concept of an action of L on R by derivations. (Here the comultiplication $k[L] \to k[L] \otimes_k k[L] \cong k[L \times L]$ is induced by the diagonal homomorphism of Lie algebras $L \to L \times L$; it takes $d \in L$ to $1 \otimes d + d \otimes 1$. We remark that in a general bialgebra, an element x such that $\mathbf{m}(x) = 1 \otimes x + x \otimes 1$ is called *primitive*.)

The concepts of bialgebra and Hopf algebra are discussed from this point of view in [**21**, §§1-7]; see also [**134**]. This viewpoint also provides a natural context for formalizing and generalizing Galois theory; see [**39**], [**74**], [**187**]. The idea that $\otimes_k$-coalgebras are the natural objects for "parametrizing" various algebraic structures is developed extensively in [**75**] (the reader of which needs to be familiar, however, with cartesian closed categories [**122**, §4.6] or [**110**] and indexed categories [**143**] and [**123**]). See also [**144**]. For the analog of this idea in more general categories than those of modules and algebras, see [**66**].

We remark that for C a $\otimes_k$-coalgebra, and R, S k-algebras, a k-bilinear map $C \times R \to S$ can be regarded as a k-linear map $R \to \mathbf{Mod}_k(C, S) \cong \hat{C} \hat{\otimes} S$. When we make this translation, we find that the condition (43.12) on the original map becomes the condition that the new map $R \to \hat{C} \hat{\otimes} S$ be a homomorphism of k-algebras. (Unfortunately, if we apply this translation to an *action* $C \times R \to R$ for C a bialgebra, though it makes the role of the coalgebra structure of C formally simpler, it makes that of the algebra structure less so.)

Some universal constructions: If M is a $\otimes_k$-coalgebra, then its coalgebra structure induces a coalgebra structure on the tensor k-algebra $k\langle M\rangle$ and on the symmetric algebra $k[M]$, making these into bialgebras which are universal among bialgebras, respectively commutative bialgebras, with coalgebra maps of M into them. The structure on $k\langle M\rangle$ has the property that an action of this bialgebra on a k-algebra R is equivalent to a system of k-module maps $R \to R$ parametrized (in the sense discussed above) by the coalgebra M; actions of $k[M]$ correspond to such systems satisfying the condition that all these endomorphisms commute with one another. As a cosemigroup in $\mathbf{Comm}^1_k$, $k[M]$ represents the functor associating to every commutative k-algebra A the whole multiplicative semigroup of $\mathbf{Mod}_k(M, A)$ under convolution. Incidentally, if the comultiplication on M is not assumed counital, the comultiplications on $k\langle M\rangle$ and $k[M]$ used in the above constructions can be modified slightly so that they become counital, with the unique augmentation annihilating M for counit. In this case, assuming k a field, $k\langle M\rangle$ and $k[M]$ will admit coinverse operations (making them Hopf algebras) if and only if the linearly compact k-algebra $\hat{M}$ is Jacobson radical, equivalently, pro-nilpotent.

The above constructions $k[M]$ and $k\langle M\rangle$ can be called the free (commutative and noncommutative) bialgebras on the coalgebra M. There is a dual concept of the *cofree* bialgebra on an *algebra*; see [**138**], and cf. [**67**].

We noted earlier that if k is a field, duality of vector spaces gives an equivalence between Hopf algebras over k and linearly compact Hopf algebras over k. In particular, *cocommutative* Hopf algebras correspond to *commutative* linearly compact Hopf algebras; and these can be identified with cogroup objects in the category of linearly compact commutative k-algebras. Now to the algebraic geometer, a linearly compact commutative k-algebra A has a "formal spectrum", Spf(A), which, if connected, looks like the "germ" of an algebraic variety. Hence a cogroup object in the category of such algebras will correspond to a "germ of an algebraic group". There is, in fact, a result on Hopf algebras corresponding to the geometric idea that the "germ" of a Lie group is determined by its Lie algebra; namely, if k has characteristic 0, every connected cocommutative Hopf algebra over k is the universal enveloping algebra $k[L]$ of a Lie algebra L, yielding an equivalence between the categories of such Hopf algebras, and of Lie algebras [**183**, §1.30]. (For a graded version of this result, the Milnor-Moore Theorem, see [**157**, Appendix B, Theorem 4.5, p.286].)

A standard reference for the general theory of Hopf algebras and bialgebras is [**183**]; a brief introduction is [**108**]; for the theory of actions of Hopf algebras see [**134**]. Certain classes of noncommutative noncocommutative Hopf algebras have recently achieved popularity under the name "quantum groups"; cf. [**177**], [**62**]. Although, as noted there, the relation of this concept to quantum mechanics is rather distant, an application of Hopf algebras to that field is proposed in [**125**].

§44. The Witt vector construction.

We shall sketch in this section the construction, alluded to in §40.2, of a representable functor $W\colon \mathbf{Comm}^1 \to \mathbf{Comm}^1$ with the property that the *additive* group of $W(S)$ $(S \in \mathbf{Comm}^1)$ is functorially isomorphic to the *multiplicative* group of formal power series

$$1 + \xi_1 t + \xi_2 t^2 + \dots \qquad (\xi_i \in S). \tag{44.1}$$

In particular, this construction will have the property that the ring $W(S)$ is without additive torsion if S is without nilpotent elements, regardless of the characteristic of S.

Let us begin heuristically by assuming S an integral domain, and considering a special class of formal power series (44.1), namely those which have only finitely many nonzero terms, and which split into linear factors:

$$1 + \xi_1 t + \dots + \xi_m t^m = (1 - x_1 t) \dots (1 - x_m t) \qquad (x_i \in S). \tag{44.2}$$

Consider two such series,

$$\begin{aligned} 1 + \dots + \xi_m t^m &= \textstyle\prod_{i=1,\dots,m} (1 - x_i t), \\ 1 + \dots + \eta_n t^n &= \textstyle\prod_{j=1,\dots,n} (1 - y_j t), \end{aligned} \tag{44.3}$$

and let f be any polynomial in two indeterminates with integer coefficients. Then if we form the power series

$$\textstyle\prod_{i \le m,\, j \le n} (1 - f(x_i, y_j) t) = 1 + \zeta_1 t + \dots + \zeta_{mn} t^{mn}, \tag{44.4}$$

we see that the coefficients ζ_h are symmetric polynomials in the x_i's, and likewise in the y_i's. It is easily deduced from standard facts about symmetric polynomials [**115**, Theorem IV.6.1, p.191], [**89**, Proposition V.2.20] that each ζ_h can be written as a polynomial in the ξ's and η's; moreover, we see by degree considerations that this polynomial involves only those ξ_i, η_j with $i \le h{\cdot}\deg_x f$, $j \le h{\cdot}\deg_y f$. These polynomials will in general depend on the m and n of (44.3).

Note that we can also write the power series (44.3) as products of larger numbers of terms, m' and n', by introducing on the right-hand sides the required number of trivial factors $(1 - 0t)$. Moreover, if the polynomial f satisfies

(44.5) $$f(0, y) = 0 = f(x, 0),$$

then we see that the value of the left-hand side of (44.4) will be unaffected by this change. Thus, in this case, the expression for each coefficient ζ_h of (44.4) in terms of the coefficients of the series (44.3) must be independent of m and n (at least for $m \geq h \cdot \deg_x f$, $n \geq h \cdot \deg_y f$, so that the set of coefficients on which ζ_h depends does not vary with m and n). Hence fixing such an f, and letting m and n go to infinity, we get a binary operation $*$ on formal power series, which determines each coefficient of the new power series as a polynomial in finitely many of the coefficients of the two given series, and which, when applied to a pair of series of the form (44.3) (for arbitrary m and n) gives (44.4).

Putting this binary operation to the side for the moment, suppose we symbolize (44.2) as

$$[x_1] + \ldots + [x_n].$$

This looks like an expression for an element of the semigroup ring $\mathbf{Z}\, S^{\mathrm{mult}}$ of the multiplicative semigroup of S. More precisely (since factors $(1 - 0t)$ make no difference in (44.2)), we can think of it as like an element of the modified semigroup ring $(\mathbf{Z}\, S^{\mathrm{mult}})_0$, where the functor $(\mathbf{Z}\, -)_0$, defined on the variety $\mathbf{Semigp}^z$ of semigroups with 0, is the left adjoint to the underlying multiplicative-semigroup-with-0 functor $\mathbf{Ring}^1_k \to \mathbf{Semigp}^z$, and is constructed by forming the ordinary semigroup ring $\mathbf{Z}\, S$, and dividing by the ideal $[0]\mathbf{Z}$. Note that multiplication of formal power series (44.3) corresponds to addition in $(\mathbf{Z}\, S^{\mathrm{mult}})_0$.

Returning to our binary operation $*$, let us fix $f(x, y)$ as the polynomial xy, which clearly satisfies (44.5). Then we see that $*$, which takes elements (44.3) to (44.4), will correspond to *multiplication* in $(\mathbf{Z}\, S^{\mathrm{mult}})_0$. Now formal power series of the form (44.2) (with m unbounded) are "dense" in arbitrary formal power series (44.1), i.e., the coefficients ξ_i do not satisfy identically any nontrivial polynomial relations. We can deduce that the binary operation $*$ we have constructed is associative and commutative, and satisfies the distributive law with respect to multiplication of formal power series, because these laws hold on a dense set. (Indeed, we can apply Corollary 41.4, with $k = \mathbf{Z}$, $F(S) = \{$formal power series (44.1)$\}$, and $A = \{$formal power series (44.2) over $\mathbf{Z}\}$.) Denoting by $W(S)$ the set of power series (44.1), and renaming the operations of multiplication and $*$ on such series "addition" and "multiplication", we see that the power series 1 and $1 - t$ will be a zero and a unit for this structure, and that the operation of multiplicative inverse of formal power series, though it does not preserve our dense subset (44.2), will be an "additive inverse". Thus we get commutative ring structures on the sets $W(S)$.

Within $W(S)$, the set of series of the form (44.2) behaves like the set of elements of $(\mathbf{Z}\, S^{\mathrm{mult}})_0$ in which the coefficients of all basis elements are nonnegative. This subset is not a subring, but if we pass to power series of the more general form $\prod_{i \leq m} (1 - x_i t) / \prod_{j \leq n} (1 - y_j t)$, it is immediate that these do form a subring of $W(S)$, and that if S is an integral domain, this subring is isomorphic to the semigroup ring $(\mathbf{Z}\, S^{\mathrm{mult}})_0$. Thus, the functor W can be thought of as a kind of "representable approximation" to the functor $S \mapsto (\mathbf{Z}\, S^{\mathrm{mult}})_0$.

If S is a field, then the larger class of elements of $W(S)$ given by all power series over S with constant term 1 that represent rational functions (but with numerator and denominator not necessarily splitting over S) can be identified with a subring of $(\mathbf{Z}\,K^{\mathrm{mult}})_0$, where K is the algebraic closure of S, consisting of those elements of this semigroup ring which are invariant under the action of $\mathrm{Gal}(K/S)$, and in which the coefficient in $\mathbf{Z}$ of each $x \in K$ is divisible by the inseparability degree of x over S.

A general phenomenon in the study of group rings is the occurrence of idempotents. These arise from subgroups of finite order, which in the above setting means finite groups of roots of unity. Suppose ω is a primitive nth root of unity in an *extension* K of the ring S, and let $e = [1] + [\omega] + \ldots + [\omega^{n-1}]$. Then $e^2 = ne$, hence if we can divide by n, the resulting element $n^{-1}e$ will be idempotent. This element corresponds to the formal power series

$$(44.6) \qquad \prod_i (1 - \omega^i t)^{n^{-1}} \;=\; (1 - t^n)^{n^{-1}}.$$

If n is invertible in S, this is evidently a well-defined formal power series, not merely over K but over S, giving an idempotent element of $W(S)$. These idempotents, taken together as n ranges over the positive integers invertible in S, lead to a direct-product decomposition of $W(S)$. (An infinite family of idempotents does not always yield an infinite direct product decomposition of a ring; topological structure is involved here. We will not go into details; the remainder of this section is thus a sketch.) If S is a $\mathbf{Q}$-algebra, so that we have the largest possible family of such idempotents, then the ring $W(S)$ decomposes into a direct product of rings isomorphic to S; in other words, W restricted to $\mathbf{Comm}^1_{\mathbf{Q}}$ is a countable direct product of copies of the identity functor on this category. The projection maps with respect to this decomposition, restricted to the semigroup ring $(\mathbf{Z}S^{\mathrm{mult}})_0$, are the homomorphisms $\pi_n\colon (\mathbf{Z}S^{\mathrm{mult}})_0 \to S$ $(n = 1, 2, \ldots)$ determined by the equations $\pi_n([s]) = s^n$ $(s \in S)$. That the resulting map

$$(44.7) \qquad \pi = (\pi_1, \pi_2, \ldots)\colon\ W(S) \;\to\; S \times S \times \ldots$$

is bijective is essentially Newton's result that the elementary symmetric functions in a finite set of variables can be expressed in terms of the power-sum ("moment") functions, using polynomials with rational coefficients! If S is any commutative integral domain of characteristic 0, (44.7) is at least an embedding, so the idempotent elements of $W(S)$ can be described as those elements whose images under π are strings of 1's and 0's. (The reader might calculate the images under π of the idempotents (44.6).)

Over $\mathbf{Comm}^1_{\mathbf{Z}/p\mathbf{Z}}$, we have idempotents corresponding to all integers n relatively prime to p; these turn out to decompose $W(S)$ into a direct product of countably many copies of a ring $W_p(S)$, called the ring of *Witt vectors* over S. The functor $W_p\colon \mathbf{Comm}^1_{\mathbf{Z}/p\mathbf{Z}} \to \mathbf{Comm}^1$ has the remarkable property of taking any perfect field S of characteristic p to the unique complete discrete valuation ring A having maximal ideal generated by the prime p, and residue field $A/pA \cong S$. Thus, it takes $\mathbf{Z}/p\mathbf{Z}$ to the ring of p-adic integers (showing that this local ring satisfies the leftmost condition in the diagram of Proposition 42.6, as mentioned earlier).

For detailed developments of this functor, see [**36**, §IX.9.1] or [**136**, §26]. In the latter reference, another interpretation of the coordinates of $W_p(S)$ is described when S is a perfect field such as $\mathbf{Z}/p\mathbf{Z}$, as digits in a modified "base p expansion" of elements of this ring. Cf. also [**37**], [**83**, Chapter III], [**84**].

(The term "Witt ring" also has an unrelated meaning in the theory of quadratic forms; cf. [**49**, p.322].)

Another interesting construction based on formal power series (which we shall not discuss here) is the representable functor $\mathbf{Comm}^1 \to \mathbf{Group}$ taking a ring S to the group of formal power series of the form $t + \xi_2 t^2 + \xi_3 t^3 + \ldots$ $(\xi_i \in S)$ under *formal composition* (substitution of one power series into another [**103**]).

§45. The co-ring of integral polynomials.

A polynomial with rational coefficients $f(x_1, \ldots, x_n) \in \mathbf{Q}[x_1, \ldots, x_n]$ is called *integral* if it assumes values in $\mathbf{Z}$ for all n-tuples of arguments in $\mathbf{Z}$. A familiar example of such a polynomial in one indeterminate which does not lie in $\mathbf{Z}[x]$ is $\binom{x}{2} = x(x-1)/2$. Let us begin by establishing some well-known facts:

LEMMA 45.1. *The ring* $\mathrm{Int}[x]$ *of integral polynomials in one indeterminate* x *is free as an additive group on the basis*

$$\{\binom{x}{0}, \binom{x}{1}, \binom{x}{2}, \ldots\},$$

where $\binom{x}{d} = x(x-1)\ldots(x-d+1)/d!$.

The ring $\mathrm{Int}[x_1, \ldots, x_n]$ *of integral polynomials in* n *indeterminates* $x_1, \ldots, x_n$ *is generated by its subrings* $\mathrm{Int}[x_m]$ $(m = 1, \ldots, n)$, *and is naturally isomorphic to their tensor product:*

$$\mathrm{Int}[x_1, \ldots, x_n] \cong \mathrm{Int}[x_1] \otimes_{\mathbf{Z}} \ldots \otimes_{\mathbf{Z}} \mathrm{Int}[x_n].$$

PROOF. First assertion: It is well-known that the indicated binomial coefficient functions are integral polynomials. Note that the polynomial $\binom{x}{d}$, evaluated at $0, 1, \ldots, d$, has values $0, 0, \ldots, 0, 1$. It follows that for any integers $a_0, \ldots, a_d$, one can find a linear combination of $\binom{x}{0}, \ldots, \binom{x}{d}$ with integer coefficients having values precisely $a_0, \ldots, a_d$ at $0, \ldots, d$. (One first determines the coefficient of $\binom{x}{0}$, then that of $\binom{x}{1}$, etc..) Thus if $f \in \mathbf{Q}[x]$ is an integral polynomial of degree d, we can find a linear combination g of these binomial coefficient functions such that $f - g$ has zeros at $0, \ldots, d$. Since the only polynomial of degree $\leq d$ with $> d$ zeros is 0, $f = g$.

We could prove the second assertion by using a generalization of this argument to show that $\mathrm{Int}[x_1, \ldots, x_n]$ is spanned by all products $\binom{x_0}{d_0} \ldots \binom{x_n}{d_n}$. However, an induction on n, using little more than the fact that $\mathrm{Int}[x_1, \ldots, x_{n-1}]$ is a free $\mathbf{Z}$-module, will be more instructive.

Suppose $f \in \mathrm{Int}[x_1, \ldots, x_n]$. Then we see that for each $r \in \mathbf{Z}$, we will have

$f(x_1, \dots, x_{n-1}, r) \in \mathrm{Int}[x_1, \dots, x_{n-1}]$, and that the resulting set-map $\mathbf{Z} \to \mathrm{Int}[x_1, \dots, x_{n-1}]$ is given by a "polynomial in r with rational coefficients"; i.e., that if B is a $\mathbf{Z}$-basis of $\mathrm{Int}[x_1, \dots, x_{n-1}]$ (which we inductively assume to exist), then for each $b \in B$, the integer giving the coefficient of b in the expression for $f(x_1, \dots, x_{n-1}, r)$ in terms of that basis varies with r as a polynomial with rational coefficients. Thus, these coefficient polynomials belong to $\mathrm{Int}[x_n]$. Also, if we take a finite subset B_0 of B whose span includes all integral polynomials in $x_1, \dots, x_{n-1}$ of degree $\leq \deg(f)$, we see that only for members of B_0 can this coefficient-polynomial be nonzero. This shows that $\mathrm{Int}[x_1, \dots, x_n]$ is a finite sum of products of members of $\mathrm{Int}[x_1, \dots, x_{n-1}]$ and $\mathrm{Int}[x_n]$. The generation part of the desired assertion follows. That the map from the tensor product to the ring of integral polynomials in n indeterminates is one-to-one may be seen from the corresponding fact about the full polynomial algebras over $\mathbf{Q}$, and the flatness of our rings of integral polynomials as $\mathbf{Z}$-modules. □

Now if $f(x) \in \mathrm{Int}[x]$, it is clear that $f(x+y)$ and $f(xy)$ belong to $\mathrm{Int}[x, y]$, and we have just seen that this can be identified with $\mathrm{Int}[x] \otimes \mathrm{Int}[y]$. We easily deduce

LEMMA 45.2. $\mathrm{Int}[x]$ *admits a unique co-ring structure with co-operations satisfying*

$$\mathbf{a}(x) = x^{\lambda} + x^{\rho}, \qquad \mathbf{m}(x) = x^{\lambda} x^{\rho}. \tag{45.3}$$

The co-additive inverse, cozero and counit co-operations of this functor are characterized by

$$\mathbf{i}(x) = -x, \qquad \mathbf{z}(x) = 0, \qquad \varepsilon(x) = 1. \quad □ \tag{45.4}$$

Note that this characterization of the co-ring structure says that the inclusion

$$\mathbf{Z}[x] \subseteq \mathrm{Int}[x] \tag{45.5}$$

is a morphism of co-rings with respect to the natural co-ring structure on $\mathbf{Z}[x]$ (the structure defined by (45.3) and (45.4)).

Let us denote by Bi (for "binomial") the functor $\mathbf{Comm}^1 \to \mathbf{Comm}^1$ represented by $\mathrm{Int}[x]$ with this co-ring structure. Then (45.5) gives a natural morphism

$$e\colon \mathrm{Bi} \to \mathrm{Id}_{\mathbf{Comm}^1}.$$

We leave to the interested reader the verification of some of the curious properties of the functor Bi:

EXERCISE 45.6. *Let* $S \in \mathbf{Comm}^1$.

(i) *Show that if* $\mathbf{Q} \subseteq S$, *then* $e(S)$ *is an isomorphism. (I.e.,* Bi *takes every* $\mathbf{Q}$*-algebra to itself.)*

(ii) *Show that if the additive group of* S *is torsion-free, then* $e(S)$ *is one-to-one, and* $\mathrm{Bi}(S)$ *can be identified with the set of elements of* S *which are carried into* S *by all the binomial coefficient functions* $\binom{x}{i}$, *regarded as polynomial maps*

$S \otimes \mathbf{Q} \to S \otimes \mathbf{Q}$.

(iii) *Show that if* S *is the ring* $\mathbf{Z}[i]$ *of Gaussian integers, or the polynomial ring* $\mathbf{Z}[t]$, *then* $\mathrm{Bi}[S] = \mathbf{Z} \subseteq S$. *(Hint: to show that a Gaussian integer* $a+bi$ *with* $b \neq 0$ *is not in* $\mathrm{Bi}(S)$, *let* p *be a prime which does not divide* b *and which remains prime in the Gaussian integers, and show that* $\binom{a+bi}{p}$ *is not a Gaussian integer.) To how large a class of rings* S *can you generalize this result?*

(iv) *Show that if* S *is any of* $\mathbf{Z}[½]$, $\mathbf{Z}[½, i]$ *or* $\mathbf{Z}[½, t]$, *then* $\mathrm{Bi}(S)$ *is the subring* $\mathbf{Z}[½]$.

(v) *Show that if* S *is* any *integral domain of characteristic* $p > 0$, *then* $\mathrm{Bi}(S)$ *can be identified with the ring of p-adic integers. (Idea: show that a homomorphism* $\mathrm{Int}[x] \to S$ *is determined by the images of* x, $\binom{x}{p}$, $\binom{x}{p^2}$, ..., *and that the only possible values of these images are* $0, \ldots, p-1$. *To show that the ring structure on these strings of elements is that of the p-adic integers, examine the natural map* $\mathrm{Bi}(\mathbf{Z}) \to \mathrm{Bi}(\mathbf{Z}/p\mathbf{Z})$.)

(vi) *Deduce that for any nonzero* S, *the ring* $\mathrm{Bi}(S)$ *is of characteristic* 0.

(vii) *Show that for* W *the functor described in the preceding section,* Bi *may be identified with the least equational subfunctor* $F \subseteq W$ *such that* $F(\mathbf{Z})$ *contains the image of* $\mathbf{Z}$ *in* $W(\mathbf{Z})$ *(i.e., the least subring of* $W(\mathbf{Z})$).

(viii) *Show that for any prime* p, *the restriction of* Bi *to* $\mathbf{Comm}^1_{\mathbf{Z}/p\mathbf{Z}}$ *factors through* $\mathbf{Bool}^1$ *as in* (40.9).

Incidentally, the ring $\mathrm{Int}[x]$ is non-Noetherian. For example, if we let I_r denote the ideal consisting of those integral polynomials that take on even values at all multiples of 2^r, then $I_0 \subseteq I_1 \subseteq \ldots$ forms an infinite strictly increasing chain, as may be seen by looking at the elements $\binom{x}{2^n}$. This may be one reason why the above co-ring has received little attention.

In this and preceding sections, we have considered some interesting representable functors $\mathbf{Comm}^1_k \to \mathbf{Comm}^1$. The authors do not know whether any essentially new phenomena arise when one considers representable functors $\mathbf{Comm}^1_k \to \mathbf{Ring}^1$. It is not clear how to pose this question formally. A large class of representable functors $\mathbf{Comm}^1_k \to \mathbf{Ring}^1$ that we would call "not essentially new" (relative to commutative-ring-valued representable functors) are those that can be factored

$$\mathbf{Comm}^1_k \to \mathbf{Comm}^1_K \to \mathbf{Ring}^1, \tag{45.7}$$

where $K \in \mathbf{Comm}^1$, the first arrow is an arbitrary representable functor (e.g., any of the constructions considered in the above sections), and the second arrow is induced by a counital coassociative $\otimes_k$-coalgebra (equivalently, it consists of the inclusion of $\mathbf{Comm}^1_K$ in $\mathbf{Ring}^1_K$, followed by a representable functor from this category to $\mathbf{Ring}^1$; cf. Theorem 29.5). But there are representable functors that do not appear to be of this precise form, but are closely related. For instance, if k is of characteristic p, the functor associating to every $S \in \mathbf{Comm}^1_k$ the subring of $M_2(S\langle\langle x, y\rangle\rangle) \times M_2(W_p(S))$ consisting of pairs of upper triangular matrices whose

(1, 1) entries have the same images under the natural maps

$$(45.8) \qquad S\langle\langle x, y\rangle\rangle \xrightarrow{\text{constant term}} S \xleftarrow{\pi_1 \text{ (see (44.7))}} W_p(S)$$

is easily seen to be representable, but probably cannot be factored as in (45.7). Another example is the functor $S \mapsto \begin{pmatrix} S[[t]] & S \\ 0 & W_p(S) \end{pmatrix}$, where S is made an $(S[[t]], W_p(S))$-bimodule via essentially the two maps (45.8). Still another class of constructions are those obtained via Boolean rings, as described in §40.4 above. So though the authors do not know of any representable functors $\mathbf{Comm}^1_k \to \mathbf{Ring}^1$ that do not arise in straightforward ways from $\mathbf{Comm}^1$-valued functors, neither do we have a conjecture on how to express all functors of the former sort in terms of the latter.

Co-ring objects in categories of commutative rings also arise in algebraic topology; see [**159**].

§46. Generalized integral polynomials.

The fact that the ring of integral polynomials is strictly larger than the ring of polynomials with integer coefficients can be looked at as a consequence of a special fact about the ring $\mathbf{Z}$, namely that it has homomorphic images which are finite rings. Let us, for simplicity, prove this relationship in the context of unique factorization domains, though more general statements can be given.

LEMMA 46.1. *Let k be a unique factorization domain, K its field of fractions, and $\mathrm{Int}_k[x]$ the ring of elements of $K[x]$ which assume values in k at all arguments in k. Noting that any element $f(x) \in K[x]$ can be written $g(x)/a$ $(g(x) \in k[x],\ a \in k)$ in such a way that the element a has no nonunit factor in common with all the coefficients of g, let us call this element $a \in k$ (well-defined up to associates) the* denominator *of f.*

Then the following conditions on an irreducible element $p \in k$ are equivalent:

(i) *There exist elements of $\mathrm{Int}_k[x]$ having denominator divisible by p.*

(ii) *The ring $k/(p)$ is finite.*

PROOF. If $f(x) = g(x)/a \in \mathrm{Int}_k[x]$ has denominator a divisible by p, then the image of g in $(k/(p))[x]$ is a nonzero polynomial which assumes the value zero at every argument. But a nonzero polynomial over an integral domain can have only finitely many roots, so (ii) must hold. Conversely, assuming (ii), there exists a nonzero polynomial over $k/(p)$ having all elements of that ring as roots. If we lift this to a polynomial g over k, then the value of g at every argument is divisible by p, so $g/p \in \mathrm{Int}_k[x]$. □

Hence, if k is a unique factorization domain, such as $\mathbf{Q}[t]$, which has no finite homomorphic images, the only "integral" polynomials over k are the polynomials actually having coefficients in k.

However, we can extend the theory of integral polynomials to the study of algebras defined by conditions that their elements take on integral values at more general specified sets of arguments, and for appropriate choices of these sets, we can get interesting constructions even when k has no finite images. We shall also be able to prove in this context a version of the result that $\mathrm{Int}[x]$ is a co-ring.

We begin with

LEMMA 46.2. *Let k be a principal ideal domain, K its field of fractions, R a commutative K-algebra, and A a subset of $\mathbf{Comm}^1_K(R, K)$ such that*

$$\bigcap_{a\in A} \ker(a) = \{0\}. \tag{46.3}$$

Let

$$\mathrm{Int}_A(R) = \bigcap_{a\in A} a^{-1}(k).$$

(Note that this need not span R as a K-vector-space.) Then

(i) *For every finite-dimensional k-subspace $V \subseteq R$, the k-module $\mathrm{Int}_A(R) \cap V$ is free.*

(ii) *Suppose S is another commutative K-algebra, and B a subset of $\mathbf{Comm}^1_K(S, K)$ such that*

$$\bigcap_{b\in B} \ker(b) = \{0\}. \tag{46.4}$$

Let us identify $A\times B$ with the induced family of homomorphisms $\{a\otimes b\colon R\otimes_K S \to K \mid a\in A,\ b\in B\}$. Then the natural map

$$\mathrm{Int}_A(R) \otimes_k \mathrm{Int}_B(S) \to \mathrm{Int}_{A\times B}(R\otimes_K S) \tag{46.5}$$

is an isomorphism.

PROOF. To verify the parenthetical remark that $\mathrm{Int}_A(R)$ may not span R as a K-vector-space, take $k = \mathbf{Z}$, $R = \mathbf{Q}[x]$, and $A =$ the set of *all* homomorphisms $R \to \mathbf{Q}$.

Let us now prove the positive assertions.

(i) From (46.3) and the finite-dimensionality of V, it follows that we can find a finite subset $A_0 \subseteq A$ such that $V \cap \bigcap_{a\in A_0} \ker(a) = \{0\}$. This means that the map $V \to K^{A_0}$ induced by the elements of A_0 is an embedding. This map takes $\mathrm{Int}_A(R) \cap V$ into the free k-module k^{A_0}. So as a module over the principal ideal domain k, $\mathrm{Int}_A(R) \cap V$ is embeddable in a free module, hence is free.

(ii) The three rings appearing in (46.5) are subrings of the K-algebras R, S and $R\otimes_K S$ respectively. Since k is a principal ideal domain, and these k-modules are torsion-free, they are flat, so in particular, the left-hand side embeds in $R\otimes_K S$, so (46.5) is one-to-one. To prove surjectivity, consider an element $t\in \mathrm{Int}_{A\times B}(R\otimes_K S)$. This will be contained in $V\otimes S$ for some finite-dimensional subspace $V\subseteq R$. By (i), $\mathrm{Int}_A(R)\cap V$ is a free k-module on some basis $\{r_1, \dots, r_m\}$; let us extend this to a K-basis $\{r_1, \dots, r_n\}$ of V $(n \geq m)$, and write

$$t = r_1\otimes s_1 + \dots + r_n \otimes s_n, \tag{46.6}$$

with $s_1, \dots, s_n \in S$. Now by definition of $\mathrm{Int}_{A\times B}(R\otimes_K S)$, for every $(a, b)\in A\times B$ the element $(a\otimes b)(t) = a(r_1)b(s_1) + \dots + a(r_n)b(s_n)$ belongs to k. This is equivalent to saying that for each $b\in B$ the element $r_1 b(s_1) + \dots + r_n b(s_n)\in R$ belongs to $\mathrm{Int}_A(R)$, hence to $\mathrm{Int}_A(R)\cap V$. By choice of $r_1, \dots, r_n$, this means $b(s_1), \dots, b(s_m)$ belong to k, and $b(s_{m+1}), \dots, b(s_n)$ are 0. Letting

b range over B, the first of these conditions tells us that $s_1, \ldots, s_m \in \mathrm{Int}_B(S)$, while by (46.4) the second implies that $s_{m+1}, \ldots, s_n = 0$. Hence t is the image of $r_1 \otimes s_1 + \ldots + r_m \otimes s_m \in \mathrm{Int}_A(R) \otimes_k \mathrm{Int}_B(S)$, proving the surjectivity of (46.5). □

We can now establish our generalization of Lemma 45.2.

PROPOSITION 46.7. *Let k be a principal ideal domain, K its field of fractions, R a commutative K-algebra, F the functor $\mathbf{Comm}^1_K \to \mathbf{Set}$ that it represents, and A a subset of $F(K)$ satisfying* (46.3). *Then any co-operation on $R \in \mathbf{Comm}^1_K$ such that A is closed under the induced operation on $F(K)$ induces a co-operation on $\mathrm{Int}_A(R) \in \mathbf{Comm}^1_k$ making the obvious diagram based on the inclusion map $\mathrm{Int}_A(R) \to R$ commute. Moreover, for any family of co-operations on R such that A is closed under the induced operations on $F(K)$, the sets of identities in these operations satisfied by the whole functor F, by the particular algebra $F(K)$, by its subalgebra A, and by the algebra-valued functor G on $\mathbf{Comm}^1_k$ represented by $\mathrm{Int}_A(R)$ are the same.*

SKETCH OF PROOF. Consider an n-ary co-operation

$$\mathbf{f}\colon\ R \to R \otimes_K \ldots \otimes_K R. \tag{46.8}$$

If A is closed under the induced n-ary operation on $\mathbf{Comm}^1_K(R, K)$, we see that $\mathbf{f}$ restricts to a homomorphism of k-subalgebras

$$\mathbf{f}_A\colon\ \mathrm{Int}_A(R) \to \mathrm{Int}_{A\times \ldots \times A}(R \otimes_K \ldots \otimes_K R). \tag{46.9}$$

By the preceding Lemma, the codomain of this homomorphism can be identified with $\mathrm{Int}_A(R) \otimes_k \ldots \otimes_k \mathrm{Int}_A(R)$, allowing us to translate (46.9) into a co-operation on $\mathrm{Int}_A(R) \in \mathbf{Comm}^1_k$. The final assertion follows from Corollary 41.4, since A is dense both in F and in G. □

Even for $k = \mathbf{Z}$ and $R = \mathbf{Q}[x]$, this result gives us new constructions. For example, it shows that the ring of those $p(x) \in \mathbf{Q}[x]$ whose value at every argument 2^i $(i = 0, 1, \ldots)$ is an integer (a much larger ring than $\mathrm{Int}[x]$) may be made a cosemigroup, because the powers of 2 form a semigroup under multiplication.

For a similar example not based on $\mathbf{Z}$, let L be any field, let k be the polynomial ring $L[t]$, so that $K = L(t)$, let $R = K[x]$, and take for A the set of homomorphisms $R \to K$ carrying x to a nonnegative power of t. The resulting ring $\mathrm{Int}_A(R)$ has a cosemigroup structure in $\mathbf{Comm}^1_k$ for the same reason as the preceding example. Moreover, it is strictly larger than $k[x]$ even if the field L is infinite (so that k has no finite homomorphic images), because there are irreducible elements modulo which the set of powers of t yields only finitely many residues; namely, t and all prime divisors of polynomials $t^n - 1$. Some examples of elements of this ring are $(x-1)x/(t-1)t$, and $(x-1)(x-t)x/(t-1)^2t^3$.

If L is an arbitrary field containing a finite field E, and we again take $k = L[t]$ (so that $K = L(t)$) and $R = K[x]$, and this time let A denote the set of k-algebra homomorphisms $R \to K$ carrying x to a member of $E[t]$, then we see that $\mathrm{Int}_A(R)$ is a *co-ring*, which contains, for instance, $(x^{\mathrm{card}(E)} - x)/t$.

Let us look again at the case $k = \mathbf{Z}$, and the ring which we called $\mathrm{Int}[x]$ in the preceding section, and can now write $\mathrm{Int}_{\mathbf{Z}}(\mathbf{Q}[x])$, for brevity identifying the set of homomorphisms $\mathbf{Q}[x] \to \mathbf{Q}$ given by evaluation at elements of $\mathbf{Z}$ with $\mathbf{Z}$ itself. We may note that the ring co-operations of this object, generated by (45.3) and (45.4), are not the only co-operations on $\mathbf{Q}[x]$ inducing operations under which $\mathbf{Z}$ is closed; clearly, for every integral polynomial f in n variables, it is also closed under the n-ary operation induced by the co-operation $\mathbf{Q}[x] \to \mathbf{Q}[x_1, \dots, x_n]$ taking x to $f(x_1, \dots, x_n)$. As we have seen, the set of such polynomials is generated under the ring operations by the 1-variable binomial coefficient polynomials, so to get this full costructure, it is enough to enrich the set of ring co-operations on $\mathrm{Int}_{\mathbf{Z}}(\mathbf{Q}[x])$ with the "co-binomial-coefficient" functions. Thus, again writing Bi for the functor represented by $\mathrm{Int}_{\mathbf{Z}}(\mathbf{Q}[x])$, the sets $\mathrm{Bi}(S)$ admit, in addition to the ring operations, unary "binomial coefficient" operations $\binom{-}{i}$ $(i = 0, 1, 2, \dots)$. Integral domains of characteristic 0 closed under these operations arise in the theory of nilpotent groups, where they are called *binomial domains* [**195**, §§10-11], [**78**, p.44 bottom = **79**, p.446]; so Bi may be regarded as a representable binomial-ring valued functor.

In the context of Lemma 46.2, one might ask whether we can strengthen conclusion (i) to say that $\mathrm{Int}_A(R)$ is a free k-module. In general, we cannot. For instance, let X be an infinite set, R the product ring K^X, and A the set of homomorphisms $R \to K$ given by evaluation at all the elements of X. Then the hypothesis of the Lemma holds, but $\mathrm{Int}_A(R)$ is the product ring k^X, which is not in general free as a k-module. However, from the result proved in Lemma 46.2(i) for finite-dimensional V, one can deduce that the same result holds for *countable*-dimensional V, hence $\mathrm{Int}_A(R)$ is a free k-module if R is a countable-dimensional K-algebra, e.g., $K[x]$.

There is an obvious parallelism between the results of this section and those of §41; it would be interesting to see whether these can be brought into some common framework.

§47. Representative functions and linearly recursive sequences.

We noted in §43 that if R is a bialgebra or Hopf algebra over a field k, then $\hat{R}$ is a "linearly compact bialgebra or Hopf algebra" over k; but that the $\hat{\otimes}$-comultiplication $\mathbf{m}\colon \hat{R} \to \hat{R} \hat{\otimes} \hat{R}$ of this structure is not simply a $\otimes_k$-comultiplication with some additional properties, because its image will not necessarily lie in the subspace $\hat{R} \otimes \hat{R}$ of $\hat{R} \hat{\otimes} \hat{R}$. (Elements of $\hat{R} \hat{\otimes} \hat{R}$ can be thought of as certain formal infinite sums of terms $s \otimes t$; the subspace $\hat{R} \otimes \hat{R}$ consists of those elements that are finite sums of such terms.) In this situation, Hopf algebraists define $R^\circ \subseteq \hat{R}$ to be the space of elements of R whose images under the comultiplication do lie in $\hat{R} \otimes \hat{R}$. This is easily shown to be a subalgebra (and to be closed under the coinverse operation, if any). A pleasant surprise is that it is a sub-$\otimes_k$-coalgebra as well, i.e., that $\mathbf{m}$ carries R° not merely into $\hat{R} \otimes \hat{R}$, but into $R^\circ \otimes R^\circ$, making R° an ordinary bialgebra or Hopf algebra. The proof, which we shall not give here, uses coassociativity in the same way as does the proof of the Fundamental Theorem on coalgebras ((25.48), (32.1)). (Alternatively, one can, as in [**183**], *define* R° to be the largest subspace of $\hat{R}$ such that $\mathbf{m}(R^\circ)$ lies in $R^\circ \otimes R^\circ$. The existence of this subspace is straightforward, without any assumption of coassociativity, and it will automatically be a subbialgebra. When one makes this choice, the result stated above translates to a useful criterion for an element of $\hat{R}$ to belong to R° [**183**, Prop. 6.0.3].)

A related situation is the following. Let M be a group or semigroup, and $\tilde{M}$ any set of k-valued functions on M which forms a k-algebra under pointwise operations, and is closed under right and left translation by members of M. (The right translate of $r \in \tilde{M}$ by $y \in M$ means the function $x \mapsto r(xy)$; left translation is defined analogously. As examples, $\tilde{M}$ might be the algebra of *all* k-valued functions on M; or, if M is a Lie group and k the field of real or complex numbers, $\tilde{M}$ might be the subalgebra of *continuous* functions.) Let $M^\circ \subseteq \tilde{M}$ be the subspace consisting of all functions $r \in \tilde{M}$ whose right translates span a finite-dimensional subspace of $\tilde{M}$. These are called the *representative functions* on M (probably because of their importance in representation theory, where they are the functions lying in finite-dimensional subrepresentations of the representation of the group or semigroup M by translations on the k-vector space $\tilde{M}$). It is easy to see that M° forms a subalgebra of the k-algebra $\tilde{M}$.

Now given $r \in M^\circ$, let $\{s_1, \ldots, s_n\}$ be a basis for the space spanned by all right translates of r. Then we have

$$(47.1) \qquad r(xy) = s_1(x)t_1(y) + \ldots + s_n(x)t_n(y) \qquad (x, y \in M)$$

for unique k-valued functions $t_1, \ldots, t_n$ on M. Clearly, (47.1) also says that every *left* translate of r is contained in the span of $t_1, \ldots, t_n$. In particular, the span of thc left translates of r is finite-dimensional; thus, the definition of M° is effectively left-right symmetric. Moreover, from the linear independence of $s_1, \ldots, s_n$ it follows that the span of the left translates of r is the whole space spanned by $t_1, \ldots, t_n$. It follows from what we have said that the functions $s_1, \ldots, s_n$ and $t_1, \ldots, t_n$ themselves belong to M°. (Note that the argument that gives this statement uses the *associativity* of M to show that a right translate of a right translate is a right translate, and similarly for left translates. This is analogous to the use of the hypothesis of coassociativity in the Fundamental Theorem on coalgebras, referred to above.)

The relations (47.1) allow us to associate to each $r \in M^\circ$ a well-defined element $s_1 \otimes t_1 + \ldots + s_n \otimes t_n \in M^\circ \otimes_k M^\circ$, and one finds that this co-operation makes the commutative k-algebra M° a bialgebra. If M is not merely a semigroup but a group, and $\tilde{M}$ is closed under the operation $r \mapsto r'$ defined by $r'(x) = r(x^{-1})$, then this induces a coinverse operation as well, making M° a Hopf algebra.

An interesting illustrative case is that in which M is the additive group of integers, and $\tilde{M}$ the k-algebra of all k-valued functions on M. The representative functions are then known as *linearly recursive sequences*, because if f is such a function, and n is the least value such that the translates of f by the integers $0, \ldots, n$ are k-linearly dependent, then the unique dependence relation among these translates becomes a *linear recursion relation* for f, i.e., an identity

$$(47.2) \qquad c_0 f(i) + \ldots + c_n f(i+n) = 0 \quad \text{holding for all } i \in \mathbf{Z},$$

with c_0 and c_n nonzero, which allows one to recursively compute the full sequence f from any n successive terms. A well-known example of a linearly recursive sequence is the sequence of Fibonacci numbers, satisfying the relation $f(i)+f(i{+}1)-f(i{+}2)=0$.

Note that on the space spanned by the translates of a sequence f satisfying (47.2), the *shift operator* T will satisfy $P(T)=0$, where P is the polynomial

$$P(x) = c_0 + \dots + c_n x^n. \tag{47.3}$$

If we pass to the algebraic closure K of k, we can put the restriction of the shift operator T to the n-dimensional space of solutions to (47.2) into Jordan canonical form, and use this decomposition to express f in closed form.

The resulting statement is simplest if $P(x)$ has n distinct roots $\rho_1,\dots,\rho_n$ in K. In that case, the solutions of (47.2) over K are all linear combinations of a set of eigenvectors for T, which we can take to be the n geometric progressions $(\dots,\rho_h^{-1},\rho_h^0,\rho_h^1,\rho_h^2,\dots)$ $(h=1,\dots,n)$. The solutions of (47.2) in the original field k are thus those K-linear combinations of these sequences which take values in k. For instance, the Fibonacci sequence in $\mathbf{Q}$ can be written $f(i)=(\tau^i-\tau'^i)/(\tau-\tau')$, where $\tau,\tau'=(1\pm\sqrt{5})/2$ are the roots of $1+x-x^2$. (Here the denominator $\tau-\tau'$ is equal to $\sqrt{5}$. This way of getting an expression for the Fibonacci sequence, by finding eigenvectors of the shift operator, is an entertaining example for an undergraduate linear algebra class.)

The opposite extreme is the case $P(x)=(x-1)^n$. If $\operatorname{char}(k)=0$, the solutions of (47.2) are then precisely the polynomial functions of degree $<n$. A statement valid in arbitrary characteristic is that the space of solutions is spanned by the images in k of the n $\mathbf{Z}$-valued binomial coefficient sequences $\binom{i}{0},\dots,\binom{i}{n-1}$. In characteristic p, one also finds that the union over all n of the space of sequences that satisfy the recursion relation $(T-1)^n=0$ is equal to the union over all m of the space of sequences periodic with period p^m.

In the general case, where $P(x)$ has roots $\rho_1,\dots,\rho_m$, of multiplicities $n_1,\dots,n_m$, the general solution to (47.2) has the form $e_1(i)\rho_1^i+\dots+e_m(i)\rho_m^i$, where each sequence $e_h(i)$ is an arbitrary k-linear combination of the sequences $\binom{i}{0},\dots,\binom{i}{n_h-1}$.

If k is not assumed a field, but a general commutative Noetherian ring, let us call a k-valued function on a group or semigroup M representative if the k-module generated by its right translates is finitely generated. (We assume k Noetherian so that every member of this module will again satisfy this condition. Actually, we will soon restrict attention to principal ideal domains.) Let us write M_k° for the ring of representative k-valued functions on M. Note that any homomorphism $k\to L$ of Noetherian rings, applied pointwise to functions on M, will carry M_k° into M_L°.

Let us now assume k a principal ideal domain, and let K be its field of fractions. Then we claim that M_k° consists precisely of those members of M_K° which are k-valued. Indeed, "$\subseteq$" is clear, and "$\supseteq$" may be seen by applying Lemma 46.2(i), with $R=M_K^\circ$ and A the set of evaluations at elements of M, and noting that a free k-submodule of a finite-dimensional K-vector-space is finitely generated.

Proposition 46.7 is also applicable to this situation. Indeed, being a commutative bialgebra or Hopf algebra over K, M_K° represents a functor G from $\mathbf{Comm}_K^1$ to **Semigp** or **Group**. The map $M \to G(K)$ taking each $x \in M$ to the evaluation-at-x map is a homomorphism of semigroups or groups (this is precisely the content of (47.1)), so the set of these evaluation maps is a subsemigroup or subgroup of $G(K)$. Hence, by that Proposition, a bialgebra or Hopf algebra structure over k is induced on the k-subalgebra $M_k^\circ = \mathrm{Int}_M(M_K^\circ)$ of k-valued representative functions.

In general, M_k° will not span M_K° over K. For instance, for k still a principal ideal domain, and $M = \mathbf{Z}$, the K-vector-space spanned by M_k° consists of those K-valued linearly recursive sequences f such that all eigenvalues (in an algebraic closure of K) of the action of the shift operator on the vector space span of the orbit of f under translations are *integral* over k, and have inverses integral over k; equivalently, such that the polynomial (47.3) can be taken to have all coefficients in k, and first and last coefficients invertible. (Idea of proof: Let f lie in the K-vector space spanned by M_k°. Then $f = c^{-1}g$ for some $c \in k-\{0\}$, $g \in M_k^\circ$, hence the k-module A spanned by the orbit of f under translations is finitely generated. Since A is a *cyclic* module over the k-algebra generated by T and T^{-1}, the k-algebra of operators on A generated by the restrictions of T and T^{-1} to A is module-isomorphic to A, hence is finitely generated as a k-module; but it has the form $k[x, x^{-1}]/(P(x))$. So the latter algebra is finitely generated as a k-module, which translates to the asserted integrality condition on the roots of P, equivalently, to the asserted invertibility condition on the first and last coefficients of P.)

For general M, we can characterize the functor represented by the bialgebra or Hopf algebra M_k° by a universal property:

PROPOSITION 47.4. *Let k be a field, or more generally, a principal ideal domain, M an object of* **Group** *or* **Semigp**, *M° the commutative Hopf algebra or bialgebra of all representative k-valued functions on M, and F the functor from* $\mathbf{Comm}_k^1$ *to* **Group** *or* **Semigp** *represented by M°.*

Then there is a natural homomorphism of groups or semigroups $M \to F(k)$, and F is universal (initial) among representable functors given with such homomorphisms. Equivalently, if we regard F as a representable functor from $\mathbf{Comm}_k^1$ *to the variety of (semi)groups given with homomorphisms of M into them, then F is initial among such functors.*

SKETCH OF PROOF. We have noted that the map taking each $x \in M$ to the k-algebra homomorphism $M^\circ \to k$ given by evaluation at x constitutes a group or semigroup homomorphism $M \to F(k)$. Now suppose G is any representable functor from $\mathbf{Comm}_k^1$ to **Group** or **Semigp**, say with representing k-algebra S, and comultiplication $\mathbf{m}_S$, and we are given a homomorphism

$$\theta:\ M \ \to\ G(k). \tag{47.5}$$

Observe that an element $a \in S$ induces a k-valued function on $G(k)$, and hence, via (47.5), on M; the latter function can be described as carrying $x \in M$ to $\theta(x)(a) \in k$. From the fact that (47.5) is a semigroup homomorphism, it follows that if $\mathbf{m}_S(a) = b_1 \otimes c_1 + \ldots + b_n \otimes c_n$, then $\theta(xy)(a) = \theta(x)(b_1)\theta(y)(c_1) + \ldots + \theta(x)(b_n)\theta(y)(c_n)$, showing that $\theta(-)(a)$ is a representative function on M, and that the map $a \mapsto \theta(-)(a)$ is a $\otimes_k$-coalgebra homomorphism $S \to M°$ as well as an algebra homomorphism. One verifies that the corresponding morphism of functors $F \to G$ is the unique morphism forming a commuting triangle with the maps from M to $F(k)$ and $G(k)$. The interpretation of this result in the final sentence is immediate. □

There is an analog of Proposition 47.4 for functions on a *ring* M:

PROPOSITION 47.6. *Let* k *be a field, or more generally, a principal ideal domain,* M *an associative unital ring, and* $M°$ *the set of k-valued functions* r *on* M *whose orbits under the action of the semigroup of right* affine translation *maps,* $x \mapsto xy+z$ $(y, z \in M)$ *are finite-dimensional; equivalently, such that* r *is representative with respect to the* addition *of* M, *and* r *and all its additive translates are representative with respect to the* multiplication *of* M.

Then $M°$ *is a co-ring object of* $\mathbf{Comm}^1_k$, *representing a functor* $F\colon \mathbf{Comm}^1_k \to \mathbf{Ring}^1$ *having a distinguished ring homomorphism* $M \to F(k)$, *and initial among representable functors given with such homomorphisms; equivalently, initial among representable functors* $\mathbf{Comm}^1_k \to M\text{-}\mathbf{Ring}^1$.

If the ring M *is commutative, then* F *is commutative-ring valued, and hence is also initial in* $\mathbf{Rep}(\mathbf{Comm}^1_k, \mathbf{Comm}^1_M)$ *and in* $\mathbf{Rep}(\mathbf{Comm}^1_k, \mathbf{Ring}^1_M)$.

SKETCH OF PROOF. Let us first verify the asserted equivalence of the two criteria for membership in $M°$. Because the affine maps $M \to M$ form a semigroup, an affine translate of an affine translate is an affine translate; from this we easily see that the first condition implies the second. Conversely, assuming the latter condition, we see that the set of multiplicative translates of additive translates of r will have finitely generated span; and these are precisely the affine translates of r. (In verifying this last fact, one must be careful with order of composition of translation operators on functions, this being contravariant in the corresponding operations on M!)

We see that $M°$ consists of certain functions representative with respect to both the addition and the multiplication of M, and is carried into $M° \otimes M°$ by each of these co-operations. It is also clearly closed under the unary co-operation induced by the additive inverse operation on M, and admits zeroary co-operations "0" and "1", namely evaluation at the corresponding elements of the ring M. By Corollary 41.4, the operations on F induced by these co-operations satisfy the identities of M, so F is ring-valued, and if M is commutative, commutative ring-valued. The canonical ring homomorphism $M \to F(k)$ is described as in the preceding Proposition, and again the proof of its universal property is straightforward. The final assertion is clear. □

DEFINITION 47.7. *If* M *is a ring and* k *a principal ideal domain, a function* $M \to k$ *whose orbit under the semigroup of right affine translations of* M *is finitely generated will be called* affinely representative.

(Since, as noted in the above Proposition, this condition can be expressed in terms of additive representativity, and multiplicative representativity of additive translates; and since, as we saw at the beginning of this section, multiplicative representativity, though defined asymmetrically, is really left-right symmetric, the condition of affine representativity is also left-right symmetric, despite the above formulation.)

QUESTION 47.8. *In the situation of the preceding Proposition, is it true that*

$$M^\circ_{\text{affine}} = M^\circ_{\text{add}} \cap M^\circ_{\text{mult}}, \tag{47.9}$$

where the three terms of this equation denote the modules of affinely representative, additively representative, and multiplicatively representative functions on M, *respectively? Equivalently, if a function* r *belongs to the right hand intersection, does every additive translate of* r *also belong to* M°_{mult}?

Note that if r is representative with respect to addition, then it follows from the distributive law in the ring M that every *multiplicative* translate of r is again representative with respect to addition. Hence if $r \in M^\circ_{\text{add}} \cap M^\circ_{\text{mult}}$, the span of the additive translates of the multiplicative translates of r will be finite-dimensional. These are the translates of r by the maps $x \mapsto (x+z)y$ $(y,\ z \in S)$. Unfortunately, the set of such maps is not in general closed under composition. (Its closure is the semigroup of all affine maps.) This set of maps will, of course, contain those maps $x \mapsto xy+z$ such that z is a left multiple of y, hence any additive translate of r has a finitely generated module of translates under multiplication by the *invertible* (or even left invertible) elements of M. Thus, when M is a division ring, we have an affirmative answer to the preceding question.

If we take M in the above Proposition to be the ring $\mathbf{Z}$, the resulting co-ring $\mathbf{Z}^\circ_k$ will represent a functor initial among all representable functors $\mathbf{Comm}^1_k \to \mathbf{Ring}^1$. Writing k-valued functions on $\mathbf{Z}$ as sequences of elements of k, let us characterize the sequences comprising this co-ring.

LEMMA 47.10. *Let* $(\ldots, f(-1), f(0), f(1), \ldots)$ *be a sequence of elements of a principal ideal domain* k. *Then the following conditions are equivalent:*

(i) f *is affinely representative.*

(ii) f *is linearly recursive (i.e., is representative with respect to the additive group structure of* $\mathbf{Z}$), *and all roots (in an algebraic closure of the field of fractions of* k) *of the polynomial* P *associated with* f *(the minimal polynomial* (47.3) *satisfying* (47.2)) *are* roots of unity.

(iii) *There exists a positive integer* d *(depending on* f) *such that for every integer* r, *the function* $f_{d,r}$ *defined by* $f_{d,r}(i) = f(di+r)$ *is given by a polynomial in* i *with coefficients in* k.

If k *has positive characteristic, these conditions are also equivalent to:*

(iv) f *is periodic.*

SKETCH OF PROOF. (i)$\Rightarrow$(ii). Suppose f satisfies (i), let $P(x)$ be the polynomial referred to in (ii), and let ρ be a root of P in the algebraic closure L of the field of fractions of k. We wish to show that ρ is a root of unity. Note that f is also linearly recursive as a sequence of elements of L, and that its associated polynomial in this context is still P, which has ρ as a root. So in proving the second half of (ii), we may assume without loss of generality that k is a field, and that ρ lies in k. Hence we can find in the space spanned by the additive translates of f an eigenvector of the shift operator T with eigenvalue ρ. Adjusted by a scalar, this will be precisely the sequence of powers of ρ; hence the latter sequence again has finite-dimensional span under the affine operations of $\mathbf{Z}$. But the translate of this sequence by the affine operation of multiplication by an integer n is the sequence of powers of ρ^n. From the properties of Vandermonde determinants, we know that the sequences of powers of distinct elements of k are linearly independent; hence only finitely many powers of ρ can be distinct. Since ρ is not 0 (for by definition, P has nonzero constant term), it is a root of unity, as claimed.

(ii)$\Rightarrow$(iii). Key ideas: Choose m such that all roots of P are mth roots of unity. Then we see that the polynomials (47.3) associated to the functions $f_{m,r}(i)$ ($r \in \mathbf{Z}$) have only the root 1, hence these functions lie in the span of the binomial coefficient functions. It is also not hard to show that for any n and r, the polynomial $\binom{n!x+r}{n} \in \mathbf{Q}[x]$ actually lies in $\mathbf{Z}[x]$, hence its values form a genuine polynomial sequence over any base ring.

(iii)$\Rightarrow$(i) is straightforward.

Finally, if k has positive characteristic, every polynomial is periodic of period p, so (iii) and (iv) are equivalent. □

Let us call a sequence f satisfying condition (iii) above *periodically polynomial*. We have shown

PROPOSITION 47.11. *Let k be any principal ideal domain. Then the ring of periodically polynomial k-valued sequences becomes a co-ring in* $\mathbf{Comm}^1_k$ *under the co-operations arising by regarding these sequences as additively and multiplicatively representative functions on* $\mathbf{Z}$. *The functor F represented by this co-ring is initial among representable functors* $\mathbf{Comm}^1_k \to \mathbf{Ring}^1$, *hence also among representable functors* $\mathbf{Comm}^1_k \to \mathbf{Comm}^1$. □

A variant of the above argument gives us the following negative result:

PROPOSITION 47.12. *Let k be a field of positive characteristic, or more generally, a principal ideal domain having a homomorphic image of positive characteristic. Then there is no nontrivial representable functor* $\mathbf{Comm}^1_k \to \mathbf{Ring}^1_{\mathbf{Q}}$.

PROOF. The initial object of $\mathbf{Rep}(\mathbf{Comm}^1_k, \mathbf{Ring}^1_{\mathbf{Q}})$ is represented by the ring of all affinely representative k-valued functions $\mathbf{Q} \to k$. We shall show that under our hypotheses on k, the only such functions are the constants, so that our initial representable functor is trivial. Since this functor admits a map to every other representable functor $\mathbf{Comm}^1_k \to \mathbf{Ring}^1_{\mathbf{Q}}$, and since a unital ring admitting a homomorphism from the trivial ring is trivial, we will have the asserted result.

We begin with the case where k is a field of characteristic p. Let $r\colon \mathbf{Q} \to k$

be affinely representative. Then by the method of proof of Lemma 47.10 (ii)⇒(iii)⇒(iv), we see that for all $c \in \mathbf{Q}$ the sequence of values $(\dots, r(-c), r(0), r(c), \dots)$ is periodic. Now as we vary c, the resulting sequences span a finite-dimensional space, hence they have some common period n. This implies that for every rational number c, we have $r(nc) = r(0)$. Since every rational number can be written nc, this says the function r is constant, as required.

Next, suppose k is a principal ideal domain which is not a field, but has an irreducible element π such that $k/\pi k$ is a field of positive characteristic. Suppose $r: \mathbf{Q} \to k$ were a nonconstant affinely representative function. Subtracting a constant function, we can assume $r(0) = 0$, and dividing by a constant in k, we can assume that π does not divide all the $r(c)$. Hence, composing with the residue map $k \to k/\pi k$, we get a nonconstant affinely representative $k/\pi k$-valued function, contradicting the result just proved. □

Let us digress slightly and record some characterizations of the class of representative functions on the *additive semigroup* $\mathbf{N}$ of nonnegative integers – largely because we think the simple result given below should be better known than it is. (Many mathematicians seem to know a few, but not all of the equivalences listed.) Note that the result is stated without reference to the bialgebra structure we have been discussing. For simplicity, k is assumed a field.

Although the distinction between left and right shift was not of fundamental importance in discussing sequences indexed by $\mathbf{Z}$, since the two shift operators generate the same group, it is so for functions on $\mathbf{N}$; hence let us note that the semigroup of additive translation operators on $k\mathbf{N}$ is generated by the *left shift* operator T, which acts by lopping off the first term of a sequence.

LEMMA 47.13. *Let k be a field, and $f = (f(0), f(1), \dots)$ a sequence of elements of k indexed by the nonnegative integers. Then the following conditions are equivalent.*

(i) *The k-vector space spanned by the orbit of f under iterated left shifts is finite-dimensional.*

(ii) *f is linearly recursive; i.e., there exist constants $c_0, \dots, c_n \in k$ with $n \geq 0$, $c_n \neq 0$ such that for every $i \geq 0$, $c_0 f(i) + \dots + c_n f(i+n) = 0$.*

(iii) *Regarded as a sequence of elements of the algebraic closure L of k, f lies in the L-algebra generated under termwise operations by the geometric progressions, the binomial coefficient functions, and the functions with finite support. Equivalently, after finitely many terms, f agrees with a (unique) function of the form $i \mapsto e_1(i)\rho_1^i + \dots + e_m(i)\rho_m^i$, where the ρ's are nonzero elements of L, and the e's are linear combinations over L of binomial coefficient functions.*

(iv) *The generating function of f is a rational function. That is, the formal power series $\Sigma\, f(i)\, x^i \in k[[x]]$ represents a member of $k(x) \subseteq k((x))$.*

SKETCH OF PROOF. We shall show (iii) ⇔ (i) ⇔ (ii) ⇔ (iv).

In the first double implication, the forward direction holds because the subspace of sequences satisfying (iii) determined by fixing any finite family of ρ_i's, any bound d on the degrees of the binomial coefficient functions used in forming the e_i's, and any bound j on the location of the last term in a summand of finite support, is finite-dimensional and closed under T. One gets the reverse implication by using the Jordan canonical form of the operator T on a finite-dimensional T-invariant space of sequences. (The "finite support" summand comes from the eigenvalue 0, which did not occur for two-sided sequences.)

The second double implication is immediate. (Note that the definition given in (ii) of a linear recursion relation for a one-sided sequence does not require $c_0 \neq 0$.)

The last double implication comes from the observation that for any $P \in k[x]$, applying $P(T)$ to a sequence corresponds to multiplying the corresponding formal power series by $P(x^{-1})$, and then discarding terms involving negative powers of x. Thus, a sequence is *annihilated* by $P(T)$ if and only if the corresponding formal power series can be written $Q(x^{-1})/P(x^{-1})$ for some $Q(x) \in xk[x]$. □

(The idea of the last step is curious – one may say that one is regarding the set of all one-sided k-valued sequences simultaneously as the formal power series ring $k[[x]]$, and as the $k[x^{-1}]$-module quotient $k((x))/x^{-1}k[x^{-1}]$ of the formal Laurent series ring $k((x))$ by its submodule $x^{-1}k[x^{-1}]$; and noting that to be annihilated by a nonzero member of $k[x^{-1}]$ under the second interpretation is the same as to be the power series of a rational function under the first.)

We note the contrasting situation for formal power series in several commuting indeterminates: $(1 - xy)^{-1} \in k[[x, y]]$ is rational, but the corresponding function on the free abelian semigroup on two generators, the Kronecker delta function $\delta_{i,j}$, is not representative. On the other hand, for power series in several *noncommuting* indeterminates, one has a result as in the one-indeterminate case; see [**50**, Prop. 3] and other works referred to there.

Some algebraic properties of the ring of linearly recursive sequences are studied in [**116**]. For some diverse results on such sequences see [**148**].

Although we began this section by regarding the restricted dual $R°$ of a bialgebra R, and the space $M°$ of representative functions on a semigroup M, as *analogous* constructions, the latter is actually a case of the former. If M is a semigroup, then as we noted in 43.11, the semigroup algebra $R = kM$ is a bialgebra under a comultiplication taking x to $x \otimes x$ for all $x \in M$. The linear dual $\hat{R}$ of this algebra can be identified with the space $\tilde{M}$ of all k-valued functions on M, and one finds that the restricted dual $R° \subseteq \hat{R}$ is precisely the bialgebra of representative functions $M° \subseteq \tilde{M}$.

Note that the condition of representativity on elements of $\tilde{M}$ is thus determined by the $\hat{\otimes}$-comultiplication of that object, which arises from the *multiplication* of the semigroup ring kM. Hence the natural *comultiplication* on the semigroup ring, while important in that it leads to the (pointwise) multiplication on $M°$, which is part of the universal bialgebra structure we have discussed, is not involved in the characterization of representative functions.

Now the polynomial algebra $k[x] = k\mathbf{N}$ admits, in addition to the comultiplication taking x to $x \otimes x$, another important cosemigroup co-operation, which takes x to $x \otimes 1 + 1 \otimes x$. (This is in fact the *coaddition* of the representing object of the forgetful functor $\mathbf{Comm}^1_k \to \mathbf{Comm}^1$, but all we care about at the moment is that it makes $k[x]$ a bialgebra.) By the above observation, the restricted dual $(k\mathbf{N})°$ of the resulting bialgebra will again consist of the linearly recursive sequences, but with a different multiplication. This turns out to correspond to "divided powers" multiplication of formal power series. (Let us recall this concept: if we write members of the polynomial ring $\mathbf{Q}[x]$ in terms of the basis of elements

$$x^{(i)} = x^i/i!$$

instead of the x^i, we find that the multiplication operation still has integer structure constants,

$$x^{(i)}x^{(j)} = \binom{i+j}{i} x^{(i+j)}.$$

This allows us to construct over arbitrary k an algebra with a basis of elements $x^{(i)}$ $(i = 0, 1, \ldots)$ which multiply by the above law, called the ring of polynomials with *divided powers*, and its completion, the ring of formal power series with divided powers.)

The *termwise* multiplication of sequences that we have considered so far, when we regard these as the sequences of coefficients of formal power series, is called *Hadamard multiplication* of formal power series.

Finally, ordinary multiplication of formal power series is dual to a coassociative $\otimes_k$-co-operation $k[x] \to k[x] \otimes_k k[x]$ which, unfortunately, does not combine with ordinary multiplication of polynomials to give a bialgebra structure, because it is not a homomorphism with respect to that multiplication: Like one of the co-operations mentioned above, it takes x to $x \otimes 1 + 1 \otimes x$; but it takes x^2 to $x^2 \otimes 1 + x \otimes x + 1 \otimes x^2 \neq (x \otimes 1 + 1 \otimes x)^2$. However, it becomes a homomorphism, and so yields a bialgebra structure, if we replace $k[x]$ by the algebra of polynomials with divided powers. For details see [**146**].

We have noted that both Hadamard and divided-power multiplications are compatible with the standard comultiplication on the dual of the semigroup ring $k\mathbf{N}$, and so, in particular, induce ring structures on the subspace of linearly recursive sequences on N. Curiously, this space is also closed under ordinary formal power series multiplication, as is seen from our characterization of the generating functions of linearly recursive sequences as the rational power series. The facts that it is closed under ordinary and divided-powers multiplication may be translated to say that when we form the product of two formal power series over a field of characteristic 0, writing this either as

$$(\Sigma f_i x^i)(\Sigma g_i x^i) = \Sigma h_i x^i \quad \text{or as} \quad (\Sigma f_i x^i/i!)(\Sigma g_i x^i/i!) = \Sigma h_i x^i/i!,$$

then in each case,

(47.14) $\quad (f_i)$ and (g_i) linearly recursive $\Rightarrow$ (h_i) linearly recursive.

The subrings of $k[[x]]$ determined by the above linear recursiveness conditions are, respectively, the ring of rational power series, and the ring of power series satisfying linear differential equations with constant coefficients. It would be interesting to know for what other sequences c_i the power series of the form $\Sigma f_i x^i/c_i$ have the property (47.14).

The constructions of this section are the last of the examples of bialgebras and Hopf algebras we shall introduce here. Some interesting classes of bialgebras arising from *combinatorial* structures are discussed in [**105**]. (However, the reader of that paper must watch out for typographical errors, misstatements, gaps in proofs, etc..) In this context, antipodes (coinverses), when they exist, correspond to inversion formulae (such as Möbius inversion); cf. [**172**], [**173**], [**77**]. A curious variant of the concept of bialgebra is introduced in [**105**, §XII], in which the condition that the comultiplication be a ring homomorphism is replaced by the condition that it be a derivation.

Possibly related to the ideas of those papers are those of [**58**] and [**59**], which show that the Witt ring functors W and W_p sketched in the preceding section are the $G = \hat{\mathbf{Z}}$ and $G = \hat{\mathbf{Z}}_{(p)}$ cases of a construction definable for any profinite group G. The way that G enters into this construction reduces to the combinatorics of the system of its open subgroups, their conjugacy relations, and their indices in one another. Cf. also [**132**].

§48. A last tantalizing observation on idempotents.

Let $\mathbf{N}_{>0} = [x]$ denote the additive semigroup of *positive* integers; equivalently, the free semigroup *without* neutral element on one generator x, and $R = k\mathbf{N}_{>0} = [k][x]$ its semigroup algebra over a principal ideal domain k. Then $\hat{R}$ can again be thought of as the set of all one-sided sequences of elements of k, though indexed by the positive rather than the nonnegative integers; and the subalgebra R° of representative functions will (except for indexing) again be the set of one-sided linearly recursive sequences, as characterized in Lemma 47.13 (since the inclusion or exclusion of one translation, the identity, does not affect finite generation of the k-module spanned by the translates of a sequence f).

From Proposition 47.4 we can see that the functor F represented by the bialgebra R° will be initial among semigroup-valued functors G on $\mathbf{Comm}^1_k$ given with a distinguished element $x \in G(k)$. It follows that the condition Idp.-in-n.e.($\mathbf{Comm}^1_k$, **Semigp**) is equivalent to the statement that the semigroup $F(k)$ contains an idempotent!

When this holds, can we find an explicit homomorphism $R^\circ \to k$ giving such an idempotent element?

Let us sketch how to do so when k is a field. We may write

$$R^\circ = R_{fs} \oplus R_{bg},$$

where $R_{fs} \subseteq R^\circ$ denotes the ideal of sequences with finite support, and R_{bg} the subalgebra of "binomial-geometric" sequences, i.e., sequences which, over the algebraic closure of k, have the form $j \mapsto e_1(j)\rho_1^j + \ldots + e_m(j)\rho_m^j$, as in Lemma 47.13(iii). Since R_{fs} is an ideal and R_{bg} is a subalgebra, the above decomposition makes R_{bg} a retract of R° as a k-algebra. From the facts that R_{fs} and R_{bg} are closed under translation, and that the cosemigroup structure is defined in terms of translation by (47.1), it follows that the above retraction respects this cosemigroup structure, so R_{bg} becomes a cosemigroup. We see that it can be identified with the cosemigroup of representative functions on the whole group $\mathbf{Z}$.

Now it is easy to show that any element of the ideal $R_{fs} \subseteq R^\circ$ with support in the first n coordinates is annihilated by any element of $F(k)$ which can be written in that semigroup as a product of $n{+}1$ factors; so, since an idempotent i satisfies $i = i^{n+1}$, in looking for idempotent elements of $F(k)$ it suffices to look for such elements in $F_{bg}(k)$, where F_{bg} is the retract of F represented by R_{bg}. That functor, being isomorphic to the functor represented by the ring of representative functions on $\mathbf{Z}$, is group-valued, so it has, indeed, a (unique) idempotent, namely evaluation at $0 \in \mathbf{Z}$.

We conclude that our present functor F has a unique idempotent, which as a homomorphism $R^\circ \to k$ can be described as taking a linearly recursive sequence on the positive integers, "cleaning" it by subtracting its finite-support summand, then extending it to a linearly recursive sequence on $\mathbf{Z}$, and evaluating this at 0. This explicit construction gives us another proof of Lemma 42.2. (The retraction of R° to R_{bg} is constructed for a different purpose in [**116**, §2], where it is denoted α.)

What about the case where k is not a field, but some other discrete valuation ring (= local principal ideal domain) satisfying Idp.-in-n.e.($\mathbf{Comm}^1_k$, **Semigp**)? Writing K for the field of fractions of k, we know that the K-algebra $(\mathbf{N}_{>0})^\circ_K$ of representative K-valued functions on $\mathbf{N}_{>0}$ contains the k-algebra $(\mathbf{N}_{>0})^\circ_k$ of representative k-valued functions on the same semigroup; and we might expect that the homomorphism $e_K\colon (\mathbf{N}_{>0})^\circ_K \to K$ which is the unique idempotent element of $F_K(K)$, would, when restricted to $(\mathbf{N}_{>0})^\circ_k$, give a homomorphism $e_k\colon (\mathbf{N}_{>0})^\circ_k \to k$ supplying the desired idempotent element of $F_k(k)$. But, in fact, the restriction of e_K to $(\mathbf{N}_{>0})^\circ_k$ is not k-valued. This may be seen by taking a nonzero nonunit $c \in k$, and forming the sequence whose nth term is c^{n-1}. This is clearly linearly recursive, but the value of the extended sequence at 0 is $c^{-1} \notin k$. (Indeed, if such a simple method worked, it could not distinguish between rings like the p-adics that satisfy Idp.-in-n.e.($\mathbf{Comm}^1_k$, **Semigp**), and other discrete local rings, such as $\mathbf{Z}_{(p)}$, which are non-Henselian, and so do not.)

However, by studying an explicit description of the p-adic integers, or a power-series ring, as a ring of strings of elements of a field, and applying the idea of Corollary 42.5(i), and the above description of the unique idempotent in the field case, one should at least be able to get results on particular cases of

PROBLEM 48.1. *Describe (if possible as explicit k-algebra homomorphisms) the idempotent element of* $F(k)$, *where* k *is a discrete valuation ring satisfying* Idp.-in-n.e.($\mathbf{Comm}^1_k$, **AbSemigp**) *and* F *is the functor represented by the bialgebra of representative k-valued functions on* $\mathbf{N}_{>0}$.

We have restricted this question to discrete valuation rings because our development of representative functions was limited to principal ideal domains. But even results in this limited case should provide some insight into the condition Idp.-in-n.e..

In the above problem, we should logically have said "Describe the idempotent element *or elements* of $F(k)$". But it will be shown in [**26**] that if $\mathbf{V}$ is any variety, I its initial object, and F initial among representable functors from $\mathbf{V}$ to abelian semigroups with one distinguished element, then $F(I)$ contains *at most* one idempotent.

CHAPTER IX.

Representable functors on categories of Lie algebras.

This will be a short chapter, consisting of some results obtained by a bit of trickery from those of earlier chapters, some counterexamples, and some questions. Though we shall get our positive results using our earlier results on functors on categories $\mathbf{Ring}_k$, which were true independent of the characteristic of k, we shall encounter important differences between the characteristic 0 and characteristic p cases here.

§49. Generalities and conventions.

Let $k \in \mathbf{Comm}^1$, and recall that $\mathbf{Lie}_k$ denotes the category of Lie algebras over k. We shall denote by $\mathrm{Br}\colon \mathbf{Ring}_k \to \mathbf{Lie}_k$ the functor taking an associative k-algebra S to the Lie algebra having the same underlying k-module as S, and Lie brackets given by commutator brackets in the ring operations of S. (This functor was written L in Theorem 25.31, but here we want to leave that symbol free to denote a Lie algebra.) The left adjoint of Br will be written $L \mapsto [k][L]$. (The bracket around the k is our somewhat arbitrary notation, introduced in §12, for the associative *nonunital* k-algebra generated by a family – mnemonically, ''generated over k, but with the k taken away''. The bracket around the L, on the other hand, represents the fact that when L is free as a k-module, this algebra is built out of the elements of a k-basis for L in a way similar to the way a commutative polynomial ring $k[X]$ is built out of the generating set X, despite the noncommutativity in the present situation.) We will refer to $[k][L]$ as the (nonunital) *universal enveloping algebra* of L, though some authors restrict the term ''enveloping algebra'' to cases where the canonical map $L \to \mathrm{Br}([k][L])$ (the unit of the adjunction) is an embedding. We recall that this map is indeed an embedding if k is a field, or more generally if L is free (or even flat) as a k-module, and also whenever k is a $\mathbf{Q}$-algebra [**43**].

The forgetful functor $\mathbf{Lie}_k \to \mathbf{Mod}_k$ also has a left adjoint, a functor $\mathbf{Mod}_k \to \mathbf{Lie}_k$ which we shall write $M \mapsto [k]\{M\}$, calling the latter object the *universal Lie algebra* on the k-module M. Taking left adjoints of the chain of functors

$$\mathbf{Ring}_k \xrightarrow{\mathrm{Br}} \mathbf{Lie}_k \xrightarrow{\text{forget}} \mathbf{Mod}_k \xrightarrow{\text{forget}} \mathbf{Set},$$

we get the chain

$$\mathbf{Set} \xrightarrow{\text{free}} \mathbf{Mod}_k \xrightarrow{[k]\{-\}} \mathbf{Lie}_k \xrightarrow{[k][-]} \mathbf{Ring}_k.$$

Hence by Lemma 5.9, the composites in this chain give the adjoints of the composites in the preceding chain. Thus, the universal Lie algebra construction followed by the universal enveloping algebra construction gives the tensor algebra construction:

$$[k][[k]\{M\}] \cong [k]<M>, \tag{49.1}$$

while *pre*composing the universal Lie algebra functor with the free module functor yields the free Lie algebra functor. Since the free module functor is easy to describe, but the free Lie algebra on a set is quite complicated ([**174**, §4.5]), the universal Lie algebra functor is also complicated; fortunately, we shall not need a full description of it. Let us note some useful things that *can* be said about it. Since the identities defining a Lie algebra are homogeneous, a Lie algebra $[k]\{M\}$ $(M \in \mathbf{Mod}_k)$ has a natural grading, and since the only Lie identities of degree 1 are those of k-modules, its component of degree 1 is precisely M. (One can also give a nice description of its component of degree 2: the only additional Lie identities occurring in that degree are those of *bilinearity* and *anticommutativity*, so this component can be identified with the exterior square $\Lambda^2 M$.) Under any conditions that insure that it embeds in its universal enveloping algebra (in particular, under the conditions recalled above from [**43**]), we can describe $[k]\{M\}$ as the Lie subalgebra of the tensor algebra $[k]<M>$ generated by M.

The free Lie algebra on a set X or $\{x, y, \dots\}$ will be written $[k]\{X\}$, respectively $[k]\{x, y, \dots\}$. There will be no danger of confusion with our notation for the universal Lie algebra on a module as long as we use different symbols for modules and free generating sets. (Cf. our notation $[k]<X>$ or $[k]<x, y, \dots>$ for free associative algebras, and $[k]<M>$ for tensor algebras.)

Let us now observe, as we did for $\mathbf{Ring}_k$, that we have a natural class of representable functors $\mathbf{Lie}_k \to \mathbf{Ab}$. Namely, for any k-module M, the functor which we may loosely write

$$L \mapsto \mathbf{Mod}_k(M, L)$$

and more formally as the composite

$$\mathbf{Lie}_k \xrightarrow{\text{forget}} \mathbf{Mod}_k \xrightarrow{\mathbf{Mod}_k(M,\,-)} \mathbf{Ab} \tag{49.2}$$

is representable, with representing Lie algebra $[k]\{M\}$, coaddition

$$\mathbf{a}_M\colon [k]\{M\} \to [k]\{M\}^{\rho} \amalg [k]\{M\}^{\lambda} \cong [k]\{M^{\lambda} \oplus M^{\rho}\}$$

determined by

$$\mathbf{a}_M(x) = x^{\lambda} + x^{\rho} \qquad (x \in M), \tag{49.3}$$

additive coinverse taking $x \in M$ to $-x$, and coneutral element necessarily taking all elements to 0. Again it is natural to ask whether every representable functor from Lie algebras to abelian groups has this form. The results of Chapter III suggest that any positive results on this question might apply more generally to representable functors from $\mathbf{Lie}_k$ to the variety $\mathbf{AbSemigp}^e$ of abelian semigroups with neutral element. We will obtain such results, under appropriate hypotheses, in the next section.

Incidentally, in this chapter we use the symbol $\amalg$ to denote coproducts in both

Lie$_k$ and **Ring**$_k$; the objects to which the operator is applied will determine which is meant. We will at times write $\amalg_{\mathbf{Lie}_k}$ or $\amalg_{\mathbf{Ring}_k}$ for emphasis.

§50. Abelian-group-valued and ring-valued functors: positive results.

Suppose F: **Lie**$_k \to$ **AbSemigp**^{e} is a representable functor, with representing Lie algebra R. If we compose on the right with Br: **Ring**$_k \to$ **Lie**$_k$, we get a representable functor **Ring**$_k \to$ **AbSemigp**^{e}, represented by $[k][R]$, with co-operations $\mathbf{a}$ and ε induced by those of R. By Theorem 13.15, we must have

$$(50.1) \qquad [k][R] = [k]\langle M\rangle$$

where

$$(50.2) \qquad M = \{x \in [k][R] \mid \mathbf{a}(x) = x^{\lambda} + x^{\rho}\}.$$

Does (50.1) imply that

$$(50.3) \qquad R \cong [k]\{M\}$$

holds? This is not known. For k a field, it is equivalent to the open question of whether a Lie algebra R such that $[k][R]$ is a free associative algebra must be a free Lie algebra. In any case, the most that a positive answer to this question would tell us is that there existed *some* isomorphism (50.3), which would not say that R contained the particular copy of M defined by (50.2). ($k\langle M\rangle$ admits a large group of automorphisms as a k-algebra, and R could be the Lie algebra generated by thc image of M under any of these.) Thus we still could not conclude that the coaddition of R satisfied (49.3) for a free generating submodule $M \subseteq R$.

To get further results, let us assume that the natural map $R \to \mathrm{Br}([k][R])$ is an embedding, so that we may regard R as a Lie subalgebra of $[k][R] \cong [k]\langle M\rangle$.

Let us now consider the derived unary co-operation on R corresponding to the derived operation of doubling (multiplication by 2) on abelian semigroups. This will be the restriction to R of the "co-doubling" map of $[k]\langle M\rangle$, which takes each $x \in M$ to $2x$. (This description of the co-doubling operation of $[k]\langle M\rangle$ follows immediately from its definition as the composite of the coaddition $\mathbf{a}$ with the codiagonal map (id, id): $[k]\langle M\rangle \amalg [k]\langle M\rangle \to [k]\langle M\rangle$.) Clearly, this map acts on the nth homogeneous component of $[k]\langle M\rangle$ as multiplication by 2^n; i.e., as a k-module endomorphism it is "diagonalizable", with eigenvalues 2, 4, 8, ... It follows that if the differences between these integers are all invertible in k, then any k-submodule of $[k]\langle M\rangle$ invariant under co-doubling must be homogeneous under the grading of the tensor algebra.

The necessary and sufficient condition for the differences between the powers of 2 (or between the powers of any other integer > 1) to be invertible in k is that k be a $\mathbf{Q}$-algebra. (It is sufficient because all these differences are invertible in $\mathbf{Q}$. To get necessity, note that if k is not a $\mathbf{Q}$-algebra, some positive integer d is not invertible in k. Since there are infinitely many distinct powers of 2, two of these are congruent modulo d; the difference between these will be noninvertible in k.)

Thus, if k is a $\mathbf{Q}$-algebra, our R is a *homogeneous* k-submodule of $[k]\langle M\rangle$. Note that the condition that k be a $\mathbf{Q}$-algebra subsumes the other assumption we have made, that the natural map $R \to [k][R] = [k]\langle M\rangle$ is an embedding.

We shall combine the above observations with

LEMMA 50.4. *Let* $k \in \mathbf{Comm}^1$ *be arbitrary, let* $M \in \mathbf{Mod}_k$, *and suppose* R *is a* homogeneous *sub-Lie-algebra of* $[k]\langle M\rangle$, *such that the natural map*

$$(50.5) \qquad [k][R] \to [k]\langle M\rangle$$

is an isomorphism. Then R *is precisely the sub-Lie-algebra of* $[k]\langle M\rangle$ *generated by* M.

PROOF. The surjectivity of (50.5) means that R generates $[k]\langle M\rangle$ as an associative algebra. Hence, as R is homogeneous, its degree-1 component must be all of M, hence R contains the Lie subalgebra of $[k]\langle M\rangle$ generated by M. Let us call that Lie subalgebra S. Assuming R larger than S, let i be the least integer such that the homogeneous component R_i is larger than S_i. Let N be the nonzero k-module R_i/S_i, and let us regard this as an associative algebra under the zero multiplication, and thus as a Lie algebra under the zero bracket operation. We see that the k-linear map $R \to N$ which acts on the ith component of R as the quotient map $R_i \to R_i/S_i$, and on all other components as 0 is a Lie algebra homomorphism, hence it induces an associative algebra homomorphism $[k][R] \to N$ which is zero on M but not on all of R. But this is impossible, since $[k][R] = [k]\langle M\rangle$ is generated by M. □

As we have noted, the hypotheses of the above Lemma hold if k is a commutative $\mathbf{Q}$-algebra, R a co-$\mathbf{AbSemigp}^e$ object of $\mathbf{Lie}_k$, and $M \subseteq [k][R]$ the submodule of elements on which the coaddition behaves as in (49.3). Recalling that the first of those hypotheses also implies that the sub-Lie-algebra of $[k]\langle M\rangle$ generated by M has the form $[k]\{M\}$, we get $R \cong [k]\{M\}$, and obtain an exact description of our co-$\mathbf{AbSemigp}^e$ object, which effectively says that the functor it represents has the form (49.2). As in §13, we note that the only morphisms among such co-$\mathbf{AbSemigp}^e$ objects are those induced by morphisms among the generating modules "M". We deduce

THEOREM 50.6 (cf. Theorem 13.15). *Let* $k \in \mathbf{Comm}^1_{\mathbf{Q}}$. *Then the functors*

$$\mathbf{Mod}_k{}^{\mathrm{op}} \xrightarrow{M \mapsto (49.2)} \mathbf{Rep}(\mathbf{Lie}_k, \mathbf{Ab}) \xrightarrow{\text{forget} \circ -} \mathbf{Rep}(\mathbf{Lie}_k, \mathbf{AbSemigp}^e)$$

are equivalences. For M *a* k*-module, the representing coalgebras for the functors* $\mathbf{Lie}_k \to \mathbf{Ab}$ *and* $\mathbf{Lie}_k \to \mathbf{AbSemigp}^e$ *induced by* M *both have underlying Lie algebra* $[k]\{M\}$. *The coaddition in each case takes* $x \in M$ *to* $x^\lambda + x^\rho$, *the coneutral element takes all elements to* 0, *and the coinverse operation of the co-*$\mathbf{Ab}$ *structure takes* $x \in M$ *to* $-x$. □

Let us now look, as in §23, at bilinear maps among such functors. Let $k \in \mathbf{Comm}^1_{\mathbf{Q}}$, and consider three k-modules M_1, M_2, M_3, and the $\mathbf{Ab}$-valued functors F, G, H they induce. A homomorphism of Lie algebras

$$(50.7) \qquad \mathbf{m}\colon [k]\{M_3\} \to [k]\{M_1\} \amalg_{\mathbf{Lie}_k} [k]\{M_2\} \cong [k]\{M_1 \oplus M_2\}$$

determines a morphism $m: F \times G \to H$ as set-valued functors, and we ask when this morphism will be *bilinear* with respect to the **Ab**-structures of these functors. The quickest way to get the answer is to first verify that m will be bilinear if and only if the same is true of the morphism among functors $\mathbf{Ring}_k \to \mathbf{Ab}$ obtained by composing with Br: $\mathbf{Ring}_k \to \mathbf{Lie}_k$. "Only if" is clear, while "if" follows from the fact that, since k is a $\mathbf{Q}$-algebra, every k-Lie algebra L is embeddable in an associative algebra $[k][L]$. Hence Proposition 23.7 tells us that m will be bilinear if and only if (50.7) carries M_3 into

$$(M_1 \otimes_k M_2 \ \oplus \ M_2 \otimes_k M_1) \ \cap \ [k]\{M_1 \oplus M_2\}$$

$$\text{(intersection taken within } [k]\langle M_1 \oplus M_2 \rangle\text{).}$$

It is not hard to see that this intersection is the image of $M_1 \otimes_k M_2$ under the map

$$x \otimes y \ \mapsto \ [x, y] \ = \ x \otimes y - y \otimes x,$$

and so is isomorphic to $M_1 \otimes_k M_2$.

(We observed parenthetically earlier that the degree-2 component of $[k]\{M\}$ is isomorphic to $\Lambda^2 M$. The image of $M_1 \otimes_k M_2$ in $[k]\{M_1 \oplus M_2\}$ under the above map corresponds to the middle summand of the formula

$$\Lambda^2(M_1 \oplus M_2) \ \cong \ (\Lambda^2 M_1) \ \oplus \ (M_1 \otimes_k M_2) \ \oplus \ (\Lambda^2 M_2),$$

which is indeed the part of $\Lambda^2(M_1 \oplus M_2)$ bilinear in M_1 and M_2.)

Thus, a bilinear morphism $F \times G \to H$ is determined by a linear map $M_3 \to M_1 \otimes_k M_2$. Specializing to the case $M_1 = M_2 = M_3$, and again stating our result first in the case of k a field, where things assume a more transparent form due to the duality between vector spaces and linearly compact vector spaces, and then in the general case, we get

THEOREM 50.8 (cf. Theorems 23.9, 25.1, 28.11). *Let k be a field of characteristic* 0. *Then any representable functor F*: $\mathbf{Lie}_k \to \mathbf{Ab}$ *can be written $L \mapsto A \mathbin{\hat{\otimes}} L$, where A is a linearly compact k-vector-space. Bilinear multiplications m on F correspond bijectively to continuous k-bilinear multiplications $*$ on A, via the formula*

(50.9) $\qquad m(as, bt) \ = \ (a*b)[s, t] \quad \text{for } a, b \in A, \ s, t \in L.$

More generally, for any $k \in \mathbf{Comm}^1_{\mathbf{Q}}$, the general representable functor F: $\mathbf{Lie}_k \to \mathbf{Ab}$ *can be written $L \mapsto A \circ L$ (cf. §28) where $A \in \mathbf{Mod}_k{}^{\mathrm{op}}$, and any bilinear multiplication m on F is again induced by a bilinear map $*$: $A \times A \to A$ (defined as in §28) via* (50.9). □

Let us now look at the conditions for such bilinear operations to yield various sorts of ring structures. Note that we can never get nontrivial *unital*-ring valued functors: $\mathbf{Lie}_k$ has a unique zeroary operation, so no representable functor on that category can admit more than one zeroary operation (cf. §11). Even if we do not ask for unitality, we shall see presently that some cases are degenerate; e.g., that there are no very interesting representable functors to associative rings. However, rather than proving those results from Theorem 50.6, we shall obtain them (later in this section) in a way that bypasses that Theorem's restriction on characteristic.

On the other hand, we clearly have a very nondegenerate representable functor $\mathbf{Lie}_k \to \mathbf{Lie}$, the forgetful functor. Let us use the above Theorem to determine the most general such functor when $k \in \mathbf{Comm}^1_{\mathbf{Q}}$. For simplicity we will assume k a field.

PROPOSITION 50.10. *Let k be a field of characteristic 0. Then any representable functor $\mathbf{Lie}_k \to \mathbf{Lie}$ has the form*

$$L \mapsto A \hat{\otimes} L \tag{50.11}$$

where A is a linearly compact nonunital commutative associative k-algebra. This construction induces an equivalence of categories

$$\mathbf{LCpComm}_k \xrightarrow{A \mapsto (50.11)} \mathbf{Rep}(\mathbf{Lie}_k, \mathbf{Lie}). \tag{50.12}$$

PROOF. Let $F: \mathbf{Lie}_k \to \mathbf{Lie}$ be representable. By Theorem 50.8, as an additive-group valued functor, F has the desired form (50.11) for some linearly compact k-vector-space A, while its multiplication has the form (50.9) for some continuous multiplication $*$ on A. To determine the conditions that $*$ must satisfy to make all the rings $A \hat{\otimes} L$ Lie rings, let us, as in §25, write down the Lie identities for such an object, substituting for each of the variables the product of an arbitrary element of A with an indeterminate in a free Lie algebra L. Thus, taking $L = [k]\{s, t\}$, the condition that the multiplication (50.9) be anticommutative says that for all $a, b \in A$,

$$(a*b)[s, t] \ + \ (b*a)[t, s] \ = \ 0. \tag{50.13}$$

Since $[s, t] = -[t, s] \neq 0$, (50.13) is equivalent to the condition $a*b - b*a = 0$, i.e., $*$ must be commutative. Likewise, to find necessary and sufficient conditions for m to satisfy the Jacobi identity (25.29), we take for L the free Lie algebra $[k]\{s, t, u\}$; then the condition is that for all $a, b, c \in A$,

$$(a*(b*c))\,[s, [t, u]] \ + \ (b*(c*a))\,[t, [u, s]] \ + \ (c*(a*b))\,[u, [s, t]] \ = \ 0.$$

Now in $[k]\{s, t, u\}$ the elements $[s, [t, u]]$ and $[t, [u, s]]$ are k-linearly independent (as may be checked by mapping this free Lie algebra into the free associative algebra $[k]\langle s, t, u\rangle$), while by the Jacobi identity, $[u, [s, t]]$ is the negative of their sum. Making this substitution in the above equation, and extracting the coefficients of those two linearly independent elements, we get the conditions $a*(b*c) - c*(a*b) = 0$ and $b*(c*a) - c*(a*b) = 0$, each of which, after application of commutativity, asserts the associativity of $*$. □

To see directly that the construction (50.11) carries Lie algebras to Lie rings, note first that given a Lie algebra L over k, the tensor product of L over k with any $A \in \mathbf{Comm}^1_k$ is a Lie algebra over A; this is just the operation of extending scalars from k to A. In particular it is a Lie ring. It is not hard to see that this conclusion is not lost on dropping the condition that A be unital, adding the condition that it be linearly compact, and then taking the completion of the tensor product.

(The same reasoning shows that if $\mathbf{V}$ is *any* subvariety of $\mathbf{NARing}$ defined by multilinear identities, and $\mathbf{V}_k$ the subvariety of $\mathbf{NARing}_k$ defined by the same identities, then for any $A \in \mathbf{LCpComm}_k$, the operation $A \hat{\otimes} -$ gives a representable functor $\mathbf{V}_k \to \mathbf{V}$. Similarly, if $\mathbf{V}$ is a subvariety of $\mathbf{NARing}^1$, or more generally a variety of rings with one or more zeroary operations, and is definable by multilinear identities, one finds that the construction of completed tensor product with any *unital* linearly compact commutative associative k-algebra carries $\mathbf{V}_k$ to $\mathbf{V}$.)

We now come to two very restrictive results, one saying that the only representable functors from Lie algebras to associative rings are those based on "associativity by default", the other making a similar statement for functors to Jordan rings. We will get these by a reduction to functors on $\mathbf{Ring}_k$ that does not go via Theorem 50.8 (namely, part (i) of the Proposition below), allowing us to use a weaker hypothesis on k than in that Theorem.

PROPOSITION 50.14. *Let* $k \in \mathbf{Comm}^1$ *be such that every Lie algebra over* k *embeds in its universal enveloping algebra. (E.g., let* k *be a* $\mathbf{Q}$*-algebra or a field.) Let* $F\colon \mathbf{Lie}_k \to \mathbf{NARing}$ *be a representable functor. Then*

(i) *If* k *is a field, there exists a linearly compact* k*-vector space* A *and a continuous* k*-bilinear operation* $*$ *on* A *such that, writing* $F'\colon \mathbf{Lie}_k \to \mathbf{NARing}$ *for the functor carrying* L *to* $A \hat{\otimes} L$ *with multiplication defined by* (50.9), *we have* $F\mathrm{Br} \cong F'\mathrm{Br}$ *in* $\mathbf{Rep}(\mathbf{Ring}_k, \mathbf{NARing})$ *(though we do not assert that* $F \cong F'$*). For general* k *satisfying our hypothesis, the corresponding statement holds with a construction of the form* $A \circ L$ *(cf. §28) in place of* $A \hat{\otimes} L$.

(ii) F *is associative-valued if and only if it satisfies the two identities saying that all* 3*-fold products are* 0.

(iii) *If* 2 *is invertible in* k, *then* F *is Jordan-valued if and only if it satisfies the commutative identity, and the five identities saying that all* 4*-fold products are* 0.

PROOF. The discussion with which we began this section shows that if we ignore the multiplication on the objects $F(L)$, regarding F as a functor $\mathbf{Lie}_k \to \mathbf{Ab}$, then there exists $F'\colon \mathbf{Lie}_k \to \mathbf{Ab}$ of the form described, such that $F\mathrm{Br} \cong F'\mathrm{Br}$. (We spoke there in terms of representing objects; here is the argument in terms of functors: By Theorem 13.15, $F\mathrm{Br} = EU$, where U is the forgetful functor $\mathbf{Ring}_k \to \mathbf{Mod}_k$ and E a representable functor $\mathbf{Mod}_k \to \mathbf{Ab}$; i.e., a functor which we may write $N \mapsto A \hat{\otimes} N$ with $A \in \mathbf{LCpMod}_k$ if k is a field, and $N \mapsto A \circ N$ with $A \in \mathbf{Mod}_k{}^{\mathrm{op}}$ in the general case. Now $U = V\mathrm{Br}$ where V is the forgetful functor $\mathbf{Lie}_k \to \mathbf{Mod}_k$, so defining $F' = EV$, we get $F\mathrm{Br} \cong F'\mathrm{Br}$ as functors to $\mathbf{Ab}$.) The key to completing the proof of (i) will be to show that the multiplicative structure on $F\mathrm{Br} \cong F'\mathrm{Br}$ induced by the given multiplication on F is also induced by some bilinear multiplication on F'. Once we have this, it will follow by the argument used to get Theorem 50.8 that this multiplication has the form (50.9).

(Note the slight abuse of notation: until we have gotten the above result, F' will merely denote an $\mathbf{Ab}$-valued functor, while $F'\mathrm{Br}$ denotes an $\mathbf{NARing}$-valued functor; namely the $\mathbf{Ab}$-valued composite of F' and Br, furnished with the multiplication induced via its isomorphism with the $\mathbf{NARing}$-valued functor $F\mathrm{Br}$.

Also note that what we are trying to prove for $F'\mathrm{Br}$ we already know for $F\mathrm{Br}$, namely that its multiplication arises from a multiplication on its left-hand factor. But since F may not be of the form (49.2), we cannot apply directly to this functor the argument that we intend eventually to apply to F'.)

Since F' has the form (49.2), Theorem 25.1 (if k is a field), respectively Theorem 28.11 (in the general case) tells us that the multiplication on $F'\mathrm{Br}$ has the form

$$(50.15) \qquad m(as, bt) = (a^*b)(st) + (b_*a)(ts),$$

for some bilinear maps $^*, {}_*: A \times A \to A$.

We now claim that $F'\mathrm{Br}$, when applied to any *commutative* associative k-algebra, gives a ring with zero multiplication. Indeed, if S is a commutative associative k-algebra, and S' the algebra with the same underlying vector space, but zero multiplication, then clearly

$$(50.16) \qquad \mathrm{Br}(S) \cong \mathrm{Br}(S') \quad \text{in } \mathbf{Lie}_k,$$

hence

$$F'\mathrm{Br}(S) \cong F\mathrm{Br}(S) \cong F\mathrm{Br}(S') \cong F'\mathrm{Br}(S') \qquad \text{in } \mathbf{NARing},$$

where the first and last steps use our isomorphism $F\mathrm{Br} \cong F'\mathrm{Br}$, while the middle step uses the fact that the multiplication on $F\mathrm{Br}$ arises from a multiplication on F, so that the Lie isomorphism (50.16) induces a ring isomorphism on applying F. Applying (50.15) to the last term in this chain, we see that since S' has zero multiplication, so does $F'\mathrm{Br}(S')$. Hence so does the first term, $F'\mathrm{Br}(S)$, as claimed.

Let us now apply this conclusion to a product $m(as, bt)$, where s, t are the indeterminates in a commuting polynomial ring $k[s, t]$, and a, b are arbitrary elements of A. Applying (50.15), setting the resulting expression equal to zero, and extracting the coefficient of $st = ts$, we get

$$(50.17) \qquad a^*b + b_*a = 0.$$

(Contrast this with the condition noted in Lemma 25.27 for m to be anticommutative: $a^*b + a_*b = 0$. Loosely, one equation is the condition for m to be commutator-like "on the right", the other "on the left".)

This allows us to express m in terms of the single operation *. Renaming this operation $*$ for symmetry, (50.15) becomes

$$(50.18) \qquad m(as, bt) = (a*b)(st-ts) \qquad (m \text{ the multiplication of } F'\mathrm{Br}).$$

But $st-ts$ is the bracket $[s,t]$ of $\mathrm{Br}(S)$; hence (50.18) says that if we give F' the multiplication (50.9) determined by our operation $*$ on A, then composition with Br yields our multiplication m on $F'\mathrm{Br}$, under which this functor is isomorphic to $F\mathrm{Br}$. This completes the proof of (i).

To get (ii) from (i), we recall from Theorem 25.25 that if the multiplication m of $F'\mathrm{Br}$ is associative, then the range of each of * and $_*$ in (50.15) is in the two-sided annihilator of the other. By (50.17), this says that * has its range in its own two-sided annihilator, i.e., writing it as $*$, that it satisfies the identities

$a*(b*c) = (a*b)*c = 0$. From (50.18) we see that this implies that all 3-fold products under the multiplication m of $F'\mathrm{Br}$ are zero. Moreover, by our hypothesis on k, every Lie algebra L embeds in a Lie algebra of the form $\mathrm{Br}(S)$, hence $F(L)$ embeds in $F\mathrm{Br}(S) \cong F'\mathrm{Br}(S)$, hence it, too, has all 3-fold products zero, as claimed. The converse is clear.

Finally, to get (iii) from (i), suppose 2 is invertible in k, and m satisfies the identities of Jordan algebras. In particular, m is commutative. Writing m as in (50.18), this says $*$ is anticommutative. Further, Theorem 25.40 says that all associations (without change of ordering) of any four-fold product of elements under $* = *$ will be equal; we claim that together with anticommutativity, this implies that their common value is zero. Indeed, the following computation, by alternating applications of four-term associativity and anticommutativity, is essentially the standard verification that an anticommutative associative algebra has cube zero, with an extra factor a thrown in because we only have associativity for four-fold products:

$$\begin{aligned} a*(b*(c*d)) &= -a*((c*d)*b) = -a*(c*(d*b)) = a*(c*(b*d)) \\ &= a*((c*b)*d) = -a*((b*c)*d) = -a*(b*(c*d)). \end{aligned}$$

Comparing the first and last terms, and recalling that 2 is invertible in k, we conclude that the above product is zero, hence so are all associations of $a*b*c*d$ in A. As in (ii), this implies the corresponding result for the multiplication of F, and again, the converse is immediate. □

Since the above result says that all representable functors $\mathbf{Lie}_k \to \mathbf{Ring}$ and $\mathbf{Lie}_k \to \mathbf{Jordan}$ actually land in much smaller varieties, it is reasonable to ask whether all representable functors $F\colon \mathbf{Lie}_k \to \mathbf{NARing}$ also land in some proper subvariety. Of course, if k satisfies the hypotheses of Proposition 50.14 and has characteristic p, then from part (i) of that Proposition we see that the additive structure of F must satisfy $px = 0$. But the next Exercise shows (for a much wider class of domain varieties than $\mathbf{Lie}_k$) that we are not forced into any proper subvariety of the variety defined by this condition.

In the definition of "monomial identity" below, the word "bracketing" refers to arrangement of parentheses, not to a "bracket" operation.

EXERCISE 50.19. *Let k be a field, and $\mathbf{V}$ any subvariety of $\mathbf{NARing}_k$ satisfying no multilinear* monomial *identity, i.e., no identity of the form*

$$(x_1 \ldots x_n) = 0$$

where $(x_1 \ldots x_n)$ is a bracketing of $x_1 \ldots x_n$. Show that there exists a representable functor $\mathbf{V} \to \mathbf{NARing}_k$ whose values are not contained in any proper subvariety of $\mathbf{NARing}_k$. Hence, dropping the k-module structure, if $\operatorname{char} k = 0$, *we get a representable functor $\mathbf{V} \to \mathbf{NARing}$ whose values are not contained in any proper subvariety of $\mathbf{NARing}$, while if* $\operatorname{char} k = p$, *we get a representable functor $\mathbf{V} \to \mathbf{NARing}_{\mathbf{Z}/p\mathbf{Z}}$ whose values are not contained in any proper subvariety of $\mathbf{NARing}_{\mathbf{Z}/p\mathbf{Z}}$.*

Hint: show that if $S \in \mathbf{V}$ satisfies no multilinear monomial identity, then the tensor product of S with the free nonassociative k-algebra on countably many *indeterminates* $a_1, a_2, \ldots$ *satisfies no k-algebra identity, and that this tensor product embeds in a "nonassociative formal power series algebra over* S", *which is the value at* S *of a representable functor on* $\mathbf{V}$.

Note that if $\mathrm{char}\ k = p$, then for a given variety $\mathbf{V} \subseteq \mathbf{NARing}_k$, there may or may not exist representable functors $\mathbf{V} \to \mathbf{NARing}$ which take on values not of characteristic p. For $\mathbf{V} = \mathbf{Comm}_k$, (and if $k = \mathbf{Z}/2\mathbf{Z}$, also for $\mathbf{V} = \mathbf{Bool}$), we have seen that there exist such functors, while when $\mathbf{V} = \mathbf{Ring}_k$ or $\mathbf{Lie}_k$, we know from Theorem 13.15 and Proposition 50.14(i) that there do not.

The computation made in the preceding exercise has an interesting strengthening in the case $\mathbf{V} = \mathbf{Lie}_k$: we claim that the free object of $\mathbf{NARing}_k$ referred to in the "Hint" can be replaced by a free object of $\mathbf{Ring}_k$; in other words, that tensor products of Lie algebras and associative algebras in general satisfy no nontrivial identities. Though this does not have any striking consequences for the theory of representable functors that we are aware of, it seems of independent interest, and is developed in the next exercise, after some combinatoric preparation.

EXERCISE 50.20. *Let* n *be a positive integer. If* π *is a permutation of* $\{1, \ldots, n\}$, *let us define equivalence relations* $\mathrm{eq}_0(\pi) \subseteq \mathrm{eq}_1(\pi) \subseteq \ldots$ *on* $\{1, \ldots, n\}$ *as follows: We take for* $\mathrm{eq}_0(\pi)$ *the discrete (least) equivalence relation, and assuming inductively that* $\mathrm{eq}_{i-1}(\pi)$ *is an equivalence relation under which each equivalence class is a block of successive integers, whose image under* π *also forms a block of successive integers, we let* $\mathrm{eq}_i(\pi)$ *be the least equivalence relation which contains* $\mathrm{eq}_{i-1}(\pi)$, *and puts two blocks from the latter relation into the same equivalence class if they are adjacent to one another (i.e., contain integers differing by* 1) *and also have adjacent images under* π.

Now suppose that w *is a Lie monomial formed by inserting brackets in some way in the ordered string* $x_1, \ldots, x_n$. *Suppose we evaluate* w *in the free associative ring* $[\mathbf{Z}]\langle x_1, \ldots, x_n\rangle$, *interpreting the Lie brackets as commutator brackets, and say that a permutation* π *of* $1, \ldots, n$ *is "associated with* w" *if one of the* 2^{n-1} *monomials occurring in the expansion of* w *is* $x_{\pi(1)} \cdots x_{\pi(n)}$.

(i) *Let* π_{alt} *be the permutation associated with* w *obtained by starting from the outermost bracket of* w, *and alternately taking first and second terms of commutators at successive levels of brackets (i.e., selecting from every commutator* $[u, v] = uv - vu$ *the term* uv *if the bracket occurs at the top level or an even number of steps down from the top, and* vu *if it occurs an odd number of steps from the top). Show that* $\mathrm{eq}_1(\pi_{\mathrm{alt}})$ *is the relation joining only those pairs of indices* $i, i+1$ *that occur on indeterminates in "innermost" brackets* $[x_i, x_{i+1}]$ *in* w *(so that any indeterminate which occurs bracketed with a larger expression forms a singleton equivalence class), and give a similar characterization of* $\mathrm{eq}_i(\pi_{\mathrm{alt}})$ *for general* i. *Letting* $\pi_{\mathrm{alt}'}$ *denote the permutation obtained in the same way, but reversing the roles of even and odd levels, show that* $\mathrm{eq}_i(\pi_{\mathrm{alt}'}) = \mathrm{eq}_i(\pi_{\mathrm{alt}})$ *for all* i.

(ii) *Deduce that from either of the permutations* $\pi_{\mathrm{alt}}, \pi_{\mathrm{alt}'}$ *one can reconstruct the bracketing of* w.

(iii) *Show that for every* π *associated with* w, $\mathrm{eq}_i(\pi) \supseteq \mathrm{eq}_i(\pi_{\mathrm{alt}})$ *for all* i, *with strict inclusion for some* i *if* $\pi \neq \pi_{\mathrm{alt}}, \pi_{\mathrm{alt}'}$.

(iv) *Deduce that given any nonempty family* W *of bracketings of* $x_1 \ldots x_n$, *if from all the words appearing in the commutator expansions of elements* $w \in W$, *we choose one word* $x_{\pi(1)} \ldots x_{\pi(n)}$ *so as to minimize the system of equivalence relations* $(\mathrm{eq}_1(\pi), \ldots, \mathrm{eq}_n(\pi))$ *(under the product of the inclusion orderings), then* $x_{\pi(1)} \ldots x_{\pi(n)}$ *appears in the expansion of exactly one member of* W.

(v) *Deduce that for any* $k \in \mathbf{Comm}^1$, *the distinct bracketings of the ordered string* $x_1 \ldots x_n$ *in the free Lie algebra* $[k]\{x_1, \ldots, x_n\}$ *give linearly independent elements, and that the k-submodule spanned by these elements is a k-module direct summand in this algebra.*

(vi) *Deduce that in the tensor product* $[k]<a_1, \ldots, a_n> \otimes_k [k]\{x_1, \ldots, x_n\}$ *of the free associative algebra on* n *generators with the free Lie algebra on* n *generators, the elements* $a_1 \otimes x_1, \ldots, a_n \otimes x_n$ *satisfy no n-linear relations (i.e., that the family of elements formed by* permuting *and* bracketing *these* n *elements in all possible ways, and evaluating the resulting monomials in this tensor product algebra, is linearly independent).*

(vii) *Deduce that the class of tensor products* $A \otimes_k L$ *of associative k-algebras* A *and Lie k-algebras* L *does not lie in any proper subvariety of* $\mathbf{NARing}_k$.

It would be interesting to know whether the pair $(\mathbf{Ring}_k, \mathbf{Lie}_k)$ is minimal for the property established in (vii) above. We note that the same argument clearly works with Lie algebras replaced by Jordan algebras (or, indeed, special Jordan algebras).

Let us end this section by observing that, just as the concept of an associative k-algebra (for $k \in \mathbf{Comm}^1$) is generalized by that of a K-ring (for $K \in \mathbf{Ring}^1$), one could generalize the concept of a Lie algebra over a commutative ring k by considering, for $K \in \mathbf{Ring}^1$, objects L consisting of a K-bimodule with an operation $[\ ,\ ]$ satisfying the identities satisfied by commutator brackets and the bimodule structure in associative K-rings. Alternatively, one could fix a Lie algebra E and consider the variety of Lie algebras given with actions of E on them by derivations – or the variety of Lie algebras given with homomorphisms of E into them. (Note that this last variety, which has a forgetful functor into the preceding one, has nonzero zeroary operations, eliminating one source of restrictions on representable functors that we noted for $\mathbf{Lie}_k$.) The questions we have considered above for varieties $\mathbf{Lie}_k$ might be studied in any of these more general contexts.

§51. Counterexamples in characteristic p.

It is a familiar fact that Lie algebras are not as well-behaved in characteristic p as in characteristic 0, and that for many purposes, the more natural concept in characteristic p is that of a *p-Lie algebra* (or *restricted Lie algebra*). This is an object which has, in addition to the operations of a Lie algebra, a unary operation, written $(-)^p$ (or, often, $(-)^{(p)}$ or $(-)^{[p]}$), satisfying identities mimicking those of the pth power operation on associative k-algebras with respect to commutator brackets [**100**]. We shall show here that, in contrast to Theorem 50.6, when k is a field of characteristic p not every representable functor $\mathbf{Lie}_k \to \mathbf{Ab}$ has the form

(49.2), precisely by taking advantage of the lack of a pth power operation. We shall then discuss what positive result might be true for functors on categories of p-Lie algebras.

Let k be any field of positive characteristic p. We have seen that if R is a co-**Ab** object of $\mathbf{Lie}_k$, then $[k][R]$ has the form $[k]\langle M\rangle$ for some k-vector space M, equivalently, is a free associative k-algebra $[k]\langle X\rangle$, where X may be chosen so that the coaddition of R is the restriction of the "standard" coaddition on this free algebra, given by $x \mapsto x^\lambda + x^\rho$ $(x \in X)$; but we cannot say that R will be a homogeneous Lie subalgebra of $[k]\langle X\rangle$, hence it may fail to contain X.

So to get our example, let us start with the free associative algebra on two indeterminates, $[k]\langle x, y\rangle$, with the standard coaddition

(51.1) $$\mathbf{a}(x) = x^\lambda + x^\rho, \qquad \mathbf{a}(y) = y^\lambda + y^\rho,$$

and let R be the Lie subalgebra generated by the elements

$$x \qquad \text{and} \qquad z = y - x^p.$$

Clearly, these elements are also a pair of free generators of $[k]\langle x, y\rangle$ as an associative algebra, hence the Lie subalgebra R that they generate is a free Lie algebra on these generators.

We claim that R is closed under the coaddition $\mathbf{a}$ on $[k]\langle x, y\rangle$ given by (51.1), i.e., is carried by $\mathbf{a}$ into $R^\lambda \amalg_{\mathbf{Lie}_k} R^\rho$. That the element $\mathbf{a}(x) = x^\lambda + x^\rho$ lies in $R^\lambda \amalg_{\mathbf{Lie}_k} R^\rho$ is clear. To show the same for $\mathbf{a}(z)$, let us recall that, as k is of characteristic p, if we write the pth power of a sum of two elements of any associative k-algebra as the sum of their pth powers plus a noncommutative polynomial f in these elements,

(51.2) $$(u+v)^p = u^p + v^p + f(u, v),$$

then this polynomial f can be written purely in terms of commutator brackets. (This is a basic fact underlying the definition of p-Lie algebra, [**100**, Def. 4, pp.187-188].) Hence

$$\begin{aligned}\mathbf{a}(z) &= \mathbf{a}(y - x^p) = (y^\lambda + y^\rho) - (x^\lambda + x^\rho)^p \\ &= y^\lambda + y^\rho - (x^\lambda)^p - (x^\rho)^p - f(x^\lambda, x^\rho) \\ &= z^\lambda + z^\rho - f(x^\lambda, x^\rho),\end{aligned}$$

which does indeed lie in $R^\lambda \amalg_{\mathbf{Lie}_k} R^\rho$.

From the facts that on $[k]\langle x, y\rangle$, $\mathbf{a}$ is coassociative and cocommutative, the unique zeroary co-operation ε is a co-neutral element for $\mathbf{a}$, and the map $\mathbf{i}$: $x \mapsto -x$, $y \mapsto -y$ is a coinverse for $\mathbf{a}$, the same properties follow immediately for the restrictions of these co-operations to the Lie subalgebra R. Thus $(R, \mathbf{a}, \mathbf{i}, \varepsilon)$ is a co-**Ab** object of $\mathbf{Lie}_k$.

The functor F represented by this object can be described as taking a Lie algebra L to the abelian group with underlying set $L \times L$, and addition

(51.3) $$(\xi, \zeta) + (\xi', \zeta') = (\xi + \xi', \zeta + \zeta' - f(\xi, \xi')).$$

(By abuse of notation, we are using the same symbol f for the associative polynomial defined by (51.2), and the Lie polynomial which, when evaluated in commutator brackets in characteristic p, gives that polynomial.)

There is another convenient way of describing this functor F. Given a Lie algebra L, embed L in *any* associative k-algebra E (which may, but need not, be its universal enveloping algebra). Then it is not hard to verify that we may make the identification

(51.4) $F(L) = \{(\xi, \zeta+\xi^p) \mid \xi, \zeta \in L\} \subseteq E\times E$, under the restriction of the ordinary additive structure of $E\times E$ to this subset.

Note that given a homomorphism of Lie algebras $a\colon L \to L'$, we can find a homomorphism of appropriately chosen associative algebras $b\colon E \to E'$ extending a. We can then characterize $F(a)$ as acting by applying b coordinatewise to elements of the form $(\xi, \zeta+\xi^p)$ $(\xi, \zeta \in L)$ in $E\times E$.

We shall use the above description of F, rather than that of (51.3), in what follows.

Let us compare F with the functor $G\colon \mathbf{Lie}_k \to \mathbf{Ab}$ taking every Lie algebra to the direct product of two copies of its underlying additive group; i.e., the functor represented by the Lie subalgebra $[k]\{x, y\} \subseteq [k]\langle x, y\rangle$ under the coaddition (51.1). Clearly, we have an isomorphism of functors $\mathbf{Ring}_k \to \mathbf{Ab}$

(51.5) $$F\mathrm{Br} \cong G\mathrm{Br}.$$

(This can be seen either from the way we constructed F, or from (51.4).) From (51.5) it is easily deduced that if F is isomorphic to *some* functor of the form (49.2), then it is isomorphic to G. Also, from the fact that the representing objects for both F and G are free Lie algebras on two generators, we see that if we write U for the forgetful functor $\mathbf{Ab} \to \mathbf{Set}$, then

(51.6) $$UF \cong UG.$$

Moreover, for each Lie algebra L, we claim that

$$F(L) \cong G(L)$$

as abelian groups. Indeed, $F(L)$ and $G(L)$ both satisfy the identity $pa = 0$ (since each embeds in the additive group of an associative k-algebra, and k has characteristic p), and by (51.6) they have the same cardinality, hence they are isomorphic.

Despite these three facts, we claim that F and G are *not* isomorphic as functors.

To see this, suppose that

$$i\colon F \to G$$

were an isomorphism. Let us apply each of F, G to the free Lie algebra $M = [k]\{t\}$ on one generator t, which is one-dimensional, with basis-element t, and which we shall regard as a Lie subalgebra of the free associative algebra in one indeterminate, $[k][t]$. Then $i(M)$ will carry $(t, t^p) \in F(M) \subseteq [k][t]\times[k][t]$ (cf. (51.4)) to an element $(\alpha t, \beta t) \in G(M) = M\times M$ with α, $\beta \in k$ not both zero; hence by composing i on the left with an automorphism of G (coming from an element of $\mathrm{GL}_2(k)$) we can assume without loss of generality that $i(M)(t, t^p) = (t, 0)$. Using the universal property of the free associative algebra $[k][t]$, we deduce that for any element ξ of any Lie algebra L contained in an associative algebra E, we have $i(L)(\xi, \xi^p) = (\xi, 0)$. But now take such a Lie algebra L containing two elements ξ and ξ' such that $f(\xi, \xi') \neq 0$. On the one hand, we see that in $F(L) \subseteq E \times E$,

$$(\xi, \xi^p) + (\xi', \xi'^p) \;=\; (\xi+\xi',\ \xi^p+\xi'^p) \;\neq\; (\xi+\xi', (\xi+\xi')^p).$$

But on the other hand, the images of (ξ, ξ^p) and (ξ', ξ'^p) under $i(L)$ satisfy

$$(\xi, 0) + (\xi', 0) \;=\; (\xi+\xi', 0).$$

Comparing these two relations, we get a contradiction to the assumption that i is a morphism of functors.

Perhaps a more satisfying way of verifying that F is not isomorphic to G is to observe that given an object of $\mathbf{Lie}_k$ and a family of endomorphisms of this object, each of our functors yields an object of **Ab** with a corresponding family of endomorphisms; thus it will suffice to display a Lie algebra L with endomorphisms α and β such that the induced systems $(F(L), F(\alpha), F(\beta))$ and $(G(L), G(\alpha), G(\beta))$ are nonisomorphic. Let E be the associative k-algebra $[k]\langle q, r \mid rq = r^2 = 0\rangle$. (Thus, the relations say that left multiplication by r is the zero operator.) We easily verify that E has k-basis $\{q, q^2, \ldots; r, qr, q^2r, \ldots\}$. (If we wish, we may also impose the relations $q^3 = q^2$, $q^2r = qr$, which say that left multiplication by q is idempotent, getting a version of E which is 4-dimensional as a vector space.) Let L be the Lie subalgebra of E generated by q and r; we find that this is spanned by the elements

$$q^i r \ (i \geq 0) \ \text{ and } \ q. \tag{51.7}$$

It is easy to see that E admits k-algebra endomorphisms α and β defined by

$$\alpha(q) = q+r, \qquad \beta(q) = 0,$$
$$\alpha(r) = r, \qquad \beta(r) = r.$$

Writing out the effects of α and β on our basis, one finds that the k-module endomorphism $(\beta-1)(\alpha-1)$ of E annihilates the subset (51.7) which we have noted spans L. (Indeed, $\alpha-1$ annihilates all the elements $q^i r$, and takes q to r which is annihilated by $\beta-1$.) Hence $(G(\beta)-1)(G(\alpha)-1)$ annihilates all of $G(L)$. On the other hand, note that for any $i > 0$, the element q^i of E (which does not belong to L unless $i = 1$), is carried by $\alpha-1$ to $q^{i-1}r$, and if $i > 1$, this is taken by $\beta-1$ to $-q^{i-1}r \neq 0$. Hence, as the action of $(F(\beta)-1)(F(\alpha)-1)$ corresponds to the coordinate-wise action of $(\beta-1)(\alpha-1)$ on the additive group of elements $(\xi, \zeta+\xi^p)$ $(\xi, \zeta \in L)$, and this takes (q, q^p) to $(0, -q^{p-1}r) \neq 0$, we have $(F(\beta)-1)(F(\alpha)-1) \neq 0$, establishing the nonisomorphism of $(F(L), F(\alpha), F(\beta))$ and $(G(L), G(\alpha), G(\beta))$.

Note that if we denote by $U\colon \mathbf{Lie}_k \to \mathbf{Ab}$ the underlying additive group functor on Lie algebras, then the functor F we have constructed can be regarded as an extension of U by U, determined by a cocycle (in the sense of group cohomology) given by the polynomial f. This suggests

PROBLEM 51.8. *Let k be a field of positive characteristic p.*

(i) *Describe* $\mathrm{Ext}(U, U)$, *where U is the underlying additive group functor* $\mathbf{Lie}_k \to \mathbf{Ab}$.

Actually, the above question is ambiguous unless one has an affirmative answer to at least the $\mathbf{V} = \mathbf{Lie}_k$ *case of*

(ii) *If* $\mathbf{V}$ *is a variety, is every extension of one representable* $\mathbf{Ab}$*-valued functor on* $\mathbf{V}$ *by another (as* $\mathbf{Ab}$*-valued functors) again representable?*

Returning to functors on $\mathbf{Lie}_k$,

(iii) *Does every representable functor* $\mathbf{Lie}_k \to \mathbf{Ab}$ *with finitely generated representing Lie algebra R have a "composition series", whose factors are copies of the functor U? (If so, can anything of the sort be said without the condition of finite generation on the representing object?)*

We end this section by considering possible results on representable functors on the variety $\mathbf{Lie}_k^{(p)}$ of p-Lie algebras; for this we will assume familiarity with the definition and elementary properties of these objects [**100**, §§V.7-8].

We note that the commutator-brackets functor Br: $\mathbf{Ring}_k \to \mathbf{Lie}_k$ factors in an obvious way

$$\mathbf{Ring}_k \to \mathbf{Lie}_k^{(p)} \to \mathbf{Lie}_k,$$

and that the change of variables $(\xi, \zeta) \mapsto (\xi, \zeta+\xi^p)$, which for the F, G of the preceding example yields the isomorphism $F\mathrm{Br} \cong G\mathrm{Br}$, can already be performed on the composites of F and G with the second factor above. (Equivalently, the cocycle determining this extension becomes a coboundary once one has a pth power operation.) Thus, we know no counterexample to

QUESTION 51.9. *If k is a field of characteristic p, does every representable functor* $\mathbf{Lie}_k^{(p)} \to \mathbf{Ab}$ *have the form*

$$\mathbf{Lie}_k^{(p)} \xrightarrow{\text{forget}} \mathbf{Mod}_k \xrightarrow{\mathbf{Mod}_k(M,-)} \mathbf{Ab} \qquad (51.10)$$

for some $M \in \mathbf{Mod}_k$ *?*

One might hope to get a counterexample by a construction such as $L \mapsto \{\xi \in L \mid \xi^p = 0\}$, which uses the pth power map, and not just the k-vector space structure. But we find that the sets given by this construction are not in general closed under addition: two elements ξ and ξ' may both be in the kernel of $(\ )^p$, but their sum will not be if $f(\xi, \xi') \neq 0$.

But if we remove the condition that k be a field, we can indeed cook up a counterexample based on this idea. Note that if k is a commutative integral domain that is not a field, there exists a nonzero k-module N such that $N \otimes_k N = 0$. (The easiest example to name is the k-module K/k, where K is the field of fractions of k.) We see that for any k-linear maps u, v of such an N into an object $L \in \mathbf{NARing}_k$, the product in L of any element of $u(N)$ and any element of $v(N)$ must be 0. In particular, if this L is a Lie algebra, then for all

$x \in N$ we have $f(u(x), v(x)) = 0$, hence if $L \in \mathbf{Lie}_k^{(p)}$, we have $(u(x)+v(x))^p = u(x)^p + v(x)^p$. Thus, if two k-module homomorphisms $N \to L$ both have image in the kernel of the pth power map, so does their sum; hence the construction

$$(51.11) \qquad L \mapsto \{u \in \mathbf{Mod}_k(N, L) \mid \forall x \in N,\ u(x)^p = 0\}$$

is $\mathbf{Ab}$-valued.

We can prove that the functor (51.11) is not of the form (51.10) if we further assume that N admits a one-to-one map $\pi\colon N \to N$ which respects addition and satisfies $\pi(cx) = c^p\pi(x)$ $(c \in k)$. (In the case $N = K/k$ mentioned above, the pth-power operation of K induces a function π on N satisfying this identity, and if k is integrally closed in K, this π is one-to-one.) Given such N and π, let us define two p-Lie algebras L_1 and L_2, each having N as its underlying k-module, each having bracket operation equal to zero (necessarily!), but with L_1 having 0 as its "pth power" operation, while L_2 has π in this role. Since L_1 and L_2 have the same k-module structure, any functor of the form (51.10) must give isomorphic abelian groups when applied to L_1 and L_2. But the functor (51.11) carries L_2 to zero because π is one-to-one, while its value at L_1 contains the identity map of N, hence is nonzero; so the functor (51.11) cannot be of the form (51.10).

§52. Functors $\mathbf{Lie}_k \to \mathbf{Group}$.

Let k be a field of characteristic 0.

Should we expect there to be representable functors $\mathbf{Lie}_k \to \mathbf{Group}$, other than those taking values in $\mathbf{Ab}$, i.e., other than those based on the additive group structure of a Lie algebra?

A naive answer might be, "Certainly! Lie algebras are intimately connected with nonabelian groups". However, when one looks at this intimate connection, one sees that the usual universal construction that obtains a Lie *group* from a finite-dimensional real Lie algebra uses the algebra as the "infinitesimal part" of the group, and "generates" the group from it. This looks like a *left* adjoint construction, while representable functors are right adjoints. (In fact, the above construction is left adjoint to the tangent space construction going from Lie groups to Lie algebras.) The sparsity of representable functors from $\mathbf{Lie}_k$ to associative rings, established in Proposition 50.14, in contrast to the abundance of such constructions on $\mathbf{Ring}_k$ and $\mathbf{Comm}_k$, also seems a negative indication.

The authors were therefore surprised to discover that there is a rich class of representable functors $\mathbf{Lie}_k \to \mathbf{Group}$. To describe these, let us recall some background from Lie group theory. If G is a Lie group, and L the associated Lie algebra (with underlying vector space the tangent space to G at the identity), then each $x \in L$ induces a one-parameter subgroup of G, written $t \mapsto \exp(tx)$ $(t \in \mathbf{R})$. The resulting map $\exp\colon L \to G$ gives an analytic bijection between a neighborhood of $0 \in L$ and a neighborhood of the identity $e \in G$. Hence the group operations of G correspond to analytic operations defined in a neighborhood of 0 in L. It is easy to characterize the operations of L corresponding to the neutral element and inverse map of G:

$$(52.1)\qquad \begin{aligned} e &= \exp(0) \\ \exp(x)^{-1} &= \exp(-x). \end{aligned}$$

Less trivially, the power series expansion of the operation corresponding to the multiplication of G turns out to be expressible in terms of the Lie algebra structure of L; namely

$$(52.2)\qquad \exp(x)\exp(y) = \exp(C(x,y))$$

where $C(x,y)$ is given by the *Campbell-Hausdorff* formula [**174**, §§4.7-4.8] [**100**, §V.5], a power series whose first few terms are shown in

$$(52.3)\qquad C(x,y) = (x+y) + \tfrac{1}{2}[x,y] + \tfrac{1}{12}([x,[x,y]]+[y,[y,x]]) + \ldots .$$

We cannot, of course, apply this formula to elements x and y of a Lie algebra L over our arbitrary field k of characteristic 0, since we cannot form infinite sums. But note that if instead we take for x and y formal power series over such an L with zero constant term:

$$(52.4)\qquad x = x_1 t + x_2 t^2 + \ldots, \qquad y = y_1 t + y_2 t^2 + \ldots \qquad (x_i, y_i \in L),$$

then we *can* apply (52.3) to get another such power series, since for each n, (52.3) has only finitely many terms that can contribute to the coefficient of t^n in $C(x,y)$. The resulting operation is not a polynomial in x and y within the power series algebra, but if we regard a power series as a sequence of elements of L, then each component of $C(x,y)$ is a Lie polynomial in the components of x and y, so we have a representable functor **Lie**$_k$ $\to$ **Group**. Intuitively, x and y in (52.4) can be thought of as formal germs of curves originating at the identity in a Lie group G having Lie algebra L, and (52.3) as the formula for the germ of their pointwise product.

Once we have this example, we can see how to construct many more: groups of truncated formal power series, subdirect products of groups of formal power series, etc.. A construction subsuming all these is the following: Choose a fixed radical linearly compact commutative associative k-algebra A. (In the example just described, this was $[k][\![t]\!]$.) Note that for a linearly compact associative algebra, the condition of being radical is equivalent to that of being pro-nilpotent (i.e., an inverse limit of nilpotent algebras) because linearly compact associative algebras are pro-finite-dimensional, and finite-dimensional radical algebras are nilpotent. Hence for such A, the construction taking a general Lie algebra L to $A \mathbin{\hat{\otimes}} L$ carries Lie algebras to pro-nilpotent Lie algebras. Now on a nilpotent Lie algebra, (52.3) is a well-defined operation, and it follows that one can also make sense of it on a pro-nilpotent Lie algebra. Moreover, the fact that in the classical case of a finite-dimensional real Lie algebra, this formula gives an associative multiplication in a neighborhood of e is easily shown to imply that associativity follows formally from (52.3) and the Lie identities; we conclude that the induced operation on $A \mathbin{\hat{\otimes}} L$ is associative. We likewise find that $x \mapsto 0$ and $x \mapsto -x$ are a neutral element and an inverse operation with respect to this operation; thus we get a functor

$$(52.5) \qquad \mathbf{Lie}_k \xrightarrow{A \hat{\otimes} -} \mathbf{ProNlptLie}_k \xrightarrow{\text{Campbell-Hausdorff}} \mathbf{Group}.$$

This construction is developed in detail in [**83**, §14], although there, it is not viewed as assigning to each radical linearly compact commutative algebra A a functor $\mathbf{Lie}_k \to \mathbf{Group}$, but as associating to each Lie algebra L a functor $\mathbf{RadLCpComm}_k \to \mathbf{Group}$ (or, what is the same thing, a functor $\mathbf{NlptComm}_k \to \mathbf{Group}$), called a "formal group law". From either point of view, the group associated to $L \in \mathbf{Lie}_k$ and $A \in \mathbf{RadLCpComm}_k$ can be described as the group of maps from the "formal spectrum" Spf(A) of A (a "germ of a variety") to a certain formal group scheme ("germ of a Lie group") $G(L)$. This $G(L)$ is, in fact, the formal spectrum of the dual of the universal enveloping algebra of L, which we referred to briefly near the end of §43. Its underlying linearly compact commutative algebra has the form of a formal power series algebra, namely $[k][[\hat{L}]]$, where $\hat{L}$ is the vector-space dual of L.

QUESTION 52.6. *If* k *is a field of characteristic* 0, *is the functor*

$$\mathbf{RadLCpComm}_k \xrightarrow{A \mapsto (52.5)} \mathbf{Rep}(\mathbf{Lie}_k, \mathbf{Group})$$

an equivalence of categories? (Is it at least surjective on isomorphism classes of objects?)

It might be possible to study the above question by the same method used in §50, namely, pre-composing a general $F \in \mathbf{Rep}(\mathbf{Lie}_k, \mathbf{Group})$ with Br: $\mathbf{Ring}_k \to \mathbf{Lie}_k$, and applying Theorem 33.7 (in place of Theorem 13.15).

We know still less about

PROBLEM 52.7. *Suppose* k *is a field of positive characteristic* p.

(i) *Characterize representable functors* $\mathbf{Lie}_k \to \mathbf{Group}$.

(ii) *Characterize representable functors* $\mathbf{Lie}_k^{(p)} \to \mathbf{Group}$.

Concerning (i), we remark that p does not occur in any denominators in the Campbell-Hausdorff formula before the degree-p terms; hence (52.5) makes sense whenever A satisfies the nilpotence condition $A^p = 0$, the result being a functor from $\mathbf{Lie}_k$ into the variety of groups nilpotent of degree p. Perhaps one can get more general functors using some version of "divided powers", customized to the coefficients of the Campbell-Hausdorff formula rather than to those of the exponential function.

CHAPTER X.

Multilinear algebra of representable functors on k-$\mathbf{Ring}^1$.

In earlier chapters, we characterized *bilinear* maps among representable **Ab**-valued functors on categories of associative rings, in order to study representable ring-valued functors on these categories. In §53 below we discuss multilinear maps, and in §54 higher-degree (quadratic etc.) maps among such functors, mainly for their own interest.

In §55, we develop a little of the theory of higher-degree maps among *modules*, where there are complications not present in the abelian group case. With this as background, we recall in §56 the definitions of certain varieties involving quadratic operations which arise in the theory of Jordan algebras, and sketch how one might study representable functors from associative rings to such objects.

Proofs which are either similar to proofs in earlier chapters, or establish tangential results, will often be omitted or sketched.

§53. Multilinear maps, and "tensor products" of representable functors.

Let $k \in \mathbf{Ring}^1$. The characterization of bilinear maps among representable functors $k\text{-}\mathbf{Ring}^1 \to \mathbf{Ab}$ gotten in Proposition 23.7 generalizes without difficulty to

PROPOSITION 53.1. *Let* $F_1, \dots, F_n,\ G \in \mathbf{Rep}(k\text{-}\mathbf{Ring}^1, \mathbf{Ab})$, *with representing objects* $k<M_1>, \dots, k<M_n>,\ k<N>$, *and cogroup structures as in Theorem 13.15. Then the multilinear morphisms* $m\colon F_1 \times \dots \times F_n \to G$ *are induced by those k-ring homomorphisms* $k<N> \to k<M_1 \oplus \dots \oplus M_n>$ *which carry* N *into the sum, over all permutations* π *of* $\{1, \dots, n\}$, *of the bimodules* $M_{\pi(1)} \otimes_k \dots \otimes_k M_{\pi(n)}$. □

A consequence is

COROLLARY 53.2. *Let* $F_1, \dots, F_n \in \mathbf{Rep}(k\text{-}\mathbf{Ring}^1, \mathbf{Ab})$, *with representing objects* $k<M_1>, \dots, k<M_n>$, *and cogroup structures as in Theorem 13.15. Then there exists a* $G \in \mathbf{Rep}(k\text{-}\mathbf{Ring}^1, \mathbf{Ab})$ *with a* universal *multilinear morphism*

$$m\colon\ F_1 \times \dots \times F_n \ \to\ G.$$

This object G *is represented by* $k<N>$, *where*

$$N \ = \ \bigoplus_{\pi \in S_n} M_{\pi(1)} \otimes_k \dots \otimes_k M_{\pi(n)},$$

and the universal morphism m *is induced by the inclusion of* N *in* $k<M_1 \oplus \dots \oplus M_n>$. □

The reader should verify that despite the contravariant relation between bimodules and representable functors, the universal property of this G, like that of a tensor product of modules, makes it *initial* among functors admitting such multilinear maps. Hence, we shall write this G as $F_1 \otimes \ldots \otimes F_n$.

However, this "tensor product" operation on representable **Ab**-valued functors is not associative! Indeed, if $n = 3$ and we compare $F_1 \otimes F_2 \otimes F_3$ (the functor with a universal trilinear map into it) with $(F_1 \otimes F_2) \otimes F_3$, we see that the former is represented by the tensor ring on a sum of $|S_3| = 6$ tensor products of the bimodules M_1, M_2 and M_3, while the latter only involves $|S_2| \cdot |S_2| = 4$ such summands, namely, those four in which M_1 and M_2 appear adjacent to one another. Similarly, the object representing $F_1 \otimes (F_2 \otimes F_3)$ is the tensor ring on the sum of the four tensor products in which M_2 and M_3 are adjacent; thus, this functor is not, in general, isomorphic to $(F_1 \otimes F_2) \otimes F_3$.

Why does the proof that for abelian groups A_1, A_2, A_3, there is a natural isomorphism

$$(53.3) \qquad (A_1 \otimes A_2) \otimes A_3 \cong A_1 \otimes A_2 \otimes A_3$$

taking the element $(a_1 \otimes a_2) \otimes a_3$ to $a_1 \otimes a_2 \otimes a_3$ $(a_i \in A_i)$ not go over to such functors? Given $F_1, \ldots, F_n$ as in Corollary 53.2 and elements $a_i \in F_i(S)$ $(i = 1, \ldots, n)$ for some $S \in k\text{-}\mathbf{Ring}^1_1$, let us write $a_1 \otimes \ldots \otimes a_n \in (F_1 \otimes \ldots \otimes F_n)(S)$ for the image of $(a_1, \ldots, a_n) \in (F_1 \times \ldots \times F_n)(S)$ under our universal multilinear map. Taking $n = 3$ as in (53.3), we find that there is, as in the abelian group case, a multilinear map

$$F_1 \otimes F_2 \otimes F_3 \to (F_1 \otimes F_2) \otimes F_3$$

which for each S carries each element $a_1 \otimes a_2 \otimes a_3$ to $(a_1 \otimes a_2) \otimes a_3$. This follows immediately from the observation that the composite of universal maps of functors

$$F_1 \times F_2 \times F_3 \to (F_1 \otimes F_2) \times F_3 \to (F_1 \otimes F_2) \otimes F_3$$

is trilinear. However, when we try to construct a map $(F_1 \otimes F_2) \otimes F_3 \to F_1 \otimes F_2 \otimes F_3$ carrying $(a_1 \otimes a_2) \otimes a_3$ to $a_1 \otimes a_2 \otimes a_3$, the argument used in the proof of (53.3), which begins "For each value of a_3, the product $a_1 \otimes a_2 \otimes a_3$ is bilinear as a function of a_1 and a_2, hence each choice of a_3 induces a linear map $- \otimes a_3 \colon A_1 \otimes A_2 \to A_1 \otimes A_2 \otimes A_3$", does not make sense, since an element a_3 lying in $F(S)$ for one S does not simultaneously lie in the groups $F(T)$ for $T \neq S$, and so does not determine a morphism of *functors* $F_1 \times F_2 \to F_1 \otimes F_2 \otimes F_3$.

The functor $(F_1 \otimes F_2) \otimes F_3$ will in fact be a homomorphic image of $F_1 \otimes F_2 \otimes F_3$, which, we see, can be characterized as universal among **Ab**-valued representable functors G with trilinear maps of $F_1 \times F_2 \times F_3$ into them which admit *factorizations* through bilinear maps on $F_1 \times F_2$:

$$F_1 \times F_2 \times F_3 \xrightarrow{\text{bilinear} \times 1} H \times F_3 \xrightarrow{\text{bilinear}} G.$$

Note that the bimodules generating the tensor rings representing $(F_1 \otimes F_2) \otimes F_3$ and $F_1 \otimes (F_2 \otimes F_3)$ respectively, regarded as subbimodules of

$k<M_1 \oplus M_2 \oplus M_3>$, intersect in $(M_1 \otimes_k M_2 \otimes_k M_3) \oplus (M_3 \otimes_k M_2 \otimes_k M_1)$. One can deduce that the tensor k-ring

$$k < (M_1 \otimes_k M_2 \otimes_k M_3) \oplus (M_3 \otimes_k M_2 \otimes_k M_1) >$$

represents an **Ab**-valued functor G which has a trilinear map $F_1 \times F_2 \times F_3 \to G$ universal for the property that it can be factored *both* through a bilinear map on $F_1 \times F_2$ *and* through a bilinear map on $F_2 \times F_3$:

$$F_1 \times F_2 \times F_3 \xrightarrow{\text{bilinear} \times 1} H_{12} \times F_3 \xrightarrow{\text{bilinear}} G,$$

$$F_1 \times F_2 \times F_3 \xrightarrow{1 \times \text{bilinear}} F_1 \times H_{23} \xrightarrow{\text{bilinear}} G.$$

Generalizing these observations, let us define a *chain-multilinear* map $F_1 \times \ldots \times F_n \to G$ of functors $k\text{-}\mathbf{Ring}^1 \to \mathbf{Ab}$ to mean an n-linear map which for each $i < n$ can be factored using a bilinear map $F_i \times F_{i+1} \to H_{i,\,i+1}$ (for some representable functor $H_{i,\,i+1}$), followed by an $(n-1)$-linear map into G. Then for $F_1, \ldots, F_n$ and G as in Proposition 53.1, a homomorphism $k<N> \to k<M_1 \oplus \ldots \oplus M_n>$ will induce a chain-multilinear morphism of functors if and only if it carries N into the subbimodule $(M_1 \otimes_k \ldots \otimes_k M_n) \oplus (M_n \otimes_k \ldots \otimes_k M_1)$ of the latter ring. From this we see that every multilinear map $F_1 \times \ldots \times F_n \to G$ decomposes uniquely as a sum of $n!/2$ maps, each of which is chain-multilinear with respect to a particular ordering of the functors F_i.

The tensor ring

$$k < (M_1 \otimes_k \ldots \otimes_k M_n) \oplus (M_n \otimes_k \ldots \otimes_k M_1) >$$

represents a functor having a universal chain-multilinear map of $F_1 \times \ldots \times F_n$ into it; we might call this functor the "chain tensor product" of the F_i, and denote it $F_1 \ominus \ldots \ominus F_n$. (Note that two-fold chain tensor products are simply two-fold tensor products.) The construction $\ominus$ is not associative, but one can write down formulas for expanding iterated expressions in these operations; the simplest example is

$$F_1 \ominus (F_2 \ominus F_3) \cong F_1 \ominus F_2 \ominus F_3 \oplus F_1 \ominus F_3 \ominus F_2.$$

We end this section by translating the above concepts into formulations in terms of dual objects, starting as usual with the case of algebras over a field.

PROPOSITION 53.4. *Let k be a field, let $F_1, \ldots, F_n$, $G \in \mathbf{Rep}(\mathbf{Ring}_k^1, \mathbf{Ab})$, and let these functors be described as the operations of completed tensor product with linearly compact k-vector-spaces $A_1, \ldots, A_n$ and B respectively. Then the n-linear maps $m\colon F_1 \times \ldots \times F_n \to G$ correspond bijectively to systems of $n!$ continuous n-linear maps*

$$\mu_\pi\colon A_{\pi(1)} \times \ldots \times A_{\pi(n)} \to B \qquad (\pi \in S_n)$$

via the formula

$$m(a_1 s_1, \ldots, a_n s_n) = \Sigma_{\pi \in S_n} \mu_\pi(a_{\pi(1)}, \ldots, a_{\pi(n)})\, s_{\pi(1)} \ldots s_{\pi(n)}.$$

The morphism m is chain multilinear *if and only if all μ_π are zero other than the two for which π is respectively the identity permutation and the permutation $\pi(i) = n+1-i$.*

Hence, the tensor product $F_1 \otimes \ldots \otimes F_n$ *of our given* n *functors* $\mathbf{Ring}_k^1 \to \mathbf{Ab}$ *is the operation of forming completed tensor products with the linearly compact vector space*

$$\bigoplus_{\pi \in S_n} A_{\pi(1)} \hat{\otimes} \ldots \hat{\otimes} A_{\pi(n)},$$

and the chain tensor product $F_1 \ominus \ldots \ominus F_n$ *is the operation of forming completed tensor products with*

$$A_1 \hat{\otimes} \ldots \hat{\otimes} A_n \;\oplus\; A_n \hat{\otimes} \ldots \hat{\otimes} A_1 .$$

For general $k \in \mathbf{Ring}^1$, *we have the analogous descriptions of n-linear and chain n-linear maps, n-fold tensor products, and n-fold chain tensor products of functors* k-$\mathbf{Ring}^1 \to \mathbf{Ab}$, *using the category* k-$\mathbf{Bimod}^{\mathrm{op}}$ *and the definition of multilinear maps among its objects given in §28, in place of* $\mathbf{LCpMod}_k$ *and continuous multilinear maps among its objects.* □

Let us remark that though the functor we have written $F_1 \otimes F_2$ is initial among *representable* functors with bilinear maps of $F_1 \times F_2$ into them, it is not in general initial among all **Ab**-valued functors given with such bilinear maps; the functor with this property is just

$$S \mapsto F_1(S) \otimes F_2(S)$$

(tensor product as abelian groups). Hence, in a context where not necessarily representable functors are also being considered, the above displayed functor might more properly be called the tensor product of F_1 and F_2, and the construction we have described in this section, their tensor product *as representable functors*. The latter is, of course, initial among representable functors k-$\mathbf{Ring}^1 \to \mathbf{Ab}$ given with morphisms from the above generally nonrepresentable functor into them.

The authors have not examined multilinear maps among representable **Ab**-valued functors on other categories. Fox [**68**] constructs a "tensor product" in the category of commutative cocommutative Hopf algebras over a field k, although some adjustment is needed to compare his ideas with ours: He looks at such Hopf algebras as **Ab**-objects in the category of cocommutative $\otimes_k$-coalgebras, rather than as co-**Ab**-objects in the category of commutative k-algebras. However, by duality, the structures he considers may be regarded as co-**Ab**-objects in the category of commutative *linearly compact* k-algebras, so that his construction gives a "tensor product" on representable functors $\mathbf{LCpComm}_k^1 \to \mathbf{Ab}$. He also constructs an internal hom functor for Hopf algebras.

§54. Higher-degree maps of Ab-valued functors.

A multilinear map on abelian groups is a map that is *linear* in each variable. A generalization of a linear map is a map which "behaves like a polynomial of degree n" for some n.

There are, in fact, two important versions of the latter concept, one modeled on the behavior of *homogeneous* polynomial functions of degree n on vector spaces, the other on that of general polynomial functions of degree $\leq n$ on such spaces. Let us begin with the latter more general concept.

DEFINITION 54.1 (cf. [**145**]). *Let* n *be a nonnegative integer, and* A, B *abelian groups. Then a set-map* $f\colon A \to B$ *is said to be* of degree $\leq n$ *if it satisfies the identity in* $n+1$ *variables* $a_0, \dots, a_n \in A$

$$\sum_{\varepsilon_0, \dots, \varepsilon_n \in \{0,1\}} (-1)^{\varepsilon_0 + \dots + \varepsilon_n} f(\varepsilon_0 a_0 + \dots + \varepsilon_n a_n) = 0. \tag{54.2}$$

A map which is of degree $\leq n$ *for some* n *is said to be* of finite degree (*or in many works, "polynomial"*).

Let us reformulate the above definition. For any set-function $f\colon A \to B$ between abelian groups, and any $a \in A$, define the function $\Delta_a f\colon A \to B$ by

$$(\Delta_a f)(x) = f(x+a) - f(x).$$

Then (54.2) says that $(\Delta_{a_0} \dots \Delta_{a_n} f)(0) = 0$ for all $a_0, \dots, a_n$. (The left-hand side of this equation differs from that of (54.2) by a factor of $(-1)^n$, but that clearly makes no difference.) Note that if we write $h_{a_1, \dots, a_n} = \Delta_{a_1} \dots \Delta_{a_n} f$, then this condition says that for each choice of $a_1, \dots, a_n$ we have $(\Delta_{a_0} h_{a_1, \dots, a_n})(0) = 0$ for all a_0, i.e., that each $h_{a_1, \dots, a_n}$ is a constant function. This in turn implies that for every choice of $a_0, \dots, a_n$, the function $\Delta_{a_0} \dots \Delta_{a_n} f = \Delta_{a_0} h_{a_1, \dots, a_n}$ is *identically* zero, not merely zero at $x = 0$.

It is not hard to show that a real-valued function f on $A = \mathbf{Z}^r$ or $\mathbf{Q}^r$ has degree $\leq n$ in the sense of Definition 54.1 if and only if it is given by a polynomial of degree $\leq n$ with real coefficients, and that the same is true of *continuous* real-valued functions on $A = \mathbf{R}^r$. For general A and f, the "only if" part of this statement can fail. For instance, the "real part" function on $\mathbf{C}$ or on $\mathbf{Z}[i]$ has degree ≤ 1 under the above definition, but is not given by a polynomial. However, the "if" direction holds in a very general form; namely, it is not hard to prove, by induction on n,

(54.3) If $h\colon A^d \to B$ is a d-linear map of abelian groups, and $d \leq n$, then the map $a \mapsto h(a, \dots, a) \in B$ is of degree $\leq n$. Hence, so is any map gotten by summing maps having this form for various $d \leq n$.

Let us now consider maps among functors. To avoid confusion, we shall be more explicit here than in the last section about the distinction between maps among **Ab**-valued functors and maps among their associated underlying-set valued functors.

PROPOSITION 54.4. *Let* $k \in \mathbf{Ring}^1$, *and let* $F, G\colon k\text{-}\mathbf{Ring}^1 \to \mathbf{Ab}$ *be representable functors, represented by* $k\langle M\rangle$ *and* $k\langle N\rangle$, *with co-operations as in Theorem 13.15. Let* $U\colon \mathbf{Ab} \to \mathbf{Set}$ *be the underlying set functor, let*

$$f\colon UF \to UG$$

be any morphism between these set-valued functors, and let us denote the corresponding k-ring homomorphism by

$$\mathbf{f}\colon k\langle N\rangle \to k\langle M\rangle.$$

Then for any nonnegative integer n, *the following conditions are equivalent:*

(i) *For every* $S \in k$-**Ring**^{1}, *the map* $f(S)\colon UF(S) \to UG(S)$ *is of degree* $\leq n$ *as a map of abelian groups, in the sense of Definition 54.1.*

(ii) *There exist multilinear morphisms* $h_d\colon F^d \to G$ *for* $d = 0, \ldots, n$, *such that* f *is given by* $f(a) = h_0 + h_1(a) + \ldots + h_n(a, \ldots, a)$.

(iii) $\mathbf{f}$ *carries* N *into the subbimodule*

$$k \oplus M \oplus \ldots \oplus M^{\otimes n} \subseteq k\langle M\rangle.$$

Sketch of Proof. (i) $\Leftarrow$ (ii) $\Leftrightarrow$ (iii) are clear (cf. (54.3)); we shall complete the proof by showing (i) $\Rightarrow$ (iii).

To do this, let S be the object with a universal $(n+1)$-tuple of elements in $F(S)$,

$$S = \coprod_{n+1 \text{ copies}} k\langle M\rangle = k\langle M^{(0)} \oplus \ldots \oplus M^{(n)}\rangle,$$

and let $a_0, \ldots, a_n \in F(S)$ be the elements of this $(n+1)$-tuple; thus $a_i\colon k\langle M\rangle \to S$ is the k-ring homomorphism which acts on M by its isomorphism with $M^{(i)}$. We observe that the sum in the abelian group $F(S)$ of any subset of this set of maps acts on M by a "diagonal" map into the sum of the corresponding subbimodules of S; in particular $a_0 + \ldots + a_n$ acts on M by the full $(n+1)$-fold diagonal map $M \to M^{(0)} \oplus \ldots \oplus M^{(n)}$.

Now suppose that $\mathbf{f}$ does not satisfy (iii); this means that for some $n' > n$, the $M^{\otimes n'}$ component of $\mathbf{f}|N\colon N \to k\langle M\rangle = \oplus_i M^{\otimes i}$ is nonzero.

Observe that the degree-n' component of the tensor ring S is a direct sum of $(n+1)^{n'}$ copies of $M^{\otimes n'}$. When we evaluate $f(a_0 + \ldots + a_n)$ on N, we get, for the degree-n' component, the full $(n+1)^{n'}$-fold diagonal copy of the degree-n' component of $\mathbf{f}|N$. This has nonzero components in all $(n+1)^{n'}$ of these summands, including those summands which involve all $n+1$ of the $M^{(i)}$. On the other hand, none of the terms $f(\varepsilon_0 a_0 + \ldots + \varepsilon_n a_n)$ with at least one ε_i equal to 0 have nonzero components in any of the latter summands, hence when the left-hand side of (54.2) is evaluated, there are no terms in that sum to cancel any of these components of the summand $f(a_0 + \ldots + a_n)$. Hence that sum is nonzero, showing that when (iii) fails, (i) fails. □

We shall, of course, call a morphism $f\colon UF \to UG$ that satisfies the above equivalent conditions a "morphism of degree $\leq n$". An interesting consequence of the above Proposition is that for F, G, M, N as in the hypothesis, if N is finitely generated as a bimodule, then every morphism of functors $f\colon UF \to UG$ is of finite degree.

We now turn to maps "homogeneous of degree n". Embarrassingly, there is no system of identities for set-maps among abelian groups giving a satisfying version of this concept. For example, if k is a perfect field of characteristic $p>0$, then as maps between additive groups, the pth power map and the identity map have the same properties (both being isomorphisms of the group structure), but as polynomials, one is homogeneous of degree 1, and the other of degree p.

However, if we do not insist on characterizing homogeneous morphisms by *identities*, we can describe a class of morphisms among our representable **Ab**-valued functors that clearly deserve to be so called. Namely, it is straightforward to verify

PROPOSITION 54.5. *With the same hypotheses and notation as in Proposition 54.4 (a morphism* $f\colon UF \to UG$, *where* $F, G \in \mathbf{Rep}(k\text{-}\mathbf{Ring}^1, \mathbf{Ab})$, *represented by* $\mathbf{f}\colon k\langle N\rangle \to k\langle M\rangle$, *and a nonnegative integer* n), *the following conditions are equivalent:*

(i) f *can be written as the composite of the diagonal morphism* $F \to F^n$ *and an* n*-linear morphism* $m\colon F^n \to G$ (*cf.* (54.3)).

(ii) $\mathbf{f}$ *carries* N *into* $M^{\otimes n} \subseteq k\langle M\rangle$.

If we call a morphism f *with these equivalent properties* internally homogeneous *of degree* n, *then any morphism internally homogeneous of degree* d *is of degree* $\leq n$ (*i.e., satisfies the equivalent conditions of Proposition 54.4*) *for all* $n \geq d$, *and conversely, any morphism of degree* $\leq n$ *can be written* uniquely *as a sum, over the nonnegative integers* $d \leq n$, *of morphisms internally homogeneous of degree* d. $\square$

Note that if, in the above Proposition, we write $\mathbf{m}\colon k\langle N\rangle \to k\langle M^{(1)} \oplus \ldots \oplus M^{(n)}\rangle$ for the ring-homomorphism corresponding to the n-linear morphism m of (i), where $M^{(1)}, \ldots, M^{(n)}$ are copies of M, then this morphism will be determined by a bimodule map $N \to \bigoplus_{\pi \in S_n} M^{(\pi(1))} \otimes_k \ldots \otimes_k M^{(\pi(n))}$. If we likewise regard $\mathbf{f}$ as determined by a bimodule map $N \to M^{\otimes n}$, then the latter bimodule map is obtained from the former by dropping the superscripts (the distinctions among the n arguments of m) and summing the resulting $n!$ maps $N \to M^{\otimes n}$. It follows that given f satisfying (i), there is a great deal of nonuniqueness in the choice of this m. One can make m unique by requiring that $\mathbf{m}$ carry N into the single summand $M^{(1)} \otimes_k \ldots \otimes_k M^{(n)}$; but that condition does not have any nice interpretation in terms of the behavior of our functor. (In fact, if $k \cong k^{\mathrm{op}}$, then it is impossible to characterize that condition category-theoretically, because it is not invariant under the self-equivalence of $k\text{-}\mathbf{Ring}^1$ carrying each k-ring S to S^{op}.) If 2 is invertible in k, however, we can make m unique by requiring it to be chain-multilinear, and symmetric with respect to the permutation of the variables interchanging i and $n+1-i$. More pleasantly, if $n!$ is invertible in k, we can make m unique by requiring that it be fully symmetric in its n arguments.

We remark that for ordinary maps among abelian groups (as distinct from maps among abelian-group valued functors), the analog of the last statement of the above Proposition (that maps decompose into homogeneous summands) is false: The map $\mathbf{Z} \to \mathbf{Z}$ given by $n \mapsto n(n+1)/2$ is of degree ≤ 2, but is not a sum of homogeneous maps $\mathbf{Z} \to \mathbf{Z}$. (Of course, when looked at as a map $\mathbf{Z} \to \frac{1}{2}\mathbf{Z}$, it *is* a sum of homogeneous maps. This illustrates another way to obtain maps $A \to B$ of degree $\leq n$ between abelian groups, generalizing (54.3): construct as in (54.3) such a map into an overgroup of $B' \supseteq B$, then show that its image actually lies in B.)

If our base-ring k is a $\mathbf{Q}$-algebra, one *can* characterize morphisms homogeneous of a given degree by identities. Namely, it is easy to verify

PROPOSITION 54.6. *Let F, G, f and n again be as in the hypotheses of Propositions 54.4 and 54.5, and suppose now that $k \in \mathbf{Ring}^1_{\mathbf{Q}}$. Let d be any integer $\neq 0, \pm 1$. Then the following conditions are equivalent:*

(i) *f is internally homogeneous of degree n.*

(ii) *f satisfies the identity $f(dx) = d^n f(x)$.* □

(The proof uses the fact that the differences between d^n and all other powers of d are invertible in $\mathbf{Q}$. More generally, for any rational number $d/e \neq 0, \pm 1$, the internally homogeneous morphisms of degree n are characterized by the identity $d^n f(ex) = e^n f(dx)$.)

If $k \in \mathbf{Ring}^1_{\mathbf{Z}/p\mathbf{Z}}$, one has a weaker result: using the behavior of f under multiplication of its argument by integers, one can characterize by identities those morphisms the degrees of whose nonzero internally homogeneous components all belong to a given *residue class* modulo $p-1$, and one can decompose any morphism uniquely as a sum, over these residue classes, of such morphisms. (This is consistent with our earlier observation that degrees 1 and p *cannot* be distinguished by identities in characteristic p.) Using the value of f at 0, one can also characterize by identities the morphisms internally homogeneous of degree 0 (constant morphisms), and the complementary group of morphisms having zero constant term. Thus one gets a grading of the additive group of morphisms $F \to G$ by a semigroup isomorphic (though noncanonically) to the multiplicative semigroup of $\mathbf{Z}/p\mathbf{Z}$. We leave the details to the interested reader.

Again we have a translation of the results of the preceding Propositions in terms of the dual objects:

PROPOSITION 54.7. *Let k be a field and let F, $G \in \mathbf{Rep}(\mathbf{Ring}^1_k, \mathbf{Ab})$. Let these functors be described as the operations of completed tensor product with linearly compact k-vector-spaces A and B respectively, and let n be a nonnegative integer. Then morphisms f: $UF \to UG$ internally homogeneous of degree n in the sense of Proposition 54.5 correspond bijectively to continuous n-linear maps*

(54.8) $$\mu: A^n \to B$$

(dual to the maps $\mathbf{f}|N$: $N \to M^{\otimes n}$ of Proposition 54.5(ii)*) via the formula*

$$f(\Sigma_{i \in I}\, a_i s_i) = \Sigma_{i_1, \ldots, i_n \in I}\, \mu(a_{i_1}, \ldots, a_{i_n}) s_{i_1} \ldots s_{i_n}.$$

(This formula is valid for arbitrary finite sums and even convergent infinite sums on the left. However, the values at sums of n terms suffice to determine the function f, given the continuity and degree-n conditions.)

For general $k \in \mathbf{Ring}^1$ and F, $G \in \mathbf{Rep}(k\text{-}\mathbf{Ring}^1, \mathbf{Ab})$, we have the analogous statement, with the concepts of multilinear algebra for k-$\mathbf{Bimod}^{\mathrm{op}}$ defined in §28 replacing those for linearly compact vector spaces. □

As in the preceding section, one can get *universal* examples of the sorts of maps we have characterized: Given any representable $\mathbf{Ab}$-valued functor F on

k-**Ring**[1], represented by $k\langle M\rangle$, the class of representable functors G given with morphisms $f\colon F \to G$ of degree $\leq n$, respectively internally homogeneous of degree n, has an initial object, represented by $k\langle k \oplus M \oplus \ldots \oplus M^{\otimes n}\rangle$, respectively by $k\langle M^{\otimes n}\rangle$.

We have discussed in this section functions of one variable. We can, of course, get a common generalization of the ideas of this and the preceding section by considering, for morphisms $UF_1 \times \ldots \times UF_n \to UG$, the property of having degrees in certain of the variables bounded by given integers, and/or being internally homogeneous of certain degrees in certain of the variables. The key to reducing the study of the behavior in the jth variable, for a given j, to the study of one-variable morphisms is, essentially as in §23, to use for each j the identification

$$k\langle M_1 \oplus \ldots \oplus M_n\rangle \;\cong\; k^{(j)}\langle M_j'\rangle,$$

where

$$\begin{aligned} k^{(j)} &= k\langle M_1 \oplus \ldots \oplus M_{j-1} \oplus M_{j+1} \oplus \ldots \oplus M_n\rangle, \\ M_j' &= k^{(j)} \otimes_k M_j \otimes_k k^{(j)} \qquad \text{(a } k^{(j)}\text{-bimodule).} \end{aligned}$$

The results come out entirely as one would expect; for instance, f has degree $\leq n_j$ in its jth variable if and only if the image of N under $\mathbf{f}$ has nonzero component only in those tensor products of the M_i in which M_j occurs $\leq n_j$ times.

§55. Higher-degree maps between modules.

In this section, we will not study representable functors at all; rather, we shall discuss how one must adjust the concept of a higher-degree map of abelian groups to get an appropriate definition of a higher-degree map of *modules*, and note some properties of these maps, thus paving the way for the study of higher-degree maps of module-valued representable functors. However, we will not undertake the latter study, but leave it to the interested reader.

Recall that the condition that a map $f\colon A \to B$ of abelian groups be of degree $\leq n$ says that for all $a_0, a_1, \ldots, a_n \in A$, $\Delta_{a_0}\Delta_{a_1} \ldots \Delta_{a_n} f = 0$, which is equivalent to saying that for all $a_1, \ldots, a_n$, the function $h_{a_1, \ldots, a_n} = \Delta_{a_1} \ldots \Delta_{a_n} f$ is *constant*. When this is true, let us write the constant value of this latter map as a function of $a_1, \ldots, a_n$, setting

$$h(a_1, \ldots, a_n) \;=\; (\Delta_{a_1} \ldots \Delta_{a_n} f)(0), \tag{55.1}$$

Then it is not hard to show that this function h is *symmetric* and *multilinear* as a map of abelian groups.

Now if A and B are *modules* over a commutative ring k, and we want to define the condition that f be "a *k-module* map of degree $\leq n$", it is natural to require that (55.1) be multilinear as a k-module map. However, this alone is only a half-way measure, for the property so defined is preserved under adding to f any map which has degree $\leq n-1$ as a map of *abelian groups*, since all such maps are annihilated by $\Delta_{a_1} \ldots \Delta_{a_n}$. So, for instance, though the above condition says, as we would wish, that the map $\mathbf{C} \to \mathbf{C}$ given by $z \mapsto \bar{z}^2$ is not of degree 2 *as a map of modules over the complex numbers*, it accepts $z \mapsto z^2 + \bar{z}$ as such a map. Hence we shall add to our definition below an inductive condition, (55.5) (which the reader can verify does exclude the above example).

Since we are using induction, we should be clear about what our terms mean when $n = 0$. Note that *every* map $A^0 \to B$ (corresponding to a choice of an element of B) is, vacuously, 0-linear. It follows that for $n = 0$, condition (55.4) below is vacuous.

DEFINITION 55.2 (cf. [**150**]). *Let* $k \in \mathbf{Comm}^1$, *let* $A, B \in \mathbf{Mod}_k$, *and let* n *be a nonnegative integer. Then a set-map* $f\colon A \to B$ *will be said to be of degree* $\leq n$ as a map of k-modules *if and only if it satisfies*

(55.3) *f has degree $\leq n$ as a map of abelian groups,*

(55.4) *The map $h\colon A^n \to B$ defined by* (55.1) *is k-multilinear; and*

(55.5) *If $n > 0$, then for all $c \in k$, the map $x \mapsto f(cx) - c^n f(x)$ is of degree $\leq n-1$ as a map of k-modules* (*this condition being assumed defined for $n-1$ by induction*).

We will call f homogeneous *of degree n as a map of k-modules if it satisfies* (55.3), (55.4), *and the following strengthening of* (55.5):

(55.6) *For all $c \in k$ and $x \in A$, $f(cx) - c^n f(x) = 0$.*

The final part of this definition may seem questionable, in view of our earlier statement that there is no satisfactory characterization of homogeneous polynomial maps via identities. We are here, in a sense, "giving in", and using "homogeneous of degree n" to describe the closest thing to the class of homogeneous polynomial functions that we *can* characterize by identities. Insofar as k is larger than its subring $\mathbf{Z}1_k$, we can at least say that condition (55.5) is stronger than the corresponding condition on f as a map of abelian groups. Nevertheless, our definition still has some unsatisfying properties; e.g., if k is a finite field, a nonzero map can be homogeneous of two different degrees.

If A, B are k-modules and $K \in \mathbf{Comm}^1_k$, a map $A \to B$ which is homogeneous of degree n as a map of k-modules may not extend uniquely to a homogeneous map of K-modules $K \otimes_k A \to K \otimes_k B$. For instance, if $k = \mathbf{Z}/p\mathbf{Z}$ and K is an infinite extension field of k, then the infinite family of maps $K \times K \to K$ homogeneous of degree $p+1$ given by $(x, y) \mapsto \alpha x y^p + (1-\alpha) x^p y$, as α ranges over K, all have the same restriction to $(\mathbf{Z}/p\mathbf{Z}) \times (\mathbf{Z}/p\mathbf{Z})$, since $x^p = x$, $y^p = y$ hold identically in $\mathbf{Z}/p\mathbf{Z}$. Thus, this common map $(\mathbf{Z}/p\mathbf{Z}) \times (\mathbf{Z}/p\mathbf{Z}) \to \mathbf{Z}/p\mathbf{Z}$ has nonunique extension to a homogeneous degree-$p+1$ map of K-modules. For a case where there is no extension, take $k = \mathbf{Z}$, $A = \mathbf{Z} \times \mathbf{Z}$, $B = \mathbf{Z}$, $f(x, y) = (x^p y - x y^p)/p$, and K an extension ring of $\mathbf{Z}$ such that K/pK is a proper extension field of $\mathbf{Z}/p\mathbf{Z}$.

In general, it seems that such pathologies occur for homogeneous maps of degree n when k has nontrivial homomorphic images of cardinality $< n$. We shall not attempt to prove that when k has no such homomorphic images, one gets unique extensions (we do not know whether this is true), but we shall prove below a known special case of this statement, namely that when $n = 2$, one always has unique extensions. We shall use "quadratic" as a familiar abbreviation for "of degree 2". We first need a calculation concerning quadratic maps of abelian groups.

LEMMA 55.7. *Suppose* $f\colon A \to B$ *is a homogeneous quadratic map of* $\mathbf{Z}$*-modules, and define the bilinear map* $h\colon A \times A \to B$ *by*

$$h(x, y) = (\Delta_x \Delta_y f)(0). \tag{55.8}$$

Then for $x_1, \ldots, x_n \in A$, *one has*

$$f(\Sigma_i\, x_i) = \Sigma_i\, f(x_i) + \Sigma_{i<j}\, h(x_i, x_j). \tag{55.9}$$

Hence the same formula holds for a sum over an infinite totally ordered index set, with all but finitely many summands zero.

SKETCH OF PROOF. The cases $n = 0$ and $n = 1$ are trivial, and the case $n = 2$ reduces to the definition of h. The case of general n is proved inductively, writing $f(\Sigma_{i\leq n}\, x_i)$ as $f((\Sigma_{i<n}\, x_i) + x_n)$, expanding using the cases 2 and $n-1$, and appealing to the bilinearity of h, (55.4). The assertion of the last sentence clearly follows. □

We can now establish

LEMMA 55.10 ([**102**, Lemma on p.I.8], cf. [**164**, Prop. II.1]). *Let* $k \in \mathbf{Comm}^1$ *and let* $f\colon A \to B$ *be a homogeneous quadratic map of k-modules. Then for any* $K \in \mathbf{Comm}^1_k$, *there is a unique homogeneous quadratic map of K-modules* $f'\colon K \otimes_k A \to K \otimes_k B$ *making the diagram*

$$\begin{array}{ccc} A & \xrightarrow{\quad f \quad} & B \\ \downarrow & & \downarrow \\ K \otimes_k A & \xrightarrow{\quad f' \quad} & K \otimes_k B \end{array} \tag{55.11}$$

commute (where the vertical arrows are the maps $x \mapsto 1 \otimes x$).

SKETCH OF PROOF. The map $h(x, y)$ defined by (55.8) is by assumption bilinear over k. From the fact that linear maps (module homomorphisms) have unique extensions under base-change, one can deduce that the same is true of multilinear maps, hence there is a unique K-bilinear map

$$h'\colon (K \otimes_k A) \times (K \otimes_k A) \to K \otimes_k B$$

extending h. (Here we are using the word "extending" to mean "making the obvious base-extension diagram commute", even when the base-extension maps are not necessarily embeddings.) Now if $f'\colon K \otimes_k A \to K \otimes_k B$ is any homogeneous quadratic map of K-modules extending f, then the function $(K \otimes_k A) \times (K \otimes_k A) \to K \otimes_k B$ taking (x, y) to $(\Delta_x \Delta_y f')(0)$ will be K-bilinear and extend h, hence it must be precisely the above map h'.

Let us now write the general element of $K \otimes_k A$ as $\Sigma_{x \in A}\, c_x \otimes x$, where the c_x are members of K, all but finitely many equal to 0. Then choosing an arbitrary total ordering "<" on the underlying set of A, and again supposing that f' is a homogeneous quadratic map of K-modules extending f, we can use Lemma 55.7 to expand $f'(\Sigma_{x \in A}\, c_x \otimes x)$ in terms of the values of f' and h' on the elements $c_x \otimes x$. We may then use the K-quadratic and K-bilinear properties of f' and h' respectively to pull out the coefficients c_x, and write the result in terms of the original f and h. Thus, we get the formula

$$(55.12) \qquad f'(\Sigma_{x \in A}\, c_x \otimes x) \;=\; \Sigma_{x \in A}\, c_x^2 \otimes f(x) \;+\; \Sigma_{x<y \in A}\, c_x c_y \otimes h(x, y).$$

Hence if such an extended quadratic map f' exists, it is unique; and our proof will be complete if we can show that (55.12) indeed yields a well-defined homogeneous quadratic map.

To do this, let us first form the free K-module $K|A|$ on the underlying set of the k-module A, writing $[x]$ for the generator corresponding to $x \in A$. Thus, every element of this module may be written uniquely as

$$(55.13) \qquad \Sigma_{x \in A}\, c_x [x]$$

with all but finitely many c_x equal to 0. Let $f''\colon K|A| \to K \otimes_k B$ denote the map taking (55.13) to the right-hand side of (55.12):

$$(55.14) \qquad f''(\Sigma_{x \in A}\, c_x [x]) \;=\; \Sigma_{x \in A}\, c_x^2 \otimes f(x) \;+\; \Sigma_{x<y \in A}\, c_x c_y \otimes h(x, y).$$

This is clearly a homogeneous quadratic map on the free module $K|A|$, because of the homogeneous quadratic coefficients, c_x^2 and $c_x c_y$.

Now $K \otimes_k A$ can be described as the quotient of $K|A|$ by the K-submodule generated by the elements $\Sigma_{x \in A}\, r_x [x]$, as we go through all equations

$$(55.15) \qquad \Sigma_{x \in A}\, r_x x \;=\; 0 \qquad (r_x \in k)$$

holding in A. (This is not the standard description of $K \otimes_k A$, but it is not hard to verify.) Hence to show that (55.12) gives a well-defined map, it will suffice to show that for every element (55.13) of $K|A|$, every relation (55.15) holding in A, and every element $q \in K$, we have

$$(55.16) \qquad f''(\Sigma\, c_x [x]) \;=\; f''(\Sigma\, (c_x + q r_x)\, [x]).$$

Once this is established, the resulting map (55.12) will be homogeneous quadratic because (55.14) is.

Let us observe that

(55.17) (55.16) holds whenever q and all the c_x lie in k,

because by hypothesis, our original map f is a homogeneous quadratic map of k-modules. Now when we expand the two sides of the general case of the desired equation (55.16) using (55.14), the terms quadratic in the system of coefficients c_x are the same on both sides, while the sum of the terms on the right not involving any c_x may be seen to be 0 by applying (55.17) to the case where all $c_x = 0$, and $q = 1$. We also claim that in the terms involving the c_x's linearly, the coefficient of each c_{x_0} on the right comes to 0. This may be seen by applying

(55.17) to the situation where q and c_{x_0} are 1, and all the other c_x are 0. Hence (55.16) holds, completing the proof of the Lemma. □

We have mentioned that the above result fails for maps of higher degree n, but that we suspect it may become true if one adds the hypothesis that k has no nontrivial homomorphic images of cardinality $<n$. If this hypothesis is insufficient, then a stronger one, such as that k has n elements whose pairwise differences are invertible (so that one can form an invertible $n\times n$ Vandermonde determinant) might prove useful. Let us record here a result with an even stronger hypothesis, part (b) of the next Lemma, which has the virtue of being very easy to prove. (This gives, in particular, an easy proof of Lemma 55.10 in the case where 2 is invertible in k.) As before, we begin with a general calculation, part (a) below.

LEMMA 55.18. *Let* $k\in \mathbf{Comm}^1$ *and suppose* $f\colon A\to B$ *is a map of k-modules homogeneous of degree* n. *Then*

(a) *For any* $x\in A$ *and* $c_1,\dots,c_n\in k$ *one has*

$$(\Delta_{c_1x}\cdots\Delta_{c_nx}f)(0) \;=\; n!\,c_1\dots c_n f(x).$$

(b) *If* $n!$ *is invertible in* k, *then for any* $K\in \mathbf{Comm}^1_k$, *there is a unique map* $f'\colon K\otimes_k A\to K\otimes_k B$ *homogeneous of degree* n *as a map of K-modules, which extends* f *(in the sense of* (55.11)).

SKETCH OF PROOF. (a) If we expand $(\Delta_{c_1x}\cdots\Delta_{c_nx}f)(0)$, we get a sum of terms $\pm f(cx)$ $(c\in k)$, and by (55.6), $f(cx)=c^nf(x)$. Hence we can extract from all these terms the common right factor $f(x)$; the effect is then that the Δ operators are being applied to the nth-power function on k. Precisely, writing this function $c^n=s_n(c)$, we get

$$(\Delta_{c_1x}\cdots\Delta_{c_nx}f)(0) \;=\; (\Delta_{c_1}\cdots\Delta_{c_n}s_n)(0)f(x).$$

Now since $(\Delta_{c_1}\dots\Delta_{c_n}s_n)(0)$ is multilinear in the c_i, when we expand it, any terms that do not involve all n of the c_i must drop out. But the only terms that do involve them all are those summands in the expansion of $(c_1+\dots+c_n)^n$ in which each c_i appears once. There are $n!$ such summands, giving the indicated formula.

(b) As in the proof of the preceding Lemma, the function h given by $h(x_1,\dots,x_n)=(\Delta_{x_1}\dots\Delta_{x_n}f)(0)$, being k-multilinear, extends uniquely to a K-multilinear function $h'\colon (K\otimes_k A)\times\dots\times(K\otimes_k A)\to K\otimes_k B$. The definition $f'(x)=(1/n!)\,h'(x,\dots,x)$ gives a K-module map $K\otimes_k A\to K\otimes_k B$ homogeneous of degree n. Part (a) (with $c_1=\dots=c_n=1$, and the definition of h applied to the left-hand side) shows that this map extends f, and (applying it to f') that it is the unique homogeneous map which can do so. □

For nonhomogeneous maps, even the analog of Lemma 55.10 fails: When $k=\mathbf{Z}/2\mathbf{Z}$ and $K\in\mathbf{Comm}^1_k$, the identity map of the k-module k can be extended to K by any of the maps $x\mapsto \alpha x+(1-\alpha)x^2$ $(\alpha\in K)$, all of which have degree ≤ 2. However, one gets plausible nonhomogeneous analogs to the conjectures we have made regarding extension of homogeneous maps if one replaces "n" by "n+1" in the hypotheses of these conjectures (e.g., as the lower bound on cardinalities of homomorphic images, or the size of a family of elements assumed to have invertible pairwise differences). We may, in fact, hope that under these modified hypotheses, every map of degree $\leq n$ decomposes uniquely as a sum of homogeneous maps of degrees $d \leq n$, reducing us to the homogeneous case. (An equivalent statement would be that under the appropriate hypotheses, a map $f: A \to B$ of degree $\leq n$ can be written $f(x) = g(1, x)$, where $g: k \times A \to B$ is a homogeneous map of degree n.)

Let us record here one more counterexample, showing that if we assume (55.3) and (55.6), but not (55.4), unique extension can fail even when k is an infinite field. Let k be an infinite field of characteristic p, let $K = k[\varepsilon \mid \varepsilon^p = 0]$, and let θ be the k-algebra endomorphism of K carrying ε to 0, which we see has fixed subalgebra k. Note that the Frobenius (p-th power) endomorphism of K also annihilates ε, and so carries K into k; hence we have the identity $\theta(a^p) = a^p$ $(a \in K)$. Now let A be the free k-module of rank p, and B the free k-module of rank 1. Then the map $A \to B$ given by

$$(a_1, \dots, a_p) \mapsto a_1 \dots a_p$$

extends to two distinct maps $K \otimes_k A \to K \otimes_k B$ satisfying (55.3) and (55.6): the map given by the same formula, and the map $(a_1, \dots, a_p) \mapsto \theta(a_1 \dots a_p)$.

Finally, we mention a phenomenon that one may or may not regard as pathological. Let k be a commutative integral domain of characteristic 0 in which p is not invertible. It follows easily from the divisibility properties of binomial coefficients that if elements $a, b \in k$ are congruent modulo pk, then their pth powers are congruent, not only modulo pk, but modulo p^2k. Consequently, the pth power map $k \to k$ induces a map $k/pk \to k/p^2k$ homogeneous of degree p. It follows that for A a (k/pk)-module, the universal k-module B admitting a homogeneous k-module map $A \to B$ of degree p will not in general be a (k/pk)-module; in particular, it will not coincide with the (k/pk)-module universal for the same property.

55.19. *Notes on the literature.* N. Roby [**164**] defines a "polynomial law" $A \to B$ of k-modules to mean a morphism of functors $\mathbf{Comm}^1_k \to \mathbf{Set}$

$$f: U(- \otimes_k A) \to U(- \otimes_k B)$$

(where U is the underlying-set functor), thus avoiding the question of existence and uniqueness of extensions under change of base ring by an elegant fiat. (If one applies such a morphism f to an element $\Sigma_1^r t_i \otimes x_i \in k[t_1, \dots, t_r] \otimes_k A$, where the t's are indeterminates and $x_1, \dots, x_r \in A$ arbitrary elements, and notes that functoriality allows one to compute from the resulting element the image under f of any element $\Sigma_1^r c_i \otimes x_i \in K \otimes_k A$ for $c_i \in K$, $K \in \mathbf{Comm}^1_k$, one sees from the form of this computation that "polynomial law" is a reasonable name.) Note that the functors $U(- \otimes_k A)$ are not representable, so their study is of a different flavor from the topic of the present work. Nonetheless, these areas have in common the feature that functoriality allows one to apply a morphism of the sort being studied to a more or less "universal" element, and thus obtain a "formula" for the action of this morphism on a general element, from which, among other properties, unique decomposition of morphisms into homogeneous components is easily deduced.

A. Prószyński calls a polynomial law (in Roby's sense) which satisfies (55.6) an "m-form", while he gives the name "m-application" to a *set-map* among modules satisfying (55.3), (55.4) and (55.6), i.e., to what we have called a homogeneous module-map of degree m. He then studies the extension-under-change-of-base-ring question as the problem of determining the kernel and cokernel of the natural map from the additive group of m-forms $A \to B$ to that of m-applications with the same range and domain. (See **[152]** and references given there, in particular **[64]** and **[150]**. The most complete development, up to 1987, is **[151]**.) A surprising result of his work is that there are identities satisfied by all polynomial laws of degree m which are not implied, as conditions on set-maps among k-modules, by the identities (55.3), (55.4) and (55.6); however, the latter identities *are* sufficient if all residue fields of k are either infinite, or prime fields. Failure to satisfy these identities is thus among the obstructions to extending homogeneous module maps under change of base-ring.

For some other work on polynomial maps, see **[98]**, **[42]**, which show that if a map of vector spaces over a field is polynomial of degree $\leq n$ on affine subspaces of small dimension (in most cases, on lines), then it is polynomial of degree $\leq n$ globally.

We know of no work on not-necessarily-homogeneous polynomial maps of modules, in the sense of Definition 55.2, over a general commutative base ring k. But we have not made a thorough literature-search; thus it may well be that some of the results conjectured above are known to be true, or false.

In connection with the condition that k have finite homomorphic images of various sizes, we mention an open question of Ralph McKenzie (personal communication): Is the condition that a commutative ring k admit a homomorphism to $\mathbf{Z}/2\mathbf{Z}$ equivalent to a first-order condition on k? (The class of rings k with this property is easily shown to be closed under ultraproducts, but it is not clear whether its complement is; we suspect not.) One may ask the same question with 2 replaced by an arbitrary prime p. In the same spirit, one can ask whether the property of having a nonzero commutative homomorphic image is a first-order property on associative unital rings.

§56. Functors to generalized Jordan rings.

As we noted in §25, the classical concept of a Jordan algebra may be thought of as modeled on the properties of an associative algebra S under the anticommutator operation $(x, y) = xy + yx$. Now if 2 is invertible in the base-ring k, one can express the operation

$$U_x(y) \;=\; xyx$$

of S (called the "triple product") in terms of this anticommutator, as $\frac{1}{2}((x,(y,x)) - (y, \frac{1}{2}(x,x)))$. Conversely, *without* any restriction on k, one can recover the anticommutator operation from the triple product and the unit element 1: $(x, y) = U_{x+y}(1) - U_x(1) - U_y(1)$. Thus, given the element 1, the operation U gives the same information as the bilinear operation $(\ ,\ )$ if 2 is invertible in k, and *as much or more* information if 2 is not invertible. It turns out, in fact, that in the latter case, the theory modeled on the anticommutator alone is poorly behaved, while there is a good theory modeled on the properties of the triple product.

Thus, one defines a *unital Jordan triple system* over $k \in \mathbf{Comm}^1$ (often called by specialists, extending the classical term, a *Jordan algebra* over k) to mean a k-module J given with a map $U: J \times J \to J$, homogeneous *quadratic* in the first variable, and *linear* in the second, and a zeroary operation 1, subject to appropriate identities; namely, using the abbreviation

$$L_{x,y}(z) = U_{x+z}(y) - U_x(y) - U_z(y),$$

the three identities

$$\begin{aligned} U_1(z) &= z, \\ L_{x,y}(U_x(z)) &= U_x(L_{y,x}(z)), \\ U_{U_x(y)}(z) &= U_x(U_y(U_x(z))), \end{aligned} \tag{56.1}$$

together with additional identities saying that (56.1) continues to hold after extension of the base ring to any $K \in \mathbf{Comm}^1_k$ [**102**]. (From the fact that U is homogeneous of degrees 1 and 2 in its two variables, it is easily deduced using Lemma 55.10 that it extends to a unique operation on $K \otimes_k J$ with the same homogeneity properties. On the other hand, the second and third identities in (56.1) are of degrees 3 and 4 in x, so it is not similarly automatic that these will continue to hold after base extension. The conditions for this to be true are a set of identities obtained from these identities by "polarization".)

To make the subject more complicated, there is also a series of weakenings of this concept. If we drop the zeroary operation "1", but retain the quadratic operation corresponding to $x \mapsto U_x(1)$, written $x \mapsto x^2$ and related to U by a (long) list of identities, we get what is called a "quadratic Jordan algebra $(J;\ U,\ (\ )^2)$". One might weaken the structure next by replacing the squaring operation with the classical Jordan multiplication, $(x, y) = (x+y)^2 - x^2 - y^2$, but the resulting objects $(J;\ U,\ (\ ,\ ))$ do not appear to have a name. Rather, the next weakening is the object $(J;\ U)$, called a "nonunital Jordan triple system", with its own set of defining identities. Finally, there is a concept called a "Jordan pair" [**119**], a type of two-sorted algebra, a model for which is the structure obtained by taking a nonunital Jordan triple system $(J;\ U)$, making two copies J^+ and J^- of J, and (roughly speaking) letting U operate "between" these, rather than carry either into itself. This has the virtue of making isomorphic certain kinds of related Jordan algebras (called "isotopes") that are not isomorphic but "by rights ought to be"; Jordan pairs also arise in the study of Lie algebras. (The reader searching the literature should be aware that there is an unrelated concept also called a "Jordan triple system", the name of which refers to Jordan canonical forms of matrices.)

Let us now sketch how the results of preceding sections might be applied to the study of representable functors from k-rings to unital Jordan triple systems (to begin with, over $\mathbf{Z}$), though without trying to work out the details.

Let $k \in \mathbf{Ring}^1$, and let

$$F\colon\ k\text{-}\mathbf{Ring}^1 \to \mathbf{JordanTrip}^1$$

be a representable functor. Thus, the underlying additive-group valued functor of F will be represented by a tensor algebra $k\langle M\rangle$, and can be written

$$S \mapsto A \bigcirc S,$$

where $A = \hat{M} \in k\text{-}\mathbf{Bimod}^{\mathrm{op}}$. A morphism $U\colon F \times F \to F$ of degree ≤ 2 in the first variable and linear in the second corresponds to a homomorphism

$$\mathbf{u}\colon\ k\langle M\rangle \to k\langle M^{\lambda} \oplus M^{\rho}\rangle$$

which carries M into

$$(56.2)\qquad \begin{aligned}(M^{\lambda} \otimes M^{\lambda} \otimes M^{\rho}) \oplus (M^{\lambda} \otimes M^{\rho} \otimes M^{\lambda}) \oplus (M^{\rho} \otimes M^{\lambda} \otimes M^{\lambda})\\ \oplus (M^{\lambda} \otimes M^{\rho}) \oplus (M^{\rho} \otimes M^{\lambda}) \oplus M^{\rho}\end{aligned}$$

If we add the homogeneity condition that for all integers d,

$$(56.3)\qquad U_{dx}(y) = d^2 U_x(y),$$

this turns out to say that the component of the image of the map $\mathbf{u}$ in the last term of (56.2) is 0, and that its components in the two preceding terms are annihilated by multiplication by 2. In terms of the dual object A, then, the operation U on F, subject to this homogeneity condition, is determined by three trilinear maps

$$\alpha_1, \alpha_2, \alpha_3\colon\ A \times A \times A \to A,$$

together with two bilinear maps

$$\beta_1, \beta_2\colon\ A \times A \to A \quad \text{both annihilated by } 2,$$

via the formula

$$(56.4)\qquad \begin{aligned} U_{\Sigma_i a_i s_i}(bt) &= \Sigma_{i,j}\, \alpha_1(a_i, a_j, b) s_i s_j t + \Sigma_{i,j}\, \alpha_2(a_i, b, a_j) s_i t s_j \\ &+ \Sigma_{i,j}\, \alpha_3(b, a_i, a_j) t s_i s_j + \Sigma_i\, \beta_1(a_i, b) s_i t + \Sigma_i\, \beta_2(b, a_i) t s_i. \end{aligned}$$

The unit of our Jordan-triple valued functor will, of course, have the form

$$(56.5)\qquad 1 = \varepsilon 1_S$$

for some $\varepsilon \in_k A$. By substituting (56.4) and (56.5) into the identities defining Jordan triple systems, one can now, in principle, get necessary and sufficient conditions on α_1, α_2, α_3, β_1 and β_2 for the algebras given by our functor to satisfy these identities. The authors have not attempted this computation; we leave it, and the corresponding problems for the other versions of Jordan systems, to workers having more insight into these structures. Recalling Theorem 25.40, we suspect that all such objects will turn out to be induced in the natural way by associative-ring valued functors; but surprises are possible.

Let us note that the homogeneity condition (56.3) led to a conclusion close to, but weaker than, the condition that the morphism U be "internally homogeneous of degree 2" in its first variable. This is an instance of the weakness of identities in defining the concept of homogeneous map. We do not know whether consideration of the full set of Jordan identities would eliminate the terms β_1 and β_2 of (56.4), or whether these represent a genuine "exotic" component which can occur in the structure of a Jordan-triple valued representable functor.

The variety of p-Lie algebras is another class of algebras involving a higher-degree operation, and the study of representable functors

$$k\text{-}\mathbf{Ring}^1 \to \mathbf{Lie}^{(p)}_{\mathbf{Z}/p\mathbf{Z}}$$

would similarly be of interest.

Finally, what about functors to Jordan triple systems and p-Lie algebras over a general commutative ring? We noted at the end of §14 that it was straightforward to obtain from our description of $\mathbf{Rep}(k\text{-}\mathbf{Ring}^1, \mathbf{Ab})$ a description of categories such as $\mathbf{Rep}(k\text{-}\mathbf{Ring}^1, \mathbf{Mod}_L)$ where $L \in \mathbf{Comm}^1$, and to generalize our characterizations of bilinear maps among objects of the former category, and the resulting descriptions of representable functors into various categories of rings, to characterizations of L-bilinear maps, and representable functors into categories of L-algebras. However, the study of higher-degree maps among $\mathbf{Mod}_L$-valued functors may not be so straightforward; we leave this project also, and its application to functors from $k\text{-}\mathbf{Ring}$ to varieties such as $\mathbf{JordanTrip}^1_L$ and $\mathbf{Lie}^{(p)}_L$, to the interested reader.

Wisbauer [**199**] considers polynomial laws in the sense of Roby (55.19 above) from one not-necessarily-associative algebra to another, subject to various identities. (The norm function from a finite-dimensional associative algebra to its base ring, together with the identity saying that it respects multiplication, is a simple example.) See Ziplies [**204**], and papers cited there, for related work, in which divided powers play an important role. One might likewise examine maps among our representable functors $k\text{-}\mathbf{Ring}^1 \to \mathbf{NARing}$ satisfying such identities.

CHAPTER XI.

Directions for further investigation.

We have noted many open problems in the preceding chapters. In this last chapter we discuss some topics for investigation which were not mentioned earlier, either because they were not closely tied to results we were presenting, or because they would have interrupted the development.

We remark that the locations of *all* open questions referred to in this work are listed in the index, under "questions (open), conjectures, etc.".

§57. Other varieties of algebras.

Obviously, there is a vast set of questions one can generate by substituting the names of any two interesting varieties $\mathbf{V}$ and $\mathbf{W}$ into the problem, "Describe the category of all representable functors from $\mathbf{V}$ to $\mathbf{W}$". The authors will be happy if the present work provokes a flowering of investigation of such questions. In this section, we shall mainly point out questions of this form related, in one way or another, to those examined in earlier chapters.

In studying representable functors from categories $\mathbf{V} = k\text{-}\mathbf{Ring}$ or $k\text{-}\mathbf{Ring}^1$ to various categories $\mathbf{W}$, we found that a few general results (the descriptions of representable functors $\mathbf{V} \to \mathbf{Ab}$ and of bilinear maps among these) allowed us to handle a wide class of cases; so wide, in fact, that we had to swear off discussing certain kinds of structure that are easy to handle but tedious to mention, namely, module and bimodule structure on objects of $\mathbf{W}$, in order to keep from drowning in details. On the other hand, small excursions from the class of *domain* categories $\mathbf{V}$ that we studied carry us into terra incognita.

A natural sort of generalization is to take for $\mathbf{V}$ a category of rings with some extra structure: an action of a group by automorphisms, a derivation, or a grading by a finite group. The three kinds of structure just mentioned can be subsumed under the heading of an *action of a fixed bialgebra* H (cf. §43.11). To formulate a generalization to this context of the class of $\mathbf{Ab}$-valued functors we discovered in Chapter III, let $k \in \mathbf{Comm}^1$, let H be a bialgebra over k, let us fix an associative k-algebra K on which H acts, and let $\mathbf{V}$ be the variety of all objects of $K\text{-}\mathbf{Ring}_k$ (or $K\text{-}\mathbf{Ring}_k^1$) given with actions of H compatible with its action on K. It seems likely that every representable functor $\mathbf{V} \to \mathbf{Ab}$ is represented by a tensor ring $[K]\langle M\rangle$ (respectively $K\langle M\rangle$), where M is a k-centralizing K-bimodule given with an H-module structure, again compatible with the action of H on K, which we extend in the natural manner to an action of H on the tensor ring. Representable functors from $\mathbf{V}$ to various categories of associative and nonassociative rings should likewise have descriptions analogous to those we obtained in Chapters V and VI.

Another important kind of structure one can have on an associative ring S is an *involution*, i.e., a unary operation $x \mapsto \bar{x}$ satisfying

$$\begin{aligned} \bar{\bar{x}} &= x, \\ \overline{xy} &= \bar{y}\,\bar{x}, \\ \overline{x+y} &= \bar{x} + \bar{y}, \end{aligned} \tag{57.1}$$

for all $x, y \in S$ [**86**]. Let $k \in \mathbf{Comm}^1$ and let $\mathbf{InvRing}^1_k$ denote the category of unital k-algebras with k-linear involution. (One could, more generally, start with a possibly noncommutative ring k, itself having an involution, and consider k-rings with involution extending the involution on k. But we shall limit ourselves to the above classical situation here for simplicity.) It seems likely that representable functors $\mathbf{InvRing}^1_k \to \mathbf{Ab}$ are again represented by tensor algebras $k\langle M\rangle$, where this time M is a k-module with an involution (a module automorphism of exponent 2) $x \mapsto \bar{x}$, which we extend to a k-algebra involution on $k\langle M\rangle$. This extended involution may be described as the unique k-module endomorphism of $k\langle M\rangle$ satisfying

$$\overline{x_1 \dots x_n} = \bar{x}_n \dots \bar{x}_1 \qquad (x_1, \dots, x_n \in M).$$

As in Chapter V, one can verify that given three representable functors of the above form, with representing objects $k\langle L\rangle$, $k\langle M\rangle$, $k\langle N\rangle$, the morphism of set-valued functors determined by a homomorphism $k\langle N\rangle \to k\langle L\rangle \amalg k\langle M\rangle$ will be bilinear with respect to the $\mathbf{Ab}$-structures on these functors if and only if it carries N into $(L \otimes_k M) \oplus (M \otimes_k L)$. But now the story changes slightly: such homomorphisms of k-algebras with involution correspond to homomorphisms $N \to (L \otimes_k M) \oplus (M \otimes_k L)$ as *modules with involution*, where the involution on this direct sum is defined to take $x \otimes y$ to $\bar{y} \otimes \bar{x}$, and $y \otimes x$ to $\bar{x} \otimes \bar{y}$ $(x \in L,\ y \in M)$. Thus, such a homomorphism is not determined by independent maps into the two summands; rather, it is determined by its composite with the projection onto either summand, say the first, which can be an arbitrary k-module homomorphism (without reference to involutions) $N \to L \otimes_k M$.

The authors have not pursued the analysis of representable functors from k-algebras with involution to algebras of various sorts to which this approach leads, but let us note a couple of well-known functors which would have to turn up. If S is a ring with involution, the sets of *symmetric* and of *skew symmetric* elements of S, i.e.,

$$\{x \in S \mid \bar{x} = x\}, \qquad \text{respectively} \qquad \{x \in S \mid \bar{x} = -x\}$$

are closed under anticommutator and commutator brackets respectively. Hence they yield subfunctors of the standard underlying-set-preserving functors to **Jordan** and to **Lie**. Recall also, as noted in §40.3, that the transpose map on $n \times n$ matrices over a commutative ring S gives an antiautomorphism (in fact, an involution) of the matrix ring, but that this does not work if S is noncommutative. If S is a not necessarily commutative ring *with involution*, however, the operation of transposing a matrix and applying the involution entrywise *is* an involution of $M_n(S)$, and with the help of this construction we can define several representable group-valued functors on $\mathbf{InvRing}_k^1$ which have analogs among representable functors on $\mathbf{Comm}_k^1$, but not among representable functors on $\mathbf{Ring}_k^1$; for example, the construction of the group of $n\times n$ *orthogonal* matrices. (Cf. §40.3. Actually, one has competing terminologies to generalize from here: Matrices over the complex numbers are called ''orthogonal'' if they have this property relative to the trivial (identity) involution of $\mathbf{C}$, but ''unitary'' if they have the same property with respect to complex conjugation.)

Is there a common generalization of the structures given by an action of a bialgebra and by an involution? We suspect that the proper concept would be that of an action of a ''$\mathbf{Z}/2\mathbf{Z}$-graded bialgebra'' H (appropriately defined), in which the *even* elements $a\in H_0$ act on products by the usual sort of law, i.e.,

$$a(xy) \;=\; \Sigma\, b_i(x)c_i(y),$$

where $\mathbf{m}(a)=\Sigma\, b_i\otimes c_i$ and all the b_i and c_i are also even, while the *odd* elements $a\in H_1$ act by laws of the form

$$a(xy) \;=\; \Sigma\, c_i(y)b_i(x),$$

where the b_i and c_i are all odd. (We shall return to this idea in §63.)

A different sort of variety for which we suspect that the analog of Theorem 13.15 should be fairly easy to prove is $\mathbf{NARing}_k$ – certainly there seems to be nothing in the structure of a general nonassociative algebra that is well enough behaved to yield an abelian group structure, other than the underlying additive structure! Likewise, we expect that the only functors from this variety to various sorts of rings are the ''obvious'' (in many cases degenerate) ones. E.g., assuming k a field for simplicity, we expect that the general bilinear multiplication on such a functor should have exactly the same description (25.5) as for functors on associative k-algebras; that such an operation will be associative if and only if all three-fold products are zero, will be a Lie multiplication if and only if it is anticommutative (i.e., if and only if $x^*y=-y_*x$) and all three-fold products are zero, and will be Jordan if and only if it is commutative and all four-fold products are zero. Likewise, we see no way to get representable functors from $\mathbf{NARing}_k$ to groups other than those that factor through the functors to associative rings just described.

One might be tempted to conjecture that it is only for especially well-behaved varieties of k-algebras, such as $\mathbf{Comm}_k$, that one can get any $\mathbf{Ab}$-valued representable functors other than those that arise from the additive structure. Perhaps such a result is true, but if so, it must be based on a fairly broad concept of ''especially well-behaved'' variety. Indeed, suppose for simplicity that k is a commutative ring in which 2 is invertible, so that the multiplication of any object of $\mathbf{NARing}_k$ is the sum of a commutative and an anticommutative multiplication, which are completely independent. Then if we let $\mathbf{V}$ be the subvariety of $\mathbf{NARing}_k$ determined by the identity saying that the commutative part of the multiplication is associative, with no restriction on the anticommutative part (or with some arbitrary set of identities on that part, and possibly even identities relating both parts, so long as these do not lead to restrictions on the commutative part), then the functor $\mathbf{V} \to \mathbf{Comm}_k$ taking an algebra S to the same algebra with only the commutative part of the multiplication will be surjective on isomorphism classes of objects. Hence the rich and exotic classes of representable functors from the latter variety into various varieties of groups and rings induce equally varied classes of representable functors on $\mathbf{V}$. Similarly, if $\mathbf{V}$ is a variety satisfying the condition that the anticommutative part of its multiplication is a Lie operation, then all representable functors $\mathbf{Lie} \to \mathbf{Group}$ (which we have seen to form a rich class) induce functors $\mathbf{V} \to \mathbf{Group}$.

Another construction based on a similar idea: Suppose α is an invertible element of k, and $\mathbf{V}$ the subvariety of $\mathbf{NARing}_k$ determined by the identities saying that the sum of the commutative part of the multiplication and α times the anticommutative part is associative. Then the functor $\mathbf{V} \to \mathbf{Ring}_k$ taking an algebra S to the algebra with the same underlying k-module, but with multiplication given by this linear combination of the commutative and anticommutative parts of the multiplication of S, is an *equivalence* of categories. In particular, the theory of representable functors on $\mathbf{V}$ is equivalent to that of representable functors on $\mathbf{Ring}_k$.

Turning to associative rings, there is an important chain of varieties lying between $\mathbf{Comm}^1_k$ and $\mathbf{Ring}^1_k$, which forms the main subject of the theory of *rings with polynomial identity*. Namely, for each positive integer n, we have the variety of k-algebras defined by the identities satisfied by $n \times n$ matrices over members of $\mathbf{Comm}^1_k$. If k is a field, it is a standard result that every infinite prime k-algebra (k-algebra in which a product of nonzero two-sided ideals is nonzero), and in particular, every noncommutative k-algebra without zero-divisors, generates either the whole variety $\mathbf{Ring}^1_k$, or a member of this countable chain of subvarieties. Cf. [**167**, Corollary 6.1.46′, p.107], [**147**, Proposition 20.3 and Corollary b, pp.403-404], [**149**, Lemma III.2.1, p.66].

We have no idea what to expect of representable functors from these varieties to $\mathbf{Ab}$. The theory of the identities of these varieties for $n > 1$ is much more complicated than that of $\mathbf{Ring}^1_k$ or $\mathbf{Comm}^1_k$, but we do not know whether these complicated sets of identities lead to anything as simple as a nonstandard $\mathbf{Ab}$-valued representable functor.

In our discussion of "rings with additional operations" above, the additional operations were k-linear. A nonlinear operation arises when we consider associative (nonunital) algebras which are their own *Jacobson radicals*. This condition is equivalent to saying that every element x has a quasiinverse, i.e., an element $\tilde{x}$ (unique if it exists) such that

$$(57.2) \qquad x\tilde{x} + x + \tilde{x} \;=\; 0 \;=\; \tilde{x}x + x + \tilde{x},$$

equivalently, such that when one embeds S in a ring with unit, $1+\tilde{x}$ is a multiplicative inverse to $1+x$. Associative k-algebras S given with a unary operation $\tilde{\ }$ satisfying (57.2) clearly form a variety, $\mathbf{RadRing}_k$; it would be of interest to examine representable functors from this variety to abelian groups, rings, etc.. No representable functors to $\mathbf{Ab}$ different from those that occur on $\mathbf{Ring}_k$ are apparent, but neither is it apparent how the proof of Theorem 13.15 might be made to work in this category, since elements of coproducts in the variety of radical algebras cannot in general be written as ordinary polynomials in elements of the given algebras.

Another sort of nonlinear operations which we might assume given on domain categories of nonunital rings are *divided power* operations [**164**]. (We mention this only in the nonunital case because a *unital* ring in which 1 has divided powers becomes a **Q**-algebra, and the divided power operations are then expressible in terms of the **Q**-algebra structure.)

§58. Some miscellaneous remarks.

58.1. *The right context for Theorem 13.15.* Most of the proof of our fundamental result, Theorem 13.15, was achieved by formally manipulating statements about homomorphisms among bimodules and their tensor products. This suggests that these bimodules might be replaced by objects of a more general abelian category **A** furnished with an associative bilinear bifunctor $\otimes: \mathbf{A} \times \mathbf{A} \to \mathbf{A}$. Given such **A** and $\otimes$, suppose we define $\mathbf{Ring}(\mathbf{A}, \otimes)$ to be the category of objects $S \in \mathbf{A}$ equipped with associative multiplications $m: S \otimes S \to S$. (Cf. the concept of *monoid object* in a *monoidal category* [**122**, pp.4, 166].) Then under appropriate hypotheses, one should be able generalize our theorem to say that every representable functor $\mathbf{Ring}(\mathbf{A}, \otimes) \to \mathbf{Ab}$ can be written

$$\mathbf{Ring}(\mathbf{A}, \otimes) \xrightarrow{\text{forget}} \mathbf{A} \xrightarrow{\mathbf{A}(M, -)} \mathbf{Ab}$$

for a unique object $M \in \mathbf{A}$, this functor being represented by an object $<M> \in \mathbf{Ring}(\mathbf{A}, \otimes)$ "freely generated" by M. What the best hypotheses on **A** are for this undertaking we leave for others to explore. A result of this nature would very likely yield as special cases many of the characterizations of representable functors on categories of k-rings and k-algebras with additional linear structure suggested at the beginning of the preceding section. (The case of rings with involution, however, seems to require a distinct construction, $\mathbf{InvRing}(\mathbf{A}, \otimes)$, where **A** is a *symmetric* monoidal abelian category.)

M. Takeuchi (personal communication) has raised essentially the same question in another way. He notes that Theorem 13.15 and its variants in §14 can be regarded as showing how to recover the categories k-**Bimod** etc. from their categories of ring objects in the above sense, and asks in what generality this can be done. Cf. [**185**], [**186**].

58.2. *Rings with coaction.* In listing sorts of rings with extra structure in §57, we mentioned "gradings by finite groups" without saying why we needed the groups to be finite; let us examine this point now. To give a grading of a k-algebra S by a group or semigroup G is clearly equivalent to specifying the projection maps π_g of S onto the homogeneous components S_g $(g \in G)$. These projections can be any $|G|$-tuple of k-linear operations on S which sum to the identity (in the sense that on every $x \in S$, only finitely many have nonzero value, and these sum to x), and which satisfy the obvious conditions describing how the π_g compose, and how they act on products in S. If G is finite, we can, as noted, encode such a family of maps as an action on S of a certain bialgebra constructed from G; but if G is infinite, the definition of such an action cannot encode the fact that every element of S is to have zero image under all but finitely many projections, and that the finitely many nonzero images should sum to the original element. Likewise, the desired product law,

$$\pi_g(xy) = \Sigma_{ef=g}\, \pi_e(x)\pi_f(y)$$

will in general involve an infinite summation, and so cannot be encoded by the comultiplication of a bialgebra. So we cannot associate to an infinite group or semigroup G a bialgebra H such that a grading of a k-algebra S by G is equivalent to an action $H \otimes_k S \to S$ of H on S.

One can, however, associate to every group or semigroup G a bialgebra H' such that a grading of S by G is equivalent to a map $S \to H' \otimes_k S$ satisfying appropriate identities; as noted in §43.11, this is called a *coaction* of H' on S. (See [**21**, §6] for further discussion of the relation between such actions and coactions.) Thus the category of all k-algebras S with a coaction of a given bialgebra H, though it is not a variety (since a coaction is not in general equivalent to a family of operations on S), is another interesting variant of the category of all k-algebras, on which one may study representable functors.

Of course, the concept of a bialgebra H itself involves structure given by a map $H \to H \otimes_k H$, which is not an operation in the sense of universal algebra; so the category of all bialgebras over k, along with its variant, the category of Hopf algebras over k, and their various subcategories, are further interesting "not quite an algebra" cases to study.

58.3. *Topological algebras.* Let $k \in \mathbf{Comm}^1$, and let $\mathbf{LTopComm}_k^1$ be the category of *linearly topologized* commutative k-algebras, i.e., commutative k-algebras given with topologies making the algebra operations continuous, and having neighborhood bases of 0 consisting of k-submodules. In this category, let R be the polynomial algebra $k[x_0, \ldots, x_n, \ldots]$, topologized so that a neighborhood basis of 0 is given by the chain of ideals

$$(x_0, \ldots, x_n, \ldots) \supseteq (x_1, \ldots, x_n, \ldots) \supseteq (x_2, \ldots, x_n, \ldots) \supseteq \ldots .$$

We see that R represents the functor $\mathbf{LTopComm}_k^1 \to \mathbf{Set}$ carrying a linear topological k-algebra S to the set of all sequences $(\xi_0, \xi_1, \ldots)$ of elements of S which converge to 0, uniformly with all their multiples. We may identify this set with the set of formal power series $\xi_0 + \xi_1 t + \ldots$ whose coefficients have the same convergence property, and this set of power series forms a commutative ring; so the set-valued functor represented by R becomes a representable ring-valued functor F: $\mathbf{LTopComm}_k^1 \to \mathbf{Comm}^1$.

Let us now consider $\mathbf{Comm}_k^1$ to be embedded in $\mathbf{LTopComm}_k^1$ as the subcategory of discrete algebras. The only sequences in a discrete algebra which converge to 0 are those with only finitely many nonzero terms, and it follows that for $S \in \mathbf{Comm}_k^1$, $F(S) \cong S[t]$, even though, as noted in §25.20, the polynomial algebra functor is *not* representable as a functor on $\mathbf{Comm}_k^1$ itself. This approach, of studying functors on a category of discrete algebras as restrictions of representable functors on a larger category of topological algebras, has long been used by algebraic geometers; cf. [**135**] and works cited there.

Remarks: One way we saw the nonrepresentability of the polynomial-ring functor on $\mathbf{Comm}_k^1$ was by noting that this functor does not respect infinite products. Our present functor on $\mathbf{LTopComm}_k^1$, being representable, does respect arbitrary products, but an infinite product in $\mathbf{LTopComm}_k^1$ of discrete algebras is not discrete, and hence the value of F on this object does not coincide with its value on the corresponding product in the subcategory $\mathbf{Comm}_k^1$.

Our choice of the category of linearly topologized algebras, rather than, say, the larger category of *all* topological algebras, or the smaller category of topological algebras having a neighborhood basis of 0 consisting of ideals, i.e., residually discrete topological algebras, or of *complete* topological algebras of any of these sorts; and likewise, our choice of R as having a family of indeterminates that converge to 0 *uniformly with all their multiples*, were somewhat arbitrary: many variants of these choices also yield ring-valued representable functors which, when restricted to $\mathbf{Comm}_k^1$, give the polynomial-ring functor.

58.4. *What is an ordered group?* The theory of ordered groups deals with a large number of sorts of objects. A *partially ordered group* is a group G given with a partial ordering $\leq$ on its underlying set, such that for all $a, b, c \in G$,

$$a \leq b \;\Rightarrow\; ac \leq bc, \quad \text{and} \tag{58.5}$$

$$a \leq b \;\Rightarrow\; ca \leq cb. \tag{58.6}$$

If only (58.5) (respectively, only (58.6)) is assumed, then G is *right* (respectively, *left*) partially ordered. If the ordering $\leq$ is total, then G is called a *linearly ordered* (respectively, a right or left linearly ordered) group. If, rather, the partial ordering $\leq$ makes G a lattice, G is said to be *lattice ordered* (or right or left lattice ordered). Groups G that are right, left, or two-sided lattice or linearly *orderable* are also considered; here the pairs of right and left orderability conditions coincide, but are not equivalent to the corresponding two-sided conditions. (However, *one-sided linear* orderability is equivalent to *embeddability* in a *two-sided lattice-orderable* group, and also to embeddability in the group of order-preserving permutations of a totally ordered set [**71**].)

These categories of groups are, to various extents, reasonable domains for the study of representable functors. (The categories of groups with linear orderings least so, because they do not admit coproducts. On the other hand, the categories of one or two-sided linearly *orderable* groups are quasivarieties, and the categories of lattice-ordered groups are varieties if one takes the morphisms to be the group homomorphisms respecting meet and join.) But our main reason for mentioning these concepts here is to bring up a question which has long puzzled the first author, concerning the larger framework in which they should be looked at. It is well-known that, say, topological groups can be regarded as *group objects* in the category of topological spaces. But we *cannot* regard partially ordered groups as group objects in the category of partially ordered sets, or lattice ordered groups as group objects in the category of lattices. The reason is that the inverse operation of such groups is not order-preserving, hence is not a morphism of the proposed base category.

One might try to get around this by making partially ordered sets, lattices, etc.

into categories with two sorts of morphisms, "even" (order-preserving) and "odd" (order-reversing), and defining a group object in such a category to have "even" multiplication and "odd" inverse. However, derived operations such as $(x, y) \mapsto y^{-1}xy$ are still not morphisms of either sort, making the resulting concepts inconvenient.

Perhaps the best way to approach the situation is to treat the class of partially ordered sets in terms of a pair of categories, $\mathbf{C} \subseteq \mathbf{C}'$ having the same objects (the partially ordered sets), but where the morphisms of $\mathbf{C}$ are the isotone maps, while those of $\mathbf{C}'$ are all the set-maps. Noting that the inclusion of $\mathbf{C}$ in $\mathbf{C}'$ respects products, one may characterize partially ordered groups as group objects of $\mathbf{C}'$ whose multiplication morphisms lie in $\mathbf{C}$. One might make similar definitions using other pairs of categories $\mathbf{C} \subseteq \mathbf{C}'$, and see whether general results of interest can be obtained in this context.

58.7. *The bifunctor viewpoint.* A number of results proved in this work have had the following form. Given varieties of algebras $\mathbf{B}$ and $\mathbf{C}$, we have found a category $\mathbf{A}$ (often another variety), and a way of constructing from an object of $\mathbf{A}$ a co-$\mathbf{C}$ object of $\mathbf{B}$, which yields an equivalence between $\mathbf{A}$ and the category of such objects; equivalently, a contravariant equivalence

$$\mathbf{A}^{\mathrm{op}} \simeq \mathbf{Rep}(\mathbf{B}, \mathbf{C}). \tag{58.8}$$

Now a functor from $\mathbf{A}^{\mathrm{op}}$ to functors $\mathbf{B} \to \mathbf{C}$ is equivalent to a bifunctor

$$\mathbf{A}^{\mathrm{op}} \times \mathbf{B} \to \mathbf{C}. \tag{58.9}$$

It might be worth examining the properties of our constructions from this point of view.

Another sort of question involving bifunctors and representable functors is considered in [**65**], which studies categories $\mathbf{C}$ admitting monoidal structures $\otimes$ (i.e., coherently associative bifunctors $\otimes : \mathbf{C} \times \mathbf{C} \to \mathbf{C}$ with neutral objects), such that for every object S, the functors $S \otimes -$ and $- \otimes S$ have right adjoints (equivalently, are left adjoints of representable functors). It is shown that the only such bifunctor on $\mathbf{Ab}$ is the ordinary tensor product, and that there are no such bifunctors on certain other categories, such as $\mathbf{Group}$ and $\mathbf{Ring}^1$. However, the observations in the first part of §57 above, looked at from the point of view of §58.1, suggest that the classification of such functors on categories of *modules* may be interesting. G. M. Kelly (personal communication) suggests that the methods of [**65**], [**90**] and the papers referred to there may be useful in such an investigation.

58.10. *Many-sorted algebras.* We noted in §34 that many-sorted algebras are in general as well-behaved as one-sorted algebras, but are liable to be omitted from introductory developments of universal algebra for reasons of brevity. The same can be said of the theory of representable functors to varieties of such algebras. A representable functor from a category $\mathbf{C}$ to a variety $\mathbf{V}$ of I-sorted algebras (algebras with I-tuples of underlying sets) will be determined by an I-tuple of objects R_i of $\mathbf{C}$, together with co-operations mapping these objects into appropriate coproducts of each other, satisfying appropriate coidentities. Note that when we characterized *bilinear maps* among $\mathbf{Ab}$-valued representable functors on

k-**Ring**^{1}, our result could have been described as a characterization of representable functors from k-**Ring**^{1} to the variety of objects consisting of three sets A_1, A_2, A_3, given with an abelian group structure on each, and a bilinear map $A_1 \times A_2 \to A_3$ among these groups.

§59. Prevarieties.

We have proved various results to the effect that there are no nontrivial representable functors from a certain variety $\mathbf{A}$ to another variety $\mathbf{B}$. We have also seen that some varieties admit a great abundance of representable functors to others, while others are much more restricted in this respect.

This suggests that there is some sort of hierarchy among the varieties of algebras, perhaps a preordering, such that one can go from $\mathbf{A}$ to $\mathbf{B}$ by a nontrivial representable functor if and only if $\mathbf{A} \geq \mathbf{B}$.

The idea, in this formulation, is easily disposed of! Let us write $\mathbf{Ab}_n$ for $\mathbf{Mod}_{\mathbf{Z}/n\mathbf{Z}}$. Then if p and q are distinct primes, one has nontrivial representable functors from $\mathbf{Ab}_p$ to $\mathbf{Ab}_{pq}$, and from $\mathbf{Ab}_{pq}$ to $\mathbf{Ab}_q$, but none from $\mathbf{Ab}_p$ to $\mathbf{Ab}_q$. (This may be seen from the characterization of representable functors among varieties of modules mentioned in the discussion following Corollary 8.16, which is proved by an easy extension of the argument we used to characterize representable functors $\mathbf{Ab} \to \mathbf{Ab}$ in §21; see [**24**, §9.7] for details.)

We see that the problem is that all representable functors $\mathbf{Ab}_p \to \mathbf{Ab}_{pq}$ land in a subvariety of $\mathbf{Ab}_{pq}$ on which every representable functor to $\mathbf{Ab}_q$ is trivial. This suggests that we should define $\mathbf{A} \geq \mathbf{B}$ to mean that there exists a representable functor $\mathbf{A} \to \mathbf{B}$ whose image is not contained in any proper subvariety.

But this relation, too, fails to be transitive. For instance, the image of the forgetful functor

$$\mathbf{Mod}_{\mathbf{Q}} \to \mathbf{Ab} \tag{59.1}$$

lies in no proper subvariety of $\mathbf{Ab}$, and likewise the image of the functor

$$A \mapsto \{x \in A \mid px = 0\}: \ \mathbf{Ab} \to \mathbf{Ab}_p \tag{59.2}$$

lies in no proper subvariety of $\mathbf{Ab}_p$. But there is no nontrivial functor $\mathbf{Mod}_{\mathbf{Q}} \to \mathbf{Ab}_p$; in particular, the composite of the above two functors is trivial.

The difficulty this time is that the class of abelian groups which the functor (59.2) takes to the trivial group, namely, the abelian groups without p-torsion, is not a subvariety. It is described by the condition

$$(\forall x) \ px = 0 \ \Rightarrow \ x = 0,$$

which, as we recalled in §9.2, is what is known as a *universal Horn sentence*. This class of groups is thus a *quasivariety*, and the forgetful functor (59.1) has trivial composite with the functor (59.2) because its image lies in this proper *subquasivariety* of $\mathbf{Ab}$.

It is, in fact, not hard to verify

(59.3) If $F: \mathbf{A} \to \mathbf{B}$ is a representable functor between varieties of algebras, whose representing object is finitely presented, and $\mathbf{C}$ is a subvariety (or more generally, a subquasivariety) of $\mathbf{B}$, then $F^{-1}(\mathbf{C})$ is a subquasivariety of $\mathbf{A}$.

(The reader may either think through the argument now, or wait to see this result approximated in Lemma 59.5 below, and then precisely recovered in Corollary 59.10(ii).) Thus, if we define $\mathbf{A} \geq \mathbf{B}$ to mean that $\mathbf{B}$ has no proper subquasivariety into which the images of all representable functors $\mathbf{A} \to \mathbf{B}$ with finitely presented representing objects fall, then the relation $\geq$ will be transitive.

The need to assume a finitely presented representing object in (59.3) comes from the fact that a Horn sentence, by definition, can have only a finite conjunction of equations to the left of the implication sign. So let us define a *generalized* (i.e., possibly infinite) universal Horn sentence to mean a condition of the form

$$(59.4) \qquad (\forall (x_i)_{i\in I})\, (\textstyle\bigwedge_{j\in J} (f_j(x_i) = g_j(x_i)) \;\Rightarrow\; u(x_i) = v(x_i)),$$

where the index-sets I and J may be infinite, and where each symbol $f_j(x_i)$, $g_j(x_i)$, $u(x_i)$, $v(x_i)$, denotes a term in the I-tuple $(x_i)_{i\in I}$ (necessarily involving only finitely many of the x_i's). (Note that the family of identities on the left-hand side of a universal Horn sentence or generalized universal Horn sentence *may* be empty; thus, the class of such sentences includes the identities.) A class of algebras defined by a system of generalized universal Horn sentences is known as a *prevariety*. (The word "quasivariety" is occasionally used for this more general concept, e.g., in [**1**].)

It is easily shown that a class of algebras of a given type is a prevariety if and only if it is closed under passing to products and subalgebras. (Key step for "if": If $\mathbf{A}$ is a class closed under these operations, and S an algebra of the given type, but not belonging to $\mathbf{A}$, then there must be some pair of distinct elements $s, s' \in S$ that cannot be separated by any homomorphism of S into a member of $\mathbf{A}$. The generalized universal Horn sentence having to the left of the implication sign the relations in any presentation of S, and to the right the equation $s = s'$, where by s, s' we here mean expressions for these elements in terms of the generators in that presentation, thus holds for all objects of $\mathbf{A}$, but not for S. Hence, generalized universal Horn sentences which hold in $\mathbf{A}$ exclude all objects not lying in $\mathbf{A}$.) A prevariety is a quasivariety if and only if it is closed under direct limits, equivalently, if and only if it is closed under ultraproducts. (Sketch of proof: It is clear that a quasivariety is closed under direct limits. Every ultraproduct is a direct limit of products, so a prevariety closed under direct limits is closed under ultraproducts. Now suppose $\mathbf{A}$ is a class of algebras closed under ultraproducts, and (59.4) is any generalized universal Horn sentence such that none of the stronger ordinary universal Horn sentences obtainable from (59.4) by weakening the conjunction of equations on the left to the conjunction of some finite subset holds for all algebras in $\mathbf{A}$. Then by finding a counterexample in $\mathbf{A}$ to each ordinary universal Horn sentence so obtained, and forming an appropriate ultraproduct, one gets a counterexample in $\mathbf{A}$ to (59.4). Hence every generalized universal Horn sentence which *does* hold in $\mathbf{A}$ is a consequence of some stronger ordinary universal Horn sentence holding in $\mathbf{A}$; so if $\mathbf{A}$ is a prevariety, it is a quasivariety.)

A prevariety is a reflective subcategory of any variety containing it. In fact, the subprevarieties of any variety can be characterized as the reflective subcategories closed under passing to subalgebras; equivalently, as those reflective subcategories that can be defined by sets of *universally quantified* predicates in first-order language with infinite conjunctions. (An example of a reflective subcategory which is not a prevariety, equivalently, is not closed under passing to subalgebras, equivalently, requires existential conditions in any axiomatization, is the image of **Group** in $\mathbf{Semigp}^e$, i.e., the full subcategory whose objects are the semigroups in which every element is invertible.)

Now that we have the concept of prevariety, we can prove the following result, simpler than (59.3). We shall sketch two proofs: a quick one based on the above characterization of prevarieties in terms of closure under products and subalgebras, and another which is longer, but is more constructive from the point of view of the generalized universal Horn sentences involved.

LEMMA 59.5. *Let* $F: \mathbf{A} \to \mathbf{B}$ *be any representable functor among prevarieties, and* $\mathbf{C}$ *any subprevariety of* $\mathbf{B}$. *Then* $F^{-1}(\mathbf{C})$ *is a subprevariety of* $\mathbf{A}$.

SKETCH OF FIRST PROOF. Representable functors carry products to products and subalgebras to subalgebras. It follows that if $\mathbf{C}$ is closed under these constructions in $\mathbf{B}$, then $F^{-1}(\mathbf{C})$ is closed under them in $\mathbf{A}$.

SKETCH OF SECOND PROOF. Suppose (59.4) is one of a family of generalized universal Horn sentences that define $\mathbf{C}$ as a subprevariety of $\mathbf{B}$. Let the representing object in $\mathbf{A}$ for the given functor F be presented by a family of generators

(59.6) $$(w_h)_{h\in H}$$

and relations

(59.7) $$(p_k(w_h) = q_k(w_h))_{k\in K}.$$

The condition on an object A of $\mathbf{A}$ saying that $F(A)$ satisfies (59.4) is a condition on I-tuples (x_i) of elements of $F(A)$ (for I as in (59.4)), that is, on I-tuples of H-tuples of elements of A such that each of these H-tuples satisfies the K-tuple of equations (59.7). The left-hand side of (59.4) for $F(A)$ is a J-tuple of equations in this I-tuple of elements of $F(A)$; each of these equations comprises an H-tuple of equations in A; and the conclusion of (59.4) is another H-tuple of such equations.

Hence, each generalized universal Horn sentence (59.4) yields an H-tuple of generalized universal Horn sentences in an $(H\times I)$-tuple of variables each having for hypothesis the conjunction of an $(I\times K)\cup(J\times H)$-tuple of equations in A, such that (59.4) holds in $F(A)$ if and only if this family of sentences holds in A. By so translating each member of our system of generalized universal Horn sentences defining $\mathbf{C}$, we get a system of such sentences determining $F^{-1}(\mathbf{C})$. □

For an example where $\mathbf{A}$, $\mathbf{B}$ and $\mathbf{C}$ are varieties, but where the prevariety $F^{-1}(\mathbf{C})$ is not a variety or even a quasivariety, let $F: \mathbf{Ab} \to \mathbf{Mod}_{\mathbf{Q}}$ be the functor

Ab(**Q**, –). Then the abelian groups whose images under F lie in the trivial variety are those having no nonzero "completely divisible" elements; i.e., those that satisfy the generalized Horn sentence

$$(\forall\, x_1, x_2, \ldots)\ \ ((x_1 = 2x_2) \wedge \ldots \wedge (x_{i-1} = i\,x_i) \wedge \ldots)\ \Rightarrow\ (x_1 = 0).$$

From Lemma 59.5 we can deduce the result we have been aiming for:

COROLLARY 59.8. *For prevarieties of algebras* **A** *and* **B**, *let us write* $\mathbf{A} \geq \mathbf{B}$ *if* **B** *has no proper subprevariety containing the images of all representable functors* $\mathbf{A} \to \mathbf{B}$. *Then* $\geq$ *is a preorder on the class of all prevarieties of algebras (of all types).*

If $\mathbf{A} \geq \mathbf{B}$ *under this preordering, and* **B** *admits a nontrivial representable functor to a prevariety* **C**, *then so does* **A**. □

Incidentally, though we are assuming, as always, that our algebras have "small" underlying sets (in older language: underlying *sets* rather than proper classes), and small sets of operations, and though it is easy to see that every variety or quasivariety is determined by a small set of identities, respectively universal Horn sentences (for, up to naming of variables, there *is* only a small set of sentences of either sort), we cannot likewise say that every prevariety is determined by a small set of generalized Horn sentences. In fact, whether this is true is known to depend on the whether one assumes a set-theoretic statement called "Vopěnka's principle" [**1**]. Whether or not one assumes this principle, the *class of subprevarieties* of a given variety, unlike the class of subvarieties or subquasivarieties thereof, can easily fail to have the cardinality of a small set. For instance, if α is any cardinal, let H_α be the generalized universal Horn sentence for an object $S \in \mathbf{Ring}^1$ saying

(59.9) For every α-tuple of elements of S, $(x_i)_{i\in\alpha}$, if all the x_i have the same square, and if $x_i x_j = 0$ for distinct i and j, then the common value of the elements x_i^2 is 0.

It is easy to verify by example that distinct sentences H_α and H_β in this large family define distinct prevarieties.

(Incidentally, because the subprevarieties of a variety can form a large class, it is not evident whether the condition $\mathbf{A} \geq \mathbf{B}$ defined in the preceding Corollary is equivalent to the statement that there is a *single* representable functor $\mathbf{A} \to \mathbf{B}$ whose image lies in no proper subvariety of **B**.)

Let us note, however, that the constructive proof of Lemma 59.5 gives systems of generalized Horn sentences whose sizes one *can* generally bound:

COROLLARY 59.10 (to second proof of Lemma 59.5). *In the context of Lemma 59.5, if the object representing* F *can be presented by* α *generators and* β *relations, and if the subprevariety* **C** *is determined within* **B** *by a system of* γ *generalized universal Horn sentences, each involving* δ *variables, and having a conjunction of* ε *equations on the left, then the subprevariety* $F^{-1}(\mathbf{C})$ *is determined within* **A** *by a system of* $\alpha\gamma$ *generalized universal Horn sentences, each involving* $\alpha\delta$ *variables, and having a conjunction of* $\beta\delta + \alpha\varepsilon$ *equations on the left. (Here* α, β, γ *are cardinals, while* δ *and* ε *may either be cardinals,* *or conditions "$<\eta$" for cardinals η, e.g., "finitely many". In the latter case, symbols like $\alpha\delta$ should be interpreted as "the product of α with a cardinal $<\eta$" etc..)*

In particular,

(i) *If the object representing F is free, and $\mathbf{A}$ and $\mathbf{C}$ are varieties, then $F^{-1}(\mathbf{C})$ is a variety.*

(ii) *If the object representing F is finitely presented, and $\mathbf{A}$ and $\mathbf{C}$ are quasivarieties, then $F^{-1}(\mathbf{C})$ is a quasivariety.*

(iii) *If $\mathbf{A}$ and $\mathbf{C}$ are determined by small sets of generalized universal Horn sentences, then so is $F^{-1}(\mathbf{C})$.* □

EXERCISE 59.11. (a) *Show that in the hypothesis of the preceding Corollary, the condition "can be presented by α generators and β relations" may be weakened to "is a retract of an object which can be presented by α generators and β relations". Deduce that if the representing object for F is* projective (*a retract of a free object*), *then F^{-1} takes varieties to varieties.*

(b) *On the other hand, show that even if the object representing F is free, F^{-1} may fail to take quasivarieties to quasivarieties. (Suggestion: Let F: $\mathbf{Ab} \to \mathbf{Mod}_{\mathbf{Z}[t]}$ be the functor associating to an abelian group A the additive group of all strings $(a_1, a_2, \ldots)$, with t acting by carrying this string to $(2a_2, 3a_3, \ldots)$. Let $\mathbf{C}$ be the quasivariety of $\mathbf{Z}[t]$-modules in which $t-1$ has zero annihilator.)*

EXERCISE 59.12. *Show that two prevarieties equivalent as categories will be equivalent under the preordering of Corollary 59.8, but that the converse is not true. (Hint for second part: compare $\mathbf{Mod}_{\mathbf{Z}/p\mathbf{Z}}$ and $\mathbf{Mod}_{(\mathbf{Z}/p\mathbf{Z})\times(\mathbf{Z}/p\mathbf{Z})}$.)*

EXERCISE 59.13. (a) *Determine the preordering of Corollary 59.10 on the family of five varieties $\mathbf{Comm}^1$, $\mathbf{Bool}^1$, $\mathbf{Ring}^1$, $\mathbf{Ring}$ and $\mathbf{Semigp}^e$, assuming that Lemma 43.7 holds for $k = \mathbf{Z}$ as well as for fields (as conjectured immediately following that Lemma).*

(b) *Determine, under the same assumption, the preordering described following* (59.3) *above, for the same five varieties. (Key fact: a finitely presented commutative ring is residually finite.)*

EXERCISE 59.14. (a) *Show that the class of all subprevarieties of $\mathbf{Comm}^1$ (like that of $\mathbf{Ring}^1$) does not have the cardinality of a small set. (Hint: if x is a nilpotent element of a ring R, and $y \neq 0$, then $xyR \neq yR$. Consider chains of such ideals.)*

On the other hand,

(b) *Show that the variety $\mathbf{Bool}^1$ has no nontrivial proper subprevarieties.*

In studying representable functors on prevarieties $\mathbf{A}$, one needs to know how to form coproducts in such categories. The general recipe is: take the coproduct in a variety containing $\mathbf{A}$, then impose on this coproduct the relations needed to force the resulting object back into $\mathbf{A}$. For instance, in the subquasivariety of $\mathbf{Comm}^1_k$ defined by the sentence

$$(\forall x)\ x^2 = 0 \ \Rightarrow\ x = 0,$$

(the quasivariety of commutative k-algebras without nonzero nilpotent elements), one may construct a coproduct $A \amalg B$ by taking the tensor product $A \otimes_k B$, and dividing out by its nil radical. This radical may be nontrivial if k is a field of nonzero characteristic.

The representable functors on a prevariety may behave quite differently from those on the variety it generates. For instance, let k be a field, and **A** the subprevariety of $\mathbf{Ring}_k^1$ defined by the condition that all idempotent elements commute with one another. This clearly contains all objects of $\mathbf{Ring}_k^1$ without zero-divisors, in particular, all free associative k-algebras $k\langle X\rangle$, hence **A** generates $\mathbf{Ring}_k^1$ as a variety. But we see that the set of idempotent elements of any object of **A** is a Boolean ring, under the same operations as in $\mathbf{Comm}_k^1$ (see 40.4), giving nontrivial representable functors from **A** to a great many varieties that admit no such functors from $\mathbf{Ring}_k^1$.

Let us note that given a representable functor $F\colon \mathbf{A} \to \mathbf{B}$ among prevarieties, not only are inverse images under F of subprevarieties of **B** subprevarieties of **A**, but the class of objects of **B** embeddable in members of the *image* of F is a subprevariety. This follows from the fact that representable functors preserve products. If **A** is a *quasivariety* and the representing object for F is finitely presented, then F will preserve ultraproducts, hence this subprevariety will be a quasivariety. (For example, taking for F the forgetful functor from groups to semigroups, we deduce that the class of semigroups embeddable in groups is a quasivariety. An elegant set of universal Horn sentences describing this quasivariety was found by A. I. Mal'cev [**126**]; cf. [**46**, §VII.3], [**127**].)

If the representing object R for F is *not* finitely presented, but the prevariety **A** is defined by a small set of generalized universal Horn sentences, one can still get bounds on the sizes of the generalized universal Horn sentences needed to characterize objects of **B** embeddable in members of the image of F. But the argument is more laborious: Let $G\colon \mathbf{B} \to \mathbf{A}$ be the left adjoint of F. Then $B \in \mathbf{B}$ can be embedded in an object of the form $F(A)$ if and only if the universal map $\eta(B)\colon B \to FG(B)$ is one-to-one. Now two elements $b,\ b' \in B$ fall together under $\eta(B)$ if and only if, when we construct $G(B)$ by gluing together copies of R (as in the proof of Theorem 8.14), the images of the two copies corresponding to b and b' coincide, element by element. If we have a set of α generators for R, this gives α equations to be satisfied in $G(B)$; each of these equations, if it holds, must be deducible from certain relations involving b, b' and other elements of B, using repeated applications of the relations defining R and the identities and generalized universal Horn sentences defining **A**. If we list all ways a family Y of relations in an object B can so force the equality of the images of two elements b and b' (with each element of Y being used), and for each such way, write down a generalized universal Horn sentence for objects of **B** saying that if these relations hold then $b = b'$, we see that this set of sentences defines the desired subprevariety of **B**. Clearly, one can bound the sizes of the sentences needed in terms of the sizes of the generalized Horn sentences defining **A**, and then bound the number of sentences of that size. But we shall not go through the details here.

We remark that when we defined the relation $\mathbf{A} \leq \mathbf{B}$ used above in terms of representable functors $\mathbf{A} \to \mathbf{B}$ not landing in any common proper subprevariety, we could have refined things a little, and gotten a stronger, if not so simple result. Namely, if we define a *degeneracy subprevariety* of a variety $\mathbf{B}$ to mean the inverse image under a representable functor $\mathbf{B} \to \mathbf{Set}$ of the trivial variety of sets (defined by the identity $x = y$), equivalently, a class of algebras determined by the condition that some system of equations have at most one solution, then it is easy to see that the inverse image of any degeneracy prevariety under any representable functor is a degeneracy prevariety, so that the relation "there is no proper *degeneracy* subprevariety of $\mathbf{B}$ containing the images of all representable functors $\mathbf{A} \to \mathbf{B}$" is sufficient to give the conclusion of Corollary 59.8. An example of a proper subprevariety – in fact, a proper subvariety – which is *not* contained in any proper degeneracy subprevariety is $\mathbf{Ab}_p \subseteq \mathbf{Ab}_{p^2}$, since every nonzero object of $\mathbf{Ab}_{p^2}$ has a nonzero homomorphic image in $\mathbf{Ab}_p$. Hence, by the indicated extension of Corollary 59.8, if $\mathbf{Ab}_{p^2}$ admits a nontrivial representable functor to a prevariety $\mathbf{C}$, then so does $\mathbf{Ab}_p$. (For an example that seems to have the same flavor, but turns out the other way, the subprevariety of $\mathbf{Ring}^1_k$ or $\mathbf{Comm}^1_k$ consisting of k-algebras without nontrivial nilpotent elements *is* contained in – in fact, is – a degeneracy prevariety, namely the prevariety of k-algebras admitting at most one homomorphism from $k[\varepsilon \mid \varepsilon^2 = 0]$.)

One can define other interesting preorderings on categories of algebras based in rather different ways on the existence of representable functors; for instance the relation "there exists a *faithful* representable functor $\mathbf{A} \to \mathbf{B}$", or the relation which arose in our study of idempotents in semigroup-valued functors in §42, "there exists a representable functor $\mathbf{A} \to \mathbf{B}$ carrying the initial object of $\mathbf{A}$ to the initial object of $\mathbf{B}$".

§60. ⊛-algebras and ⊛-coalgebras.

Recall that given $k \in \mathbf{Comm}^1$, the definition of a *bialgebra* over k (§43) is built out of two definitions, that of an associative k-algebra and that of a coassociative $\otimes_k$-coalgebra, each of which can be defined and studied independent of the other. Now bialgebras arise as a generalization of co-$\mathbf{Semigp}^e$ objects of $\mathbf{Comm}^1_k$, and by definition, the latter have underlying $\mathbf{Comm}^1_k$ objects, which in particular are associative k-algebras, thus explaining the presence of the first kind of structure. What is not as obvious is why co-$\mathbf{Semigp}^e$ objects of $\mathbf{Comm}^1_k$ should have underlying $\otimes_k$-coalgebra objects. One can look on this as an accident – a result of the fact that the underlying k-module of a coproduct of objects of $\mathbf{Comm}^1_k$ depends only on the k-module structures of these objects, and is in fact their tensor product over k.

When we looked at the category k-$\mathbf{Ring}$ of nonunital k-rings (where $k \in \mathbf{Ring}^1$), we found that there, too, the underlying k-bimodule of a coproduct $\coprod_{\alpha \in A} S_\alpha$ could be described solely in terms of the k-bimodule structures of the S_α; we recalled such a description in (12.2):

$$\coprod_{\alpha \in A} S_\alpha \;=\; \bigoplus_{(\alpha_1, \ldots, \alpha_n)} S_{\alpha_1} \otimes_k \cdots \otimes_k S_{\alpha_n},$$

where the direct sum is over all strings

(60.1) $(\alpha_1, \dots, \alpha_n)$ with $n \geq 1$, all $\alpha_i \in A$, and $\alpha_i \neq \alpha_{i+1}$ $(1 \leq i \leq n-1)$.

Although this is certainly not as neat a construction as that of the tensor product of modules, let us make

Definition 60.2. *Let* $k \in \mathbf{Ring}^1$. *Then for any family of k-bimodules* $(M_\alpha)_{\alpha \in A}$, *we define the k-bimodule*

(60.3) $$\circledast_{\alpha \in A} M_\alpha = \bigoplus_{(\alpha_1, \dots, \alpha_n)} M_{\alpha_1} \otimes_k \dots \otimes_k M_{\alpha_n},$$

where the direct sum is taken over all strings (60.1). *For finite families* $M_1, \dots, M_r$, *we shall also write this construction* $M_1 \circledast \dots \circledast M_r$.

For each $\beta \in A$, *we shall write* $i^\beta: M_\beta \to \circledast_{\alpha \in A} M_\alpha$ *for the inclusion as k-bimodules, where we identify* M_β *with the summand in* (60.3) *indexed by the length-one string* (β), *and* $\varepsilon^\beta: \circledast_{\alpha \in A} M_\alpha \to M_\beta$ *for the projection onto that summand.*

Clearly, for any index-set A the construction

$$\circledast_{\alpha \in A}: (k\text{-}\mathbf{Bimod})^A \to k\text{-}\mathbf{Bimod}$$

is functorial. Let us note two degenerate cases: when the index-set A is empty, this is the functor giving the zero bimodule, and when A is a singleton $\{\alpha\}$, it is isomorphic to the identity functor, via the morphisms i^α and ε^α.

It is also not hard to verify that to every map of sets $f: A \to B$, and A-tuple of k-bimodules $(M_\alpha)_{\alpha \in A}$, we can associate an isomorphism of bimodules

(60.4) $$\circledast_{\beta \in B} (\circledast_{f(\alpha)=\beta} M_\alpha) \cong \circledast_{\alpha \in A} M_\alpha,$$

in a way which on the one hand is functorial in the M_α, and on the other hand respects composition of maps of index-sets; where the first of these assertions means that given bimodule homomorphisms $M_\alpha \to N_\alpha$ $(\alpha \in A)$, one has the commuting diagram

(60.5) $$\begin{array}{ccc} \circledast_{\beta \in B} (\circledast_{f(\alpha)=\beta} M_\alpha) & \xrightarrow{\sim} & \circledast_{\alpha \in A} M_\alpha \\ \downarrow & & \downarrow \\ \circledast_{\beta \in B} (\circledast_{f(\alpha)=\beta} N_\alpha) & \xrightarrow{\sim} & \circledast_{\alpha \in A} N_\alpha, \end{array}$$

and the second means that given maps of index-sets $A \xrightarrow{f} B \xrightarrow{g} C$, the diagram formed from four instances of (60.4):

(60.6) $$\begin{array}{ccc} \circledast_{\gamma \in C} (\circledast_{g(\beta)=\gamma} (\circledast_{f(\alpha)=\beta} M_\alpha)) & \xrightarrow{\sim} & \circledast_{\gamma \in C} (\circledast_{gf(\alpha)=\gamma} M_\alpha) \\ \downarrow \wr & & \downarrow \wr \\ \circledast_{\beta \in B} (\circledast_{f(\alpha)=\beta} M_\alpha) & \xrightarrow{\sim} & \circledast_{\alpha \in A} M_\alpha. \end{array}$$

commutes.

We shall henceforth restrict attention to finite families of bimodules, indexed by sets of the form $\{1, \ldots, r\}$. We note that the isomorphisms (60.4) allow us to express the functors $\circledast_{i=1,\ldots,r}$ for $2<r<\infty$ in terms of the 2-variable functor $-\circledast-$. This bifunctor will be coherently associative and commutative [**122**, §VII.2]: the associativity and commutativity isomorphisms

$$(M_1 \circledast M_2)\circledast M_3 \cong M_1 \circledast M_2 \circledast M_3 \cong M_1 \circledast (M_2 \circledast M_3) \quad \text{and} \quad M_1 \circledast M_2 \cong M_2 \circledast M_1$$

are the instances of (60.4) corresponding to the obvious maps of index-sets

$$\{1,2\} \leftarrow \{1,2,3\} \rightarrow \{1,2\} \quad \text{and} \quad \{1,2\} \rightarrow \{1,2\},$$

and the required coherency conditions follow from the commutativity of (60.6). We likewise see, using the two maps $\{1\} \to \{1,2\}$ and the fact that under each of these, one element has empty inverse image, that the zero bimodule is a two-sided "neutral element" for $\circledast$. In the language of [**122**], these observations show that the bifunctor $\circledast$ (like $\otimes_k$ for $\mathbf{Mod}_k$, and like $\oplus$ for any abelian category) makes k-**Bimod** a *symmetric monoidal category*.

It is thus natural to make

DEFINITION 60.7. *A* ⊛-algebra *in* k-**Bimod** *will mean a pair* (M, μ), *where* $M \in k$-**Bimod**, *and* μ *is a bimodule map* $M \circledast M \to M$ *(or as we shall often write it,* $M^\lambda \circledast M^\rho \to M$*). This ⊛-algebra will be called* ⊛-associative *(or simply* associative *when there is no danger of ambiguity) if the two maps* $M \circledast M \circledast M \to M$ *obtained in the obvious ways from* μ *coincide. It will be called* ⊛-commutative (commutative *when there is no danger of ambiguity) if the map* μ *is invariant under precomposition with the automorphism of* $M^\lambda \circledast M^\rho$ *interchanging the* λ *and* ρ *arguments, and* ⊛-unital (*or* unital) *if* $\mu i^\lambda = \mathrm{id}_M = \mu i^\rho$.

We similarly define a ⊛-coalgebra *to mean a pair* $(M, \mathbf{m})$ *with* $M \in k$-**Bimod** *and* $\mathbf{m}$ *a bimodule map* $M \to M^\lambda \circledast M^\rho$, *and let the duals of the properties of the preceding paragraph define the conditions of* ⊛-coassociativity, ⊛-cocommutativity *and* ⊛-counitality *on such objects (the last condition meaning* $\varepsilon^\lambda \mathbf{m} = \mathrm{id}_M = \varepsilon^\rho \mathbf{m}$).

(The reader should beware confusing the conditions of associativity etc. on μ in the above Definition with the properties of $\circledast$ with the same names, mentioned in the preceding paragraph. It is, of course, the fact that $\circledast$ has those properties that allows one to *define* these conditions on a ⊛-operation μ.)

We shall generally suppress the symbol $\otimes$ in writing decomposable elements of the components $\ldots \otimes_k M^\lambda \otimes_k M^\rho \otimes_k \ldots$ of $M^\lambda \circledast M^\rho$, and write such elements as "products" $\ldots x_i^\lambda x_{i+1}^\rho \ldots$; the same will apply in ⊛-products of more than two copies of M. The ⊛-product of r copies of M will be written $M^{(1)} \circledast \ldots \circledast M^{(r)}$; thus, for $r = 2$, the superscripts $^\lambda$ and $^\rho$ will be regarded as abbreviations for $^{(1)}$ and $^{(2)}$. We shall not take advantage of the option of dropping the prefix "⊛-" as allowed by the above Definition; we have included this option for the convenience of possible future workers.

We may now observe that large parts of the arguments by which the structure of a cocommutative, coassociative, possibly counital co-operation on an object of $\mathbf{Ring}_k^1$ was analyzed in §§13 and 17 did not involve the algebra structure, but only the bimodule structure of the coproduct and the coidentities assumed for the comultiplication map, and so could have been formulated as results on

⊛-coalgebras! If the first author felt no qualms about devoting several more years of his research time to the completion of this book, he might experiment with rewriting those sections from this point of view. However, aside from his lack of time, there is a good argument for leaving them as they are, so as not to add an additional layer of abstraction to this first exposition of the topic.

Let us examine the concept of a *⊛-algebra* as defined above. A general (not necessarily ⊛-associative) ⊛-algebra (M, μ) is a rather baroque object: the multiplication μ specifies in an arbitrary manner two k-multilinear operations $M^n \to M$ for each $n > 0$:

$$(x_1, x_2, \ldots, x_n) \mapsto \mu(x_1^\lambda x_2^\rho \ldots) \qquad \text{and} \qquad (x_1, x_2, \ldots, x_n) \mapsto \mu(x_1^\rho x_2^\lambda \ldots).$$

If we assume μ is ⊛-unital, that assumption determines the two operations with $n = 1$, and if we assume it ⊛-commutative, this reduces the two operations for each $n > 1$ to a single operation; but we still have infinitely many independent operations.

The condition of ⊛-associativity, however, cuts things down considerably. For convenience, let us consider it in conjunction with ⊛-unitality.

THEOREM 60.8. *Let* $M \in k$-**Bimod**. *Then a ⊛-unital ⊛-associative ⊛-algebra operation* μ *on* M *is determined by the two operations*

$$(x, y) \mapsto \mu(x^\lambda y^\rho) \qquad \textit{and} \qquad (x, y) \mapsto \mu(x^\rho y^\lambda),$$

which we shall denote

$$(x, y) \mapsto x^* y \qquad \textit{and} \qquad (x, y) \mapsto x_* y$$

respectively. These can be any two k-bilinear operations satisfying the four associativity conditions

$$(60.9) \qquad \begin{array}{ll} x^*(y^*z) = (x^*y)^*z, & x^*(y_*z) = (x^*y)_*z, \\ x_*(y^*z) = (x_*y)^*z, & x_*(y_*z) = (x_*y)_*z. \end{array}$$

SKETCH OF PROOF. Let μ be a ⊛-unital ⊛-associative ⊛-algebra operation on M. We first want to show how to express all the multilinear operations comprising μ in terms of * and $_*$. It will be most elegant to do still more. The condition that μ is ⊛-unital and ⊛-associative is equivalent to saying that it belongs to a family of operations

$$\mu^r\colon\ M^{(1)} \circledast \ldots \circledast M^{(r)} \to M \qquad (r = 0, 1, 2, \ldots)$$

(where $M^{(1)}, \ldots, M^{(r)}$ are copies of M), with μ^0 the zero map, μ^1 the identity, and $\mu^2 = \mu$, which satisfy a generalized associativity law, saying that for

$$r = r_1 + \ldots + r_s,$$

μ^r is equal to the composite

$$M^{(1)} \circledast \ldots \circledast M^{(r)} \xrightarrow{\mu^{r_1} \circledast \ldots \circledast \mu^{r_s}} M^{(1)} \circledast \ldots \circledast M^{(s)} \xrightarrow{\mu^s} M,$$

where the left-hand map acts by first using (60.4) to identify $M^{(1)} \circledast \ldots \circledast M^{(r)}$ with

$$(M^{(1)}\circledast\ldots\circledast M^{(r_1)})\circledast\ \ldots\ \circledast(M^{(r_{s-1}+1)}\circledast\ldots\circledast M^{(r_s)}),$$

and then combining the first r_1 copies of M under μ^{r_1}, the next r_2 copies under μ^{r_2}, etc..

From the above properties, and the unitality law for μ, we can deduce that these maps will also satisfy the unitality laws

$$(60.10)\qquad \mu^r\,i^{(m)} = \mathrm{id}_M,$$

for $m \leq r$, and that they are "consistent", in the sense that for $i_1, \ldots, i_n \leq r$,

$$(60.11)\qquad \mu^r(x_1^{(i_1)}\ldots\, x_n^{(i_n)}) = \mu^{r+1}(x_1^{(i_1)}\ldots\, x_n^{(i_n)}).$$

Using these observations, we shall describe, for any $r > 0$ and any $i_1, \ldots, i_n \in \{1, \ldots, r\}$ with $n \geq 1$ and $i_m \neq i_{m+1}$ $(1 \leq m \leq n-1)$, how to compute the value of μ^r on a decomposable element of the summand

$$M^{(i_1)}\otimes_k \ldots \otimes_k M^{(i_n)} \subseteq M^{(1)}\circledast\ldots\circledast M^{(r)}$$

in terms of the binary operations * and $_*$. Namely, we claim that for $x_1, \ldots, x_n \in M$, the element

$$(60.12)\qquad \mu^r(x_1^{(i_1)}\ldots\, x_n^{(i_n)})$$

can be evaluated by inserting in the string $x_1 \ldots x_n$ a * between x_m and x_{m+1} if $i_m < i_{m+1}$ and a $_*$ if $i_m > i_{m+1}$, and nesting parentheses on the left (though once we have proved (60.9), it will follow that it does not matter how parentheses are placed). For $n = 1$, this assertion is just (60.10); so let $n > 1$, and assume inductively that the desired result is true for expressions of length $< n$.

There are two cases to consider. Suppose first that

$$i_{n-1} < i_n.$$

Let us write $r = r_1 + r_2$, where r_1 is any integer such that $i_{n-1} \leq r_1 < i_n$. Then the general associative law for μ noted above says that we can get the value of (60.12) by breaking $x_1^{(i_1)}\ldots\, x_n^{(i_n)}$ into substrings which alternately have the property that all their members have superscripts $\leq r_1$ and that all have superscripts $> r_1$, then applying μ^{r_1} to substrings of the first sort and μ^{r_2} to substrings of the second sort (in the latter case, after first reducing all superscripts by r_1), marking the element of M obtained from each substring of the first sort with the superscript $^\lambda$ and the element obtained from each substring of the second sort with $^\rho$, and finally, applying to the resulting string in $M^\lambda \circledast M^\rho$ the operation $\mu^2 = \mu$.

Now note that by choice of r_1, the last of the strings into which we have broken $x_1^{(i_1)}\ldots x_n^{(i_n)}$ is the single term $x_n^{(i_n)}$. By (60.10) μ^{r_2} takes this element to x_n. From this it is easily seen that the value of (60.9) is unchanged if we increase the superscript on x_n from $^{(i_n)}$ to any higher value; moreover, by (60.11) we may first pass to the operation μ^{r+1} on $M^{(1)}\circledast\ldots\circledast M^{(r+1)}$, and then increase this superscript to $^{(r+1)}$. In other words, (60.12) is equal to

$$(60.13)\qquad \mu^{r+1}(x_1^{(i_1)}\ldots\, x_{n-1}^{(i_{n-1})}x_n^{(r+1)}).$$

We can now apply the general associative law to the above expression, this time decomposing the superscript $r+1$ on the operator μ as the sum of r and 1. This groups the first $n-1$ terms of (60.13) into one string and the last term into another string, and thus expresses (60.12) as

$$\mu(\mu^r(x_1^{(i_1)} \dots x_{n-1}^{(i_{n-1})})^\lambda x_n^\rho),$$

in other words, as

$$\mu^r(x_1^{(i_1)} \dots x_{n-1}^{(i_{n-1})}) * x_n. \tag{60.14}$$

By our inductive hypothesis, the element $\mu^r(x_1^{(i_1)} \dots x_{n-1}^{(i_{n-1})})$ is obtained by inserting $*$, $_*$, and parentheses among its $n-1$ terms in the manner described earlier, and evaluating. Substituting the resulting expression into (60.14), we get the corresponding description of the element (60.12), as required.

The case $i_{n-1} > i_n$ is similar, except that at the step where we rewrite (60.12) in terms of μ^{r+1}, we first "push up" the superscripts by 1, i.e., use the general associative and unital laws to get, not (60.11), but

$$\mu^r(x_1^{(i_1)} \dots x_n^{(i_n)}) = \mu^{r+1}(x_1^{(i_1+1)} \dots x_n^{(i_n+1)}).$$

We then *lower* the final superscript i_n+1 to 1 at the point in the argument where previously we raised i_n to $r+1$.

To get (60.9), note that in proving the result just established, we could have nested parentheses on the *right* instead of on the *left*, by isolating x_1 rather than x_n. Equating the results of the "right-nesting" and "left-nesting" reductions applied to each of

$$\mu^3(x^{(1)}y^{(2)}z^{(3)}), \quad \mu^2(x^{(1)}y^{(2)}z^{(1)}), \quad \mu^2(x^{(2)}y^{(1)}z^{(2)}), \quad \mu^3(x^{(3)}y^{(2)}z^{(1)}),$$

we obtain the four desired equations (60.9).

The converse statement, that any pair of operations satisfying (60.9) yields a ⊛-unital ⊛-associative ⊛-algebra, is straightforward to verify; one defines the actions of the operations μ^r as described following (60.12) above, and notes that the general associative law then holds. (This argument takes for granted that the identities (60.9) imply that parentheses can be dropped in any monomial in the operations $*$ and $_*$. This can be formally verified by the same argument as for a single associative operation; alternatively, it can be reduced to the result for a single associative operation by Lemma 61.2 in the next section.) □

In the above situation, to further require ⊛-commutativity is clearly to take $* = {}_*$. Hence we get

COROLLARY 60.15. *A ⊛-commutative ⊛-unital ⊛-associative ⊛-algebra structure on an object* $M \in k$-**Bimod** *is equivalent to a structure on* M *of (in general nonunital, noncommutative!) associative k-ring extending the given k-bimodule structure.* □

Using the approach of §28, one can get results for ⊛-*coalgebras* parallel to the above results on ⊛-algebras. In the following statement, given a pair of maps $\mathbf{p}: M \to M^\lambda \otimes_k M^\rho$ and $\mathbf{q}: M \to M^\rho \otimes_k M^\lambda$, the "induced" map from M to a tensor product $\dots \otimes_k M^\lambda \otimes_k M^\rho \otimes_k \dots$ will mean the map obtained by the appropriate left-nested composition. E.g., the induced map

$$M \to M^\rho \otimes_k M^\lambda \otimes_k M^\rho$$

is the composite

$$M \xrightarrow{\mathbf{p}} M^\lambda \otimes_k M^\rho \cong M \otimes_k M^\rho \xrightarrow{\mathbf{q}\otimes 1} M^\rho \otimes_k M^\lambda \otimes_k M^\rho .$$

Corollary 60.16 (to proofs of Theorem 60.8 and Corollary 60.15). *A ⊛-counital ⊛-coassociative ⊛-co-operation*

$$\mathbf{m}\colon M \to M^\lambda \circledast M^\rho$$

($M \in k$-**Bimod**) *is determined by its composites with the projections of* $M^\lambda \circledast M^\rho$ *onto its direct summands* $M^\lambda \otimes_k M^\rho$ *and* $M^\rho \otimes_k M^\lambda$. *These composites can be any pair of* $\otimes_k$*-comultiplications on* M *which* (i) *satisfy the coassociativity conditions dual to* (60.9), *and* (ii) *have the property that for each* $x \in M$, *only finitely many of the induced maps* $M \to \ldots \otimes_k M^\lambda \otimes_k M^\rho \otimes_k \ldots$ *carry* x *to a nonzero value. (I.e., the system of maps of* M *into these tensor product bimodules induced by the two comultiplications gives a map of* M, *not merely into the direct product of these bimodules, but into their direct sum,* $M^\lambda \circledast M^\rho$.)

Such a ⊛-co-operation will be ⊛-cocommutative if and only if the two $\otimes_k$*-co-operations determining it are equal, as maps* $M \to M \otimes_k M$. □

Recall that we constructed our functor ⊛ so that the underlying k-bimodule of a *coproduct* of k-rings would be the ⊛-product of the underlying bimodules of those k-rings, and thus, in particular, so that a co-operation on a k-ring would have the form of a ⊛-co-operation on its underlying bimodule. Let us describe the ⊛-coalgebra structure corresponding to the coaddition $\mathbf{a}_M$ defined in §12 on a tensor ring $T = [k]\langle M\rangle$ ($M \in k$-**Bimod**). The counitality, cocommutativity and coassociativity properties of T as a co-**Ab** object of k-**Ring** mean that $\mathbf{a}_M$ makes the underlying bimodule of T a ⊛-counital ⊛-cocommutative and ⊛-coassociative ⊛-coalgebra; thus by Corollary 60.16 it is determined by a single $\otimes_k$-comultiplication. We find that this map acts on decomposable elements of T by

$$x_1 \ldots x_n \mapsto \Sigma_{1 \le i < n}\, (x_1 \ldots x_i) \otimes (x_{i+1} \ldots x_n) \in T \otimes_k T \qquad (x_1, \ldots, x_n \in M)$$

(Here we are using the tensor product sign in writing elements of $T \otimes_k T$, but juxtaposition for the internal multiplication of T.) For each r, the induced map $T \to T^{\otimes r}$ likewise takes $x_1 \ldots x_n$ to

$$\Sigma_{1 \le i_1 < \ldots < i_r = n}\, (x_1 \ldots x_{i_1}) \otimes \ldots \otimes (x_{i_{r-1}+1} \ldots x_n)$$

In particular, this map is zero if $r > n$, so condition (ii) of the preceding corollary is satisfied.

If k is a field and we restrict attention to k-rings that are k-algebras, and identify the linear dual of the k-vector-space $[k]\langle M\rangle$ with the vector space of formal tensor power series $[k]\langle\langle A\rangle\rangle$ over the linearly compact vector space $A = M^*$, then the ⊛-coalgebra structure on T induces a ⊛-algebra structure on this dual. This will be ⊛-unital, ⊛-associative and ⊛-commutative, hence it corresponds to an ordinary k-algebra structure on $[k]\ll A\gg$, which we find to be the standard k-algebra structure on formal tensor power series. Observe also that for each $r \geq 0$, the vector subspace $M \oplus \ldots \oplus M^{\otimes r}$ forms a $\circledast$-subcoalgebra of T, though not a subalgebra. The dual of this subcoalgebra is an algebra of truncated formal tensor power series.

At this point, we face the inverse of the riddle with which this section began: We have set things up so that co-operations on objects in k-**Ring** and coidentities that these satisfy correspond to certain $\circledast$-co-operations on their underlying k-bimodules and coidentities for these $\circledast$-co-operations (just as, for k commutative, co-operations on objects of $\mathbf{Comm}_k^1$ and their coidentities correspond to certain $\otimes_k$-co-operations and coidentities on their underlying k-modules); but now how do we explain the fact that a *k-ring* structure on a k-bimodule is equivalent to a $\circledast$-operation which is $\circledast$-commutative $\circledast$-unital and $\circledast$-associative, even though the k-ring structure need *not* be commutative or unital (Corollary 60.15)? This seems to mimic the $\mathbf{Comm}_k^1$ case unreasonably closely!

To make sense of this, let us note that if R is a k-ring, the $\circledast$-operation μ on its underlying k-bimodule corresponding to its k-ring structure is (after identification of $R \circledast R$ with $R \amalg R$) the *codiagonal* homomorphism of k-rings, $R^{\lambda} \amalg R^{\rho} \to R$, given by $r^{\lambda} \mapsto r \leftarrow r^{\rho}$. Now the codiagonal morphism of *any* object R of any category with pairwise coproducts is easily seen to be commutative, unital and associative. So what Corollary 60.15 tells us is that every $\circledast$-commutative $\circledast$-unital $\circledast$-associative $\circledast$-operation on a k-bimodule M arises, via the forgetful functor from k-rings to k-bimodules, from the codiagonal map associated with a unique k-ring structure on M. Though not an obvious result, this is indeed a plausible one.

§61. Further observations on $\circledast$.

Let us return to the subject of $\circledast$-unital $\circledast$-associative $\circledast$-algebras that are *not* necessarily $\circledast$-commutative, characterized in Theorem 60.8. Although these are unfamiliar objects, we shall show that they can be "embedded" in ordinary associative k-rings. To see how, let us note that given a k-ring S, and any two elements $u, d \in S$ (mnemonic for "up" and "down"), if we define two binary operations on S,

$$(61.1) \qquad a^{*}b = aub, \qquad a_{*}b = adb,$$

these will satisfy (60.9). If u and d are k-centralizing, these maps will be k-bilinear (in the strong sense we are giving that term in this book; cf. the display in Definition 10.4). Let us now prove a converse:

LEMMA 61.2. *Let* $k \in \mathbf{Ring}^1$ *and let* M *be a k-bimodule given with k-bilinear binary operations* * *and* $_{*}$ *satisfying the associativity conditions* (60.9). *Then* M *can be embedded as a k-bimodule in an associative k-ring* S, *with two k-centralizing elements* u *and* d *such that for all* $a, b \in M$, (61.1) *holds.*

Moreover, one can make S *unital, with* u, *or alternatively* d *(but unless* $^{*} = {}_{*}$, *not both) the multiplicative neutral element.*

Sketch of Proof. Given $(M, {}^*, {}_*)$, let $U = ku$ be a k-bimodule presented by a single generator u, and relations saying that u centralizes k (so that U is an isomorphic copy of the k-bimodule k), let $D = kd$ be another such bimodule, with k-centralizing generator d, and let S be the k-ring presented by the generating k-bimodule $M \oplus U \oplus D$, and the relations

$$(61.3) \qquad aub = a^*b, \qquad adb = a_*b \qquad (a, b \in M),$$

where the right-hand sides of these relations are evaluated in M using our given pair of operations.

The relations (61.3) can be thought of as specifying the multiplication of S on the products of bimodules $M \times U \times M$ and $M \times D \times M$. Clearly, the desired embedding is possible if and only if the canonical map of M into the k-ring S so presented is one-to-one. By [**19**, Theorem 6.1], this one-one-ness will hold if (in the language of that paper), all "ambiguities" in the indicated presentation are "resolvable". (That result shows that S will then be the direct sum of all tensor products of strings of M's, U's and D's not containing any substrings MUM or MDM. In particular, the natural map $M \to S$ will be an embedding, since the length-1 string M has no such substrings. We will assume familiarity with the method of [**19**, §6] for the remainder of this proof.)

The above presentation has exactly four ambiguities, corresponding to the two alternative ways of reducing members of each of the four products

$$MUMUM, \qquad MUMDM, \qquad MDMUM, \qquad MDMDM.$$

The four resolvability conditions are seen to be precisely the associativity conditions (60.9), which hold by hypothesis. Hence the canonical map $M \to S$ is an embedding, as desired.

To obtain such an embedding in which u is a multiplicative neutral element, we let S be the *unital* k-ring presented as above, except that we omit the summand U, and we replace the first equation of (61.3), $aub = a^*b$, by $ab = a^*b$. Thus, we get a presentation by the bimodule $M \oplus D$, and relations describing how to reduce MM and MDM. Again we have four ambiguities:

$$MMM, \qquad MMDM, \qquad MDMM, \qquad MDMDM,$$

and again these are resolvable precisely because of the conditions (60.9). The statement with d rather than u made to equal 1 is proved in the same way.

If we attempt to make both u and d multiplicative neutral elements, however, we get an additional ambiguity, based on the two ways of reducing members of the product MM, and this will be resolvable only if ${}^* = {}_*$. □

Turning in another direction, let us look at this concept of ⊛-associative ⊛-unital ⊛-algebra in the general context of mathematical objects with some version of an "associative unital operation". For such an object (A, μ), it is generally of interest to define the concept of an *action* of (A, μ) on an arbitrary object M. In the present situation, an action of a ⊛-algebra (A, μ) on a k-bimodule M should mean a bimodule map $\nu\colon A⊛M \to M$ such that the obvious square diagram relating μ and ν commutes, and the "left unitality" triangle for this action likewise commutes. Note that the operation ν will include maps $M \otimes_k A \otimes_k M \to M$, etc.; thus, this structure on M is more "ring-like" than the corresponding structure with $\otimes_k$ in place of $\circledast$, which is that of a *module* M over a k-algebra A.

The easiest examples of such actions are obtained by starting with a $\circledast$-associative $\circledast$-unital $\circledast$-algebra (K, μ) (whose operation μ we shall often express in terms of two bilinear maps * and $_*$ as in Theorem 60.8), and taking two subbimodules,

(61.4) $$A \subseteq K, \qquad M \subseteq K,$$

such that

(61.5) the composite maps $A \circledast A \to K \circledast K \xrightarrow{\mu} K$ and $A \circledast M \to K \circledast K \xrightarrow{\mu} K$ have images in A and M respectively.

The former composite thus constitutes a $\circledast$-algebra structure on A, and the latter an action ν, given by

(61.6) $$\begin{aligned} &\nu(\dots a_i\, x_{i+1}\, a_{i+2}\, x_{i+3} \dots) \\ &= \mu(\dots a_i^{\lambda}\, x_{i+1}^{\rho}\, a_{i+2}^{\lambda}\, x_{i+3}^{\rho} \dots) \\ &= \dots {}_* a_i {}^* x_{i+1} {}_* a_{i+2} {}^* x_{i+3} {}_* \dots \qquad (a\text{'s in } A,\ x\text{'s in } M). \end{aligned}$$

It is easily deduced from the associativity conditions (60.9) on (K, μ) that the necessary and sufficient conditions for (61.5) to hold are that

(61.7) the two sets $A{}^*A$, $A{}_*A$ both lie in A, and

(61.8) the four sets $M{}_*A{}^*M$, $A{}^*M$, $M{}_*A$, A all lie in M.

(Note that because A acts formally on the left on M, but not on the right, (61.8) does not involve products such as $A{}_*M$ and $M{}^*A$. Cf. (61.6).)

One can get generic examples of this construction by taking for K a free k-ring (or if k is commutative, a free k-algebra) on a set of generators $\{a_i \mid i \in I\} \cup \{x_j \mid j \in J\} \cup \{u, d\}$, making this a $\circledast$-algebra as in Lemma 61.2, taking for A the least subbimodule containing the a_i and satisfying (61.7), and then taking for M the least subbimodule containing the x_j and satisfying (61.8). For a simpler example, let k be any commutative integral domain, K its field of fractions, and $c \in k$ a nonzero nonunit. Being an associative k-ring, K can be regarded as a $\circledast$-commutative $\circledast$-associative $\circledast$-unital $\circledast$-algebra. We find that its k-submodules $A = ck$ and $M = c^{-1}k$ satisfy (61.7) and (61.8).

To get further examples of actions, we can start with an action determined as in (61.4)-(61.6), and find a k-subbimodule $N \subseteq M$ such that the action of A on M induces an action on M/N. Clearly, the necessary and sufficient condition for this to hold is that in every string $\dots {}^*M{}_*A{}^*M{}_*A{}^* \dots$, if we replace at least one "M" by an "N", the resulting product lies in N. We find that this will hold if and only if

(61.9) the four sets $N{}_*A{}^*M$, $M{}_*A{}^*N$, $A{}^*N$, $N{}_*A$ all lie in N.

For instance, in our above example with $A = ck$ and $M = c^{-1}k$, if we let $N = k$, we find that (61.9) holds, so we get an action v of $A = ck$ on $M/N = (c^{-1}k)/k$. Observe that this v annihilates the summands A, $(M/N) \otimes_k A$, and $A \otimes_k (M/N) \subseteq A \circledast (M/N)$ because A, $M_* A$ and $A^* M$ all lie in N, but it does not annihilate $(M/N) \otimes_k A \otimes_k (M/N)$, since μ carries $\overline{c^{-1}}\, c\, \overline{c^{-1}}$ to $\overline{c^{-1}} \neq 0$. In contrast, in any example of the form (61.6), v will clearly annihilate $\ldots a_i x_{i+1} a_{i+2} x_{i+3} \ldots$ if it annihilates a substring of this product.

We remark, though we do not have any interesting examples in mind, that, just as the operation $\otimes_k$ may be taken as the basis for the concept of a "k-linear category", so one can define the concept of a "⊛-category" $\mathbf{C}$, whose hom-objects $\mathbf{C}(A, B)$ are k-bimodules, and whose "composition" is given by bimodule homomorphisms $\mathbf{C}(B, C) \circledast \mathbf{C}(A, B) \to \mathbf{C}(A, C)$ $(A, B, C \in \mathbf{C})$. (Cf. [**122**, §VII.7].)

Let us return to the concepts of ⊛-algebra and ⊛-coalgebra per se. We have noted that if k is a field and M a ⊛-coalgebra whose underlying k-bimodule centralizes k, and can thus be considered a k-vector-space, then the linear dual space $A = M^*$ becomes a linearly compact ⊛-algebra. Suppose M is ⊛-counital and ⊛-coassociative, but not necessarily ⊛-cocommutative; thus, its operation $\mathbf{m}$ can be expressed in terms of the two components $M \to M^\lambda \otimes_k M^\rho$ and $M \to M^\rho \otimes_k M^\lambda$, and, likewise, the operation μ of the dual ⊛-algebra A can be expressed in terms of the corresponding two bilinear maps * and $_*$. Now if S is any other ⊛-associative ⊛-unital ⊛-algebra with k-centralizing underlying bimodule structure, the vector space $\mathbf{Mod}_k(M, S) \cong A \hat{\otimes} S$ will have *four* mutually associating multiplications, obtained from all pairings of one of the operations $_*$ and * of A with one of the operations of S. Likewise, in the more general context of k-bimodules $(k \in \mathbf{Ring}^1)$, we get for any ⊛-counital ⊛-coassociative ⊛-coalgebra M and any ⊛-unital ⊛-associative ⊛-algebra S a family of four "convolution" multiplications (cf. §29) on the abelian group k-$\mathbf{Bimod}(M, S)$. It is not clear whether there is a "natural" way to choose two of these four operations – or perhaps two linear combinations of them – to give a ⊛-algebra structure to this object. (One could look to the constructions of Theorem 25.25 and Chapter VII for guidance; however, the situation there is more restricted, since the ⊛-algebra structure of that S is ⊛-commutative.)

It seems likely that one should be able to develop analogs of many concepts from the theory of bialgebras and Hopf algebras using ⊛-algebras and ⊛-coalgebras in place of $\otimes_k$-algebras and $\otimes_k$-coalgebras. For instance, the objects representing the functors from k-**Ring** to **Group** discussed in §33 should correspond to "⊛-commutative ⊛-Hopf algebras".

§62. Analogs of ⊛.

We were led to look at the bifunctor ⊛ above because it had properties with respect to coproducts in k-**Ring** and the forgetful functor k-**Ring** $\to$ k-**Bimod** $(k \in \mathbf{Ring}^1)$ analogous to those that $\otimes_k$ has with respect to coproducts in $\mathbf{Comm}_k$ and the forgetful functor $\mathbf{Comm}_k^1 \to \mathbf{Mod}_k$ $(k \in \mathbf{Comm}^1)$. Let us look for further analogous situations.

There is one that closely parallels the situation we have been discussing: Replacing k-**Ring** and k-**Bimod** by **Semigp** and **Set** respectively, we get a ''nonlinear'' version of that situation. Thus, if for every index-set A we define a functor $\mathbf{Set}^A \to \mathbf{Set}$, which for convenience we shall also denote $\circledast$, by

(62.1) $$\circledast_{\alpha\in A} X_\alpha = \coprod_{(\alpha_1,\dots,\alpha_n)} X_{\alpha_1} \times \dots \times X_{\alpha_n},$$

where $\coprod$ denotes disjoint union, and $(\alpha_1,\dots,\alpha_n)$ runs over all strings as in (60.1), and if we write U for the underlying set functor $\mathbf{Semigp} \to \mathbf{Set}$, then the diagram

(62.2) $$\begin{array}{ccc} \mathbf{Semigp}\times\mathbf{Semigp} & \xrightarrow{\ \coprod\ } & \mathbf{Semigp} \\ \downarrow{\scriptstyle U\times U} & & \downarrow{\scriptstyle U} \\ \mathbf{Set}\times\mathbf{Set} & \xrightarrow{\ \circledast\ } & \mathbf{Set} \end{array}$$

commutes up to a functorial isomorphism. (The same is true for coproducts over arbitrary index-sets A, of course, but in this section we will for simplicity talk about two-fold coproducts.) It is easy to prove the analogs for this situation of the characterizations of $\circledast$-associative $\circledast$-unital possibly $\circledast$-commutative $\circledast$-algebras obtained in Theorem 60.8 and Corollary 60.15. However, though we have maps $i^\beta\colon M_\beta \to \circledast_{\alpha\in A} M_\alpha$ as in the rings-and-bimodules case, there are no corresponding maps $\varepsilon^\beta\colon \circledast_{\alpha\in A} M_\alpha \to M_\beta$, so we do not have a concept of $\circledast$-*counitality* of a $\circledast$-coalgebra, and cannot reproduce the results involving that condition. Nonetheless, properties of $\circledast$-algebras and $\circledast$-coalgebras in **Semigp** both seem worthy of study.

What about analogous constructions for unital semigroups, and unital k-rings? Here there are curious difficulties in even getting analogs of the functor $\circledast$.

Let us begin with the ring case, and, for simplicity, start with a *field* k and consider the category $\mathbf{Ring}^1_k$ of associative unital k-algebras. There is a description of the k-vector-space structure of the coproduct $S \amalg T$ of two such algebras, similar to the description of the coproduct of nonunital k-algebras: if one takes k-bases $B_S \cup \{1_S\}$ and $B_T \cup \{1_T\}$ of S and T respectively, then a k-basis for the coproduct is given by the set of all strings of elements taken alternately from B_S and B_T, with the empty string allowed and considered to represent $1_{S\amalg T}$. Let $\mathbf{Mod}^1_k$ denote the category of k-vector-spaces with a distinguished element, which we shall denote 1. Then we might hope to turn the above description of the underlying vector space of a coproduct into a functor $\circledast$ making the following diagram commute up to an isomorphism of functors:

(62.3) $$\begin{array}{ccc} \mathbf{Ring}^1_k\times\mathbf{Ring}^1_k & \xrightarrow{\ \coprod\ } & \mathbf{Ring}^1_k \\ \downarrow{\scriptstyle U\times U} & & \downarrow{\scriptstyle U} \\ \mathbf{Mod}^1_k\times\mathbf{Mod}^1_k & \xrightarrow{\ \circledast\ } & \mathbf{Mod}^1_k. \end{array}$$

Unfortunately, this approach does not work. For suppose $S,\ T,\ T' \in \mathbf{Ring}^1_k$, with vector-space bases $B_S \cup \{1_S\}$, $B_T \cup \{1_T\}$, $B_{T'} \cup \{1_{T'}\}$, let $s_0, s_1 \in B_S$, $t \in B_T$, and suppose there is a k-algebra homomorphism $f\colon T \to T'$ taking t to $1_{T'}$. Then the induced map of k-algebras $S \amalg T \to S \amalg T'$ takes $s_0 t s_1$ to $s_0 s_1$.

Hence $\mathrm{id}_{U(S)} \circledast U(f)$ must also take $s_0 t s_1$ to $s_0 s_1$; but this is impossible because $U(S)$ forgets the multiplication of S.

Surprisingly, however, there *is* a bifunctor $\circledast$ making (62.3) commute up to isomorphism; but with a different description. Indeed, fix any $k \in \mathbf{Comm}^1$ (not necessarily a field). Given $S, T \in \mathbf{Ring}_k^1$, let us define the k-module

$$N = (S/k1_S) \otimes_k (T/k1_T),$$

writing the decomposable elements thereof as $\bar{s} \otimes \bar{t}$ $(s \in S,\ t \in T)$, and now consider the k-module

$$U(S) \otimes_k U(T) \otimes_k U(k\langle N \rangle). \tag{62.4}$$

It is shown in [**55**, Proposition A1] that the k-module homomorphism from (62.4) to $U(S \amalg T)$ given by

$$s_0 \otimes t_0 \otimes (\bar{s}_1 \otimes \bar{t}_1) \dots (\bar{s}_n \otimes \bar{t}_n) \mapsto s_0 t_0 (s_1 t_1 - t_1 s_1) \dots (s_n t_n - t_n s_n) \tag{62.5}$$

is bijective. Hence if for objects $L, M \in \mathbf{Mod}_k^1$ we define

$$L \circledast M = L \otimes_k M \otimes_k U(k\langle (L/k1_L) \otimes_k (M/k1_M) \rangle), \tag{62.6}$$

then the diagram (62.3) commutes up to a functorial isomorphism given by (62.5).

The authors have not investigated whether this bifunctor $\circledast$ is associative; more precisely, whether there is an isomorphism of functors

$$- \circledast (- \circledast -) \cong (- \circledast -) \circledast - \tag{62.7}$$

that forms a commuting diagram with the associativity isomorphism for the coproduct functor on $\mathbf{Ring}_k^1$,

$$- \amalg (- \amalg -) \cong (- \amalg -) \amalg -,$$

via the maps induced by the connecting isomorphism

$$U(-) \circledast U(-) \cong U(- \amalg -)$$

of (62.3). One troubling sign is that (62.5) is not the unique choice for that connecting isomorphism. Indeed, as noted in the same Appendix to [**55**] (though in slightly different notation), the map $U(-) \circledast U(-) \to U(- \amalg -)$ given by

$$s_0 \otimes t_0 \otimes (\bar{s}_1 \otimes \bar{t}_1) \dots (\bar{s}_n \otimes \bar{t}_n) \mapsto s_0 (s_1 t_1 - t_1 s_1) \dots (s_n t_n - t_n s_n) t_0 \tag{62.8}$$

is also an isomorphism, and one can show that the isomorphisms described by (62.5) and (62.8) are not related to one another by an automorphism of the bifunctor $\circledast$. (Idea: For $S, T \in \mathbf{Ring}_k^1$, compare the two elements of $S \circledast T$ corresponding to an element of the form $s_0 t_0 (s_1 t_1 - t_1 s_1) \in S \amalg T$ under these two isomorphisms. These turn out to be

$$s_0 \otimes t_0 \otimes (\bar{s}_1 \otimes \bar{t}_1) \quad \text{and} \quad s_0 \otimes 1_T \otimes (\bar{s}_1 \otimes \overline{t_0 t_1}) - s_0 \otimes t_1 \otimes (\bar{s}_1 \otimes \bar{t}_0)$$

respectively. To get the latter from the former, we would need to know $\overline{t_0 t_1}$, the residue of $t_0 t_1$ modulo $k1_T$, and this is not determined by the structure of T as a module with distinguished element.) We do not even know whether the bifunctor $\circledast$ given by (62.4) is the essentially *unique* bifunctor making (62.3) commute up to isomorphism. (There is some "inessential" nonunicity; e.g., one can modify (62.6) by killing all or part of $L \circledast M$ whenever 1_L and/or 1_M is zero.) Finally, it is not evident whether there is any way to generalize these constructions from the case of algebras over commutative k to that of k-rings for a general associative unital ring k, i.e., to construct a bifunctor $\circledast$ making the diagram

$$\begin{array}{ccc} k\text{-}\mathbf{Ring}^1 \times k\text{-}\mathbf{Ring}^1 & \xrightarrow{\coprod} & k\text{-}\mathbf{Ring}^1 \\ \downarrow{\scriptstyle U\times U} & & \downarrow{\scriptstyle U} \\ k\text{-}\mathbf{Bimod}^1 \times k\text{-}\mathbf{Bimod}^1 & \xrightarrow{\circledast} & k\text{-}\mathbf{Bimod}^1 \end{array} \tag{62.9}$$

commute up to a (hopefully, coherently associative) isomorphism of functors. Clearly, there are many questions to be looked at!

We turn now to semigroups with neutral element. Does there exist a bifunctor $\circledast$: $\mathbf{Set}^{\mathrm{pt}} \times \mathbf{Set}^{\mathrm{pt}} \to \mathbf{Set}^{\mathrm{pt}}$ making the diagram

$$\begin{array}{ccc} \mathbf{Semigp}^e \times \mathbf{Semigp}^e & \xrightarrow{\coprod} & \mathbf{Semigp}^e \\ \downarrow{\scriptstyle U\times U} & & \downarrow{\scriptstyle U} \\ \mathbf{Set}^{\mathrm{pt}} \times \mathbf{Set}^{\mathrm{pt}} & \xrightarrow{\circledast} & \mathbf{Set}^{\mathrm{pt}} \end{array} \tag{62.10}$$

commute up to isomorphism of functors?

We shall sketch a proof that there does not! The idea will be to show that such a functor would have to look very much like the construction (62.1) that worked for nonunital semigroups, and then get a contradiction using the neutral element.

Suppose $\circledast$ were a bifunctor making (62.10) commute. We shall construct, for every finite string of alternating λ's and ρ's, $\sigma = (\dots, \lambda, \rho, \dots)$, say of length n, an "operation" $[\dots]_\sigma$, which, given pointed sets X and Y, maps the n-fold product-set $\dots \times X \times Y \times \dots$ (with X in those positions where σ has λ and Y in those where σ has ρ) into $X \circledast Y$. Since our hypothesis on $\circledast$ concerns semigroups, our construction must introduce some semigroups; so let L_σ be the semigroup with underlying set consisting of an identity element e, and an element t_i for each i such that σ has λ in the ith position, and multiplication making the t_i left zeros (i.e., satisfying $t_i t_j = t_i$ for all i, j), and similarly, let R_σ be the semigroup consisting of a neutral element e, and left zero elements t_i for those i such that the ith term of σ is ρ. (The virtue of these left zero multiplications is that they can be described in terms of the pointed set structures alone. We could equally well have used right zero multiplications, or one right, and the other left.)

Now given any two pointed sets X and Y, and elements $x_1, \dots, x_n$, with x_i taken from X if the ith term of σ is λ and from Y if the ith term of σ is ρ, consider the maps $U(L_\sigma) \to X$, $U(R_\sigma) \to Y$ of pointed sets carrying each t_i to x_i, and each neutral element e to the corresponding base-point e. These yield a map

$$U(L_\sigma \amalg R_\sigma) \cong U(L_\sigma) \circledast U(R_\sigma) \to X \circledast Y,$$

where the first step is the given functorial bijection making (62.10) commute, and the second is gotten by applying the bifunctor $\circledast$ to the two maps of pointed sets just described. We now define $[x_1: \ldots : x_n]_\sigma \in X \circledast Y$ to be the image under this map of the semigroup element

$$t_1 \ldots t_n \in L_\sigma \amalg R_\sigma.$$

It is immediate that these operations $[\ldots]_\sigma$ are functorial, i.e., that given $f\colon X \to X'$, $g\colon Y \to Y'$, one has

$$[\ldots : f(x_i) : g(x_{i+1}) : \ldots]_\sigma \;=\; (f \circledast g)([\ldots : x_i : x_{i+1} : \ldots]_\sigma).$$

Let us note one degenerate case: when σ is the unique length-zero string, $()$, the above construction gives a zeroary operation, in other words, a distinguished element $[\,]_{()} \in X \circledast Y$. It is straightforward to verify that this element is the basepoint e of the pointed set $X \circledast Y$.

We sketch the remainder of the argument, leaving the details to the interested reader.

EXERCISE 62.11. *Assume as above that* $\circledast\colon \mathbf{Set}^{\mathrm{pt}} \times \mathbf{Set}^{\mathrm{pt}} \to \mathbf{Set}^{\mathrm{pt}}$ *makes the diagram* (62.10) *commute up to functorial isomorphism, and let the operations* $[\ldots]_\sigma$ *be defined as above.*

(i) *Show that for* $X, Y \in \mathbf{Set}^{\mathrm{pt}}$, *each element of* $X \circledast Y$ *can be written in the form* $[x_1: \ldots : x_n]_\sigma$ *for a unique* σ, *and unique elements* x_i *lying in* $X - \{e\}$ *and* $Y - \{e\}$. (*Suggestion: Make* X *and* Y *into semigroups such that all nonidentity elements are left zeroes.*)

(ii) *Show that there must exist a semigroup word* w *in three variables such that for all* $S, T \in \mathbf{Semigp}^e$, $x, z \in S$, $y \in T$, *one has*

$$[x : y : z]_{(\lambda, \rho, \lambda)} \;=\; w(x, y, z),$$

where we have identified $U(S) \circledast U(T)$ *with* $U(S \amalg T)$ *by the given isomorphism of functors, and written the coprojections of* S *and* T *into* $S \amalg T$ *as inclusions.* (*Hint: consider* $[s : t : s']_{(\lambda, \rho, \lambda)}$ *in the coproduct of the two* free *semigroups with neutral element,* $\langle s, s' \rangle$ *and* $\langle t \rangle$.)

(iii) *Show that one of the three equations*

$$[x : e : z]_{(\lambda, \rho, \lambda)} = [x]_{(\lambda)} \quad or \quad [x : e : z]_{(\lambda, \rho, \lambda)} = [\,]_{()} \quad or \quad [x : e : z]_{(\lambda, \rho, \lambda)} = [z]_{(\lambda)}$$

must hold identically for all X, Y. (*Hint: Apply* (i) *to the pointed sets* $X = \{s, s', e\}$, $Y = \{e\}$, *noting that* $Y - \{e\}$ *is empty in this case.*)

(iv) *Deduce that the semigroup word* w *of* (ii) *involves at most one of* x *and* z. (*Aside: Would* (ii) *and* (iii) *imply* (iv) *if we were working with* **Group** *rather than* $\mathbf{Semigp}^e$?)

(v) *Note that the unicity statement of* (i) *implies that for all pointed sets* X *and* Y, *and every* $y \in Y - \{e\}$, *the map*

$$[- : y : -]_{(\lambda, \rho, \lambda)}\colon\; (X - \{e\}) \times (X - \{e\}) \;\to\; X \circledast Y$$

is one-to-one. Deduce that the semigroup word w *of* (ii) *involves both* x *and* z.

The contradiction between (iv) *and* (v) *shows that no functor* $\circledast$ *with the indicated properties can exist.*

On the other hand, we shall now show that the situation for *groups* resembles that for unital rings (cf. (62.5) above), but with additive commutators $[s, t] = st - ts$ replaced by multiplicative commutators

$$[g, h] = g^{-1}h^{-1}gh.$$

We will need

LEMMA 62.12. *Let G and H be groups, which we assume for notational convenience have disjoint underlying sets, and let elements of their coproduct $G \amalg H$ be written as alternating products of nonidentity elements of G and H, so that we can speak unambiguously of elements of this group as "ending" with certain strings of elements of G and H. Then for $w \in G \amalg H$, $g \in G - \{e\}$, $h \in H - \{e\}$, we have:*

(i) *If w does not end in the substring hg, then $w[g, h]$ is longer than w, and ends in the substring gh.*

(ii) *If $w = hg$, then $w[g, h] = gh$.*

(iii) *If w ends in the substring hg and has length > 2, then $w[g, h]$ is shorter than w.*

Likewise

(i′) *If w does not end in the substring gh, then $w[g, h]^{-1}$ is longer than w, and ends in the substring hg.*

(ii′) *If $w = gh$, then $w[g, h]^{-1} = hg$.*

(iii′) *If w ends in the substring gh and has length > 2, then $w[g, h]^{-1}$ is shorter than w.*

SKETCH OF PROOF. (i) and (ii) are verified by computation. (iii) follows by writing $w = vhg$, where v ends with an element of G, applying (ii), and noting that the first factor of gh contracts with the last factor of v. We get (i′)-(iii′) by noting that $[g, h]^{-1} = [h, g]$, and applying (i)-(iii) with the roles of G and H reversed. □

COROLLARY 62.13. *Let G and H be groups. Then*

(a) *The kernel K of the natural homomorphism $G \amalg H \to G \times H$ is free on $\{[g, h] \mid g \in G - \{e\}, h \in H - \{e\}\}$.*

Hence

(b) *Every element $w \in G \amalg H$ can be written uniquely as a product ghv, where $g \in G$, $h \in H$, and v is a group-theoretic reduced word in $\{[g, h] \mid g \in G - \{e\}, h \in H - \{e\}\}$. Every element can also be written uniquely $gv'h$, with g, h and v' chosen from these same sets.*

SKETCH OF PROOF. (a) The commutators $[g, h]$ clearly lie in K. Given any $w \in K$, we see from (iii) and (iii′) of the preceding Lemma that we can shorten w by repeated right multiplications by such commutators and their inverses until we get an element of length ≤ 2, i.e., of one of the forms gh, hg, g, h or e. But of these, only e lies in K, hence we have reduced w to e; i.e., w lies in the group generated by the indicated commutators.

To verify that these commutators generate K freely, one uses (i) and (i′) of the preceding Lemma to show inductively that any nontrivial *reduced group word* in these commutators with last factor $[g, h]$ represents an element of $G \amalg H$ which ends in gh, while one with last factor $[g, h]^{-1}$ represents an element which ends in hg; so no such word is equal to the identity.

To get (b), let w be any element of $G \amalg H$, and let its image in $G \times H$ be (g, h). Then $v = h^{-1}g^{-1}w$ and $v' = g^{-1}wh^{-1}$ lie in K. Applying (a) to these elements gives the desired unique expressions for w. □

Let us note that the free group K described in part (a) of the above Corollary is determined functorially by the underlying pointed sets of G and H. Indeed, the construction can be described abstractly as the functor $P\colon \mathbf{Set}^{\mathrm{pt}} \times \mathbf{Set}^{\mathrm{pt}} \to \mathbf{Group}$ taking a pair of pointed sets X, Y (with basepoints denoted e) to the free group on $(X-\{e\}) \times (Y-\{e\})$, and taking any pair of morphisms $a\colon X \to X'$, $b\colon Y \to Y'$ of pointed sets to the homomorphism of free groups that carries the free generator (x, y) to the free generator $(a(x), b(y))$ if neither component of the latter pair is a basepoint e, and to the group identity e otherwise. It is easy to check that this is indeed a functor. Now defining $\circledast\colon \mathbf{Set}^{\mathrm{pt}} \times \mathbf{Set}^{\mathrm{pt}} \to \mathbf{Set}^{\mathrm{pt}}$ by

$$X \circledast Y = X \times Y \times UP(X, Y),$$

where $U\colon \mathbf{Group} \to \mathbf{Set}^{\mathrm{pt}}$ is the forgetful functor, we see that point (b) of the above Corollary describes an isomorphism $U(- \amalg -) \cong U(-) \circledast U(-)$ making the diagram

(62.14)
$$\begin{array}{ccc} \mathbf{Group} \times \mathbf{Group} & \xrightarrow{\amalg} & \mathbf{Group} \\ \downarrow{\scriptstyle U \times U} & & \downarrow{\scriptstyle U} \\ \mathbf{Set}^{\mathrm{pt}} \times \mathbf{Set}^{\mathrm{pt}} & \xrightarrow{\circledast} & \mathbf{Set}^{\mathrm{pt}} \end{array}$$

commute (in fact, two such isomorphisms).

The questions on coherent associativity, etc., that we raised for unital k-algebras would also be worth studying in this case.

There are many other forgetful functors U among varieties for which one may ask whether there exists a bifunctor $\circledast$ making a commuting diagram, as in the cases considered above. Let us end our discussion of this topic by noting some cases where such a $\circledast$ exists, but is not particularly exotic. If U is the forgetful functor from the variety of *nonunital* commutative k-algebras, $\mathbf{Comm}_k$, to $\mathbf{Mod}_k$, then we can take $M \circledast N = M \oplus N \oplus (M \otimes_k N)$. If U is the forgetful functor from $\mathbf{Ab}$ or $\mathbf{AbSemigp}^e$ to $\mathbf{Set}$, then $\circledast$ is the direct product bifunctor on $\mathbf{Set}$. (This is the nonlinear analog of the case of the forgetful functor $\mathbf{Comm}_k^1 \to \mathbf{Mod}_k$ and the bifunctor $\otimes_k\colon \mathbf{Mod}_k \times \mathbf{Mod}_k \to \mathbf{Mod}_k$, which suggested this investigation.) If U is the forgetful functor $\mathbf{Group} \to \mathbf{Semigp}^e$, then $\circledast$ is the coproduct bifunctor on $\mathbf{Semigp}^e$. Note that in this last case, the analog of Corollary 60.15 does *not* hold: $\circledast$-commutative $\circledast$-associative $\circledast$-unital $\circledast$-algebras here correspond to arbitrary objects of $\mathbf{Semigp}^e$, which need not be groups.

§63. Tall-Wraith monads and hermaphroditic functors.

In this section and the next, we shall sketch briefly (with some changes in notation and language, and tacit correction of some minor errors) an interesting concept introduced by D. Tall and G. Wraith [**188**] (also Wraith [**202**, §9]), and point out some questions to which it leads.

Consider a representable functor $\mathbf{Comm}^1 \to \mathbf{Comm}^1$ with representing object

$$(B, \mathbf{a}, \mathbf{m}, \mathbf{i}, \mathbf{z}, \mathbf{u}),$$

where B is the underlying commutative ring, $\mathbf{a}$ and $\mathbf{m}$ are the coaddition and comultiplication, $\mathbf{i}$ is the additive coinverse, and $\mathbf{z}$ and $\mathbf{u}$ are the additive and multiplicative co-neutral-elements. We shall, as usual, often simply refer to this as "the co-ring B". Following [**188**], let us write the left adjoint to the functor represented by B as "$B\odot -$". We recall from the proof of Theorem 8.14 that for $S\in \mathbf{Comm}^1$, the ring $B\odot S$ is generated by a family of homomorphic images of the ring B, indexed by the elements of S. For $b\in B$, $s\in S$, let us write $b\odot s\in B\odot S$ for the image of b in the copy of B indexed by s.

The behavior of $b\odot s$ as a function of b is clear from this description: for each s, the map $b\mapsto b\odot s$ is a ring homomorphism $B\to B\odot S$. To describe the behavior of $b\odot s$ as a function of s, we need some further notation. Any pair of elements $s, s'\in S$ yields a pair of homomorphisms $-\odot s,\ -\odot s'\colon B\to B\odot S$, which can be thought of as equivalent to one map $B\otimes B\to B\odot S$, since the coproduct operation in $\mathbf{Comm}^1$ is $\otimes$ (i.e., $\otimes_{\mathbf{Z}}$). We shall denote this map $-\odot(s, s')\colon B\otimes B\to B\odot S$. It is now straightforward to verify that for all $s, s'\in S$,

$$\begin{aligned} b\odot(s+s') &= \mathbf{a}(b)\odot(s, s'),\\ b\odot(ss') &= \mathbf{m}(b)\odot(s, s'),\\ b\odot(-s) &= \mathbf{i}(b)\odot s,\\ b\odot 0 &= \mathbf{z}(b),\\ b\odot 1 &= \mathbf{u}(b). \end{aligned}$$

Let us note two important cases of such co-rings: The identity functor of $\mathbf{Comm}^1$ is represented by the free ring on one generator,

$$I = \mathbf{Z}[x], \tag{63.1}$$

made a co-$\mathbf{Comm}^1$ object of $\mathbf{Comm}^1$ via the obvious co-operations. And if B, B' are two co-$\mathbf{Comm}^1$ objects of $\mathbf{Comm}^1$, then the composite of the functors they represent is represented by the object $B'\odot B$ (with factors in the opposite of the order for the represented functors, i.e., in the proper order for their left adjoints), which is thus again a co-$\mathbf{Comm}^1$ object of $\mathbf{Comm}^1$.

In particular, if B is a co-$\mathbf{Comm}^1$ object of $\mathbf{Comm}^1$, then $B\odot B$ is again such an object. The authors of [**188**] now consider a co-$\mathbf{Comm}^1$ object B of $\mathbf{Comm}^1$ given with morphisms of co-$\mathbf{Comm}^1$ objects,

$$\pi\colon B\odot B\to B, \qquad \eta\colon I\to B, \tag{63.2}$$

such that π is associative and has η as a neutral element; i.e., such that if we define

(63.3) $$b \cdot c = \pi(b \odot c) \quad \text{and} \quad e = \eta(x),$$

we have the identities

(63.4) $$(b \cdot c) \cdot d = b \cdot (c \cdot d), \qquad e \cdot b = b = b \cdot e.$$

So again, we have a mathematical entity with an "associative operation" having a "neutral element". Let us recall a class of instances of that pattern which has been given a name: an endofunctor M of a category $\mathbf{C}$, given with an associative morphism of functors $MM \to M$, and a morphism $\mathrm{Id}_{\mathbf{C}} \to M$ which is a "neutral element" for that morphism, is called by categorists a *monad* or *triple*. (Tall and Wraith use the latter term, but we shall follow **[122]** and use the former.) Since each co-$\mathbf{Comm}^1$ object B of $\mathbf{Comm}^1$ induces a functor $B \odot -$: $\mathbf{Comm}^1 \to \mathbf{Comm}^1$, such co-$\mathbf{Comm}^1$ objects given with structures of the sort described in (63.2)-(63.4) correspond to certain monads on the category $\mathbf{Comm}^1$.

Tall and Wraith call a co-$\mathbf{Comm}^1$ object B of $\mathbf{Comm}^1$ a *bi-ring*, and call a bi-ring with a structure as described in (63.2)-(63.4) a *bi-ring triple*. We shall not use these terms because of the danger of confusion with *bialgebras* in the sense of Hopf algebra theory, but shall instead call their "bi-ring triples" *Tall-Wraith monad objects* (or *TW-monad* objects) in $\mathbf{Comm}^1$, and the monads $B \odot -$ that these induce *Tall-Wraith monads* (or *TW-monads*) on that category.

As we remarked in §61, when one considers associative unital operations on some sort of mathematical object, it is natural to look also at *actions* of objects with such structure on unstructured objects. We therefore define an *action* of a TW-monad object B on any $S \in \mathbf{Comm}^1$ (called in **[188]** a structure on S of "module over B" or "B-ring") to mean a ring homomorphism

(63.5) $$\kappa\colon B \odot S \to S$$

such that, if one also abbreviates $\kappa(b \odot s)$ to $b \cdot s$, one has

(63.6) $$(b \cdot c) \cdot s = b \cdot (c \cdot s), \qquad e \cdot s = s, \qquad (b, c \in B,\ s \in S).$$

Here (in contrast to §61) we will be interested in TW-monads mainly for the sake of their actions.

As noted in **[188]**, the above ideas may be carried over to any variety of algebras $\mathbf{V}$ in the sense of universal algebra. Thus, for $\mathbf{V}$ any variety, let us call a co-$\mathbf{V}$ object B of $\mathbf{V}$ given with morphisms (63.2) satisfying (63.4) a *TW-monad object* of $\mathbf{V}$, and define an *action* of such an object on a general object S of $\mathbf{V}$ to mean a morphism (63.5) satisfying (63.6). *Morphisms* of TW-monad objects of $\mathbf{V}$, and morphisms among objects with actions of a fixed such monad object B, are defined in the obvious ways.

Tall and Wraith give various examples of TW-monad objects, mostly in $\mathbf{Comm}^1$. We shall discuss in the next section several of their examples, and several others. In the remainder of this section we shall sketch some general facts not noted in **[188]**, which will, among other things, make it easier to recognize examples.

It is straightforward to verify that if B is a TW-monad object of a variety $\mathbf{V}$, then the category of objects of $\mathbf{V}$ with B-action is itself a variety $\mathbf{W}$, whose unary operations correspond to the elements of B, and generate the clone of all operations of $\mathbf{W}$ over the operations of $\mathbf{V}$.

PROPOSITION 63.7. *Let* $U: \mathbf{W} \to \mathbf{V}$ *be a functor between varieties which* preserves underlying sets. *Then the following conditions are equivalent:*

(i) $\mathbf{W}$ *can be described as the variety of objects of* $\mathbf{V}$ *given with action of some TW-monad object* B *of* $\mathbf{V}$, *and* U *as the functor which forgets this action.*

(ii) U *respects finite coproducts* (*including empty coproducts, i.e., initial objects*).

(iii) U *has a right adjoint* (*in addition to the left adjoint which it necessarily has because a functor preserving underlying sets is representable*).

SKETCH OF PROOF. Assume (i). Note that $B \odot -: \mathbf{V} \to \mathbf{V}$, being a left adjoint, respects colimits. Hence given any system consisting of a family of objects of $\mathbf{V}$, each with an action of B, and a family of morphisms among these objects respecting these actions, we get an action of B on the colimit S of this system. It is easy to verify that this colimit gives the colimit of these objects as objects of $\mathbf{V}$ with B-action, i.e., as objects of $\mathbf{W}$; in other words, that colimits of objects of $\mathbf{W}$ arise from the corresponding colimits in $\mathbf{V}$; in other words, that U respects colimits. In particular, we have (ii).

Now assume (ii). Let $<x>_{\mathbf{W}}$ denote the free object on one generator x in $\mathbf{W}$. The object $<x>_{\mathbf{W}}$ has a natural structure of co-$\mathbf{W}$ object in $\mathbf{W}$, representing the identity functor $\mathbf{W} \to \mathbf{W}$. Since this structure is given by morphisms of $<x>_{\mathbf{W}}$ into coproducts of finite families of copies of itself, and U respects finite coproducts, a co-$\mathbf{W}$ structure is also induced on $U(<x>_{\mathbf{W}})$. We claim that the functor $G: \mathbf{V} \to \mathbf{W}$ represented by this co-$\mathbf{W}$ object is a right adjoint to U.

Indeed, let us denote by $U': \mathbf{W} \to \mathbf{V}$ the left adjoint to G; we shall prove this functor naturally isomorphic to U. We know that U' is given by $S \mapsto U(<x>_{\mathbf{W}}) \odot S$ $(S \in \mathbf{W})$, and we easily verify that for each $S \in \mathbf{W}$, the map $U(S) \to U'(S)$ given by

$$s \mapsto x \cdot s \tag{63.8}$$

is a homomorphism in $\mathbf{V}$, giving us a morphism of functors $U \to U'$. Now the $\mathbf{V}$-algebra $U(<x>_{\mathbf{W}}) \odot S$ is generated by elements $w(x) \cdot s$ $(w(x) \in <x>_{\mathbf{W}},\ s \in S)$; but because of the nature of our co-$\mathbf{W}$ structure on $U(<x>_{\mathbf{W}})$, we can write $w(x) \cdot s$ as $x \cdot w(s)$. Thus, all these generators of $U(<x>_{\mathbf{W}}) \odot S$ are in the image of (63.8); hence that homomorphism is surjective. Finally, since the relations imposed on the elements $x \cdot s$ in the construction of $U(<x>_{\mathbf{W}}) \odot S \in \mathbf{V}$ are a subset of those imposed on the corresponding elements in the construction of $<x>_{\mathbf{W}} \odot S \in \mathbf{W}$, which is just S, (63.8) is one-to-one. So (63.8) is an isomorphism $U \cong U'$, so G is a right adjoint to U, completing the proof that (ii) $\Rightarrow$ (iii).

Before proving (iii) $\Rightarrow$ (i), let us recall some general facts. Given a pair of adjoint functors between arbitrary categories, $\mathbf{W} \underset{F}{\overset{U}{\rightleftarrows}} \mathbf{V}$ (with U the right adjoint and F the left), the composite functor

$$M = UF: \mathbf{V} \to \mathbf{V}$$

admits a structure of monad. (This is the motivating case of that concept. The morphism $MM \to M$ of the monad structure is gotten by composing the morphism ε of Proposition 5.3(iii) on the left with U (there called V) and on the right with F. The unit of the monad is the η of that Proposition.) In this situation, every object $U(T)$ $(T \in \mathbf{W})$ admits a natural "action" of this monad (a morphism $M(U(T)) \to U(T)$ making appropriate diagrams commute), and this gives a factorization of the functor $U\colon \mathbf{W} \to \mathbf{V}$ as

$$\mathbf{W} \to \mathbf{V}^M \to \mathbf{V}, \tag{63.9}$$

where the middle term denotes the category of objects of $\mathbf{V}$ with such action [**122**, Theorem VI.3.1, p.138]. Intuitively, this action of M on an object $U(T)$ constitutes "all the information coming from the $\mathbf{W}$-structure of T that can be expressed as additional structure on its image $U(T) \in \mathbf{V}$", and the category $\mathbf{V}^M$ tends to closely approximate $\mathbf{W}$. When it approximates the latter so well that the first arrow of (63.9) is an equivalence, one says that the functor U is "monadic" (or "tripleable". We will again follow [**122**] in using the former term, though we must warn the reader not to confuse it with the property of "having a structure of monad", which we will also be considering. The idea of this terminology is that it says U is "essentially" the forgetful functor $\mathbf{V}^M \to \mathbf{V}$ associated with a monad M.) It is well-known that this condition holds, in particular, whenever U is an underlying-set preserving functor among varieties of algebras, and F its left adjoint.

Now the functor M in this general situation, being a composite of a left adjoint functor F and a right adjoint functor U, is generally neither. But in (iii), we have assumed that U has not only a left adjoint F (because it preserves underlying sets) but also a right adjoint (i.e., that it *is* a left adjoint). Hence $UF\colon \mathbf{V} \to \mathbf{V}$ is a composite of two left adjoint functors, hence is a left adjoint functor, i.e., has the form $B \odot -$ for some co-$\mathbf{V}$ object B of $\mathbf{V}$. Since UF has a structure of monad, its representing object B will have a structure (63.2) of TW-monad object. As noted above, the fact that U preserves underlying sets implies that $UF = B \odot -$ is monadic (not to be confused with the fact that it has a structure of monad!), i.e., that when we factor U as in (63.9), with $M = UF = B \odot -$, the first arrow is an equivalence $\mathbf{W} \simeq \mathbf{V}^{B \odot -}$, establishing (i). □

The equivalence of (i) and (iii) is also proved in [**202**, Theorem 9.3].

We observed in the above proof that the free object $\langle x \rangle_{\mathbf{W}}$ has a natural co-$\mathbf{W}$ structure in $\mathbf{W}$, making it the representing object for the identity functor of that category. If we restrict attention to the co-operations corresponding to the operations of $\mathbf{V}$, this yields the co-$\mathbf{V}$ object of $\mathbf{W}$ which represents the forgetful functor U. As also noted in the proof, since coproducts are the same in $\mathbf{V}$ and $\mathbf{W}$, if we apply U to this object of $\mathbf{W}$ it retains both its co-$\mathbf{W}$ and co-$\mathbf{V}$ structures; in the latter capacity, it represents the right adjoint to UF, and hence is the object B of statement (i); in the former, it represents the right adjoint to U. Thus, our TW-monad object B, with various combinations of operations and co-operations, appears under four hats in this context!

Can we get some feeling for what TW-monads look like; specifically, what it means for an underlying-set preserving functor $\mathbf{W} \to \mathbf{V}$ between varieties of algebras to be expressible as the forgetful functor $\mathbf{V}^{B\odot -} \to \mathbf{V}$ associated with a TW-monad on $\mathbf{V}$?

If $U\colon \mathbf{W} \to \mathbf{V}$ is any underlying-set preserving functor between varieties, the variety $\mathbf{W}$ can be regarded as obtained by adding to the description of $\mathbf{V}$ some additional operations and identities; in that situation, let us call a family of additional operations on an object S of $\mathbf{V}$ such that the appropriate identities are satisfied a "$\mathbf{W}$-structure" for S. Condition (ii) of the above Proposition, which we showed equivalent to the condition (i) that we are interested in, can now be restated as

(63.10) If (S_i) is a finite family of objects of $\mathbf{V}$, each given with a $\mathbf{W}$-structure, then these $\mathbf{W}$-structures extend in a unique manner to a $\mathbf{W}$-structure on $\coprod_{\mathbf{V}} S_i$.

We shall see in the next section that this formulation is convenient for recognizing when familiar forgetful functors U arise from TW-monads. But to study from a general point of view the properties of the operations and identities of $\mathbf{W}$ that will lead to this condition holding, let us write out formally the original statement that U respects coproducts; this says that for every finite family (S_i) of objects of $\mathbf{W}$, the canonical homomorphism

$$\coprod_{\mathbf{V}} U(S_i) \;\to\; U(\coprod_{\mathbf{W}} S_i) \tag{63.11}$$

is an isomorphism.

There is an easy criterion for one half of this statement to hold; namely, (63.11) will be *surjective* for all families (S_i) if and only if

(63.12) For every derived operation w of $\mathbf{W}$, say of arity n, the variety $\mathbf{W}$ satisfies an identity of the form $w(x_1, \ldots, x_n) = v(w_1(x_{i_1}), \ldots, w_m(x_{i_m}))$, where v is an operation of $\mathbf{V}$ (of some arity m), each w_j is a *unary* operation of $\mathbf{W}$, and each $i_j \in \{1, \ldots, n\}$.

Indeed, we see the necessity of this condition by letting each S_i in (63.11) be the free algebra $\langle x_i \rangle_{\mathbf{W}}$ on one indeterminate, and writing down what it means for the element $w(x_1, \ldots, x_n)$ on the right-hand side of (63.11) to come from an element on the left-hand side. That (63.12) is also sufficient for (63.11) to be surjective is easily verified.

Note that (63.12) says, in particular, that every derived operation of $\mathbf{W}$ can be expressed in terms of operations of $\mathbf{V}$ and *unary* operations of $\mathbf{W}$, i.e., that the operations of $\mathbf{W}$ are generated over those of $\mathbf{V}$ by unary operations. (For the case of TW-monads, i.e., the case where (63.11) is *bijective*, we already know this from the discussion immediately preceding Proposition 63.7.) Now if we choose any set of "primitive" operations for $\mathbf{V}$, and some set of *unary* operations generating the operations of $\mathbf{W}$ over those of $\mathbf{V}$, then one can show by induction that (63.12) holds in general if and only if it holds whenever w is the operation obtained by following a member of the former family by a member of the latter; i.e.,

LEMMA 63.13. *Suppose* $U\colon \mathbf{W} \to \mathbf{V}$ *is an underlying-set preserving functor among varieties of algebras, and we regard* $\mathbf{W}$ *as consisting of objects of* $\mathbf{V}$ *with some additional operations, and subject to some additional identities. Then the following conditions are equivalent:*

(i) *For every finite family* (S_i) *of objects of* $\mathbf{W}$, *the map* (63.11) *is surjective.*

(ii) (a) *The operations of* $\mathbf{W}$ *are generated over those of* $\mathbf{V}$ *by unary operations, and* (b) *for each primitive operation* v *of* $\mathbf{V}$, *and each member* w *of a set of unary operations generating the operations of* $\mathbf{W}$ *over those of* $\mathbf{V}$, *the variety* $\mathbf{W}$ *satisfies an identity of the form*

(63.14)
$$wv(x_1,\dots,x_n) \;=\; v'(w_1(x_{i_1}),\dots,w_m(x_{i_m})),$$

where v' *is some derived operation of* $\mathbf{V}$, *and each* w_j *is a derived unary operation of* $\mathbf{W}$. □

(Statement (ii) above is a little sloppy on order of quantification; the precise formulation is, of course, that (i) implies (ii)(a), and that if (ii)(a) holds, and we specify any family of "primitive" operations for $\mathbf{V}$, and any family of unary operations generating the operations of $\mathbf{W}$ over those of $\mathbf{V}$, then (i) is equivalent to (ii)(b) holding for those families. Roughly, (ii)(b) says there are identities specifying how each new operation w "acts on" each existing operation v.)

The condition for the maps (63.11) to be one-to-one cannot be restated so conveniently, but informally, assuming those maps are known to be surjective, it can be thought of as saying that if we are given a family of objects of $\mathbf{V}$ with $\mathbf{W}$-structures, and take their coproduct in $\mathbf{V}$, and try to use the identities (63.14) to obtain the value of each unary operation of $\mathbf{W}$ on each element of this coproduct, then the resulting definitions will be consistent, and will give operations which indeed satisfy the identities of $\mathbf{W}$.

A special case of the situation which we have been considering is that in which the set of additional operations adjoined to $\mathbf{V}$ in forming $\mathbf{W}$ is empty, so that $\mathbf{W}$ is determined by additional *identities* only, i.e., is a subvariety of $\mathbf{V}$. Here part (iii) of Proposition 63.7 immediately gives

COROLLARY 63.15. *The inclusion of a subvariety* $\mathbf{W}$ *in a variety* $\mathbf{V}$ *is the forgetful functor associated with a TW-monad if and only if* $\mathbf{W}$ *is coreflective in* $\mathbf{V}$ (*in addition to being reflective, as a subvariety necessarily is*). □

Let us end this section by looking at TW-monads in a more general category-theoretic context.

Suppose $U\colon \mathbf{P} \to \mathbf{Q}$ is a functor among arbitrary categories, having both a left adjoint F and a right adjoint G. Then:

(i) As we have observed, the endofunctor UF of $\mathbf{Q}$ has a structure of monad, and also has a right adjoint, namely UG.

(ii) This is equivalent to saying that the latter endofunctor UG of $\mathbf{Q}$ has a structure of *comonad*, and has left adjoint UF.

(iii) Further, the endofunctor FU of $\mathbf{P}$ has (like UG), a structure of comonad;

but (like UF) it is the composite of two *left* adjoint functors, hence has a right adjoint, namely GU.

(iv) Which is to say that the endofunctor GU of $\mathbf{P}$ has a structure of monad, and has a left adjoint, FU.

Let us call any endofunctor M of any category $\mathbf{Q}$ which, like the UF of (i) above, is given with a structure of monad and has a right adjoint, a "TW-monad" on $\mathbf{Q}$. We can call its right adjoint, and the induced comonad structure thereon (cf. (ii) above) the corresponding "TW-comonad" on $\mathbf{Q}$; since adjoint functors determine one another, these are equivalent structures on $\mathbf{Q}$. Symmetrically, we might call an endofunctor of a category $\mathbf{P}$ given with a structure of comonad and admitting a right adjoint (as in (iii)) a "WT-comonad" on $\mathbf{P}$, and the structure that its right adjoint acquires (as in (iv)) the corresponding "WT-monad".

Proposition 63.7 and the discussion preceding it show that any TW-monad (equivalently, any TW-comonad) on a *variety* $\mathbf{V}$ in fact arises as above from a functor U with both a left and a right adjoint, and that this U is essentially unique if we require that its domain be a variety and that it preserve underlying sets. We do not know what may be said of WT-monads (equivalently, WT-comonads) on varieties $\mathbf{P}$; in particular, whether they all arise as above from functors $\mathbf{P} \to \mathbf{Q}$ with adjoints on both sides, and if so, whether the category $\mathbf{Q}$ can be taken to be a variety.

Let us note that if $M\colon \mathbf{Q} \to \mathbf{Q}$ is a TW-monad on any category $\mathbf{Q}$, N its right adjoint, and S an object of $\mathbf{Q}$ given with a structure of algebra with respect to M, i.e., a morphism $\kappa\colon M(S) \to S$ making appropriate diagrams commute, then the adjunction between M and N yields a morphism $S \to N(S)$ making S a coalgebra with respect to the TW-comonad N. The converse is also true, so algebras with respect to M are equivalent to coalgebras with respect to N. On the other hand, so far as we can see, the adjunction associated with a WT-monad and comonad works the wrong way for identifying an algebra structure with respect to one with a coalgebra structure with respect to the other.

It would be interesting to know how to construct or characterize the most general functor U among varieties of algebras which has both a left and a right adjoint. (The "hermaphrodites" of the title of this section.) One way to get outside the class of functors characterized in Proposition 63.7 is by composing members of that class on the right and/or the left with *equivalences* of varieties which are not underlying-set preserving.

We remark that for representable functors between varieties that do not necessarily preserve underlying sets, the implication (ii)$\Rightarrow$(iii) of Proposition 63.7 does not hold. For instance, the functor $\mathbf{Ab} \to \mathbf{Ab}$ represented by any non-free finitely generated abelian group preserves arbitrary coproducts, but not general colimits, hence does not have a right adjoint. Another example with these properties is the functor $U\colon \mathbf{Semigp} \to \mathbf{Set}$ taking every semigroup to its set of idempotent elements.

This book focuses more on representable functors than on their left adjoints, and we may note that it is the *comonad* member of a TW-monad/comonad pair that is representable, and hence fits this framework. On the other hand, its representing object B is a TW-*monad* object, and the representable functor $\mathbf{V}^{B\odot -} \to \mathbf{V}$ is "the forgetful functor associated with a TW-monad". In the next section, we will speak most often about "TW-monad objects B" and "forgetful functors associated with TW-monads", but when, at times, we look at the functor represented by B itself, and call it a TW-*comonad*, the reader should remember that this is just a third way of describing the same data.

§64. Examples of TW-monads, and further remarks.

It's time for some examples!

First, let G be any semigroup with neutral element, let $\mathbf{V}$ be any variety of algebras, let $\mathbf{V}^G$ be the variety consisting of objects of $\mathbf{V}$ given with an action of G by endomorphisms, and let $U\colon \mathbf{V}^G \to \mathbf{V}$ be the forgetful functor. Clearly, U satisfies (63.10), hence it is the forgetful functor associated with a TW-monad. We can also verify this by displaying its right adjoint: This is the functor taking any object S of $\mathbf{V}$ to the direct product $S^{|G|}$ of a $|G|$-tuple of copies of S, with G acting on the index-set of this product; the universal map $U(S^{|G|}) \to S$ is the projection to the identity component. (The left adjoint of U similarly takes S to the *coproduct* of a $|G|$-tuple of copies of S, with G again acting on the indices, and with the coprojection of S to the copy of itself indexed by e as universal map.) The associated TW-monad object B of $\mathbf{V}$ is the free object $\langle\{x_g \mid g \in G\}\rangle_{\mathbf{V}}$ on a $|G|$-tuple of generators, with TW-monad structure determined by the relations $x_g \cdot x_h = x_{gh}$, and with x_e for the e of (63.3) and (63.4).

In the opposite direction, let us note that if B is any TW-monad object of a variety $\mathbf{V}$, and $\mathbf{V}^{B\odot -}$ the variety of objects of $\mathbf{V}$ with B-action, then the set of those unary operations of $\mathbf{V}^{B\odot -}$ which act on all objects by *endomorphisms* of the $\mathbf{V}$-structure – equivalently, the set B_{end} of elements $b \in B$ whose actions $b\cdot -$ induce such endomorphisms – forms a semigroup with neutral element, yielding a functor

$$B \mapsto B_{\mathrm{end}} \tag{64.1}$$

from the category of TW-monad objects of $\mathbf{V}$ to $\mathbf{Semigp}^e$; the multiplication and neutral element of B_{end} are the restrictions of the operations (63.3) of B. As Tall and Wraith **[188]** point out, for any variety $\mathbf{V}$, the construction of the preceding paragraph, associating to a semigroup G the TW-monad object of $\mathbf{V}$ inducing the forgetful functor $\mathbf{V}^G \to \mathbf{V}$, is left adjoint to (64.1)!

Suppose, next, that k is a commutative ring and B is a TW-monad object of $\mathbf{Comm}^1_k$, and let us write

$$B_{k\text{-lin}}$$

for the set of elements of B whose actions on all objects of $(\mathbf{Comm}^1_k)^{B\odot -}$ are k-*linear*, i.e., respect the k-module structure. (Thus, $B_{\mathrm{end}} \subseteq B_{k\text{-lin}} \subseteq B$.) Then $B_{k\text{-lin}}$ can be seen to form an *associative unital k-algebra*, under the k-linear structure of B and the multiplication and unit element given by (63.3) (*not* the multiplication and unit element of B as an object of $\mathbf{Comm}^1_k$!). Of the linearity conditions characterizing $B_{k\text{-lin}}$, the condition

$$b\cdot(s+s') \;=\; b\cdot s + b\cdot s'$$

determines the form the coaddition of the co-ring B must take on elements of $B_{k\text{-lin}}$, namely,

$$\mathbf{a}(b) \;=\; b \otimes 1 + 1 \otimes b,$$

(we might also write the right-hand side $b^{\lambda} + b^{\rho}$); similarly, the conditions

$$b \cdot cs \;=\; c(b \cdot s)$$

$(c \in k)$ translate, on writing $\mathbf{s}_c$ for the co-operation of B corresponding to scalar multiplication by $c \in k$, to

(64.2)
$$\mathbf{s}_c(b) \;=\; cb.$$

The behaviors of the additive coinverse $\mathbf{i}$ and the cozero $\mathbf{z}$ on these elements are similarly determined by the linearity condition. Thus, of the co-operations of the co-ring B, it is only the comultiplication $\mathbf{m}$ and the counit $\mathbf{u}$ whose behavior on $B_{k\text{-lin}}$ contains nontrivial information.

Let us now assume k is a field. We do not know whether the cocommutative comultiplication $\mathbf{m}$ of B will always carry $B_{k\text{-lin}}$ into the subspace $B_{k\text{-lin}} \otimes_k B_{k\text{-lin}} \subseteq B \otimes_k B$, but there *will* be a largest sub-k-algebra (under $\cdot$) $(B_{k\text{-lin}})^{\circ} \subseteq B_{k\text{-lin}}$ such that

$$\mathbf{m}((B_{k\text{-lin}})^{\circ}) \;\subseteq\; (B_{k\text{-lin}})^{\circ} \otimes_k (B_{k\text{-lin}})^{\circ}.$$

(Cf. §47.) One finds that $\mathbf{m}$ and $\mathbf{u}$ make the k-algebra $(B_{k\text{-lin}})^{\circ}$ into a cocommutative counital *k-bialgebra*.

Again, we have a left adjoint to this construction: given a cocommutative counital k-bialgebra H, there is a universal TW-monad object B in $\mathbf{Comm}^1_k$ with a map of H into $(B_{k\text{-lin}})^{\circ}$; this has for underlying commutative k-algebra the symmetric algebra $k[H]$ on H, co-k-module structure determined by the condition that the elements of H should act k-linearly, comultiplication and counit obtained from the $\otimes_k$-coalgebra structure of H, and maps π and η constructed using the k-algebra structure of H. One finds that an action of this TW-monad object B on a commutative k-algebra S is equivalent to an action of the bialgebra H on S, in the sense recalled in §43.11.

Thus, we see that actions of semigroups on objects of arbitrary varieties, and actions of k-bialgebras on objects of $\mathbf{Comm}^1_k$, are both cases of actions of TW-monads. It would be of interest to investigate how much of the theory of k-bialgebras can be generalized to arbitrary TW-monad objects, in $\mathbf{Comm}^1_k$ or in general varieties. It would also be interesting to look for further analogs of the two observations "B_{end} forms a semigroup with neutral element" and "$B_{k\text{-lin}}$ forms a k-algebra"; i.e., for descriptions of the natural structure one can put on the set of elements of a TW-monad object B in a variety $\mathbf{V}$ that respect a specified part of the $\mathbf{V}$-structure. (Incidentally, Wraith [**202**] calls elements of B_{end} "primitive" elements of B; Tall and Wraith [**188**] call them "super-primitive", and call the $\mathbf{W}$-operations they determine "Adams operations".)

There is a natural generalization of the construction $B_{k\text{-lin}}$. Let k be a field and L a subfield, and within any TW-monad object B of $\mathbf{Comm}^1_k$, let us look at the set $B_{L\text{-lin}}$ of elements of B inducing L-linear operations. This will be a

k-ring containing the k-algebra $B_{k\text{-lin}}$. For example, if G is a group given with an action on our base field k by L-algebra automorphisms, and $\mathbf{W}$ is the variety of objects S of $\mathbf{Comm}_k^1$ with actions of G by ring automorphisms which, rather than *respecting* the k-module structure, satisfy the equivariance identities

$$g(cx) = g(c)g(x) \qquad (g \in G,\ c \in k,\ x \in S),$$

then (63.10) is again easily verified, so $B = \langle x \rangle_{\mathbf{W}}$ becomes a TW-monad object of $\mathbf{Comm}_k^1$; and we see that every element of G induces an element of $B_{L\text{-lin}}$, but only those which act trivially on k induce elements of $B_{k\text{-lin}}$. We suspect that the object $B_{L\text{-lin}}$ of this example can be taken as the model for some generalization (or "relativization") of the concept of a bialgebra H over k, in which one assumes that k is an L-algebra on which a fixed L-bialgebra H_0 acts, and H contains both k and H_0. (Perhaps this is related to the concept introduced in [**200**].)

There are certain constructions of TW-monads on varieties $\mathbf{Comm}_k^1$ that are special to positive characteristic. Let k be a field of characteristic p. If k is perfect (every element has a pth root) and we let $\mathbf{Comm}_{k,\varphi^{-1}}^1$ denote the variety of commutative k-algebras given with a pth-root map (an inverse to the Frobenius map $\varphi(x) = x^p$), then the forgetful functor $\mathbf{Comm}_{k,\varphi^{-1}}^1 \to \mathbf{Comm}_k^1$ has a right adjoint, the functor taking an algebra S to the inverse limit of the system of rings

(64.3) $$\ldots \overset{\varphi}{\to} S \overset{\varphi}{\to} S \overset{\varphi}{\to} S,$$

made a k-algebra in such a way that the canonical homomorphism from this inverse limit into the last term of (64.3) is a k-algebra homomorphism. The representing object, the k-algebra $k[x, x^{1/p}, x^{1/p^2}, \ldots]$, is thus a TW-monad object of $\mathbf{Comm}_k^1$.

If $k = \mathbf{Z}/2\mathbf{Z}$, the forgetful functor $\mathbf{Bool}^1 \to \mathbf{Comm}_k^1$ has a right adjoint, the Boolean-ring-of-idempotents functor. In fact, for every finite field k, the subvariety of $\mathbf{Comm}_k^1$ generated by k, which is determined by the identity

$$x^{\operatorname{card}(k)} = x,$$

is equivalent to $\mathbf{Bool}^1$ (though not by an underlying-set-preserving functor). Calling this variety $\mathbf{Bool}_k^1$, the forgetful functor $\mathbf{Bool}_k^1 \to \mathbf{Comm}_k^1$ has a right adjoint, whose representing object $k^{\operatorname{card}(k)}$ thus becomes a TW-monad object.

The following observations shed some light on the above two examples. Suppose $k \in \mathbf{Comm}^1$, and let $I = k[x]$ be the TW-monad object representing the identity functor on $\mathbf{Comm}_k^1$. This will be initial among all TW-monad objects of $\mathbf{Comm}_k^1$. (Indeed, part of the structure of any TW-monad object B is a morphism $I \to B$, and this turns out to be a morphism of TW-monad objects.) If k has characteristic p, one finds that the k-ring $I_{(\mathbf{Z}/p\mathbf{Z})\text{-lin}}$ is $k[f; \varphi]$, a skew polynomial ring in one indeterminate f, where f represents the Frobenius endomorphism φ, and twists coefficients in k by that endomorphism, i.e., satisfies

(64.4) $$fc = c^p f \qquad (c \in k).$$

(Recall that the multiplication of $B_{k\text{-lin}}$ and related constructions is not induced by the ring multiplication of B; so though $I = k[x]$ is commutative, $I_{(\mathbf{Z}/p\mathbf{Z})\text{-lin}}$

need not be. In fact, $I_{(\mathbf{Z}/p\mathbf{Z})\text{-lin}}$ consists of all polynomials of the form $\Sigma\, c_i x^{p^i}$, under addition and *composition* of polynomials.) Consequently, for any TW-monad object B of $\mathbf{Comm}_k^1$, $B_{(\mathbf{Z}/p\mathbf{Z})\text{-lin}}$ is not merely a k-ring, but a k-ring with a distinguished element f satisfying (64.4). Thus, in constructing TW-monads based on "generalized k-bialgebras" H as suggested above, one should be able to include in a presentation of such H relations involving this distinguished element. In the first case above, we in effect adjoined an inverse to f, while in the second, we imposed the relation $f^r = 1$, where $p^r = \mathrm{card}(k)$.

There are further sorts of TW-monads in "mixed characteristic", i.e., when k is a commutative ring of characteristic 0 having homomorphic images of positive characteristic. For instance, we discussed in §44 the "Witt ring" functor $W\colon \mathbf{Comm}^1 \to \mathbf{Comm}^1$, taking each commutative ring S to a ring $W(S)$ whose additive group is the multiplicative group of formal power series with constant term 1 over S. Recall that $W(S)$ has a "dense" set of elements which we wrote as formal sums $\Sigma[x_i]$ $(x_i \in S)$, and which behave approximately like members of the semigroup ring on the multiplicative semigroup of S. One finds that W has morphisms to the identity functor and to WW, characterized respectively by the conditions of taking an element $\Sigma[x_i] \in W(S)$ to $\Sigma x_i \in S$ and to $\Sigma[[x_i]] \in W(W(S))$. These yield morphisms (63.2) making the representing coalgebra B of W a TW-monad object of $\mathbf{Comm}^1$. Algebras with respect to the monad induced by B (perhaps more easily pictured as coalgebras with respect to the comonad structure on W) are called "special λ-rings" and are one of the main subjects of Tall and Wraith's paper [**188**]. To see why we have called this a mixed-characteristic phenomenon, recall from §44 that every ring $W(S)$ $(S \neq \{0\})$ has characteristic 0; hence a nontrivial ring S admitting a homomorphism $S \to W(S)$ must have characteristic 0. But as also noted there, when all primes are made invertible, the functor W splits into a direct product of copies of the identity functor. The resulting TW-monad on $\mathbf{Comm}_{\mathbf{Q}}^1$ turns out to be the one corresponding (in the sense of the first paragraph of this section) to actions of the multiplicative semigroup of the positive integers by endomorphisms; so it is only in mixed characteristic that we have a distinctive construction.

Another mixed-characteristic concept which can very likely be characterized in terms of algebras with respect to TW-monads on commutative rings is that of *binomial ring* (mentioned at the end of §46). This becomes vacuous over $\mathbf{Q}$: every object of $\mathbf{Comm}_{\mathbf{Q}}^1$ has unique binomial coefficient operations.

In fact, on the variety $\mathbf{Comm}_{\mathbf{Q}}^1$ we know of no TW-monads other than those induced by $\mathbf{Q}$-bialgebras H. It would be interesting to know whether these are the only ones, and if so, whether this result can be generalized to some characterization of TW-monads on $\mathbf{Comm}_k^1$ holding whenever k is a commutative $\mathbf{Q}$-algebra.

We also wonder whether the concept of a *coaction* of a bialgebra on a ring can be generalized, in some way analogous to the way actions of TW-monads on objects of general varieties $\mathbf{V}$ generalize actions of bialgebras on k-algebras. (Perhaps WT-monads and comonads would be relevant here?)

Turning to varieties of noncommutative rings, it should be possible to use the characterizations of representable ring-valued functors on such varieties obtained in

Chapters V and VI to completely describe the TW-monads on these varieties. We shall sketch here the approach in the case of $\mathbf{Ring}_k^1$ when k is a prime field ($\mathbf{Q}$ or $\mathbf{Z}/p\mathbf{Z}$), leaving the development of full details and generality to the interested reader. (The "prime" assumption keeps us from having to consider operations linear over a subring of k but not over k; the assumption that k is a field allows us to use again the convenient duality between $\mathbf{Mod}_k$ and $\mathbf{LCpMod}_k$. Incidentally, note that for k a prime field, or more generally, any epimorph of $\mathbf{Z}$, we have $k\text{-}\mathbf{Ring}^1 = \mathbf{Ring}_k^1$.)

From Theorem 13.15 we see that for k a prime field, every representable functor $\mathbf{Ring}_k^1 \to \mathbf{Ab}$ takes values in abelian groups having unique k-module structures. Hence representable functors $\mathbf{Ring}_k^1 \to \mathbf{Ring}_k^1$ are essentially the same thing as representable functors $\mathbf{Ring}_k^1 \to \mathbf{Ring}^1$. We have seen that every functor of the latter sort has the form

$$S \mapsto (A_0 \mathbin{\hat{\otimes}} S) \times (A_1 \mathbin{\hat{\otimes}} S)^{\mathrm{op}} \tag{64.5}$$

where A_0 and A_1 are linearly compact associative unital k-algebras. To put on the representing object B for the functor (64.5) a structure of TW-monad object, we need to give appropriate morphisms

$$\eta\colon\ I \to B \quad\text{and}\quad \pi\colon\ B \odot B \to B.$$

Since the relation between the functor (64.5) and the representing object B is contravariant, the above maps correspond to morphisms from the functor (64.5) to the identity functor on the one hand, and to the composite of (64.5) with itself on the other. Hence let us write the identity functor and this composite in the form (64.5). The former is

$$S \mapsto (k \mathbin{\hat{\otimes}} S) \times (\{0\} \mathbin{\hat{\otimes}} S)^{\mathrm{op}},$$

while composing (64.5) with itself we get

$$S \mapsto ((A_0 \mathbin{\hat{\otimes}} A_0 \times A_1^{\mathrm{op}} \mathbin{\hat{\otimes}} A_1) \mathbin{\hat{\otimes}} S) \times ((A_1 \mathbin{\hat{\otimes}} A_0 \times A_0^{\mathrm{op}} \mathbin{\hat{\otimes}} A_1) \mathbin{\hat{\otimes}} S)^{\mathrm{op}}.$$

From the first of these descriptions we see that the unit η of the TW-monad object B corresponds to an augmentation on A_0:

$$A_0 \to k, \tag{64.6}$$

and from the second, that the operation π corresponds to a pair of morphisms of linearly compact k-algebras,

$$A_0 \to A_0 \mathbin{\hat{\otimes}} A_0 \times A_1^{\mathrm{op}} \mathbin{\hat{\otimes}} A_1, \qquad A_1 \to A_1 \mathbin{\hat{\otimes}} A_0 \times A_0^{\mathrm{op}} \mathbin{\hat{\otimes}} A_1. \tag{64.7}$$

Hence, a TW-monad object of $\mathbf{Ring}_k^1$ corresponds to a pair of linearly compact associative unital k-algebras A_0 and A_1, given with morphisms (64.6) and (64.7), subject to identities corresponding to the associativity and unitality conditions on η and π.

Now the representing object B for (64.5) is a tensor algebra $k\langle M\rangle$, where M is a discrete k-vector-space, from which, we recall, the linearly compact vector space $A_0 \times A_1$ arises by dualization. The structure on $A_0 \times A_1$ we have described corresponds to a decomposition $M = M_0 \times M_1$, a coassociative counital

$\otimes_k$-coalgebra structure on each of these summands (dual to the algebra structures on the A_i), and a unital $\mathbf{Z}/2\mathbf{Z}$-graded *algebra* structure on their direct sum (dual to the maps (64.6) and (64.7)) which respects the coalgebra structures up to a "twist" (corresponding to the two occurrences of $(\ )^{\mathrm{op}}$ in (64.7)). It appears that we have here the concept of $\mathbf{Z}/2\mathbf{Z}$-graded bialgebra whose existence we conjectured in §57, actions of which on associative algebras would constitute a common generalization of actions of ordinary bialgebras, and involutions!

The above discussion shows, incidentally, that the statement, which for $\mathbf{Comm}^1_k$ we could only conjecture when $k=\mathbf{Q}$, that any TW-monad object B was generated in the canonical way by its subspace $B_{k\text{-lin}}$, is true in $\mathbf{Ring}^1_k$ for k any prime field.

Let us note that the forgetful functor $U\colon \mathbf{InvRing}^1_k \to \mathbf{Ring}^1_k$, from k-algebras with involution (§57) to ordinary k-algebras, satisfies (63.10); hence $\mathbf{InvRing}^1_k \cong (\mathbf{Ring}^1_k)^{B\odot -}$ for some TW-monad object B. We have just seen that this TW-monad object will be determined by an object $A_0 \times A_1$ with algebra and coalgebra structures as discussed above. By examining the free object on one generator in $\mathbf{InvRing}^1_k$, and its canonical co-$\mathbf{InvRing}^1_k$ structure, we easily verify that

$$A_0 = A_1 = k.$$

Here the structure maps (64.6) and (64.7) are uniquely determined by the condition of unitality, since each of A_0 and A_1 is k. The right and left adjoint of U take a k-algebra S to $S\times S^{\mathrm{op}}$ with the involution $(s,s')\mapsto(s',s)$, and to $S \amalg S^{\mathrm{op}}$ with involution $s^\lambda \leftrightarrow s^\rho$, respectively.

Note that the functor $\mathbf{Ring}^1_k \to \mathbf{Ring}^1$ represented by this B is the one which was shown in Corollary 25.22 to be *initial* in $\mathbf{Rep}(\mathbf{Ring}^1_k, \mathbf{Ring}^1)$. Assuming again that k is a prime field, this says this functor is initial in $\mathbf{Rep}(\mathbf{Ring}^1_k, \mathbf{Ring}^1_k)$, and we easily deduce that our TW-comonad is initial in the category of TW-comonads on $\mathbf{Ring}^1_k$. (Equivalently, its representing TW-monad object B is *final* among TW-monad objects in $\mathbf{Ring}^1_k$. Cf. our earlier remark that the object representing the identity functor is *initial* among TW-monad objects in any variety.) This constitutes a curious confirmation of the observation that the theory of associative rings with involution is a natural companion to ordinary associative ring theory.

As we have mentioned, it will be shown in [**26**] that for any two varieties $\mathbf{U}$ and $\mathbf{V}$, the category $\mathbf{Rep}(\mathbf{U}, \mathbf{V})$ has an initial object. If $\mathbf{V}$ is a variety and M the initial object of $\mathbf{Rep}(\mathbf{V}, \mathbf{V})$, then M has unique morphisms to MM and to $\mathrm{Id}_{\mathbf{V}}$, and it is immediate (from the uniqueness of all morphisms out of M in $\mathbf{Rep}(\mathbf{V}, \mathbf{V})$) that these make it a comonad. Hence, M is a TW-comonad, and so is initial among TW-comonads on $\mathbf{V}$. For k a general field, we do not know what the variety of algebras with respect to the initial TW-comonad of $\mathbf{Ring}^1_k$ looks like; we suspect that its objects will be k-algebras equipped not only with involutions, but also with operations acting by ring endomorphisms that are not k-algebra endomorphisms.

As a final observation on TW-monads on categories of rings and algebras, consider an arbitrary homomorphism of rings,

(64.8) $$k \to k' \quad \text{in } \mathbf{Ring}^1 \text{ or } \mathbf{Comm}^1,$$

and let us ask whether the restriction-of-scalars functor from k'-rings (or k'-algebras, or commutative k'-algebras) to k-rings (respectively k-algebras, commutative k-algebras) is the forgetful functor associated with a TW-monad on the latter category. By definition, this functor preserves underlying sets, so it suffices to determine whether it also respects finite coproducts. For varieties of *unital* rings and algebras (such as we have been discussing so far in this section), the problem is quickly settled by considering the coproduct of the empty family of objects: this gives the initial object in any variety, hence gives k' in one variety and k in the other; hence restriction of scalars via a homomorphism (64.8) other than an isomorphism never corresponds to a TW-monad on k-$\mathbf{Ring}^1$, $\mathbf{Ring}_k^1$, or $\mathbf{Comm}_k^1$. In the nonunital cases, the initial object is zero and poses no problem, so it suffices to consider whether restriction of scalars respects *pairwise* coproducts. It is not hard to show that this holds if and only if (64.8) is an epimorphism of rings. So on varieties k-$\mathbf{Ring}$, $\mathbf{Ring}_k$, and $\mathbf{Comm}_k$, we have such a TW-monad corresponding to each epimorph k' of k.

A few notes on other varieties:

On $\mathbf{Semigp}^e$, as on our varieties of rings, we obtained earlier an exact characterization of all representable endofunctors, and one should be able to use this to get a description of all TW-monad objects of this category. In addition to forgetful functors $(\mathbf{Semigp}^e)^G \to \mathbf{Semigp}^e$ of the sort described at the beginning of this section, another underlying-set-preserving functor having a right adjoint which would have to show up is the forgetful functor $\mathbf{Group} \to \mathbf{Semigp}^e$.

This example, incidentally, is related to a question asked at the end of the last section. We noted there that a functor with adjoints on both sides induces not only a TW-monad and comonad on its codomain category, but also a WT-monad and comonad on its *domain* category. In the above example, the latter monad and comonad are the identity functor of $\mathbf{Group}$. But it is easy to see that these are also the WT-monad and comonad induced *by* the identity functor of $\mathbf{Group}$, regarded as a functor with a left and right adjoint. Hence, whatever may be the answer to the question of whether WT-monads and comonads on varieties $\mathbf{W}$ are always induced by underlying-set-preserving functors U with right adjoints, we see that when such U does exist, it need not be unique.

We turn next to module categories. As noted following Corollary 8.16 (and developed in detail in [**24**, §9.7]), if we write ${}_k\mathbf{Mod}$ for the category of left modules over $k \in \mathbf{Ring}^1$, then representable functors ${}_k\mathbf{Mod} \to {}_{k'}\mathbf{Mod}$ correspond to (k, k')-bimodules. In particular, a representable functor ${}_k\mathbf{Mod} \to {}_k\mathbf{Mod}$ is determined by a k-bimodule B. If we look for a pair of morphisms (63.2) with the associativity and unitality conditions required to make B a TW-monad object, these turn out to be equivalent to a structure on B of *unital k-ring*, extending the above k-bimodule structure. Regarding B as a TW-monad object on the one hand and as a k-ring on the other, we find that $({}_k\mathbf{Mod})^{B\odot -}$ is just ${}_B\mathbf{Mod}$, and the forgetful functor $({}_k\mathbf{Mod})^{B\odot -} \to {}_k\mathbf{Mod}$ is restriction of scalars. That is, TW-monads on module varieties correspond precisely to restriction-of-scalars functors induced by arbitrary morphisms (64.8) in $\mathbf{Ring}^1$.

Wraith [**202**] treats a general variety $\mathbf{V}$ of algebras as the category of all "models" of an algebraic theory A (an algebraic theory being a clone of formal operations, clothed in category-theoretic garb) and considers *models of theories* as generalizations of *modules over rings*. He in fact motivates the concept of TW-monad in [**202**, §9] using the analogy (which we state in our language):

$$(k\text{-modules} : k\text{-bimodules} : k\text{-rings}) \;::\; (\text{objects of } \mathbf{V} : \text{co-}\mathbf{V}\text{-objects of } \mathbf{V} : \text{TW-monad objects of } \mathbf{V}),$$

which he emphasizes by calling k-rings "k-algebras", and calling the three concepts on the right-hand side "A-models", "A-bimodels" and "A-algebras". (He also calls equations (63.14) "distributivity laws".) He notes that there is a left adjoint to the forgetful functor from TW-monad objects of $\mathbf{V}$ to co-$\mathbf{V}$-objects of $\mathbf{V}$, generalizing the tensor ring construction in the ring-and-module case ([**202**, Exercise 9.3]). Another phenomenon of ring theory which generalizes nicely to TW-monads is the transitivity of the relation "K is a k-ring": if B is a TW-monad object in a variety $\mathbf{V}$, one finds that the TW-monad objects in $\mathbf{V}^{B\odot-}$ correspond to the maps of TW-monad objects from B to other TW-monad objects B' of $\mathbf{V}$. Perhaps the most fundamental difference between the module case and the general case is that for general $\mathbf{V}$, the categories of co-$\mathbf{V}$-objects of $\mathbf{V}$ and of TW-monad objects of $\mathbf{V}$ cannot be identified with varieties of algebras.

Returning to varieties of modules, let us note that restriction-of-scalars functors sometimes admit more than just a left adjoint and a right adjoint:

Exercise and Problem 64.9. *Let $f: k \to k'$ be a morphism in $\mathbf{Ring}^1$, and $U_f: {}_{k'}\mathbf{Mod} \to {}_k\mathbf{Mod}$ the corresponding restriction-of-scalars functor. Having seen that in general this functor has both a left and a right adjoint, let us ask in particular cases whether its left adjoint in turn has a left adjoint, whether its right adjoint has a right adjoint, and, if either is true, let us look further for a left adjoint to that left adjoint, a right adjoint to that right adjoint, etc..*

(i) *Show that if k' is the ring $M_2(\mathbf{Z}) = \begin{pmatrix}\mathbf{Z} & \mathbf{Z}\\ \mathbf{Z} & \mathbf{Z}\end{pmatrix}$ of 2×2 matrices over the integers, k its subring $\begin{pmatrix}\mathbf{Z} & \mathbf{Z}\\ 0 & \mathbf{Z}\end{pmatrix}$ of upper triangular matrices, and $f: k \to k'$ the inclusion map, then the chain of adjoint functors indicated above has exactly 7 terms.* (*Suggestion for making this case more intuitively understandable: Identify ${}_{k'}\mathbf{Mod}$ with $\mathbf{Ab}$, and ${}_k\mathbf{Mod}$ with the category of arrows in $\mathbf{Ab}$.*)

(ii) *Suppose that to every ring homomorphism f we associate the pair (m, n), where m, an integer or $+\infty$, is the number of functors that appear to the left of U_f in the maximal chain of adjoints, and n the number that appear to the right of U_f.* (*So by Proposition 63.7*(i)$\Rightarrow$(iii), *m and n are each ≥ 1.*)

Investigate the possible values of the pair (m, n), and the ring-theoretic properties of f that determine these values.

(iii) *Investigate the corresponding question for not-necessarily-underlying-set-preserving functors between varieties ${}_k\mathbf{Mod}$ and ${}_{k'}\mathbf{Mod}$ (recalling that an adjoint pair of functors between these varieties corresponds to a (k, k')-bimodule).*

The reader who enjoyed part (i) above might like to analyze similarly the cases where $k = \begin{pmatrix} \mathbf{Z} & \mathbf{Z} \\ 0 & \mathbf{Z} \end{pmatrix}$, $k' = \mathbf{Z}$, and f is one of the maps taking a matrix (a_{ij}) to a_{11} or to a_{22}. For some other results in the literature on chains of adjunctions, see [**111**], [**165**].

Our characterization of TW-monads on varieties of modules has a "nonlinear" analog: It is easy to show that TW-monads on **Set** correspond to semigroups G with neutral element, the associated forgetful functor having the form $\mathbf{Set}^G \to \mathbf{Set}$ ([**202**, second half of Exercise 9.1]).

An interesting case of an underlying-set preserving functor that "just fails" to satisfy Lemma 63.13(ii) is the forgetful functor from p-Lie algebras over a field k of characteristic p to ordinary Lie algebras over k. Part (a) of that condition holds, and letting w be the unary operation $(\ )^p$, we find that there exist identities of the sort required by (b) when v is each of the k-vector-space operations of a Lie algebra – but not when v is the Lie bracket operation.

If B is a TW-monad object in a variety $\mathbf{V}$, we have seen that the forgetful functor $\mathbf{V}^{B \odot -} \to \mathbf{V}$ respects both underlying sets and coproducts. This suggests that there should be a particularly close relationship between representable functors on $\mathbf{V}^{B \odot -}$ and on $\mathbf{V}$. We have also seen that the sorts of varieties of rings with additional linear structure mentioned at the beginning of this chapter can all be expressed as $\mathbf{V}^{B \odot -}$, for $\mathbf{V}$ one of the varieties of rings studied in Chapter III. So general results on representable functors on varieties $\mathbf{V}^{B \odot -}$ might be useful in the study of functors on these varieties of rings with additional structure.

§65. The Ehrenfeucht question for semigroups and associative algebras.

We shall end this work by discussing an interesting application of representable functors, and a related open question. Let us approach these ideas, not through the history of the conjecture named above, but as the first author happened to encounter them.

In the Introduction to [**118**], Jacques Lewin asks, in an aside, whether a finitely generated subalgebra of a free associative algebra must be finitely related, noting that the answer was apparently unknown even for semigroups.

There is in fact an easy counterexample to the question for semigroups, which leads to a counterexample to the ring-theoretic question as well: Within the free semigroup on two generators $\langle x, y \rangle$, consider the subsemigroup S generated by

$$x^2,\quad yx,\quad xy,\quad y. \tag{65.1}$$

These elements satisfy two infinite families of relations

$$(yx)(x^2)^n(xy) = (y)(x^2)^{n+1}(y) \quad \text{and} \quad (yx)(x^2)^n(y) = (y)(x^2)^n(xy) \quad (n \geq 0). \tag{65.2}$$

Now each of the above equations says that the common value of its two sides can be expressed in two different ways in terms of the generating set (65.1); but it is easy to verify that no *proper subword* of any of these words can be written in more than one way in terms of these generators. From this it is easy to deduce that none of the relations (65.2) is a consequence of any relations of shorter length in x and y that hold among the generators (65.1). Since any *finite* set of relations satisfied by these generators admits a bound on the lengths of the words occurring, we see that no finite set of relations holding among these elements implies all the relations (65.2); hence S is not finitely related. It is easy to deduce that the subalgebra of the free associative k-algebra $k<x, y>$ generated by the same four elements (65.1) is also not finitely related.

Consider, however, the following plausible argument for an *affirmative* answer to the semigroup case of Lewin's question:

It is known that the free semigroup on two generators can be embedded in the multiplicative semigroup of 2×2 matrices over a commutative ring. (For instance, the free group on two generators, and hence the subsemigroup which these generate, can be embedded in 2×2 matrices over the complex numbers by sending x and y to $\begin{pmatrix}1 & r\\ 0 & 1\end{pmatrix}$ and $\begin{pmatrix}1 & 0\\ s & 1\end{pmatrix}$ respectively, whenever r and s are complex numbers whose product has absolute value ≥ 4, e.g., $r = s = 2$; cf. [**120**, p.168].) Since the free semigroup on countably many generators embeds in the free semigroup on two generators, we have in particular that any free semigroup on finitely many generators embeds in 2×2 matrices over a commutative ring.

Now suppose E is any set of semigroup relations in finitely many indeterminates $x^{(1)}, \ldots, x^{(n)}$. By substituting for each semigroup indeterminate $x^{(m)}$ a generic 2×2 matrix, i.e., a matrix of distinct commutative-ring indeterminates $((x_{ij}^{(m)}))$, we get from each member of E four polynomial relations in these $4n$ indeterminates. The Noetherian property for polynomial algebras in finitely many indeterminates now implies that a finite subset of these polynomial relations entails all of them. In particular, if we take for E the set of all semigroup relations satisfied by a certain n elements $u_1, \ldots, u_n$ of a certain free semigroup of finite rank, there will be a finite subset $E_0 \subseteq E$ such that any n elements of any free semigroup which satisfy E_0 satisfy *all* the relations holding among $u_1, \ldots, u_n$.

A little thought shows, however, that this does not prove that the semigroup generated by these elements is finitely related in the variety of all semigroups, but rather, that it is finitely related in the *quasivariety* of semigroups embeddable in 2×2 matrices over commutative rings, and hence, in particular, in the class of semigroups whose elements can be separated by homomorphisms into free semigroups.

The above argument was found by A. Doohovskoy in 1971, and clarified as above by the first author of the present work, but neither of us knew that the result it established was equivalent to the then outstanding Ehrenfeucht Conjecture, arising in the theory of formal languages, and it was not published. That Conjecture was eventually settled by M. Albert and J. Lawrence [**4**] (cf. also [**5**]), using methods that ultimately (though not on the surface) reduce to 2×2 matrix representations, and by V. S. Guba [**76**], by essentially the method shown above. A more general statement was given by Stallings [**180**]. Let us state the result in an intermediate degree of generality, as

PROPOSITION 65.3 (after Stallings). *Let d be a positive integer. Then every finitely generated object in the quasivariety of semigroups embeddable in $d\times d$ matrices over commutative rings is finitely related in that quasivariety.*

Consequently, every finitely generated subsemigroup of a free semigroup is finitely related in the smaller quasivariety (or the still smaller prevariety) generated by all free semigroups.

The same results hold with "group" in place of "semigroup". □

The Ehrenfeucht Conjecture was originally posed in a different form: It asked whether every subset S of a free semigroup A on finitely many generators contains a finite subset S_0 such that two homomorphisms from A into another free semigroup agree on S if and only if they agree on S_0. This clearly asks whether the *dominion* in A (Definition 30.10 above) of the subsemigroup generated by the set S is equal to the dominion in A of some finitely generated subsubsemigroup, if we take dominions, not with respect to the variety of all semigroups, but with respect to the prevariety generated by the free semigroups. This can be rephrased as asking whether the class of dominion-closed subsemigroups of A with respect to that prevariety has *ascending chain condition.* Let us establish the equivalence of this formulation of the Conjecture to the result proved above (cf. [**52**]).

LEMMA 65.4. *Let* $\mathbf{P}$ *be a prevariety of algebras. Then the equivalent conditions* (i)-(i″) *below imply the equivalent conditions* (ii)-(ii′).

(i) *Every finitely generated object of* $\mathbf{P}$ *is finitely presented in* $\mathbf{P}$.

(i′) *Every finitely generated object of* $\mathbf{P}$ *has ascending chain condition on* $\mathbf{P}$*-congruences (congruences of the form* $\{(a,a') \mid f(a)=f(a')\}$, *where* f *is a homomorphism of the given object into an object of* $\mathbf{P}$).

(i″) *Every object of* $\mathbf{P}$ *free on finitely many generators has ascending chain condition on* $\mathbf{P}$*-congruences.*

(ii) *Every finitely generated object of* $\mathbf{P}$ *has ascending chain condition on* $\mathbf{P}$*-dominion-closed subalgebras (subalgebras which are difference-kernels of pairs of morphisms into objects of* $\mathbf{P}$).

(ii′) *Every object of* $\mathbf{P}$ *free on finitely many generators has ascending chain condition on* $\mathbf{P}$*-dominion-closed subalgebras.*

Moreover, if $\mathbf{P}$ *has the property:*

(65.5) *a direct product of two finitely generated objects is finitely generated,*

then, conversely, (ii)-(ii′) *imply* (i)-(i″).

PROOF. The equivalence of (i)-(i″), respectively of (ii)-(ii′), is straightforward, using the fact that a finitely generated algebra A is a homomorphic image of a free algebra F on finitely many generators, and that congruences or subalgebras of A correspond to certain congruences or subalgebras of F.

Let us now recall that prevarieties admit coproducts and objects presented by generators and relations. Clearly, the coproduct in $\mathbf{P}$ of two finitely generated algebras is finitely generated. Now given a $\mathbf{P}$-dominion-closed subalgebra S of A, let B_S be the object of $\mathbf{P}$ gotten by imposing on $A \amalg_{\mathbf{P}} A$ the relations identifying corresponding elements of the two copies of S. This will be a universal example of an object of $\mathbf{P}$ with two homomorphisms of A into it agreeing on S, hence since S was assumed **P**-dominion-closed, the set of elements on which these two homomorphisms agree is exactly S; that is, if we write $a^{(1)}$ and $a^{(2)}$ for the images of an element $a \in A$ under the two coprojections $A \to A \amalg_{\mathbf{P}} A$, the pair $(a^{(1)}, a^{(2)})$ will lie in the congruence on $A \amalg_{\mathbf{P}} A$ defining B_S if and only if a lies in S. Thus, we have embedded the partially ordered set of **P**-dominion-closed subalgebras of A in the partially ordered set of congruences on $A \amalg_{\mathbf{P}} A$, and the implication (i′)⇒(ii) follows.

To get the final sentence, suppose $f\colon A \to B$ is a homomorphism in **P**. We see that a pair $(a, a') \in A \times A$ will lie in the *congruence* induced by this homomorphism if and only if it lies in the *difference kernel* of the two homomorphisms $f\pi_1, f\pi_2\colon A \times A \to B$. Assuming that $A \times A$ is finitely generated whenever A is, we conclude that (ii)⇒(i′). □

Note that if X and Y are finite generating sets for objects $S, T \in \mathbf{Semigp}^e$, then $\{(x, e_T), (e_S, y) \mid x \in X, y \in Y\}$ is a finite generating set for $S \times T$. Hence the above Proposition shows that the dominion-closed-subalgebra formulation of the Ehrenfeucht Conjecture, which is condition (ii′) for the prevariety generated by free semigroups, is indeed equivalent to condition (i) for that prevariety, which we proved in Proposition 65.3. (The original formulation of the Ehrenfeucht Conjecture was stated for free semigroups with additional zeroary operations, i.e., constants, given by a subset of the free generators. But it is easy to verify that those statements are equivalent to the above simpler formulations.)

We remark that the variety **Semigp**, in contrast to $\mathbf{Semigp}^e$, does not satisfy (65.5); for example the direct product of two copies of the free semigroup on one generator x requires the infinite family of generators $\{(x, x^n), (x^n, x) \mid n>0\}$. However, even in the absence of (65.5), we have the implication (i)⇒(ii) of Lemma 65.4, hence Proposition 65.3 shows that both (i) and (ii) of that Lemma hold for the subprevariety of **Semigp** generated by the free objects.

To see that the implication (ii)⇒(i) of Lemma 65.4 may fail in the absence of (65.5), let G be a group and **P** the variety of all G-sets. Every finitely generated G-set consists of finitely many orbits, hence has ascending chain condition on subalgebras (unions of orbits), hence in particular on dominion-closed subalgebras. But congruences on the free G-set on one generator correspond to subgroups of G, hence taking a G which has an infinite ascending chain of subgroups, we see that conditions (i)-(i″) fail.

Let us make

DEFINITION 65.6. *A prevariety* **P** *satisfying the equivalent conditions* (i)-(i″) *of Proposition 65.4 will be called* locally Noetherian.

Abstracting the argument that yields Proposition 65.3, we get

PROPOSITION 65.7. *Suppose* $U\colon \mathbf{V} \to \mathbf{W}$ *is a representable functor among prevarieties of algebras, with* finitely generated *representing object. Then if* **V** *is locally Noetherian, so is the prevariety of all* **W***-algebras embeddable in algebras* $U(S)$ $(S \in \mathbf{V})$, *and hence so is any subprevariety thereof.* □

Actually, it suffices to talk about quasivarieties rather than prevarieties in these considerations, in view of

LEMMA 65.8. *A prevariety* $\mathbf{P}$ *is locally Noetherian if and only if the least quasivariety* $\mathbf{Q}$ *that contains* $\mathbf{P}$ *is.*

SKETCH OF PROOF. "If" is clear, since the property of being locally Noetherian carries over to subprevarieties. To get the converse, let $\mathbf{P}' \supseteq \mathbf{P}$ denote the prevariety determined by those generalized universal Horn sentences holding in $\mathbf{P}$ that involve only finitely many variables (but possibly an infinite conjunction of equations). We see that $\mathbf{P} \subseteq \mathbf{P}' \subseteq \mathbf{Q}$, and that $\mathbf{P}'$ contains the same *finitely generated* algebras as $\mathbf{P}$; hence $\mathbf{P}'$ is locally Noetherian if and only if $\mathbf{P}$ is. But if $\mathbf{P}'$ is locally Noetherian, then for every generalized universal Horn sentence H in finitely many variables holding in $\mathbf{P}'$, some finite subset of the conjunction on the left-hand side of H entails that full conjunction for all algebras in $\mathbf{P}'$, hence $\mathbf{P}'$ satisfies the stronger ordinary universal Horn sentence obtained from H by replacing its hypothesis by that finite conjunction. Hence $\mathbf{P}'$ is determined by a set of ordinary universal Horn sentences, hence it equals $\mathbf{Q}$, proving the latter quasivariety locally Noetherian. □

Though we saw that Lewin's original question on subalgebras of free associative algebras had a negative answer, that question now suggests

QUESTION 65.9. *Let* k *be a field. Is the subquasivariety of* $\mathbf{Ring}_k^1$ *generated by all free algebras* $k\langle X\rangle$ *locally Noetherian?*

By Proposition 65.7, we would have an affirmative answer if we could find a representable functor U, with finitely generated representing object, from some locally Noetherian quasivariety $\mathbf{V}$ into $\mathbf{Ring}_k^1$, such that every free associative k-algebra was embeddable in an algebra $U(S)$; this last condition is easily seen to be equivalent to the condition that the image of U not lie in any proper subvariety of $\mathbf{Ring}_k^1$.

Can we find such a functor? Although there are many representable functors U from varieties $\mathbf{Comm}_K^1$ to varieties $\mathbf{Ring}_k^1$, those that we know of which have finitely generated representing objects also land in proper subvarieties. Let us show that under certain restrictions on U, this is inevitable. Namely, suppose k is an infinite field, and $U\colon \mathbf{Comm}_k^1 \to \mathbf{Ring}_k^1$ is *k-linear*, in the sense that for some way of writing elements of objects $U(S)$ $(S \in \mathbf{Comm}_k^1)$ as d-tuples of elements of S (possibly subject to some relations), the k-vector space operations on elements of $U(S)$ are given by the k-vector space operations on their coordinates in S. One can deduce that the multiplication of $U(S)$ will be given by bilinear expressions in these coordinates under the multiplication of S, and hence that given elements $x_1, \ldots, x_r \in U(S)$, the dimension of the k-subspace of the k-algebra $U(S)$ spanned by all degree-n monomials in $x_1, \ldots, x_r$ is $\leq d$ times the dimension of the space spanned by all degree-n monomials in the dr coordinates of $x_1, \ldots, x_r$ in S. Note that d is here a constant, depending on U, r is a fixed value (without loss of generality 2), and what we are interested in is how the dimensions of our spaces of expressions of degree n grow with n. Now the dimension of the space spanned by the degree-n monomials in the generators $x_1, \ldots, x_r$ of a free associative algebra $k\langle x_1, \dots, x_r\rangle$ is r^n, while the dimension of the space spanned by the degree-n monomials in any finite family of elements of a commutative algebra S grows only polynomially in n. Hence a free algebra $k\langle x_1, \dots, x_r\rangle$ $(r>1)$ cannot be embedded in $U(S)$.

There are, of course, varieties $\mathbf{V}$ which admit representable functors $U: \mathbf{V} \to \mathbf{Ring}_k^1$ with finitely generated representing objects such that free algebras $k\langle X\rangle$ can be embedded in algebras $U(S)$; but the obvious examples of these – such as $\mathbf{Ring}_k^1$ itself – are not locally Noetherian.

A class of cases that suggests cautious optimism are the varieties of *differential algebras*. The simplest example is the variety $\mathbf{Comm}_k^{1,D}$ consisting of objects of $\mathbf{Comm}_k^1$ given with one k-derivation D; equivalently, with an action of the k-bialgebra having for underlying k-algebra the polynomial ring $k[D]$, with comultiplication taking D to $D \otimes 1 + 1 \otimes D$. The free object on r indeterminates $x_1, \dots, x_r$ in this variety has the k-algebra structure of a polynomial ring in infinitely many indeterminates, $D^i x_j$ $(i = 0,\ 1,\ \dots;\ j = 1, \dots, r)$, and so is not Noetherian as an ordinary k-algebra. Nevertheless, if the field k has characteristic 0, this free algebra is known to have ascending chain condition on semiprime D-invariant ideals [**163**, Theorem in §I.12], which means that the quasivariety of objects of $\mathbf{Comm}_k^{1,D}$ without nilpotents is locally Noetherian. There are also versions of this result in positive characteristic; see [**112**, §III.4].

Now what can be said about the growth rates of objects of $\mathbf{Comm}_k^{1,D}$? To estimate how fast the space spanned by the expressions obtainable from a given family of elements under n applications of an arbitrary finite family of derived operations grows with n, it is not hard to show that it suffices to study the rate of growth of the set of those monomials u in a fixed set of indeterminates and their iterated derivatives such that both the total degree of u in these elements, and the total number of applications of D within u, are each at most n. This number appears to have faster than polynomial, but slower than exponential growth. (Consider, as a rough indicator, the case of a single indeterminate x, and polynomials of degree exactly n involving exactly n applications of D. These have the form $(D^{i_1}x)\dots(D^{i_n}x)$ with $i_1 + \dots + i_n = n$, and with order of factors irrelevant. In other words, they correspond to partitions of n; and the number of such partitions is known to grow roughly as $e^{\sqrt{n}}$ [**6**, Theorem 6.3].)

Since this growth-rate is subexponential, no k-linear representable functor on this variety can have the properties needed to give an affirmative answer to Question 65.9. Similar calculations made in [**25**] exclude a large class of related varieties. Nevertheless, the above example shows that the local Noetherian property can hold (and be proved!) in cases with faster than polynomial growth, suggesting that a solution to Question 65.9 via Proposition 65.7 is not impossible.

Let us digress to note that there are interesting representable functors on $\mathbf{Comm}_k^{1,D}$ that have no obvious analogs on $\mathbf{Comm}_k^1$; for instance, the functor $\mathbf{Comm}_k^1 \to \mathbf{Lie}_k$ associating to each differential algebra $S \in \mathbf{Comm}_k^{1,D}$ the Lie algebra of derivations sD $(s \in S)$. To make this precise, we need to know how to compute from two elements $s, t \in S$ the element u such that the commutator bracket of sD and tD is uD. We find that this element is $s(Dt) - (Ds)t$; so we define our functor to take S to its underlying k-module, with the Lie bracket operation

(65.10) $$[s, t] = s(Dt) - (Ds)t.$$

A few final notes on congruences in free groups and semigroups. Let S be the free group on n generators. If C denotes the partially ordered set of congruences on S induced by homomorphisms into other free groups (rather than into direct products of free groups, as above), then from the fact that such a homomorphic image is always again a free group, and, if not isomorphic to S, is free of smaller rank, we can deduce that all chains in C have length at most n. It seems likely that a similar result is true for free semigroups, but we don't see how to prove it. Even though every free semigroup embeds in a free group, and hence every congruence on a free semigroup induced by a homomorphism into a free semigroup is the restriction of a congruence on a free group induced by a homomorphism into a free group, the restriction operation may carry incomparable congruences to comparable ones, and so create chains of congruences on free semigroups that did not exist among the group congruences.

For congruences induced by homomorphisms into *products* of free groups or semigroups, somewhat more can be proved than is indicated in the statement of Proposition 65.3. That Proposition tells us that the lattice of such congruences on the free group or semigroup on n generators has ascending chain condition, equivalently, that any chain in that lattice is reverse-well-ordered, i.e., isomorphic to the opposite of some ordinal. Using the ideal theory of polynomial rings, one can bound this ordinal by $\omega^{4n}+1$. (One can lower the coefficient of n here with a bit more work.) In the group case, infinite chains do in fact occur: e.g., if for every integer i, we let K_i denote the congruence on the free group $<x, y, z>$ induced by the endomorphism that fixes x and y, and takes z to y^{-i}, then we find that an iterated commutator of the form

$$[\ldots[[x,\, y^{j_1} z],\, y^{j_2} z],\, \ldots,\, y^{j_r} z]$$

lies in K_i if and only if $i \in \{j_1, \ldots, j_r\}$. Hence this commutator will lie in a finite intersection $K_{i_1} \cap \ldots \cap K_{i_s}$ if and only if $\{i_1, \ldots, i_s\} \subseteq \{j_1, \ldots, j_r\}$. This shows that $K_{i_1} \cap \ldots \cap K_{i_s}$ determines $\{i_1, \ldots, i_s\}$, and it follows that in an infinite chain such as $K_0 \supseteq K_0 \cap K_1 \supseteq K_0 \cap K_1 \cap K_2 \supseteq \ldots$, all steps are distinct.

On the other hand, among congruences on the free *semi*group on n generators induced by homomorphisms into products of free semigroups, the longest chains the authors know of are of length $1 + n + n(n-1)/2$.

REFERENCES

(Numbers in angle brackets at the end of each listing show pages on which the work is referred to.)

1. Jiří Adámek, *How many variables does a quasivariety need?* Algebra Universalis **27** (1990) 44-48. MR **90m**:08008. <304, 306>

2. Abraham Adrian Albert, *On the power-associativity of rings,* Summa Brasiliensis Mathematicae **2** (1948) 21-32. MR **10**, p.97. <134>

3. Abraham Adrian Albert, *Power-associative rings,* Trans. Amer. Math. Soc. **64** (1948) 552-593. MR **10**, p.349. <134>

4. M. H. Albert and John Lawrence, *A proof of Ehrenfeucht's conjecture,* Theoretical Comp. Sci. **41** (1985) 121-123. MR **87h**:68083. <342>

5. M. H. Albert and John Lawrence, *Test sets for finite substitutions,* Theoretical Comp. Sci. **43** (1986) 117-122. MR **87k**:68072. <342>

6. George E. Andrews, *The Theory of Partitions,* Addison-Wesley, Encyclopedia of Mathematics, vol. 2, 1976. MR **58** #27738. <346>

7. Michael A. Arbib and Ernest G. Manes, *Machines in a category: an expository introduction,* SIAM Review **16** (1974) 163-192. MR **50** #16156. <34>

8. Michael A. Arbib and Ernest G. Manes, *Machines in a category,* J. Pure and Applied Algebra **19** (1980) 9-20. MR **82i**:68037. <34>

9. Richard F. Arens and Irving Kaplansky, *Topological representations of algebras,* Trans. Amer. Math. Soc. **63** (1948) 457-481. MR **10**, p.7. <33, 213>

10. Reinhold Baer, *Zur Einführung des Scharbegriffs,* J. reine und angew. Math. **160** (1929) 199-207. <96>

11. Yu. A. Bakhturin, A. M. Slin'ko and I. P. Shestakov, *Nonassociative rings* (Russian), Algebra, Topology, Geometry **18** (1981) 3-72. MR **83h**:17002. <135>

12. George M. Bergman (shown as "George W. Bergman"), *Skew fields of rational functions, after Amitsur,* exposé 16 (18 pp) in *Séminaire M. P. Schützenberger, A. Lentin et M. Nivat, Problèmes mathématiques de la théorie des automates, année 1969/70,* Secrétariat mathématique, 11 Rue Pierre et Marie Curie, 75 – Paris 5, 1970. MR **43** #6260. <213>

13. George M. Bergman, *Groups acting on hereditary rings,* Proc. London Math. Soc. (3) **23** (1971) 70-82. (Corrigendum to Theorem 4.2 at (3) **24** (1972) 192.) MR **45** #293. <63>

14. George M. Bergman, *Hereditary commutative rings, and centres of hereditary rings,* Proc. London Math. Soc. (3) **23** (1971) 214-236. MR **46** #9022. <222>

15. George M. Bergman, *Hereditarily and cohereditarily projective modules,* pp. 29-62 of *Ring Theory* (proceedings of a conference held in Park City, UT, March 2-6 1971), Robert Gordon ed., Academic Press, 1972. MR **52** #13934. <222>

16. George M. Bergman, *Modules over coproducts of rings,* Trans. Amer. Math. Soc. **200** (1974) 1-32. MR **50** #9970. <204>

17. George M. Bergman, *Element chasing and exact embedding,* unpublished, 7 pp., ca. 1974. <153>

18. George M. Bergman, *Some category-theoretic ideas in algebra,* pp. 285-296 in *Proceedings of the International Congress of Mathematicians* (Vancouver, B.C., 1974), vol. 1, Ralph D. James, ed., Canadian Math. Society, 1975. MR **58** #22222. <11, 31>

19. George M. Bergman, *The diamond lemma for ring theory,* Advances in Mathematics **29** (1978) 178-218. MR **81b**:16001. <38, 86, 122, 317^2>

20. George M. Bergman, *Hyperidentities in groups and semigroups,* Aequationes Mathematicae **23** (1981) 50-65. MR **83b**:08016. <95^2, 96>

21. George M. Bergman, *Everybody knows what a Hopf algebra is,* pp. 25-48 in *Group Actions on Rings* (proceedings of a conference at Bowdoin College, Bowdoin, Maine, July 18-24, 1984), Susan Montgomery, ed., Contemporary Math. **43** (1985). MR **87e**:16024. <30, 236, 300>

22. George M. Bergman, *On the scarcity of contravariant left adjunctions,* Algebra Universalis **24** (1987) 169-185. MR **88k**:18003. <34>

23. George M. Bergman, *Co-rectangular bands and cosheaves in categories of algebras,* Algebra Universalis **28** (1991) 188-213. MR **92k**:08004. <7, 94>

24. George M. Bergman, *An Invitation to General Algebra and Universal Constructions,* Berkeley Mathematics Lecture Notes, vol. 7, U. C. Berkeley. (References are correct for Summer 1995 version, and should remain accurate for the near future.) <7^3, 9, 17, 18^2, 29^2, 82, 89, 104, 137, 303, 339>

25. George M. Bergman, *On the growth of algebras with bialgebra action,* to appear in Israel J. Math., volume in memory of S. Amitsur (preprint 6/95, 22 pp.). <346>

26. George M. Bergman, *Colimits of representable functors,* not yet written, title tentative. <119, 168, 257, 338>

27. George M. Bergman, *Algebra structures on pro-objects,* not yet written, title tentative. <127^3, 173>

28. George M. Bergman and W. Edwin Clark, *The automorphism class group of the category of rings,* J. Algebra **24** (1973) 80-99. MR **47** #210. <7>

29. George M. Bergman and P. M. Cohn, *Symmetric elements in free powers of rings,* J. London Math. Soc. (2) **1** (1969) 525-534. MR **40** #4301. <63^3, 79>

30. George M. Bergman and Samuel M. Vovsi, *Embedding rings in completed graded rings, 2. Algebras over a field,* J. Algebra **84** (1983) 25-41. MR **85i**:16001 b. <180>

31. Israel Berstein, *On cogroups in the category of graded algebras,* Trans. Amer. Math. Soc. **115** (1965) 257-269. MR **34** #6757. <6, 175, 179, 182, 184, 186>

32. Garrett Birkhoff and John D. Lipson, *Heterogeneous algebras,* J. Combinatorial Theory **8** (1970) 115-133. MR **40** #4119. <180>

33. Nicolas Bourbaki, *Éléments de Mathématique. Algèbre, Ch. 8, Modules et anneaux semi-simples,* 1958. MR **20** #4576. < 172 >

34. Nicolas Bourbaki, *Éléments de Mathématique. Topologie Générale, Ch. 1-2,* Act. Sci. Ind., vol. 858, 1940; 3rd ed., Act. Sci. Ind., vol. 1142, 1960. MR **25** #4480. < 105 >

35. Nicolas Bourbaki, *Éléments de Mathématique. Algèbre Commutative, Ch. 5-7,* Masson, 1985. MR **86k**:13001b. < 216, 221 >

36. Nicolas Bourbaki, *Éléments de Mathématique. Algèbre Commutative, Ch. 8-9,* Masson, 1985. MR **86j**:13001. < 240 >

37. P. Cartier, *Groupes formels associés aux anneaux de Witt generalisés,* Comptes Rendus Ac. Sci., Sér. A-B **265** (1967) pp.A49-A52. MR **36** #1448. < 240 >

38. Jeremiah Certaine, *The ternary operation* $(abc) = ab^{-1}c$ *of a group,* Bull. Amer. Math. Soc. **49** (1943) 869-877. MR **5**, p.227. < 96 >

39. Stephen U. Chase and Moss E. Sweedler, *Hopf Algebras and Galois Theory,* Lecture Notes in Math., vol. 97, Springer-Verlag, 1969. MR **41** #5348. < 236 >

40. T. Cheatham and E. Enochs, *The epimorphic images of a Dedekind domain,* Proc. Amer. Math. Soc. **35** (1972) 37-42. MR **46** #1784. < 223 >

41. O. Chein, H. O. Pflugfelder and J. D. H. Smith, eds., *Quasigroups and Loops: Theory and Applications,* Sigma Series in Pure and Applied Math., vol. 8, Heldermann Verlag, 1990. MR **93g**:20133. < 63 >

42. Stephen D. Cohen, *Functions and polynomials in vector spaces,* Archiv der Math. **48** (1987) 409-419. MR **89f**:11161. < 291 >

43. P. M. Cohn, *A remark on the Birkhoff-Witt Theorem,* J. London Math. Soc. **38** (1963) 197-203. MR **26** #6223. < 259, 260 >

44. P. M. Cohn, *Some remarks on the invariant basis property,* Topology **5** (1966) 215-228. MR **33** #5676. < 233 >

45. P. M. Cohn, *Free radical rings,* pp.135-145 in *Rings, Modules and Radicals (Proc. Colloq. Keszthely, 1971),* Colloquia Mathematica Societatis János Bolyai, vol. 6, North-Holland, 1973. MR **50** #13113. < 196^5, 197^9, 198^2 >

46. P. M. Cohn, *Universal Algebra,* 2nd ed., Reidel, 1981. MR **82j**:08001. < 17, 18^2, 27, 137^2, 138, 308 >

47. P. M. Cohn, *Free Rings and their Relations,* 2nd ed., London Math. Soc. Monographs, vol. 19, Academic Press, 1985. (1st ed. MR **51** #8155) MR **87e**:16006. < 63, 196, 198^5 >

48. P. M. Cohn, *Algebra,* 2nd ed., vol. 2, Wiley & Sons, 1989. (1st ed. MR **58** #26625) MR **91b**:00001. < 1, 171, 172, 233 >

49. P. M. Cohn, *Algebra,* 2nd ed., vol. 3, Wiley & Sons, 1991. MR **92c**:00001. < 1, 165, 240 >

50. P. M. Cohn and Christophe Reutenauer, *A normal form in free fields,* Canadian J. Math. **46** (1994) 517-531. MR **95j**:16031. < 254 >

51. Gary Cornell and Joseph H. Silverman, eds., *Arithmetic Geometry,* Springer-

Verlag, 1986. MR **89b**:14029. <209, 213>

52. Karel Culik, II, and Juhani Karhumäki, *Systems of equations over a free monoid, and Ehrenfeucht's Conjecture,* Discrete Mathematics **43** (1983) 139-153. MR **85b**:68025. <343>

53. B. A. Davey and Heinrich Werner, *Dualities and equivalences for varieties of algebras,* pp.101-275 in *Contributions to Lattice Theory, (Szeged, 1980),* A. P. Huhn and E. T. Schmitt, eds., Colloquia Mathematica Societatis János Bolyai, vol. 33, North Holland, 1983. MR **85c**: 08012. <32, 33>

54. Robert Davis, *Universal coalgebra and categories of transition systems,* Math. Systems Theory **4** (1970) 91-95. MR **42** #3142. <34>

55. Warren Dicks, *Meyer-Vietoris presentations over coproducts of rings,* Proc. London Math. Soc. (3) **34** (1977) 557-576. MR **56** #3059. <321^2>

56. J. Dieudonné, *Linearly compact vector spaces and double vector spaces over sfields,* Amer. J. Math. **73** (1951) 13-19. MR **9**, p.325. <105>

57. G. V. Dorofeev, *Varieties of generalized standard and generalized accessible algebras* (Russian), Algebra i Logika **15** (1976) 143-167 (transl. in *Algebra and Logic* **15** (1976) 90-104). MR **58** #16811. <135>

58. Andreas W. M. Dress and Christian Siebeneicher, *The Burnside ring of profinite groups and the Witt vector construction,* Advances in Mathematics **70** (1988) 87-132. MR **89m**:20025. <256>

59. Andreas W. M. Dress and Christian Siebeneicher, *The Burnside ring of the infinite cyclic group and its relations to the necklace algebra, λ-rings, and the universal ring of Witt vectors,* Advances in Mathematics **78** (1989) 1-41. MR **90k**:18015. <256>

60. B. Eckmann and P. J. Hilton, *Structure maps in group theory,* Fundamenta Math. **50** (1961/1962) 207-221. MR **24** #4897. <95>

61. B. Eckmann and P. J. Hilton, *Group-like structures in general categories, I-III,* Math. Ann., *I:* **145** (1962) 227-255, *II:* **151** (1963) 150-185, *III:* **150** (1963) 165-187. MR **25** #108, **27** #3681, **27** #3682. <21>

62. Edward G. Effros and Zhong-Jin Ruan, *Discrete quantum groups, I. The Haar measure,* International J. Math., **5** (1994) 681-723. MR **95j**:46089. <237>

63. Solomon Feferman, *Set theoretic foundations of category theory,* with an appendix by G. Kreisel, pp. 201-224 in *Reports of the Midwest Category Seminar, vol. III,* Lecture Notes in Math., vol. 106, Springer-Verlag, 1969. MR **40** #2727. <17>

64. Miguel Ferrero and Artibano Micali, *Sur les m-applications,* pp. 33-53 in *Colloque sur les Formes Quadratiques, 2,* Montpellier, 1977. Bull. Soc. Math. France, Mémoire, vol. 9, 1979 (MR **80h**:15016). MR **80j**:10032. <291>

65. F. Foltz, C. Lair and G. M. Kelly, *Algebraic categories with few monoidal biclosed structures or none,* J. Pure and Applied Algebra **17** (1980) 171-177. MR **82f**:18009. <302^2>

66. Thomas F. Fox, *The coalgebra enrichment of algebraic categories,*

Communications in Algebra **9** (1981) 223-234. MR **82g**:18003. <236>

67. Thomas F. Fox, *The construction of cofree coalgebras,* J. Pure and Applied Algebra **84** (1993) 191-198. MR **94e**:16004. <236>

68. Thomas F. Fox, *The tensor product of Hopf algebras,* Rendiconti dell'Istituto di matematica dell'Università di Trieste, **24** (1994) 65-71. MR **95j**:16044. <280>

69. Peter Freyd, *Algebra-valued functors in general and tensor products in particular,* Colloquium Mathematicum (Wrocław) **14** (1966) 89-106. MR **33** #4116. <3, 26, 27>

70. V. Ginzburg and M. Kapranov, *Koszul duality for operads,* Duke Mathematical Journal **76** (1995) 203-272. (Erratum to appear, regarding §2.2.) <143>

71. A. M. W. Glass and W. Charles Holland eds., *Lattice-Ordered Groups. Advances and Techniques,* Series in Mathematics and its Applications, vol. 48, Kluwer Academic Publishers, 1989. MR **91i**:06017. <301>

72. Robert Gordon, ed., Problems collection, pp. 373-381 in *Ring Theory* (proceedings of a conference held in Park City, Utah, March 2-6, 1971), Academic Press, 1972. MR **48** #8467. <62>

73. George Grätzer, *Universal Algebra,* 1st ed., Van Nostrand, 1968, 2nd ed., Springer-Verlag, 1979. MR **40** #1320, **80g**:08001. <18, 137, 138>

74. C. Greither and Bodo Pareigis, *Hopf Galois theory for separable field extensions,* J. Algebra **106** (1987) 239-258. MR **88i**:12006. <236>

75. Luzius Grünenfelder and Robert Paré, *Families parametrized by coalgebras,* J. Algebra **107** (1987) 316-375. MR **88i**:16044 <236>

76. V. S. Guba, *Equivalence of infinite systems of equations in free groups and semigroups to finite systems,* Mat. Zametki **40** (1986) 321-324. MR **88d**:20060. <342>

77. Mark Haiman and William R. Schmitt, *Incidence algebras, antipodes, and Lagrange inversion in one and several variables,* J. Combinatorial Theory, Ser. A **50** (1989) 172-185. MR **90f**:05005. <256>

78. Philip Hall, *The Edmonton Notes on Nilpotent Groups,* Queen Mary College Mathematics Notes, 1969. MR **44** #316. <246>

79. Philip Hall, *Collected Works,* K. W. Gruenberg and J. E. Roseblade, eds., Clarendon Press, 1988. MR **90b**:01008. <246>

80. Adam O. Hausknecht, *Coalgebras in categories of associative algebras,* doctoral thesis, University of California at Berkeley, 1975. <7^2>

81. Adam O. Hausknecht, *The automorphism class group of the category of rings over an arbitrary commutative base ring,* unpublished. <7>

82. Adam O. Hausknecht, *Cogroups in categories of associative rings,* title tentative, in preparation. <7, 175, 177, 178>

83. Michiel Hazewinkel, *Formal Groups and Applications,* Series in Pure and Applied Math., vol. 78, Academic Press, 1978. MR **82a**:14020. <240, 276>

84. Michiel Hazewinkel, *Twisted Lubin-Tate formal group laws, ramified Witt vectors and (ramified) Artin-Hasse exponentials,* Trans. Amer. Math. Soc. **259** (1980) 47-63. MR **81m**:14032. <240>

85. Irving Roy Hentzel and Giulia Maria Piacentini Cattaneo, *Degree three identities,* Communications in Algebra **12** (1984) 2349-2400. MR **86d**:17002. <135>

86. Israel N. Herstein, *Rings with Involution,* Chicago Lectures in Mathematics, University of Chicago Press, 1976. MR **56** #406. <296>

87. Philip J. Higgins, *Algebras with a scheme of operators,* Math. Nachrichten **27** (1963) 115-132. MR **29** #1239. <180>

88. Karl Heinrich Hofmann and Paul S. Mostert, *Elements of Compact Semigroups,* Merrill Research and Lecture Series, C. E. Merrill Books, Columbus, Ohio, 1966. MR **35** #285. <229>

89. Thomas W. Hungerford, *Algebra,* Graduate Texts in Mathematics, vol. 73, Springer-Verlag, 1974. MR **50** #6693. <237>

90. Geun Bin Im and G. M. Kelly, *A universal property of the convolution monoidal structure,* J. Pure and Applied Algebra **43** (1986) 75-88. MR **87m**:18011. <302>

91. John R. Isbell, *Uniform Spaces,* Amer. Math. Soc. Surveys, vol. 12, 1964. MR **30** #561. <105>

92. John R. Isbell, *Epimorphisms and dominions,* pp. 232-246 in *Conference on Categorical Algebra* (La Jolla, CA, 1965), S. Eilenberg et al., eds., Springer-Verlag, 1966. (Some errors in this paper concerning particular categories are corrected in [**95**].) MR **35** #105a. <17, 157^2>

93. John R. Isbell and J. M. Howie, *Epimorphisms and dominions, II,* J. Algebra **6** (1967) 7-21. MR **35** #105b. <17, 157>

94. John R. Isbell, *Epimorphisms and dominions, III,* Amer. J. Math. **90** (1968) 1025-1030. MR **38** #5877. <17, 157>

95. John R. Isbell, *Epimorphisms and dominions, IV,* J. London Math. Soc. (2) **1** (1969) 265-273. MR **41** #1774. <17, 157, 353>

96. John R. Isbell, *Epimorphisms and dominions, V,* Algebra Universalis **3** (1973) 318-320. MR **50** #2029. <17, 157>

97. John R. Isbell, *Compatibility and extensions of algebraic theories,* Algebra Universalis **6** (1976) 37-51. MR **54** #212. <32>

98. John R. Isbell, *Polynomials in modules. I. Vector spaces,* J. Algebra **112** (1988) 478-493. MR **89e**:15040. <291>

99. Nathan Jacobson, *Structure of Rings,* Amer. Math. Soc. Colloq. Pub., vol. 37, 1956, revised 1964. MR **36** #5158. <177, 194>

100. Nathan Jacobson, *Lie Algebras,* Interscience Tracts in Pure and Applied Math., vol. 10, 1962. MR **26** #1345. <121, 122, 269, 270, 273, 275>

101. Nathan Jacobson, *Structure and Representations of Jordan Algebras,* Amer. Math. Soc. Colloq. Pub., vol. 39, 1968. MR **40** #4330. <121, 123>

102. Nathan Jacobson, *Lectures on Quadratic Jordan Algebras,* Tata Institute for Fundamental Research Lectures on Mathematics and Physics, vol. 45, Bombay, 1969. MR **48** #4062. <287, 292>

103. D. L. Johnson, *The group of formal power series under substitution,* J. Austral. Math. Soc. (A) **45** (1988) 296-302. (Note correction to this paper in MR review.) MR **89j**:13021. <240>

104. Peter T. Johnstone, *Stone Spaces,* Cambridge Studies In Advanced Math., vol. 3, Cambridge University Press, 1982. MR **85f**:54002. <33, 107, 127^3, 211>

105. S. A. Joni and Gian-Carlo Rota, *Coalgebras and bialgebras in combinatorics,* Studies in Applied Mathematics **61** (1979) 93-139. MR **81c**:05002. <255, 256>

106. Daniel M. Kan, *On monoids and their dual,* Boletín de la Sociedad Matemática Mexicana (2) **3** (1958) 52-61. MR **22** #1900. <6, 33, 90>

107. Daniel M. Kan, *Adjoint functors,* Trans. Amer. Math. Soc. **87** (1958) 294-329. MR **24** #A1301. <11>

108. Irving Kaplansky, *Bialgebras,* Lecture notes, University of Chicago, 1975. MR **55** #8087. <237>

109. Elyahu Katz, *Topological groups with co-monoid structures,* Glasgow Math. J. **18** (1977) 145-152. MR **57** #510. <7>

110. G. M. Kelly, *Basic concepts of enriched category theory,* London Math. Soc. Lecture Note Series, vol. 64, Cambridge University Press, 1982. MR **84e**:18001. <236>

111. G. M. Kelly and F. W. Lawvere, *On the complete lattice of essential localizations,* Bull. Soc. Math. de Belgique **XLI** (1989) 289-319. MR **91c**:18001. <341>

112. Ellis R. Kolchin, *Differential Algebra and Algebraic Groups,* Series in Pure and Applied Math., vol. 54, Academic Press, 1973. MR **58** #27929. <346>

113. Aleksandr Gennadievich Kurosh, *Theory of Groups,* transl. by K. A. Hirsch, Chelsea Pub. Co., 1956, 1960. MR **15**, p. 501, **17**, p. 124, **18**, p. 188. <95>

114. Tsit-Yuen Lam, *A First Course in Noncommutative Rings,* Graduate Texts in Math., vol. 131, Springer-Verlag, 1991. MR **92f**:16001. <171>

115. Serge Lang, *Algebra,* Addison-Wesley, (1st ed. 1965, 2nd ed. 1984) 3rd ed. 1993. MR **33** #5416 (1st ed.), **86j**:00003 (2nd ed.). <1, 15, 233, 237>

116. Richard G. Larson and Earl J. Taft, *The algebraic structure of linearly recursive sequences under Hadamard product,* Israel J. Math. **72** (1990) 118-132. MR **92g**:16056. <254, 257>

117. Solomon Lefschetz, *Algebraic Topology,* Amer. Math. Soc. Colloq. Pub., vol. 27, 1942, reprinted 1963. MR **4**, p. 84. <105>

118. Jacques Lewin, *Free modules over free algebras and free group algebras,* Trans. Amer. Math. Soc. **145** (1969) 455-465. MR **40** #2706. <341>

119. Ottmar Loos, *Jordan Pairs,* Lecture Notes in Math., vol. 460, Springer-Verlag, 1975. MR **56** #3071. <292>

120. Roger Lyndon and Paul Schupp, *Combinatorial Group Theory,* Ergebnisse der Mathematik, vol. 89, Springer-Verlag, 1977. MR **58** #28182. <342>

121. Saunders Mac Lane, *One universe as a foundation for category theory,* pp. 192-201, *Reports of the Midwest Category Seminar, vol. III,* Lecture Notes in Math., vol. 106, Springer-Verlag, 1969. MR **40** #2731. <17^2>

122. Saunders Mac Lane, *Categories for the Working Mathematician,* Graduate Texts in Math., vol. 5, Springer-Verlag, 1971. MR **50** #7275. <1, 9, 11, 12, 15, 16^2, 17^4, 22, 27, 29, 89, 154, 236, 299, 311^2, 319, 327, 329^2>

123. Saunders Mac Lane and Robert Paré, *Coherence for bicategories and indexed categories,* J. Pure and Applied Algebra **37** (1985) 59-80. MR **86k**:18003. <236>

124. Wilhelm Magnus, Abraham Karrass and Donald Solitar, *Combinatorial Group Theory,* Interscience Publishers, 1966, Dover Publications, 1976. MR **34** #7617, **54** #10423. <95>

125. Shahn Majid, *Hopf algebras for physics at the Planck scale,* Classical and Quantum Gravity **5** (1988) 1587-1606. MR **90f**:81041. <237>

126. Anatoliy I. Mal'cev, *Über die Einbettung von assoziativen Systemen in Gruppen* (Russian, German summary), Mat. Sb. N.S. **6** (1939) 331-336. MR **2**, p. 7. <308>

127. Anatoliy I. Mal'cev, *Über die Einbettung von assoziativen Systemen in Gruppen. II.* (Russian, German summary), Mat. Sb. N.S. **8** (1940) 251-264. MR **2**, p. 128. <308>

128. Anatoliy I. Mal'cev, *On the general theory of algebraic systems* (Russian), Mat. Sb. N.S. **35** (1954) 3-20. MR **16**, p. 440. <96>

129. Ernest Gene Manes, ed., *Category Theory Applied to Computation and Control, Proceedings of the first International Symposium, San Francisco, California, Feb. 25-26, 1974,* Lecture Notes in Comp. Sci., vol. 25, Springer-Verlag, 1975. MR **51** #2816. <34>

130. Ernest Gene Manes, *Algebraic Theories,* Graduate Texts in Math., vol. 26, Springer-Verlag, 1976. MR **54** #7578. <34>

131. Ralph McKenzie, George McNulty and Walter Taylor, *Algebras, Lattices, Varieties, vol. 1.* Wadsworth and Brooks/Cole, 1987. MR **88e**:08001. <18, 96>

132. N. Metropolis and Gian-Carlo Rota, *Witt vectors and the algebra of necklaces,* Advances in Mathematics **50** (1983) 95-125. MR **85d**:05026. <256>

133. Barry Mitchell, *Rings with several objects,* Advances in Mathematics **8** (1972) 1-161. MR **45** #3524. <195>

134. Susan Montgomery, *Hopf Algebras and their Actions on Rings,* Amer. Math. Soc. Regional Conference Series in Mathematics, vol. 82, 1993. MR **94i**:16019. <236, 237>

135. Robert A. Morris and Bodo Pareigis, *Formal groups and Hopf algebras over discrete rings,* Trans. Amer. Math. Soc. **197** (1974) 113-129. MR **51** #3181. <300>

136. David Mumford, *Lectures on Curves on an Algebraic Surface,* with a section by G. M. Bergman, Annals of Math. Studies, vol. 59, Princeton University Press, Princeton, 1966. MR **35** #187. <209, 240>

137. David Mumford, *Abelian Varieties,* Oxford University Press, 1970, 1974. MR **44** #219. <213>

138. Kenneth Newman and David E. Radford, *The cofree irreducible Hopf algebra on an algebra,* Amer. J. Math. **101** (1979) 1024-1045. MR **80i**:16017. <236>

139. Warren D. Nichols, *Quotients of Hopf algebras,* Communications in Algebra **6** (1978) 1789-1800. MR **80a**:16017. <234>

140. J. Marshall Osborn, *Varieties of algebras,* Advances in Mathematics **8** (1972) 163-369. MR **44** #6775. <135>

141. J. Marshall Osborn, *What are nonassociative algebras?,* Algebras, Groups and Geometries **3** (1986) 264-285. MR **88b**:17040. <135>

142. Freddy M. J. van Oystaeyen and Alain H. M. J. Verschoren, *Non-commutative Algebraic Geometry: An Introduction,* Lecture Notes in Math., vol. 887, Springer-Verlag, 1981. MR **85i**:16006. <213>

143. Robert Paré and Dietmar Schumacher, *Abstract families and the adjoint functor theorem,* pp. 1-125 in *Indexed Categories and their Applications,* R. T. Johnstone and R. Paré, eds., Lecture Notes in Math., vol. 661, Springer-Verlag, 1978. MR **80b**:18004. <236>

144. Bodo Pareigis, *Endomorphism bialgebras of diagrams and of non-commutative algebras and spaces,* pp. 153-187 in *Advances in Hopf Algebras,* Jeffrey Bergen and Susan Montgomery, eds., Dekker Lecture Notes in Pure and Applied Algebra, vol. 158, 1994. MR **95i**:18002. <236>

145. Inder Bir S. Passi, *Polynomial maps on groups,* J. Algebra **9** (1968) 121-151. MR **38** #241. <281>

146. Brian Peterson and Earl J. Taft, *The Hopf algebra of linearly recursive sequences,* Aequationes Mathematicae **20** (1980) 1-17. MR **81j**:16012. <255>

147. Richard S. Pierce, *Associative Algebras,* Graduate Texts in Math., vol. 88, Springer-Verlag, 1982. MR **84c**:16001. <172, 298>

148. A. J. van der Poorten, *Some facts that should be better known, especially about rational functions,* pp. 497-528 in *Number Theory and Applications,* R. A. Mollin, ed., NATO ASI Series C: Mathematical and physical sciences, vol. 265, Kluwer Acad. Publishers, 1989. MR **92k**:11011. <254>

149. Claudio Procesi, *Rings with Polynomial Identities,* Series in Pure and Applied Math., vol. 17, Marcel Dekker, 1973. MR **51** #3214. <298>

150. Andrzej Prószyński, *Forms and mappings. I. Generalities,* Fundamenta Math. **122** (1984) 219-235. MR **86e**:13016. <286, 291>

151. Andrzej Prószyński, *Odwzorowania wyższych stopni,* (*Higher Order Mappings,* Polish with English and French summaries), Wyższa Szkoła Pedagogiczna, Bydgoszcz, Poland, 1987. MR **89b**:15043. <291>

152. Andrzej Prószyński, *Equationally definable functors and polynomial* *mappings,* J. Pure and Applied Algebra **56** (1989) 59-84. MR **90c**:13007. <291>

153. Heinz Prüfer, *Theorie der abelschen Gruppen. I,* Math. Z. **20** (1924) 165-187. <96>

154. Mohan S. Putcha, *On linear algebraic semigroups, I & II,* Trans. Amer. Math. Soc. **259** (1980) 457-469, 471-491. MR **81i**:20087. <209>

155. Mohan S. Putcha, *On linear algebraic semigroups, III,* Internat. J. Math. Math. Sci. **4** (1981) 667-690; correction at *ibid.* **5** (1982) 205-207. MR **83k**:20073a, b. <209>

156. Mohan S. Putcha, *Linear Algebraic Monoids,* London Math. Soc. Lecture Note Series, 133, Cambridge University Press, 1988. MR **90a**:20003. <209>

157. Daniel Quillen, *Rational homotopy theory,* Annals of Math. (2) **90** (1969) 205-295. MR **41** #2678. <237>

158. David E. Radford, *When pro-affine monoid schemes are group schemes,* J. Algebra **57** (1979) 497-501. MR **80d**:14025. <209, 224^2, 225, 234>

159. Douglas C. Ravenel and W. Stephen Wilson, *The Hopf ring for complex cobordism,* J. Pure and Applied Algebra **9** (1976/77) 241-280. MR **56** #6644. <243>

160. Michel Raynaud, *Anneaux Locaux Henséliens,* Lecture Notes in Math., vol. 169, Springer-Verlag, 1970. MR **43** #3252. <227^2>

161. Michel Raynaud, *Schémas en groupes de type* $(p, p, \ldots, p)$, Bull. Soc. Math. Fr. **102** (1974) 241-280. MR **54** #7488. <209>

162. Irving Reiner, *Maximal Orders,* London Math. Soc. Monographs, vol. 5, Academic Press, 1975. MR **52** #13910. <172>

163. Joseph Fels Ritt, *Differential Algebra,* Amer. Math. Soc. Colloq. Pub., vol. 33, 1950. MR **12**, p. 7. <346>

164. Norbert Roby, *Lois polynomes et lois formelles en théorie des modules,* Ann. Sci. École Normale Sup. (3) **80** (1963) 213-348. MR **28** #5091. <287, 290, 299>

165. R. Rosebrugh and R. J. Wood, *An adjoint characterization of the category of sets,* Proc. Amer. Math. Soc. **122** (1994) 409-413. MR **95a**:18003. <341>

166. Maxwell Rosenlicht, *Some basic theorems on algebraic groups,* Amer. J. Math. **78** (1956) 401-443. MR **18**, p.514. <213>

167. Louis H. Rowen, *Ring Theory,* vol. II, Series in Pure and Applied Math., vol. 128, Academic Press, 1988. MR **89h**:16001. <298>

168. Walter Rudin, *Fourier Analysis on Groups,* Interscience Tracts in Pure and Applied Math., vol. 12, 1962. MR **27** #2808. <108>

169. Richard D. Schafer, *An introduction to nonassociative algebras,* Series in Pure and Applied Math., vol. 22, Academic Press, 1966. MR **35** #1643. <131, 135>

170. Boris M. Schein, *On the theory of inverse semigroups and generalized grouds,* Amer. Math. Soc. Transl. (2) **113** (1979) 89-122. MR **35** #283, **80m**:00006. <96>

171. Norbert H. Schlomiuk, *On co-H-objects in the category of augmented algebras,* Bolletino dell'Unione Matematica Italiana (4) **5** (1972) 506-510. MR **47** #4897. <189>

172. William R. Schmitt, *Antipodes and incidence coalgebras,* J. Combinatorial Theory, Ser. A **46** (1987) 264-290. MR **88m**:05006. <256>

173. William R. Schmitt, *Incidence Hopf algebras,* J. Pure and Applied Alg., **96** (1994) 299-330. <256>

174. Jean-Pierre Serre, *Lie Algebras and Lie Groups,* Benjamin, 1965. MR **36** #1582. <61, 260, 275>

175. L. Silver, *Non-commutative localization and applications,* J. Algebra **7** (1967) 44-76. MR **36** #205. <78, 158>

176. L. A. Skornyakov, *Unars,* pp. 735-743 in *Universal Algebra,* B. Csákány, E. Fried, E. T. Schmidt, eds., Colloquia Mathematica Societatis János Bolyai, vol. 29, Budapest, 1982. MR **83f**:08013. <61>

177. S. P. Smith, *Quantum groups: an introduction and survey for ring theorists,* pp.131-178 in *Noncommutative Rings,* S. Montgomery and L. Small, eds., Math. Sci. Res. Inst. Publication, vol. 4, Springer-Verlag, 1992. MR **94g**:17032. <237>

178. Edwin H. Spanier, *Algebraic Topology,* McGraw-Hill, 1966. MR **35** #1007. <4>

179. John R. Stallings, *Whitehead torsion of free products,* Annals of Math. (2) **82** (1965) 354-363. MR **31** #3518. <41>

180. John R. Stallings, *Finiteness properties of matrix representations,* Annals of Math. (2) **124** (1986) 337-386. MR **88b**:20105. <342>

181. Hans H. Storrer, *Epimorphic extensions of non-commutative rings,* Commentarii Mathematici Helvetici **48** (1973) 72-86. MR **48** #342. <78>

182. A. K. Suškevič, *Theory of Generalized Groups,* Gos. Naučno-Tehn. Izdat. Ukrainy, Kharkov, 1937. <96>

183. Moss E. Sweedler, *Hopf Algebras,* Math. Lecture Note Series, Benjamin, N.Y., 1969. MR **40** #5705. <30, 126, 151, 164, 172, 231^2, 237^2, 246, 247>

184. Moss E. Sweedler, *The predual theorem to the Jacobson-Bourbaki Theorem,* Trans. Amer. Math. Soc. **213** (1975) 391-496. MR **52** #8188. <150, 154, 159, 160>

185. Mitsuhiro Takeuchi, *Equivalences of categories of algebras,* Communications in Algebra **13** (1985) 1931-1976. MR **86j**:16038. <299>

186. Mitsuhiro Takeuchi, $\sqrt{\textit{Morita}}$ *Theory,* J. Math. Soc. Japan **39** (1987) 301-336. MR **88k**:16037. <299>

187. Mitsuhiro Takeuchi, *A Hopf algebra approach to Picard-Vessiot theory,* J. Algebra, **122** (1989) 481-509. MR **90j**:12016. <236>

188. D. O. Tall and G. C. Wraith, *Representable functors and operations on rings,* Proc. London Math. Soc. (3) **20** (1970) 619-643. MR **42** #258. <326^3, 327^3, 333, 334, 336>

189. John Tate and Frans Oort, *Group schemes of prime order,* Ann. Sci. École Normale Sup. (4) **3** (1970) 1-21. MR **42** #278. <213^3>

190. Walter Taylor, *Varieties obeying homotopy laws,* Canadian J. Math. **29** (1977) 498-527. MR **55** #7891. <63>

191. Walter Taylor, *Laws obeyed by topological algebras – extending results of Hopf and Adams,* J. Pure and Applied Algebra **21** (1981) 75-98. MR **82h**:55010. <63>

192. A. I. Valitskas, *Absence of a finite basis of quasiidentities for the quasivariety of rings imbeddable in radical rings,* Algebra i Logika **21** (1982) 13-26 (Russian), Algebra and Logic **21** (1982) 8-24 (English translation). MR **84d**:16014. <198>

193. Dan Voiculescu, *Dual algebraic structures on operator algebras related to free products,* J. Operator Theory **17** (1987) 85-98. MR **88c**:46080. <7>

194. Dan V. Voiculescu, Kenneth J. Dykema, and Alexandru Nica, *Free Random Variables: a Noncommutative Probability Approach to Free Products with Applications to Random Matrices, Operator Algebras, and Harmonic Analysis on Free Groups,* Amer. Math. Soc. CRM Monograph Series, vol. 1, 1992. MR **94c**:46133. <7>

195. Robert B. Warfield, Jr., *Nilpotent Groups,* Lecture Notes in Math., vol. 513, Springer-Verlag, 1976. MR **53** #13413. <246>

196. William C. Waterhouse, *Introduction to Affine Group Schemes,* Graduate Texts in Math., vol. 66, Springer-Verlag, 1979. MR **82e**:14003. <209>

197. Alan G. Waterman, *General-valued Polarities,* doctoral thesis, Harvard University, 1971. <32>

198. Niklaus Wirth, *Algorithms + Data Structures = Programs,* Series in Automatic Computation, Prentice-Hall, 1976. MR **55** #13850. <96>

199. Robert Wisbauer, *Homogene Polynomgesetze auf nichtassoziativen Algebren über Ringen,* J. reine und angew. Math. **278/279** (1975) 195-204. MR **52** #5753. <294>

200. Gavin C. Wraith, *Hopf algebras over Hopf algebras,* Annali di Mat. Pura ed Appl. (4) **76** (1967) 149-163. MR **37** #1444. <335>

201. Gavin C. Wraith, *Algebras over theories,* Colloquium Mathematicum (Wrocław) **23** (1971) 181-190. MR **46** #231. <28>

202. Gavin C. Wraith, *Algebraic Theories,* Lecture Note Series, vol. 22, Aarhus Universitet, Matematisk Institut, Lectures Autumn 1969, revised version of notes, Feb. 1975. (MR **41** #6943 reviews the original version; the revision has far fewer errors. But note re p.49, line 6 from bottom: it is not true that "... every monad on A^b arises in this way".) <28, 326, 329, 334, 340^3, 341>

203. Oscar Zariski and Pierre Samuel, *Commutative Algebra, vol.I,* van Nostrand, 1958 and Graduate Texts in Math., vol. 28, Springer-Verlag, 1975. MR **19**, p.833, **52** #5641. <172>

204. Dieter Ziplies, *Abelianizing the divided powers algebra of an algebra,* J. Algebra **122** (1989) 261-274. MR **90j**:16003. <294>

Word and phrase index

The authors have long been frustrated with indexes to mathematical works that only note where a concept is defined, and give no information on where else it occurs. We have tried to make the indexes of this book more useful.

We quickly encountered a tangle of problems – When does a mention of a concept merit referencing? Should items be listed redundantly under various synonymous terms and permuted wordings (''adjunction'' / ''adjoint functors'' / ''functors, adjoint'', etc.), or should this redundancy be replaced by cross-referencing, or eliminated altogether? Is the referencing of very common terms such as ''ring'' useful? If a concept is referred to on each of pages m, $m+1$, ... , n, should these be joined as ''m-n'' even if some of the references are quite unrelated to others, and are separated by paragraphs not related to the concept?

We were not able to come up with general answers to these questions, but have tried to use good sense. We suspect that when we have erred, it has mostly been in the direction of including entries that we might have omitted. One negative rule we have stuck to is not to reference remarks of the form ''In this Chapter we shall show ...'', but only the pages where the results in question actually occur – unless the initial remarks are combined with some nontrivial observations.

This index, and the symbol index which follows, were prepared by inserting, in the source-files for the text, codes signaling ''at this point – is mentioned'', ''at this point an extended discussion of – begins/ends'', ''at this point – is defined'', etc.. Macro packages written by the first author collected, alphabetized and formatted the resulting information.

It is possible that occasionally, a term occurring in the top line of one page may be listed under the preceding page, because of the nature of the program `troff`, which may make the ''decision'' that a word, or in some cases a whole line, has to begin a new page only *after* our macros have recorded the current page number.

Boldface numbers indicate the pages where terms are defined, or conventions relating to them are set. We considered providing other information in similar ways: e.g., perhaps small type for brief tangential references, a raised dot after each page-number to signal the approximate height on the page at which the term occurs, etc.. We may experiment with some of these in the future, but we have preferred to keep this first trial simple.

Personal names are indexed here if results by the individuals in question are mentioned which are not contained in works cited. The pages where items in the References are cited are listed at the end of each bibliographic reference. (We have adopted this idea from [**46**].) Terms used by other authors for which we here use different wording are, if referenced, put in single quotes; e.g., 'natural transformation', for our ''morphism of functors''.

Our subject involves a number of hyphenated terms beginning ''$\otimes$-'', which are alphabetized under ''tensor''. In §§60-62 a few terms beginning ''$\circledast$-'' are introduced; we alphabetize this symbol as ''wheel'', for want of better inspiration. Other symbols with non-obvious alphabetical locations are restricted to the symbol index.

Symbol Index

This index has two functions: It is a glossary of *symbols* used, and it lists the pages where the *topics* so symbolized are treated. For instance, the entry for **Semigp**^{e} informs one that this denotes the category of semigroups with neutral element, and lists various pages, on which one may or may not find the symbol **Semigp**^{e}, but where one will always find reference to such semigroups.

Under each letter of the alphabet, the lower-case letter is followed by the upper-case letter, then Greek and miscellaneous related symbols, in a somewhat arbitrary order. (For a particularly complicated example, the order we have set up under `p` is p, P, π, Π, $\prod$, $\coprod$, φ, Φ, ψ, Ψ, though not all of these symbols actually occur.) Symbols that are not even approximately alphabetical are alphabetized by assigning them spellings; e.g., "0" and "1" are alphabetized as `zero` and `one`; "=" and related symbols such as $\cong$ are alphabetized, in an arbitrary order, under `equal`; $\otimes$ is alphabetized as `tensor`, and $\circledast$ as `wheel`. Fortunately, the reader does not have to know all the details of this system, though some symbols will doubtless require somewhat more search than others. Font-differences, and "punctuation" such as brackets, do not affect ordering unless everything else is equal.

Operator-symbols are often shown in combination with letters with which they are frequently used; e.g., $\circ$ is shown as $\hat{M} \circ N$, alphabetized under `M`; similarly, the notation for a presentation of an object by generators and relations is shown as $<X \mid Y>$, alphabetized under `XY`.

We have not found it easy to decide *which* symbols to include. In general, if a symbol is defined in one place, and used again without explanation more than a page or so away, we have included it. For some very frequently used symbols, e.g., $\prod$ and $\coprod$, only the locations of the definitions are given; the word-and-phrase index can be used to search for particular topics related to these concepts. Symbols of standard and uncontroversial usage which are not useful for locating topics are generally not included. A few symbols (such as m) have specialized uses in some places, and nonspecialized uses (e.g., as an arbitrary integer) elsewhere; only the specialized uses are recorded.

As in the word-and-phrase index, boldface page-numbers indicate pages where definitions are given.

a coaddition map $R \to R^{\lambda} \coprod R^{\rho}$, 37, 42-54, 154, 261-262, 270-271, 326, 334.

$a \in_U \ldots$ "a is a U-element of ...", **146**-153.

$\mathbf{a}_M$ canonical coaddition on $[k]<M>$, **44**-45, 53-54, 62-63, 101-102, 151, 260, 270, 281, 315.

$a \coprod b$ morphism $A \coprod B \to A' \coprod B'$ induced by $a: A \to A'$ and $b: B \to B'$, **19**.

$as \in A \hat{\otimes} S$	abbreviation for $a \otimes s$, **115**.
$a*b$, a_*b	pair of bilinear operations on a module or bimodule, **115**-143, 148-149, 178-179, 266-267, **312**-319.
$\mathbf{a}^\sigma$, $(\mathbf{aa})^\sigma$	σ-part of $\mathbf{a}$, resp. $\mathbf{aa}$, for an index-string σ, **45**-50.
Ab	variety of abelian groups, 4, 13, 16, **35**, 55, 91-94, 115, 148, 170, 177, 210, 220, 234, 260-263, 269-274, 277-294, 302-303, 305, 307, 309, 325, 332, 340.
AbBinar^e	variety of sets with an abelian not necessarily associative binary operation and a neutral element, **61**-63, 91.
AbSemigp	variety of (nonunital) abelian semigroups, **35**, 63-78, 92, 94, 224-229.
AbSemigp^e	variety of abelian semigroups with neutral element, **35**, 37-39, 42-56, 62, 94, 210, 224, 230-231, 260-262, 325.
AbSemigp^i	variety of abelian semigroups with distinguished idempotent, **64**-65, 77-78.
$\mathbf{b}$	nonlinear part of a co-operation $\mathbf{a}$ or $\mathbf{m}$, 43, 50-52, 182-190.
□□	box diagrams for E-systems, **87**-91.
$\mathbf{B^A}$	category of all (covariant) functors $\mathbf{A} \to \mathbf{B}$, **9**, 20.
$B \odot -$	in §§63-64, left adjoint to functor with representing object B, **326**-341.
B_{end}, $B_{k\text{-lin}}$	certain sets of elements in a TW-monad object B, **333**-335, 338.
Bi	functor $\mathbf{Comm}^1 \to \mathbf{Comm}^1$ represented by $\text{Int}[x]$, **241**-242, 246.
Binar^e	variety of sets with a not necessarily associative binary operation and a neutral element, 91, 182-186, 223.
Bool^1, **Bool**	varieties of unital and nonunital Boolean rings, **33**, 211-213, 242, 243, 268, 307, 308, 335.
Br(S)	Lie ring formed from the associative ring S using commutator brackets, **259**-273, 276.
$B(S)$	Boolean ring of idempotent elements of the commutative ring S, **211**-213, 335.

$\mathbf{C}^{\mathrm{aug}}$ category of augmented objects of **C**, **37**-40.

$\mathbf{Comm}_k^1$ variety of commutative associative unital k-algebras, 4, **209**-257, 297, 298, 300, 307, 309, 326-327, 333-336, 339, 345-346.

$\mathbf{Comm}^1$ variety of commutative associative unital rings, 119, 137, 300, 342.

$\mathbf{C}^{\mathrm{pt}}$ category of pointed objects of **C**, **39**-40, 322-325.

$\mathbf{C}(X, Y)$ set *or* algebra of morphisms $X \to Y$ in **C** (see various definitions), **9**, **21**, **24**.

d "diagonal, modulo the radical", *see* $(\mathbf{Z}\times\mathbf{Z})$-$\mathbf{Ring}^{\mathbf{d}}$.

$\deg(x)$ $= \mathrm{ht}(\mathbf{a}(x))$, **43**, 51-53, 83, 191-194.

Δ_a difference operator: $(\Delta_a f)(x) = f(x+a) - f(x)$, **281**, 285-289.

$\simeq$ equivalence of categories, **9**.

$\cong$ isomorphism, **9**.

E-**Syst** $\simeq$ **Rep**(**Semigp**^{e}, **Semigp**^{e})^{op}, **85**-91, 93-94, 200, 339.

ε augmentation map (including coneutral element of a coalgebra), *or* counit of an adjunction, *or* nilpotent ring element, **13**, **37**, **37**.

ε^β bimodule projection map $\circledast_{\alpha\in A} M_\alpha \to M_\beta$, **310**, 311, 320.

η unit of adjunction, **13**, 16, 230, 235, 308, 329.

f (in Chapters III-IV and §34) forgetful maps $R^{\alpha_1}\otimes_k \ldots \otimes_k R^{\alpha_h} \to R\otimes_k \ldots \otimes_k R$ etc., **46**, 70, 187.

(f, g) morphism out of coproduct, or into product, induced by two morphisms f and g, **19**.

$F_1 \ominus \ldots \ominus F_n$ chain tensor product of representable functors, **279**-280.

GL_n general linear group functor, 2-4, 29, 175, 210, 272.

Group variety of groups, 6, 16, **35**, 89, 94-98, 175-190, 210-213, 219-220, 223, 231, 247-256, 274-276, 297, 299, 302, 305, 323-325, 339, 343, 344, 347.

Gr...$^{(>0)}$ categories of objects graded by the positive integers, **181**-190.

$h(a_1, \ldots, a_n)$	$= (\Delta_{a_1} \ldots \Delta_{a_n} f)(0)$, n-linear function of $a_1, \ldots, a_n$ obtained by polarization from f, **285**-289.
$[h]$, $[h]'$	the two $\{\lambda, \rho\}$-strings of length h, **42**-53, 83-85.
$\mathrm{ht}(x)$	height of an element x of a coproduct, **42**, 83.
i	(in Chapter IV) idempotent element of a semigroup (*see also word index*), **64**-78.
i^λ, i^μ, i^ρ, etc.	coprojections $R \to R^\lambda \amalg R^\rho$, $R \to R^\lambda \amalg R^\mu \amalg R^\rho$, etc., **42**, 150, **310**, 320.
I	often: trivial functor, or initial object (*see word index*), in addition to uses for index-set, ideal, etc..
$\mathrm{Id}_{\mathbf{C}}$	identity functor of the category $\mathbf{C}$, **9**.
Idp.-in-n.e.$(\mathbf{V}, \mathbf{S})$	condition on existence of idempotent elements, 64-78, 206-207, **225**-230, 256-257, 309.
$\mathbf{I}$, $\mathbf{I}^1$	arbitrary classes of identities for objects of **NARing**, $\mathbf{NARing}^1$, **136**-142.
$\mathrm{Int}_{\ldots}[\ldots]$	rings of integral polynomials, **240**-246, 249.
$\mathbf{InvRing}^1_k$	variety of k-algebras with involution, 211, **296**-297, 338.
$J(A)$	Jacobson radical of A, 188, **194**-205, 206, 298.
Jordan, $\mathbf{Jordan}^1$	varieties of nonunital and unital Jordan rings, 6, 103, **122**-125, 127, 152, 153, 265-267, 269, 291-294, 296, 297.
k	fixed associative unital ring (sometimes with additional restrictions); *cf.* k-**Ring** *etc. and* K *below, and* characteristic, integral domain *etc. in word index*, **35**, 37, 51, 61, 63, **104**, **146**, 152, 175, 181, 188, 190, 206, 259, 277, 310, 339.
k-**Bimod**	variety of k-bimodules, 4-5, 16, 36, **41**, 43-59, 99-104, 146-172, 280, 284, 299, 309-319, 322, 339.
$(k \times k)$-$\mathbf{LCpRing}^{\mathbf{d}}_k$	category of linearly compact k-algebras in $(\mathbf{Z} \times \mathbf{Z})$-$\mathbf{Ring}^{\mathbf{d}}$ (*q.v.*), **199**-204, 234.
k-$\mathbf{GrRing}^{(>0)}$	variety of connected graded nonunital k-rings, **181**-190.
$[k][L]$	(in Chapter IX) universal enveloping algebra of the Lie algebra L (*see also word index*), **259**-260, 263, 265, 271, 276.

$[k]$<M>	nonunital tensor ring on a k-bimodule (*for* $[k]$<X> *see under* free *in word index*), **44**-45, 52-56, 62-63, 182-186, 260-263, 270, 295, 299, 315.
k<M>	unital tensor ring on a k-bimodule (*for* k<X> *see under* free *in word index*), **43**, 54, 56, 99-103, 236, 277-285, 293, 295, 296, 321, 337, 340.
$[k]\{M\}$, $[k]\{X\}$	(in Chapter IX) universal Lie algebra on the k-module M; free Lie algebra on the set X, **259**-264, 269, 270-271.
k-**Ring**	variety of associative nonunital k-rings, **36**-58, 64-67, 76-78, 103, 148, 295-300, 309, 314, 316-317, 339.
k-$\mathbf{Ring}^1$	variety of associative unital k-rings, 4, 16, **36**-40, 54-58, 64-82, 99-103, 148-173, 177, 207, 277-294, 295-300, 316, 317, 322, 337, 339.
$k[[t]]$	formal power series algebra (*see word index*).
$k\otimes_{\mathbf{Z}}k$, $K\otimes_k K$	free k-bimodule or k-centralizing K-bimodule on one generator, 146, 155, 161.
K	fixed associative unital k-algebra (in definition of K-$\mathbf{Ring}_k$ etc., *q.v.*).
K-$\mathbf{Bimod}_k$	variety of k-centralizing K-bimodules, **56**-59, 103.
K-$\mathbf{Ring}_k$	variety of nonunital k-centralizing K-rings, **56**-58, 295.
K-$\mathbf{Ring}_k^1$	variety of unital k-centralizing K-rings, **56**-58, 103, 161, 229, 250, 295.
LCp...	prefix denoting category of linearly compact topological objects, **114**.
$\mathbf{LCpComm}_k^1$, etc.	categories of linearly compact commutative associative k-algebras, 210, 264-265, 275-276, 280.
$\mathbf{LCpMod}_k$	category of linearly compact k-vector spaces (*see word index*), **106**.
$\mathbf{LCpRing}_k^1$, etc.	categories of linearly compact associative k-algebras, 6, **114**, 126-128, 177, 199-204.
$\mathbf{Lie}_k$, **Lie**	varieties of Lie algebras, Lie rings, 3, 5, 28, 103, **121**-122, 124, 126, 137, 151, 179, 235, 237, 259-276, 296, 297, 341, 346.

$\mathbf{Lie}_k^{(p)}$ variety of p-Lie ('restricted Lie') algebras, **269**, 273-274, 276, 294, 341.

$\underleftarrow{\mathrm{Lim}}$, $\underrightarrow{\mathrm{Lim}}$ limits and colimits (including inverse and direct limits), *see mainly word index*, **14**.

$\mathbf{LTopMod}_k$ category of linearly topologized k-vector spaces, **104**-114.

λ, μ, ρ superscripts used in indexing coproduct of two or three copies of R, 2, **42**, **45**, 150.

$\Lambda^2 M$ exterior square of the module M, 260, 263.

m map $R \otimes_k \dots \otimes_k R \to R$ induced by the multiplication of R, **51**-53.

$\mathbf{m}$ comultiplication (*see word index*) in a co-ring or cosemigroup, 82.

$(-)^p$ "pth power" operation of p-Lie algebra; *see* $\mathbf{Lie}_k^{(p)}$ *above*, **269**.

M° *see* R°.

$\hat{M}$ object of k-$\mathbf{Bimod}^{\mathrm{op}}$ corresponding to $M \in k$-$\mathbf{Bimod}$ (*cf.* $\hat{V}$), **146**-154.

$\hat{M} \circ N$ k-$\mathbf{Bimod}(M, N)$, **146**-153, 226, 263, 265.

$M_n(R)$ ring of $n \times n$ matrices over R, 118, 160, 210, 233.

$\mathbf{Mod}_k$ variety of k-modules (k commutative), 5, 28, 36, **55**-59, 103-114, 152, 259-274, 285-291, 294.

$\tilde{M}$ translation-invariant ring of functions on group, semigroup, or ring M, **247**-248, 254.

NARing variety of not necessarily associative rings; *see also* **Lie**, **Jordan**, $\mathbf{Ring}^{(n)}$, *etc. here, and* bilinear maps *in word index*, 5, **102**, 128-143, 149, 152, 168, 264-269, 297-298.

1_X, 1_R (through §24) identity morphism of an object X, (from §25 on) identity (unit) element of a ring R, **9**, **115**.

$(\ )^{\mathrm{op}}$ opposite (category, ring, or semigroup) (*see word index*).

$\mathbf{P}$, $\mathbf{P}^\varepsilon$ category of objects $(A, {}^*, {}_*)$, respectively $(A, {}^*, {}_*, \varepsilon)$, **136**-143.

$\prod_I A_i$ product of the A_i, **15**.

$\coprod_I A_i$ coproduct of the A_i, **15**.

φ Frobenius endomorphism, $x \mapsto x^p$ (*see word index*), **211**.

Q quasimultiplicative semigroup functor **Ring** $\to$ **Semigp**^{e}, **177**, 186-191, 194, 199, 203, 206.

$|R|$ underlying **C**-object of a **C**-based algebra or coalgebra R, **21**.

R°, M°, etc. various bialgebra constructions, all more or less describable as $\{a \mid \mathbf{m}(a) = \text{a finite sum } \Sigma\, b_i \otimes c_i\}$, **246**, **247**-257, **334**.

Rep(C, V) category of representable functors $\mathbf{C} \to \mathbf{V}$ (*see also under particular* **C** *and* **V**, *and terms such as* cogroup), **25**, 40, 54, 55, 74, 119, 155, 168, 170, 202, 205.

R_h $\{x \in R \mid \deg(x) \leq h\}$, **43**-53.

Ring variety of associative nonunital rings, 5, **36**, 120-121, 127, 137, 168, 306, 307.

$\mathbf{Ring}_k$ variety of associative nonunital k-algebras, 28, 36, **55**-58, 61-63, 115-116, 119, 120, 177, 234, 259-276, 298, 339.

$\mathbf{Ring}_k^1$ variety of associative unital k-algebras, 4, 5-6, **55**-58, 63, 95, 103, 115-143, 175-179, 190-207, 210, 214, 218, 224-228, 232, 233, 235, 242, 250-253, 298, 308, 311, 320-322, 336-339, 341, 345-346.

$\mathbf{Ring}^{(n)}$ variety of not necessarily associative rings with n-fold products associative, **125**, 132.

$\mathbf{Ring}^1$ variety of associative unital rings, 5, 13, **36**, 117-119, 127, 151, 302, 307.

R^λ, R^μ, R^ρ copies of R in $R \amalg R$, $R \amalg R \amalg R$, **42**, **45**, **82**.

R^1 *see* neutral element: adjunction of, *in word index.*

R^σ subbimodule or subset of $R^\lambda \amalg R^\rho$ or $R^\lambda \amalg R^\mu \amalg R^\rho$ determined by index-string σ, **42**-52, **83**-85.

Semigp variety of (nonunital) semigroups, **35**, 39, 75, 94, 206-207, 219, 224-230, 238, 247-257, 319, 332.

$\mathbf{Semigp}^e$	variety of semigroups with neutral element, 6, 13, **35**, 82-98, 176, 188-208, 213, 214, 231-234, 299, 305, 307, 309, 322-323, 333, 339, 341-344, 347.
Set	category of sets, **9**, 13, 34.
S^n	nth power of an object or functor S, **20**.
$\mathrm{Spf}(A)$	formal spectrum of a linearly compact commutative algebra A, 236, 276.
Sum	functor carrying graded rings etc. (regarded as many-sorted) to ungraded, **180**, 190.
Sym, SSym	prefixes for subrings of symmetric and strongly symmetric elements in coproducts, **68**-82.
$\otimes$	tensor product (used without subscript in writing tensor products of elements, submodules of tensor product modules, tensor products of functors), **41**, 51-53, 103, **278**-280, 311.
$\hat{\otimes}$	completed tensor product *or* analogous bimodule construction, **111**, **146**-147, 232-233, 263-265, 275, 279, 284.
$\otimes_k$	tensor product over k, **41**.
$\otimes_k$**-co-Ring**	category of noncounital $\otimes_k$-co-rings, **151**-173.
$\otimes_k$**-co-Ring**^{1}	category of counital $\otimes_k$-co-rings, **151**-152, 155-161, 166-173.
$\tau,\ \tau^1$	relation "– satisfies –" on $\mathbf{P}\times\mathbf{I}$, respectively $\mathbf{P}^{\varepsilon}\times\mathbf{I}^1$, **136**-138.
$U_x(y)$	Jordan "triple product", **291**-293.
$\hat{V}$	topological dual of discrete vector space (*cf.* $\hat{M}$), **106**, 226, 232-237, 246-255.
$\mathbf{V}^M,\ \mathbf{V}^{B\odot -},\ \mathbf{V}^G$	category of objects of $\mathbf{V}$ with action of monad M, of TW-monad $B\odot -$, or of semigroup G, **329**-330, **333**-335, 336, 338, 339, 341.
$\circledast$	functors on bimodules etc. mimicking behavior of coproduct of rings etc., **310**-325.
$W,\ W_p$	Witt ring construction, 210, **239**, 237-240, 242, 256, 336.

ISBN 0-8218-0495-2

9 780821 804957